FRONTIERS IN ELECTRONICS

Future Chips

SELECTED TOPICS IN ELECTRONICS AND SYSTEMS

Editor-in-Chief: **M. S. Shur**

Published

Vol. 7: Current Trends in Optical Amplifiers and Their Applications
ed. *T. P. Lee*

Vol. 8: Current Research and Developments in Optical Fiber Communications in China
eds. *Q.-M. Wang and T. P. Lee*

Vol. 9: Signal Compression: Coding of Speech, Audio, Text, Image and Video
ed. *N. Jayant*

Vol. 10: Emerging Optoelectronic Technologies and Applications
ed. *Y.-H. Lo*

Vol. 11: High Speed Semiconductor Lasers
ed. *S. A. Gurevich*

Vol. 12: Current Research on Optical Materials, Devices and Systems in Taiwan
eds. *S. Chi and T. P. Lee*

Vol. 13: High Speed Circuits for Lightwave Communications
ed. *K.-C. Wang*

Vol. 14: Quantum-Based Electronics and Devices
eds. *M. Dutta and M. A. Stroscio*

Vol. 15: Silicon and Beyond
eds. *M. S. Shur and T. A. Fjeldly*

Vol. 16: Advances in Semiconductor Lasers and Applications to Optoelectronics
eds. *M. Dutta and M. A. Stroscio*

Vol. 17: Frontiers in Electronics: From Materials to Systems
eds. *Y. S. Park, S. Luryi, M. S. Shur, J. M. Xu and A. Zaslavsky*

Vol. 18: Sensitive Skin
eds. *V. Lumelsky, M. S. Shur and S. Wagner*

Vol. 19: Advances in Surface Acoustic Wave Technology, Systems and Applications (Two volumes), volume 1
eds. *C. C. W. Ruppel and T. A. Fjeldly*

Vol. 20: Advances in Surface Acoustic Wave Technology, Systems and Applications (Two volumes), volume 2
eds. *C. C. W. Ruppel and T. A. Fjeldly*

Vol. 21: High Speed Integrated Circuit Technology, Towards 100 GHz Logic
ed. *M. Rodwell*

Vol. 22: Topics in High Field Transport in Semiconductors
eds. *K. F. Brennan and P. P. Ruden*

Vol. 23: Oxide Reliability: A Summary of Silicon Oxide Wearout, Breakdown, and Reliability
ed. *D. J. Dumin*

Vol. 24: CMOS RF Modeling, Characterization and Applications
eds. *M. J. Deen and T. A. Fjeldly*

Vol. 25: Quantum Dots
eds. *E. Borovitskaya and M. S. Shur*

Selected Topics in Electronics and Systems – Vol. 26

FRONTIERS IN ELECTRONICS

Future Chips

Proceedings of the 2002 Workshop on Frontiers in Electronics (WOFE-02)

St. Croix, Virgin Islands, USA   6 – 11 January 2002

Editors

Yoon Soo Park
Office of Naval Research, USA

Michael S. Shur
Rensselaer Polytechnic Institute, USA

William Tang
DARPA, USA

World Scientific
New Jersey • London • Singapore • Hong Kong

Published by

World Scientific Publishing Co. Pte. Ltd.
5 Toh Tuck Link, Singapore 596224
USA office: Suite 202, 1060 Main Street, River Edge, NJ 07661
UK office: 57 Shelton Street, Covent Garden, London WC2H 9HE

British Library Cataloguing-in-Publication Data
A catalogue record for this book is available from the British Library.

FRONTIERS IN ELECTRONICS: FUTURE CHIPS
Proceedings of the 2002 Workshop on Frontiers in Electronics (WOFE-02)

ISBN 981-238-222-4

This book is printed on acid-free paper.

Printed in Singapore by Mainland Press

PREFACE

Workshop On Frontiers in Electronics (WOFE-02) in St Croix, Virgin Islands, in January 2002, was the third in the series of the WOFE workshops. Over seventy leading experts from academia, industry, and government agencies came to St. Croix to report on the most recent developments in their fields and to exchange views on future trends and directions of the electronics and photonics industry. The issues they addressed ranged from system-on-chip to DNA doping, from ultra-thin SOI to electro textiles, from photonics integration on ULSI platform to wide band gap semiconductor devices and solid state lighting. Rapid pace of electronic technology evolution compels a merger of different technical areas, and WOFE-02 provided a unique opportunity for cross-fertilization of emerging fields of microelectronics, photonics, and nanoelectronics. The Workshop was informal and stimulated provocative views, visionary outlooks, and discussions of controversial issues.

The plenary session opened with a talk by Professor Yariv on periodic optical structures (which include what is now called photonic crystals). These structures allow one to control the confinement of optical radiation. DFB lasers, fiber Bragg gratings, photonic bandgap defect mode lasers, adiabatic mode filters, control of spontaneous emission, defect mode waveguides in 2D media, and coupled resonator optical waveguides are examples of applications relying on periodic optical structures.

The second plenary paper by Dr. Hee-Gook Lee et al dealt with the applications of MEMS (Micro Electro Mechanical Systems) Technology for optical telecommunication and RF telecommunication technologies, where the demands of smaller volume, higher performance, lower cost, and lower power mobile handsets with multi-band functionality require the use of RF MEMS.

In his plenary talk, Professor Shklovskii applied the ideas of low temperature physics to complicated phenomena of screening of macro ions such as DNA, proteins, lipid membranes or colloids in water solutions by multivalent counter ions. He showed that multivalent counter ions form a strongly correlated two-dimensional liquid at the macro ion surface and explained the phenomenon of charge inversion that was observed in the electrophoresis of several biological and non-biological macro ions.

The ancient pictograph in Fig. 1 shows the transition from a sleigh to a wheeled vehicle. In our view, this is a great illustration of a scientific discovery that we can compare to the transition from a transistor to an integrated circuit. One of the most interesting sessions of WOFE-02 was the session on Ultra Large Scale Integrated circuits organized by Professors Iwai and Nishi who provided a detailed summary of this session opening these Proceedings.

Fig. 1. Pictograph of wheeled vehicle from Erech, Mesopotamia, 3500 B.C

The decade of 2001 – 2010 promises to become the decade of nanotechnology and nanoelectronics. In their paper in this issue, Professor Robert Hull, Dr. Richard Martel, and Professor J. M. Xu give a brief summary of current activities in the fabrication and integration of active components with dimensions of tens of nanometers or less, with particular emphasis on nanoelectronic devices discussed at WOFE-02.

Drs. Elana Ethridge and Dick Urban organized the session on electrotextiles and flexible electronics. Their paper in this issue discusses the merging of electronics and textiles and the synergy between the electronics and textile industries. Two other papers on electrotextiles published in this issue present technical challenges and early technical results obtained in this emerging field.

Wide band gap semiconductor electronics and photonics were richly represented at WOFE-02. The paper by Professor Khan et al deals with Strain Energy Band Engineering (SEBE) approach to AlN/GaN/InN heterojunction devices. This approach (coupled with Pulsed Atomic Epitaxy (PALE) technique) improves the performance of heterojunction GaN-based devices by controlling strain and/or lattice mismatch. It was successfully used to develop novel GaN-based transistors, UV light emitting diodes, and solar blind photodetectors.

Professor E. Munoz et al describe the applications of nitride UV devices in biology, medicine, combustion control, water monitoring, and UV detection. Their paper is followed by the paper by Professor Žukauskas et al dealing with the optimization of white all-semiconductor lamps for solid-state lighting applications. Present lighting technology relies on tungsten incandescent and compact fluorescence lamps for residential use, fluorescence lamps for work environment, and sodium lamps for street lighting. Today, 21% of energy use is in lighting, and, perhaps, half of this energy or more can be saved by switching to efficient and cold solid-state lighting sources. Solid-state lighting will use visible and UV light emitting diodes that are projected to reach lifetimes exceeding 100,000 hours. From traffic lights to road signs, from automobile taillights to outdoor displays, from landscape to accent lights, solid-state light sources have already arrived as harbingers of the next lighting revolution.

The paper by Professor Ozpineci et al discusses the system impact of silicon carbide power devices. These devices have many advantages for high voltage power electronics applications such as a hybrid electric vehicle (HEV) traction drive.

Two other papers on wide band semiconductor materials and devices in this issue (by Professor Matulionis et al and Dr. Rumyantsev et al) deal with hot-phonon limited electron energy relaxation and low frequency noise phenomena in AlN/GaN heterostructures.

The terahertz session organized by Professor Jim Allen and Dr. Dwight Woolard dealt with applications of terahertz technology for sensing of chemical and biological warfare agents and with new approaches to detecting and generating terahertz radiation.

Dr. Samuels et al discuss how to discriminate between hazardous materials in the environment and ambient constituents. Naturally occurring bacteria, plant pollen, fungi, and other airborne materials make this task very challenging. The spectroscopic properties of the chemical warfare agents in the long wavelength infrared region have been successfully used for environmental sensing. However, for biological warfare agents, such radiation is much less effective, and recent studies of the spectroscopic properties of biological agents in the terahertz range might provide useful spectral signatures of these materials.

The paper by Professor Gelmont et al discusses the effect of polar-optical phonons on polar semiconductor device behavior at terahertz frequencies. Their results suggest that Bi_2S_3 Schottky diode based multipliers have potential as enhanced sources for the terahertz range.

X. Peralta et al demonstrated resonant detection of terahertz radiation by two-dimensional plasma waves in two field effect devices: a commercial field effect transistor (FET) and a double quantum well field effect transistor with a periodic grating gate. In both devices, the standing 2-D plasmon is tuned to the frequency of the THz radiation by varying the gate bias. The double quantum well field effect transistors exhibit a rich photoconductive response corresponding to spatial harmonics of the standing 2-D plasmons under the metal part of the periodic gate.

The section of this issue on advanced concepts opens with the paper by Dr. Johnstone who discusses the issue of the thermal management in optoelectronics devices. He demonstrates that integrated cooling should improve the performance of infrared lasers and detectors.

Dr. Blalock et al describe a novel multiple-gate MOS-JFET transistor that combines two different transistors, JFET and MOSFET, in a single silicon island. This device can be viewed as a four-gate transistor with each of these four gates controlling the conduction characteristics.

Professor Dorojevets considers the architecture and design of an 8-bit flux-1 superconductor Rapid Single Flux Quantum (RSFQ) microprocessor. The first single-chip superconductor FLUX-1 microprocessor has been designed in the RSFQ logic and fabricated using Josephson junction technology. This microprocessor chip contains 65,759 Josephson junctions on a 10.6 mm x 13.2 mm die with flip-chip packaging.

The paper by Dr. Refaat et al in the sensor application section of this issue reports on the spectral response measurements of quantum short wave infrared detectors. The detectors used in their study comprise InGaAs PIN diodes and InGaAsSb avalanche photodiodes (APDs) for short wave infrared applications. Dr. Nurul Abedin et al report on the noise characteristics of these compound photodetectors for 2 μm lidar/dial remote sensing applications.

The modeling and simulation section features the papers by Drs. Fossum, Fok and Pulfrey, Suris and Dmitriev, and Mayergoyz and Andrei. The paper by Professor Fossum deals with device design issues and performance of extremely scaled CMOS. The discussed simulation tools include a 2D numerical device simulator, a 1D numerical Poisson-Schrödinger solver, and a generic, physics/process-based MOSFET compact model in SPICE. Drs. Fok and Pulfrey examine power-supply noise using circuit simulation of the complete integrated circuit power net. Drs. Suris and Dmitriev discuss electron localization in ideal 2D and 3D quantum dot superlattices in a homogeneous electric field and Bloch oscillations in such structures. Finally, the paper by Drs. Mayergoyz and Andrei presents a new approach to the analysis of random dopant-induced effects in semiconductor devices. This approach based on the perturbation technique is applied to the analysis of random dopant-induced fluctuations of threshold voltages and frequency response characteristics of MOSFET devices.

On behalf of the WOFE Organizing, Program, and Steering Committees, we would like to thank all WOFE-02 participants and especially the invited contributors to this issue for making WOFE-02 a successful conference. We also gratefully acknowledge generous support of this workshop by the US Office of Naval Research, US Defense Advance Research Project Agency, NSF, US Army Research Office, and US European Army Research Office. Our special thanks go the Members of Organizing, Program, and Steering Committees, and to Session Organizers for their tireless work and inspiration.

This material is based upon work partially supported by the National Science Foundation under Grant No. 0114810.

Yoon Soo Park
Michael S. Shur
William Tang

Editors

CONTENTS

International Journal of High Speed Electronics and Systems, Vol. 12, No. 2 (2002) 207–214

OPTICAL WAVE PROPAGATION IN PERIODIC STRUCTURES

AMNON YARIV
Department of Applied Physics 128-95, California Institute of Technology
Pasadena, California 91125, USA

SHAYAN MOOKHERJEA
Department of Electrical Engineering 136-93, California Institute of Technology
Pasadena, California 91125, USA

We present a formalism for optical waveguiding in two-dimensional periodic structures by means of Bragg reflection. The theory provides analytic expressions for the field modal profiles, dispersion, and attenuation or gain in the waveguide. Line defects in two-dimensional photonic crystal slabs are used as the principal example.

Keywords: waveguides; laser amplifiers; guided waves

1. Introduction

The use of periodic structures to affect the propagation of electromagnetic waves dates back many years. As a matter of fact, the title of this paper with the exception of the word "optical" is the same as a 1946 monograph by L. Brillouin[1]. For many years the main application of periodic structures was in microwave amplifiers and oscillators where, as "slow wave structures" they were used to slow the electromagnetic wave to a point where it could be velocity-synchronized and exchange power with an electron beam. This is the basis of the operation of the traveling wave tube amplifier [2].

In the optical regime, periodic optical structures found their main applications as Bragg grating reflection filters in fibers and in distributed feedback (DFB) semiconductor lasers. These two devices, which are based on the periodic modulation of the refractive index, are now two of the main weapons in the optical communications arsenal. They are designed to operate in a regime where the wavelength is approximately twice the spatial period. In this Bragg regime, the waves are reflected strongly due to constructive interference of reflections from all the periods, and consequently, do not penetrate very far into the periodic medium[1].

The formalism applied nearly from the outset to describe these phenomena and devices was that of the coupled modes. This approach is valid when the relative periodic index modulation is small, i.e., $\Delta n \sin kx$ satisfies $\Delta n/n \ll 1$. The fundamental reason is that in this limit, the Floquet-Bloch theorem which expresses the wave propagation in the periodic medium as an infinite sum,

$$\mathbf{E} = \sum_{m=-\infty}^{\infty} \mathbf{e}_m \exp\left[i\left(\omega t - \beta_0 - m\frac{2\pi}{\Lambda}z\right)\right], \tag{1}$$

where Λ is the period, reduces to just two terms[3] corresponding to forward and backward-propagating modes.

The coupled-mode theory[4] has proven immensely useful and is the basis for most of the theoretical work as the design of actual devices. The success is due in large measure to the simplicity and the intuitive form which lends itself to invention. It is also very close in spirit to the coupled-mode formalism used to describe other phenomena such as electrooptic, acoustooptic, and phase conjugate wave interactions, and parametric amplification, as well as time-dependent perturbation theory in quantum mechanics.

In recent years, following a theoretical suggestion by Yablonovich[5] and John[6], interest was focussed anew on this area now under the guise of "photonic crystals". The original proposals were to fabricate three-dimensional periodic media with periodicities in the vicinity of the optical wavelength. Such structures would display optical properties not unlike those possessed by electrons in the periodic environment of crystals. Specifically, there would exist "forbidden gaps" of frequencies in which an incident optical wave will be strongly reflected due to Bragg resonance. The daunting problem of fabricating such 3D structures and the tasks of coupling between such structures and conventional optical waveguides led to a major shift of emphasis from 3D photonic crystals to two-dimensional structures in which optical confinement in one dimension is achieved by conventional dielectric slab optical waveguiding while a 2D array of holes provides Bragg confinement in the remaining two dimensions. In these hybrid structures, the basic distinction between the early periodic structures and photonic crystals has disappeared. One difference which is cultural rather than real is that photonic crystals have come to be associated with strong index perturbations $\Delta n/n \approx 1$, where formal solutions are not feasible. Most of the work in this field thus relies on numerical solutions*to Maxwell's equations[7]. The strong index perturbations are most often achieved by drilling holes into slab dielectric waveguides such as the regions shown in Fig. 1. The large index modulation in these cases cases causes the optical field to evanesce to insignificance in a small number of periods. This has given rise to the so-called defect

*__AY:__ I personally find myself straddling the fence. As can be seen from the list of references, I have worked in both camps. This perhaps allows me the privilege of neutrality. (The remaining alternative is someone who worked in neither field.) My opinion is that both endeavors are justified. The "photonic crystal" numerical calculations are needed because real structures are often not amenable to formal solutions. The advantage of the periodic optical structures ($\Delta n/n \ll 1$) camp is that the formal approach is what is needed for invention and seeing relationships between phenomena. This might explain why most of the "working" devices and phenomena were invented and explained using this point of view.

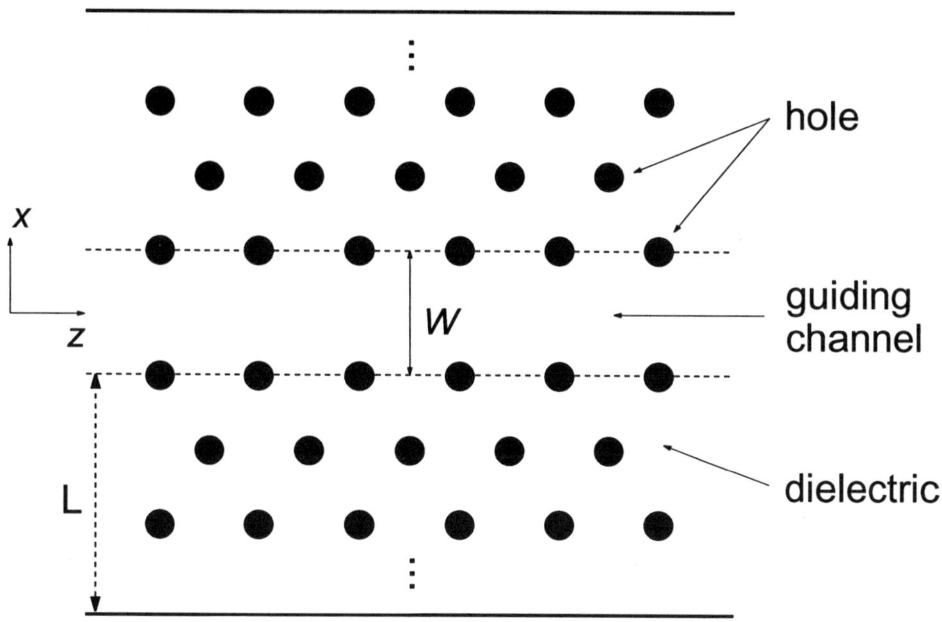

Fig. 1. A triangular array of air holes bounding a guiding channel (GC). The structure is shown in the top view.

mode resonators[8]. Such resonators are formed when the periodicity of the photonic bandgap is interrupted or perturbed. A defect mode can be produced, for example, by "replugging" one hole; a wave starting within such a defect cannot escape since it is reflected by Bragg scattering in all three directions. This confinement creates a defect mode optical resonator, which may be "pumped" (e.g., optically) to form a laser whose mode volume is $\ll \lambda^3$. It is interesting to note that the commercial DFB laser is also based on a defect mode except that here the defect is called "$\lambda/4$ slip"[9].

To illustrate the power of the analytic approach, we consider the case of optical guiding in periodic structures. The top view of such a waveguide is shown in Fig. 2. The dark dots indicate drilled holes. There have been many papers, experimental as well as numerical, dealing with such waveguides which are usually formed by leaving out i.e., "replugging" a row of holes. What has recently emerged from these is that there is a preferred region of channel width W where the transmission losses are smallest but no theoretical explanation is offered for this observation. In what follows, we will show that applying the coupled-mode formalism to this scenario affords an intuitive picture of the guiding and, in addition to explaining the preferred values of W, suggest some new experimental situations.

2. Formalism

We assume that the structure, as shown in Fig. 1, has no variations along the

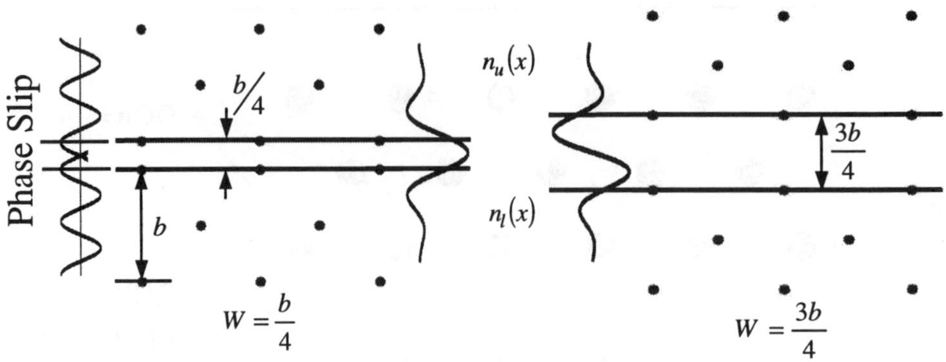

Fig. 2. A triangular array of air holes bounding a guiding channel (GC) of width W. The structure is shown in the top view, in the $x - z$ plane.

y axis (out of the plane). As functions of $\mathbf{r} = (x, z)$ and t, we look for solutions of Maxwell's equations in the guiding channel,

$$\nabla^2 \mathbf{E}(\mathbf{r}, t) - \mu\epsilon(\mathbf{r}) \frac{\partial^2 \mathbf{E}(\mathbf{r}, t)}{\partial t^2} = 0, \tag{2}$$

that represent waveguide modes. The amplitude of such a mode will decay exponentially with increasing distance away from the guiding channel, since Poynting power propagates in the the z direction. The solutions of this equation are harmonic or plane waves. Working in one (implicit) polarization, we take the solution to be of the form

$$E(\mathbf{r}, t) = \frac{1}{2} E_{\mathrm{i}} \exp\left[i\left(\omega t - k_{\mathrm{i}\perp} x - k_{\mathrm{i}\parallel} z\right)\right] + \frac{1}{2} E_{\mathrm{r}} \exp\left[i\left(\omega t + k_{\mathrm{r}\perp} x - k_{\mathrm{r}\parallel} z\right)\right] \tag{3}$$

i.e., two intersecting plane waves (Fourier components) propagating at an equal angle to the z axis. The amplitudes E_{i} and E_{r} are mutually dependent as a consequence of Bragg reflection into each other at the interfaces between the guiding channel and the periodic media.

To further analyze this, we need to evaluate the complex reflection coefficients at the two interfaces. We carry out the analysis for the upper interface, with analogous results applicable to the lower interface. Taking advantage of the 2D periodicity of the index of refraction in this region, we can expand $\Delta n^2(\mathbf{r})$ in a Fourier series,

$$\Delta n^2(x, z) = \sum_{m,n} \frac{\Delta n^2_{mn}}{2} e^{-i\left[K_{mn\perp} x + K_{mn\parallel} z\right]}, \tag{4}$$

where $K_{mn\perp} = m(2\pi/b)$ and $K_{mn\parallel} = n(2\pi/a)$ are the components of the reciprocal lattice vector in this geometry. The coefficients Δn^2_{mn} are given by

$$\Delta n^2_{mn} = \frac{1}{ab} \int_{-a/2}^{a/2} dz \int_{-b/2}^{b/2} dx\, \Delta n^2(x, z)\, e^{i\left[K_{mn\perp} x + K_{mn\parallel} z\right]}. \tag{5}$$

Applying the usual techniques of coupled-mode theory [3,10], we can derive the equations describing the spatial evolution of $E_{\rm i}$ and $E_{\rm r}$ inside the periodic structure, assuming for simplicity that $n = 0$ so that $K_{mn\|} = 0$ and $k_{{\rm i}\|} = k_{{\rm r}\|} = k_{\|}$. Introducing a coupling coefficient κ, we write

$$\begin{aligned} \cos\theta_{\rm i}\frac{dE_{\rm i}}{dx} &= \kappa^* E_{\rm r}(x)e^{i2\Delta k_\perp x}, \\ \cos\theta_{\rm r}\frac{dE_{\rm r}}{dx} &= \kappa\, E_{\rm i}(x)e^{-i2\Delta k_\perp x} \end{aligned} \tag{6}$$

where

$$\kappa \equiv \frac{i\omega}{2c}\Delta n^2_{mn} \tag{7}$$

and $\Delta k_\perp$ is the detuning from the k-vector conservation condition:

$$2\Delta k_\perp \equiv k_{{\rm i}\perp} + k_{{\rm r}\perp} - m\frac{2\pi}{b}. \tag{8}$$

On resonance, therefore, the two-dimensional propagation problem is reduced to two coupled-wave equations involving a single coordinate x and a dispersion relation,

$$k_\| = \sqrt{\left(\frac{\omega}{c}\right)^2 n_{00}^2 - \left(m\frac{\pi}{b}\right)^2} \tag{9}$$

using $k_{{\rm i}\perp} = k_{{\rm r}\perp} = m\pi/b$. In the above equation, n_{00} is defined in terms of the background refractive index, $n_{00} = n_0 + \Delta n_{00}$.

We can write the solution by defining the reflection coeffcients at the upper and lower interfaces bounding the guiding channel,

$$\begin{aligned} \hat{r} &= \left.\frac{E_{\rm r}(x)}{E_{\rm i}(x)}\right|_{x=\frac{W}{2}} = -\left(\frac{\hat{\kappa}^*_{n0}}{|\hat{\kappa}_{n0}|}\right)\tanh(|\hat{\kappa}_{n0}|L), \\ \check{r} &= \left.\frac{E_{\rm i}(x)}{E_{\rm r}(x)}\right|_{x=-\frac{W}{2}} = -\left(\frac{\check{\kappa}^*_{n0}}{|\check{\kappa}_{n0}|}\right)\tanh(|\check{\kappa}_{n0}|L), \end{aligned} \tag{10}$$

where L is the thickness of the periodic structures in the transverse (x) direction as shown in Fig. 1.

In the triangular lattice structure, we assume hole dimensions are small compared to a and b so that to a good approximation,

$$\Delta n^2(x,z) = (n_0^2 - 1)A_0\left[\delta(x)\delta(z) + \delta\left(x - \frac{b}{2}\right)\delta\left(z - \frac{a}{2}\right)\right] \tag{11}$$

where $A_0 = \pi r^2$ is the area of the hole written in terms of r, the radius of the hole. A more detailed model may be adopted in numerical recipes if necessary. We find that

$$\Delta n^2_{mn} = (n_0^2 - 1)\,\frac{A_0}{ab}\left[1 + (-1)^{m+n}\right]. \tag{12}$$

We find that $\Delta n^2_{m0} = 0$ for all odd m, and in particular, for $m = 1$. The lowest-order non-zero Fourier coefficient of $\Delta n^2(x, z)$ that defines κ according to Eq. (7) is, for $m = 2$,

$$\begin{aligned} \Delta n^2_{20} &= 2(n_0^2 - 1)\frac{A_0}{ab}, \\ \text{i.e., } \kappa \equiv \kappa_{20} &= \frac{i\omega}{c}(n_0^2 - 1)\frac{A_0}{ab}. \end{aligned} \tag{13}$$

If the longitudinal (z) phasematching is not exact, then we may maintain the same formalism by redefining the coupling coefficient,

$$\kappa' \equiv \frac{i\omega}{2c}\Delta n^2_{mn} e^{-i\Delta k_\parallel Z}\frac{\sin(\Delta k_\parallel Z)}{\Delta k_\parallel Z} \tag{14}$$

where $2\Delta k_\parallel \equiv k_{\mathrm{i}\parallel} - k_{\mathrm{r}\parallel} - n(2\pi/a)$ measures the detuning from longitudinal resonance.

To complete the solution, we use $\hat{r}$ and $\check{r}$ as boundary conditions on the field of the guiding channel at $x = W/2$ and $x = -W/2$,

$$\frac{E_{\mathrm{r}} e^{ik_\perp W/2}}{E_{\mathrm{i}} e^{-ik_\perp W/2}} = \hat{r}, \quad \frac{E_{\mathrm{i}} e^{ik_\perp W/2}}{E_{\mathrm{r}} e^{-ik_\perp W/2}} = \check{r}. \tag{15}$$

Taking the upper and lower periodic structures to be identical, so that $\hat{r} = \check{r} \equiv r$, we examine the phase condition of Eq. (15),

$$2k_\perp W - 2\theta_r = s\,2\pi, \quad s = 0, 1, 2, \ldots \tag{16}$$

where θ_r is the phase of r. This condition, which we call "transverse Bragg resonance," states that a wave returns to the starting plane after a round trip with the same phase (modulo 2π). An associated amplitude conservation condition needed for low modal propagation loss requries that $|R| \approx 1$ which is fulfilled if $\kappa L \gg 1$, and may be achieved in other ways, e.g., by providing gain in the periodic structures.

In Eq. (16), we substitute $k_\perp = 2\pi/b$ and $\theta_r = \pi/2$ as appropriate for the triangular lattice geometry. We then obtain a series of quantized widths,

$$W = \left(s + \frac{1}{2}\right)\frac{b}{2} = \frac{b}{4}, \frac{3b}{b}, \frac{5b}{4}, \ldots \tag{17}$$

In contrast with index-guided waveguides, the width of the guiding channel is quantized and is determined by b, and diffraction order m and the reflectance phase θ_r. Further, the usual practice of leaving out a row of holes[11], or an integer number of holes, will not make for a low loss waveguide. Such an approach would result in a (substantial) detuning from the Bragg resonance condition, resulting in lowering $|r|$ and increasing the modal loss of propagation. The spacing $b/4$ of the lowest-order mode is formally equivalent to the $\lambda/4$ phase slip of DFB lasers[9], except that here it involves the transverse dimension x rather than the propagation direction of Poynting power, z.

In the limit $|\kappa L| \gg 1$, we can use the fact that the field is evanescent in the periodic structures bounding the guiding channel to write the modal profiles,

$$E(x,z) = E_0\, e^{-ik_\parallel z} \begin{cases} \cos\left(\dfrac{2\pi}{b}x\right), & |x| \le \frac{b}{8}, \\ \cos\left(\dfrac{2\pi}{b}x\right) \exp\left[-|\kappa_{20}|\left(x - \dfrac{b}{8}\right)\right], & x \ge \frac{b}{8}, \\ \cos\left(\dfrac{2\pi}{b}x\right) \exp\left[-|\kappa_{20}|\left(x + \dfrac{b}{8}\right)\right], & x \le -\frac{b}{8}, \end{cases} \quad \Bigg| \quad W = \frac{b}{4} \tag{18}$$

$$E(x,z) = E_0\, e^{-ik_\parallel z} \begin{cases} \sin\left(\dfrac{2\pi}{b}x\right), & |x| \le \frac{3b}{8}, \\ \sin\left(\dfrac{2\pi}{b}x\right) \exp\left[-|\kappa_{20}|\left(x - \dfrac{3b}{8}\right)\right], & x \ge \frac{3b}{8}, \\ \sin\left(\dfrac{2\pi}{b}x\right) \exp\left[-|\kappa_{20}|\left(x + \dfrac{3b}{8}\right)\right], & x \le -\frac{3b}{8}, \end{cases} \quad \Bigg| \quad W = \frac{3b}{4} \tag{19}$$

Notomi *et al.*[12] have numerically and experimentally investigated transmission in a triangular (hexagonal) lattice, with an observed waveguiding band, in our notation, in the range $W = 0.7b$ to $W = 0.8b$, and have shown that the $W = b$ single-line missing-hole design indeed does not lead to low-loss single-mode waveguiding. Their measured range overlaps our theoretical value of $0.75b$ for the odd solution, Eq. (19); they did not report on the symmetry of the spatial distribution of the mode and did not explore waveguiding at $W = b/4$ where the even mode, Eq. (18), should exist.

In summary, we have applied a coupled mode formalism to obtain closed-form expressions for the properties of traveling wave modes in 2D dielectric waveguides confined by periodically patterned structures. Transverse Bragg resonance is critical to the definition of low-loss propagation and we have shown that the width of the waveguiding channel is quantized, and should not be an integral multiple of the transverse periodicity of the structure. In particular, leaving out a row of holes in a line-defect photonic crystal waveguide does not make for an optimized structure.

Acknowledgments

This work was supported by the U.S. Office of Naval Research. The authors thank G.T. Paloczi for the preparation of Fig. 2.

References

1. L. Brillouin, *Wave Propagation in Periodic Structures*, McGraw-Hill, New York, 1946.
2. J.R. Pierce, *Traveling-wave Tubes*, Van Nostrand, New York, 1950.
3. A. Yariv, *Optical Electronics in Modern Communications*, Oxford, New York, 1997.
4. A. Yariv, "Coupled-mode theory for guided-wave optics", *IEEE Journal of Quantum Electronics*, **QE-9** (1973) 919–933.
5. E. Yablonovich, "Inhibited spontaneous emission in solid-state physics and electronics" *Physical Review Letters* **58** (1987) 2059–2062.

6. S. John, "Strong localization of photons in certain disordered dielectric superlattices" *Physical Review Letters* **58** (1987) 2486–2489.
7. B. D'Urso, O.J. Painter, J.D. O'Brien, T.A. Tombrello, A. Yariv, and A. Scherer, "Modal reflectivity in finite-depth two-dimensional photonic-crystal microcavities", *Journal of the Optical Society of America B*, **15** (1998) 1155-1159.
8. O. Painter, R.K. Lee, A. Scherer, A. Yariv, J.D. O'Brien, P.D. Dapkus, I. Kim, "Two-dimensional band-gap defect mode laser", *Science*, **284** (1999) 1819–1821.
9. H.A. Haus and C.V. Shank, "Antisymmetric taper of distributed feedback lasers", *IEEE Journal of Quantum Electronics*, **QE-12** (1976) 532–539.
10. A. Yariv, "A coupled wave formalism for optical waveguiding by transverse Bragg reflection", accepted by *Optics Letters* (2002).
11. J.D. Joannopoulos, R.D. Meade, and J.N. Winn, *Photonic Crystals*, Princeton University Press, Princeton (1995), p. 102.
12. M. Notomi, A. Shinya, K. Yamada, J. Takahashi C. Takahashi, and I. Yokohama, "Singlemode transmission within photonic bandgap of width-varied single-line-defect photonic crystal waveguides on SOI substrates" *Electronics Letters*, **37** (2001) 293–295.

International Journal of High Speed Electronics and Systems, Vol. 12, No. 2 (2002) 215–233

MEMS TECHNOLOGY FOR ADVANCED TELECOMMUNICATION APPLICATIONS

HEE-GOOK LEE, JAE YEONG PARK, JONG UK BU, and YOUNGJOO YEE

LG Electronics Institute of Technology,
16 WooMyeon-Dong, SeoCho-Gu, Seoul, 137-724, Korea

MEMS (Micro Electro Mechanical Systems) technology is becoming a crucial enabling technology for optical and RF telecommunication applications due to the performance, cost and integration advantages. All-optical switches and silicon optical benches for optical network systems and RF MEMS components such as RF MEMS switches, tunable capacitors, high Q inductors, and thin film bulk acoustic resonators for miniaturized single chip RF transceivers are discussed. Despite all the promising aspects, practical problems to be addressed for successful commercialization are not trivial. By reviewing optical and RF MEMS research activities and some of the implementation examples, key success factors of MEMS components for the commercialization are discussed.

Keywords: Optical/wireless telecommunications; RF MEMS; optical MEMS; bulk micromachining; surface micromachining; transceiver.

1. MEMS Technology

MEMS technology enabling the batch fabrication of miniaturized mechanical structures, devices, and systems has originated from the silicon planar technology, or more specifically from the silicon based sensor technology of the 1960s. Thereafter, significant technical advances and development of infrastructure have occurred in the last decade. The technology will play important roles in many different industries including telecommunication, bio-medical electronics, automobile, power electronics, and consumer electronics. In particular, key application areas experiencing the greatest impact will be micro-displays using actuated micromirror arrays, mass/nano storage including hard disk drives/optical drives, RF/optical switching products, and sensing products such as accelerometers and IR/pressure sensors.

In this paper, the current status of optical and RF MEMS research activities in general will be reviewed. Following that, the various aspects of the MEMS development for advanced telecommunication applications will be discussed.

1.1. *MEMS fabrication: Bulk micromachining*

MEMS process technology can be divided into bulk and surface micromachining. Bulk micromachining is a technology to fabricate microstructures by etching substrates anisotropically or isotropically. Wet etching techniques are performed

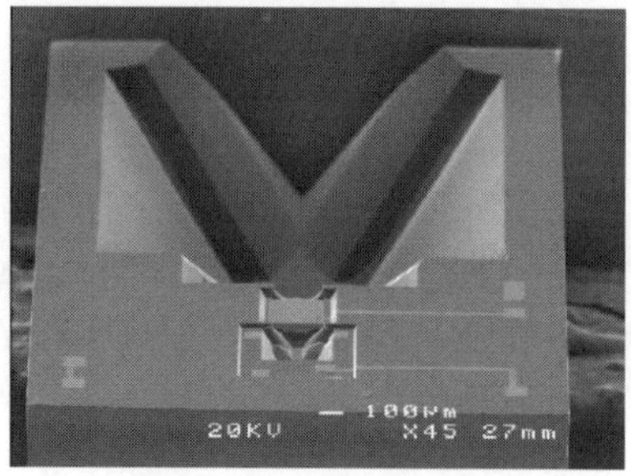

Fig. 1. SEM picture of a bulk micromachined silicon optical bench.

using chemical solutions such as KOH, EDP, and TMAH. Dry etching techniques utilize methods such as RIE, ICP, laser machining, etc.

Anisotropic etching results in different etch rates along different crystallographic directions to form V grooves, pyramids, mirrors and channels. A typical V groove formed on a (100) silicon surface has a slope of 54.74 degrees with respect to the (111) surface of the substrate. Figure 1 shows an SEM picture of a silicon optical bench consisting of different sized etch cavities. Process technology evolution has made precise control of the required dimensions achievable in production. Isotropic etching can be used to form hemispheres and the size can be controlled by varying the etch window pattern. It is possible to get polymeric micro lens by combining a silicon hemisphere mold with microinjection molding technology. The diameter of the hemispheres can be as small as 200 μm.

1.2. *MEMS fabrication: Surface micromachining*

Surface micromachining is a technology used to create microstructures on top of a substrate by depositing and etching materials such as metals, polysilicon, nitride,

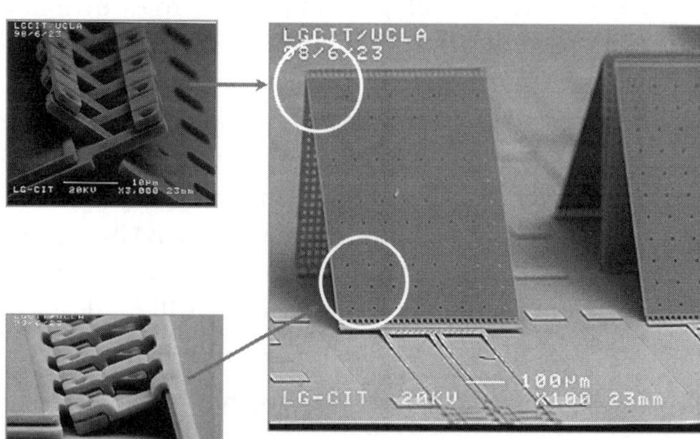

Fig. 2. SEM picture of scratch drive actuator (SDA).

(a) (b) (c)

Fig. 3. SEM pictures of (a) a 45 degree optical mirror fabricated by using SU-8 epoxy resin, which is integrated with a laser diode on a silicon substrate, (b) a suspended high Q inductor using electroplated copper and (c) a micromold, by using electroplated nickel and SU-8 epoxy resin as a mold for plating.

oxide, organic materials, etc. Various micro-components and systems comprising of sensors and actuators can be made using surface micromachining technology.

The commonly used deposition techniques are evaporation, sputtering, screen-printing, CVD, spin-casting, electroplating, etc. Dry and wet etching techniques are also employed. Representative examples made possible by surface micromachining are shown in Figs. 2 and 3. Figure 2 shows a scratch drive actuator (SDA) fabricated using polysilicon as the structural material. Several other structures constructed with surface micromachining techniques are shown in Fig. 3. They are the optical mirror, the inductor and the micro mold respectively.

2. MEMS for Optical Telecommunication

2.1. *Optical switch*

Inherent advantages of MEMS for the field of optical telecommunication can be realized by the integration of mechanical, electrical, and optical functional building blocks on a single substrate. Such integration provides precise alignment, easy assembly and lower cost.

Recently there has been growing interest on the high capacity optical cross-connect (OXC) switch for the applications in the Dense Wavelength Division Multiplexing (DWDM) systems. To implement an OXC, two- or three-dimensional configurations of the micromirrors are used. A large-scale OXC based on the electrostatic actuation has been successfully demonstrated.[1] Figure 4 shows a unit cell of the micromirror array of the large scale OXC. Architecture of the OXC is based on the actuated micromirror with two degrees of freedom (2-DOF) tilting motion. The traffic of a light signal from an input fiber is redirected and bound for its destination output node via the free space by a pair of the 2-DOF mirrors. When the node counts of the input and the output fibers are N, respectively, a set of two arrays, each with N 2-DOF individually actuated mirrors is sufficient to switch a signal from any input fiber node to any other output node. Therefore, the total

Fig. 4. An electrostatic micromirror with two degrees of freedom for all-optical cross connects (Lucent Technologies).

number of the actuated micromirrors required for this free space OXC configuration is only $2N$. However, the actuation of the 2-DOF micromirrors should be controlled in a virtually analog manner, not by "ON" and "OFF" switching. This means that the actuation resolution of each micromirror requires at least N levels while maintaining sub-micron position tolerance. Furthermore, the complexity of the control logic for the $2N$ 2-DOF micromirrors array is proportional to N^2 because each of N mirrors requires N controlled position.

As the node count of the fibers increases, the dynamic range of the actuation of the micromirror has to be increased. However, the stable operating range of the actuation by the electrostatic force should be limited by the pull-in phenomenon. The micromirror can be pulled abruptly and collapsed to the fixed counter electrode if the applied bias voltage between the two electrodes exceeds the threshold value, causing the tilting angle of the micromirror to be in an unstable pull-in position. In general, one third of the initial gap between the moveable electrode and the fixed one at null bias condition is a first order estimation of the stable actuation range of the electrostatic actuators of a basic parallel plate capacitor configuration. An alternative approach, which removes this kind of instability related to the electrostatic actuation principle, is the use of piezoelectric unimorph actuator comprised of a metal-PZT-metal sandwich structure that utilizes the rotation of the suspending micromirror around the two orthogonal axes.[2] It can be actuated upward or downward by the change in the strain of the PZT layer. Figure 5(a) shows the unit cell of the PZT actuated micromirror array. Because the maximum applied voltage for the operation range of the PZT actuated micromirrors is only 15 V, this architecture is suitable for low voltage applications.

The pitch between the micromirrors in the two-dimensional array is limited by the diameter of the optical fiber. For large scale OXC's, the micromirrors should have a large tilting angle of actuation to be able to redirect the light into any micromirrors in the increased area of the other mirror array chip. Because of the pull-in limit, the air-gap between the moveable mirror and the fixed forcer electrode should be large enough to guarantee the increase in the stable actuation characteristics. The micromirrors of the large scale OXC employed additional thermal

(a) (b)

Fig. 5. (a) SEM picture of PZT actuated 2-DOF micromirror for free space optical cross connects, (b) a photograph of packaged micromirror array for a 16 × 16 OXC. (LG Electronics).

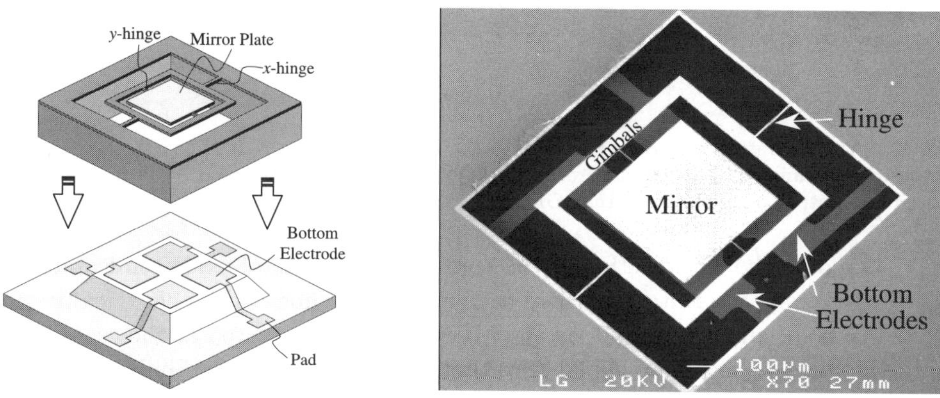

(a) (b)

Fig. 6. (a) A schematic drawing of an electrostatic 2-DOF micromirror having built-in large air-gap for large-scale free space OXC, (b) micromirror after assembly (LG Electronics).

actuators to achieve sufficiently large electrostatic actuation gap. The large gap can be facilitated by the other actuation methods such as the SDA.[3] Both the thermal actuator and the SDA make the device and its controls more complex and they require additional power consumption in addition to tilting the micromirror itself, which increase the complexity in control. A new type of electrostatic micromirror is proposed, requiring no additional actuators through choosing a large built-in air-gap by bulk micromachining.[4] The micromirror in Fig. 6 can have large actuation angles of up to ±10 degrees. The top substrate, including the suspended micromirror array, can be aligned passively by picking and placing it on the bottom substrate with the pyramidal mesas, on which the bottom electrodes have been delineated.

Several other types of optical switches have been developed based on the matrix array of the micromirrors actuated in the bi-state positions. The configuration is

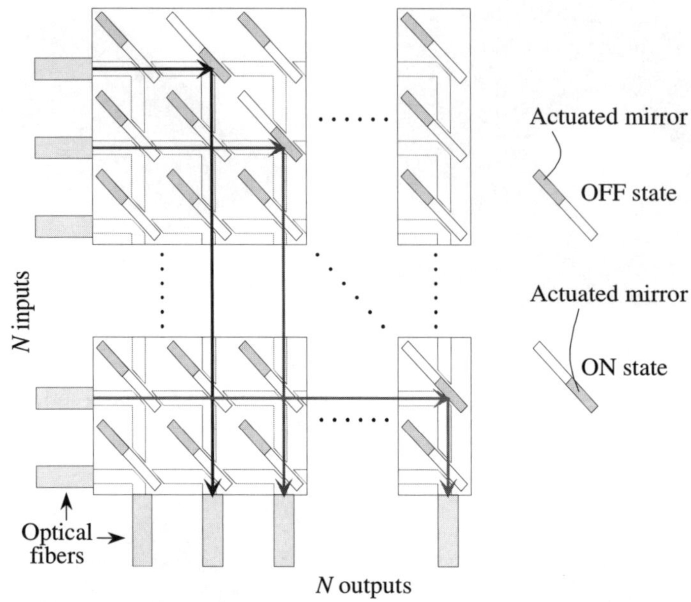

Fig. 7. An illustration of the optical cross-point switching by an $N \times N$ matrix array of the actuated micromirrors. The micromirror in the "ON state" redirects the laser beam from an input fiber to an output node. Those in the "OFF state" stay out of the light path.

very similar to the construction of the complementary metal oxide semiconductor (CMOS) cross-point switches, where a column metal line can be electrically connected to a specific row output by the MOSFET switch located at the cross point of the input and the output signal lines that one wants to switch. Similarly the micromirror at the cross point redirects the path of the light trajectory to the desired output node, as shown in Fig. 7. Because the required actuation of each micromirror for this matrix configuration is only "ON" and "OFF" bi-state characteristics, the control logic determining the states of the micromirrors is simple and the resolution or the accuracy of the actuation is not as critical as the 2-DOF micromirrors of the free space OXC's. However, the total number of the micromirrors required for N-input and N-output fabric is N^2. For this type of OXC, the cost increases rapidly as N increases. The acceptable insertion loss for OXC is approximately -7 dB, which can be realized up to 32×32 OXC with current technology. Various actuation principles for the matrix switches have been reported so far. Flip-up micromirror actuated by electro-magnetic force is another approach of the matrix switch.[5] Microfluidics utilizing the thermo-capillary force is yet another approach.[6]

2.2. *Silicon optical bench for photonic devices*

In optical module assembly, precise alignment of optical fibers to laser diodes or photo diodes is essential. Passive alignment using silicon optical bench is attractive due to the precise alignment, ease of assembly and lower cost.

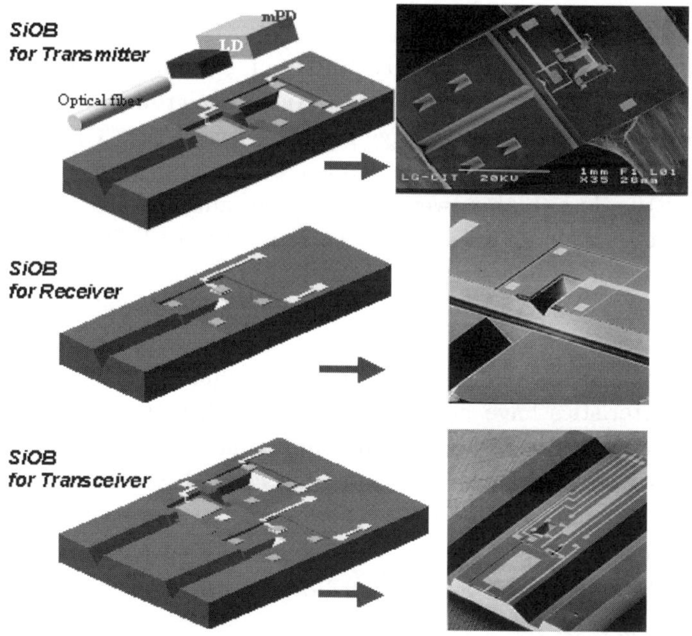

Fig. 8. Schematic drawings and photomicrographs of micromachined silicon optical benches for single chip transmitter, receiver, and transceiver applications (http://misob.lg-elite.com).

The passive alignment techniques include flip-chip bonding technology using the surface tension of molten solder, the index alignment of optical devices with the marker, and the mechanical contact alignment on the precisely fabricated planar light waveguide or micromachined silicon motherboard. Anisotropic etching of silicon substrate provides the dimension control to make a silicon optical bench shown in Fig. 8. Besides anisotropic etching, other key technologies needed to build a silicon optical bench are the patterning of metal electrodes on three-dimensional surfaces, the convex corner compensation technology, and the solder deposition technology.

3. MEMS for Wireless Telecommunication

In wireless telecommunication, the need for compact, higher performance, lower cost, and lower power mobile handsets with multi-band, multi-mode functionality is creating an opportunity for the use of RF MEMS technology. MEMS technology enables mass production of RF devices or systems through their batch fabrication. A single chip RF transceiver, one of the key elements in the wireless communication system, can be realized by fabricating micro-scale RF components and integrating them with MMIC's or RFIC's.

3.1. *RF MEMS switches*

So far, semiconductor switches such as FET's or PIN diodes are widely used in millimeter wave integrated circuits (MMIC's) and microwave circuits. However, these semiconductor switches have large insertion loss in the "ON" state and poor isolation in the "OFF" state at high frequency. The semiconductor switches also have large resistive loss, high power consumption, and low signal power handling capability. Some semiconductor switches, such as silicon FET's, can handle a high level of signal power at low frequency, but the performance degrades rapidly as operating frequency increases. GaAs metal-semiconductor field effect transistors (MESFET's)[7,8] and PIN diodes[9] operate well at high frequency, but have limited signal power handling capability. Accordingly, switching elements with better performance characteristics have been demanded to substitute these semiconductor switches. Recently developed MEMS micromechanical switches seem to have become the new switching elements. The micromechanical switches have low resistive loss, low power consumption, good isolation, and high power handling capability compared to the semiconductor switches. RF MEMS switches also have integration compatibility with other electronic components to realize a compact module assembly.

Micromechanical switches were among the first devices studied in the RF MEMS field. In early 1979, a membrane-based RF MEMS switch was fabricated on silicon.[10] Since then, many different types of RF MEMS switches with electrostatic actuation designs[11] including cantilevers[12–14] and multiple-supported or membrane structures[15–18] have been investigated. Applications of RF MEMS switches include signal routing, impedance matching networks, arrayed capacitor banks, phase shifters, adjustable gain amplifiers, etc.

The RF MEMS switches can be realized by two different approaches: metal contact[12] and capacitive coupling.[16] The switches can also be categorized into two types such as serial and shunt, as shown in Fig. 9. Most metal contact switches use the serial switch configuration, while capacitive coupling switches use the shunt switch configuration. The metal contact switches use metal-to-metal ohmic contact between two electrodes, as illustrated in Fig. 9(a). The MEMS metal contact switch shown in Fig. 10 covers a broader frequency range, but has a shorter lifetime than

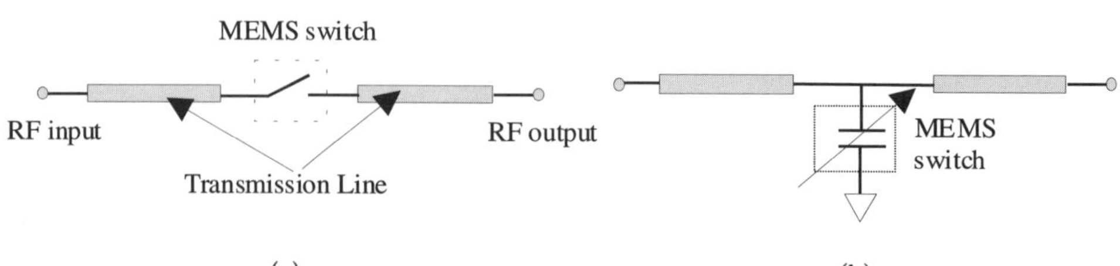

Fig. 9. Schematic drawing of (a) metal contact serial switch configuration and (b) capacitive coupling shunt switch configuration.

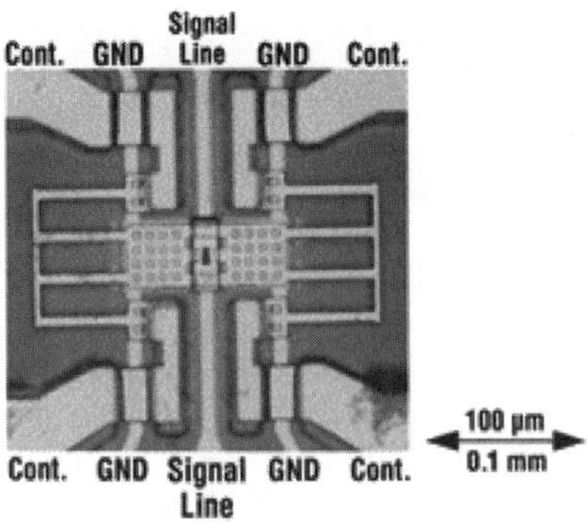

(a)

(b)

Fig. 10. SEM pictures of metal contacting resistive RF MEMS switches from (a) RSC and (b) Northeastern University.

the capacitive coupling switches. The contact material should be selected based on the contact resistance, metal sticking behavior, lifetime, and environmental and packaging compatibility.

The MEMS capacitive coupling switch has a thin dielectric film and an air gap between the two metal electrodes. Silicon nitride or oxide is commonly used as a dielectric layer in the capacitive RF MEMS switches. The capacitive switch controls electrical current or signal by using the on/off impedance ratio, while the resistive switch utilizes the mechanical connection of two isolated transmission lines. Thus, high capacitance on/off ratio is desirable to achieve high impedance ratio of the switch. When a voltage is applied between the movable plate and the transmission line of the capacitive switch, the electrostatic force pulls the movable plate down toward the dielectric layer. The dielectric layer can reduce stiction and eliminate micro welding between the movable plate and the transmission line. The actuation voltage of the micro mechanical switch can be determined by the hinge geometry, the membrane material properties, and the gap height between the movable plate and dielectric layer on top of the transmission line.

A RF MEMS shunt capacitive switch shown in Fig. 11(a) has been developed on high resistivity silicon wafers by using surface micromachining technique.[16,17] The switch has dimensions of 120 μm in width and 280 μm in length, and consists of a thin aluminum alloy membrane suspended over a dielectric film deposited on top of the bottom electrode. In the membrane "UP" state, most of the RF signal passes through the transmission line with a minimal capacitance (C_{off}) of 35 fF between the bottom signal line and the suspended movable top electrode. When an electrostatic voltage of 50 V is applied between the movable top electrode and the bottom electrode, the top electrode moves "DOWN" toward the dielectric film on the bottom electrode. This configuration has an on-off capacitance ratio of approximately between 60 and 100. Silicon nitride or oxide commonly used may

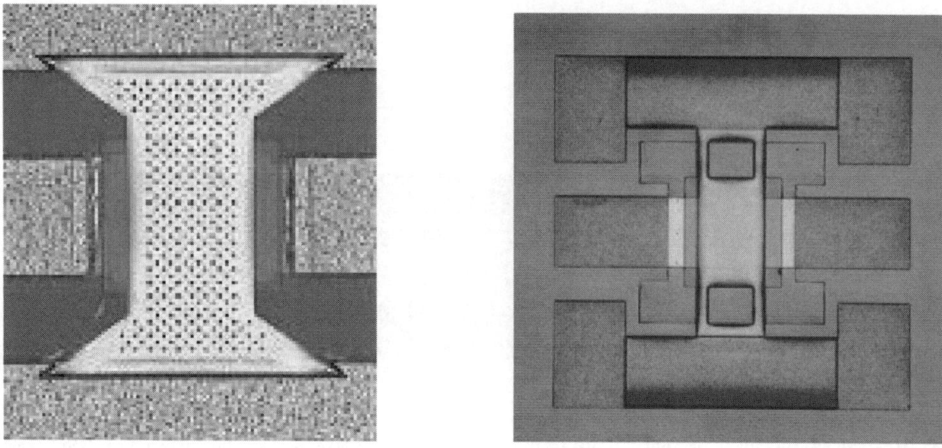

(a) (b)

Fig. 11. Photomicrographs of capacitive coupling capacitive RF MEMS switches from (a) Texas Instrument and (b) LG Electronics.

not be appropriate for making RF MEMS capacitive switches with small size, high capacitance, and high on/off ratio, due to their small relative dielectric constant at high frequencies. Thus, Strontium Titanate Oxide with high dielectric constant was utilized for fabricating RF MEMS capacitive switches shown in Fig. 11(b).[18] The switches have low insertion loss of less than 0.1 dB, high on/off ratio of 500, high isolation of 40 dB at 2 GHz, and an operation voltage of 7.5 volts. Table 1 shows performance characteristics of FET and electrostatic RF MEMS switches. As shown in Table 1, the RF MEMS switches have lower insertion loss, higher isolation, higher operating voltage, and lower switching speed than the FET switches. Thus, additional research efforts are required to reduce the operating voltage and increase switching speed of the RF MEMS switches.

Table 1. Comparison of performance characteristics of FET and electrostatic RF MEMS switches.

Parameters	GaAs FET	RSC	TI	HRL	Michigan	LG Electronics
Actuation voltage (V)	~ 1	~ 60	~ 50	~ 25	$15 \sim 20$	$5 \sim 10$
Insertion loss (dB)	2 at 6 GHz	0.2	0.15 at 10 GHz	0.2	0.6 at 22 GHz	~ 0.1
Isolation (dB)	22 at 2 GHz	32 at 10 GHz	15 at 10 GHz	40 at 12 GHz	20 at 22 GHz	42 dB at 5 GHz
Switching time (us)	0.01	2 – 5	3.5 – 5.3	20	—	5 – 10
Contact mechanism		resistive	capacitive	resistive	capacitive	capacitive

An important parameter of RF MEMS switches for commercialization is the switch lifetime, which is measured in a number of different ways depending on the requirement of target applications. Cold-switching life time refers to the measured time during the opening and closing of a switch with zero signal level through the contact, while hot-switching life time refers to that of a switch with a specified signal level through the contact. Thus, a switch lifetime can differ significantly based on different switching conditions. For general RF applications, a satisfactory hot-switched lifetime would be necessary.

3.2. *Tunable capacitors*

Tunable capacitors or varactors are widely used components in RF communication due to their tuning functionality. They are utilized in various applications including harmonic frequency generators, low noise parametric amplifiers, and frequency controllers such as voltage-controlled oscillators. Important parameters of the tunable capacitors are unbiased base capacitance, tuning ratio, equivalent series resistance or quality factor, self-resonant frequency, and device linearity in response to RF power. The value of the unbiased base capacitance depends on the circuit requirements of a specific application. Typically, the useful values range from 0.1 pF to a few pico farads for RF applications. The tuning ratio is a parameter that is a limiting factor in RF circuit design. In most applications, large tuning ratio in excess of 2:1 is demanded. The Q factor can be calculated by using the following equation, since other loss factors are negligible such as resistive loss caused by the equivalent series resistance.

$$Q = \frac{1}{\omega C R} . \tag{1}$$

As shown in the above equation, when the resistance is larger, the quality factor is small. Thus, the equivalent series resistance should be small enough to achieve high Q of the tunable capacitor. The series inductance in the equivalent circuit of the tunable capacitor should also be low enough to achieve high electrical self-resonant frequency. In most applications the tunable capacitor is required to have high linearity in response to RF signal power. Since a solid-state varactor has a large dependence on the signal power, multiple varactor devices are often used in a parallel-series configuration to increase the overall linearity. Although the parallel-series configuration of the varactors behaves similar to a single varactor in terms of capacitance tuning, the increased number of components generate issues in reliability, cost, size, weight, and power consumption. In summary, the solid-state varactors often suffer from a small tuning ratio, excessive resistive loss, a low Q caused by large series resistance, and low self-resonant frequency due to the large parasitic effects from the silicon substrate.

Recently MEMS based tunable capacitors are developed to overcome these shortcomings of the solid-state varactors. MEMS tunable capacitors tune their capacitance by adjusting device physical dimensions such as varying the gap between top and bottom electrodes or overlapping capacitive area of the interdigitated comb structures.

In 1996, a gap tuning capacitor was developed using a surface micromachining technique.[19] As shown in Fig. 12, the top electrode of the tunable capacitor supported by micromachined springs, is moved in the vertical direction over the bottom electrode by the tuning voltage between the electrodes. In the tunable capacitors with electrostatically actuated two parallel plates, the operation should be limited within one-third of the initial gap due to the equilibrium condition between the electrostatic attracting force and the restoring force of the movable electrode. Thus, this imposes a theoretical limit of 50% on the capacitance tuning range. In 1997 the tunable capacitor comprised of three-plate parallel plate system was developed to increase the theoretical MTR (maximum tuning range) limit from 50% to 100% using a multi-user MEMS polysilicon process (MUMP's).[20] In 1999, the tunable capacitor using electro-thermal actuator was fabricated by using MUMP's process and then flip-chip assembled onto an alumina substrate.[21] The device has a tuning ratio of 100% with a tuning voltage of 5 V and a quality factor of 256 at 1 GHz for the 0.102 pF, and a self-resonant frequency of 31 GHz. In 2001, an RF MEMS tunable capacitor shown in Fig. 13 was fabricated on a Quartz substrate by using a piezoelectric actuator.[22] It has a tuning ratio of approximately 200% at 6 volts and a quality factor of 210 at 1 GHz. The electrostatically actuated RF MEMS tunable capacitors described in the above can be integrated with other conventional electronic components and circuits.

In 1998, a MEMS area-tuning capacitor based on interdigitated comb structures has been demonstrated using a deep reactive ion etch of single-crystal silicon which is later coated with metal film.[23] When a tuning voltage is applied across the interdigitated comb structure, the electrostatic force of the comb fingers laterally actuates the movable portion of the device, resulting in a change of the overlapping area in the capacitor. However, the gap between the comb finger electrodes

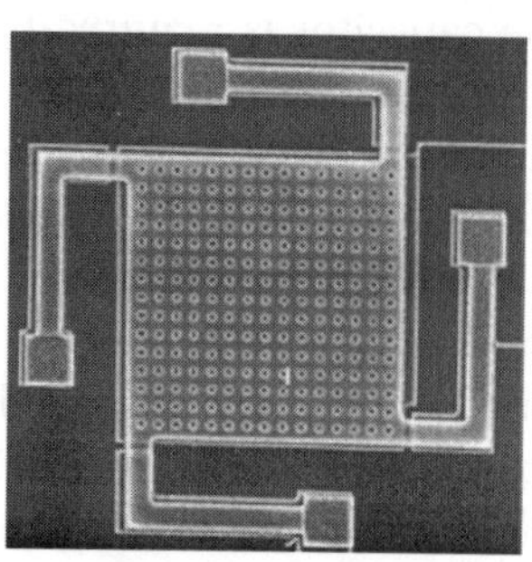

(a)

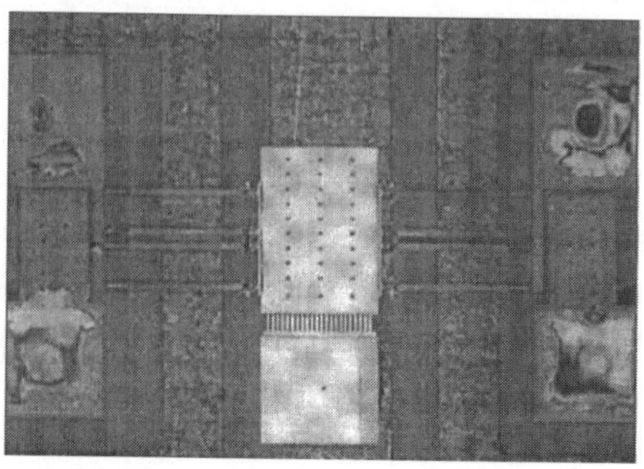

(b)

Fig. 12. Photomicrographs of RF MEMS tunable capacitors with gap tuning configuration by (a) University of California at Berkeley and (b) University of Colorado.

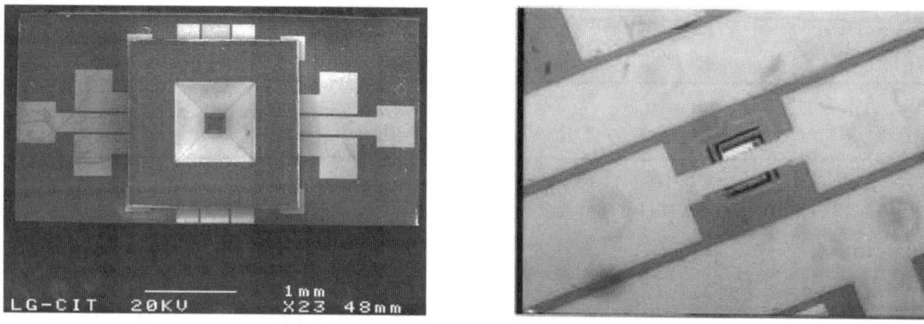

(a) (b)

Fig. 13. Photomicrographs of RF MEMS tunable capacitors with gap tuning configuration: (a) picture was taken at the front side, (b) picture was taken at the backside (from LG Electronics).

remains unchanged. Unlike the gap-tuning parallel plate tunable capacitors, there is no theoretical tuning limit for the tunable capacitors with area-tuning interdigitated comb structures. The only practical limits for the tuning range are related with the supporting spring design and the length of the comb fingers. The tunable capacitors shown in Fig. 14 were fabricated, characterized, and compared on low resistivity silicon, high resistivity silicon, and a glass substrate. The tunable capacitor fabricated on glass substrate has better performance because the resistive loss and the substrate parasitic effects are small compared to capacitor on silicon substrates. The fabricated tunable capacitor on a glass substrate has a self-resonant frequency of 5 GHz, Q of 34 at 500 MHz, and tuning range of 100% at 5 V and 200% at 14 V respectively. Table 2 shows performance characteristics of solid-state varactors and RF MEMS tunable capacitors. As shown in Table 2, the RF MEMS tunable capacitors have better performance characteristics at high frequencies than

(a) (b)

Fig. 14. (a) An SEM picture and (b) a close up view of RF MEMS tunable capacitor with area tuning configuration (from Rockwell Science Center).

Table 2. Comparison of performance characteristics of solid-state varactors and RF MEMS tunable capacitors.

Parameters	Solid-state varactors	RSC	U of Colorado	HRL	U of Cal. at Berkeley	LG Electronics
Actuation voltage (V)	~ 4	2 – 14	~ 5	80 – 200	~ 5.5	~ 6
Min. capacitance (pF)	2 at 4 V	1.8	0.9	0.035	2.11	0.1
Tuning range (%)	335	200	~ 90	185	16	200
Quality factor	350 at 0.05 GHz	34 at 0.5 GHz	256 at 1 GHz	—	62 at 1 GHz	210 at 1 GHz
Actuation mechanism	Hyper abrupt junction	Electro-static	Electro-thermal	Sliding motor	Electro-static	Electro-piezoel.

solid-state varactors. On the other hand, the high operating voltage of MEMS tunable capacitors should be further reduced for commercial applications.

3.3. *High Q inductors*

Integrated inductors with high performance are in increasingly higher demand for modern wireless communications products. They can be utilized in various applications including filters, oscillators, low-power converters, and RF front ends of wireless transceivers. Key parameters in the integrated inductors are the quality factor Q, the large inductance, and the self-resonance frequency. Resistance of metal lines and dielectric losses in the substrate are contributing to the degradation of the quality factor and the inductance value. The self-resonant frequency is limited by various fringing and parasitic capacitance between windings and a substrate. The quality factor of integrated inductors on silicon substrate ranges from 1 to 10, while that of integrated inductors suspended on the substrate ranges from 10 to 50. So far, the highest quality factor of 140 has been reported for a large suspended spiral inductor with thick copper coil on a glass substrate fabricated using surface machining techniques and polymer/metallic processes.[24] Figures 15 and 16 show integrated high Q MEMS inductors with electroplated thick copper conductor lines.

For high frequency applications, it is most appropriate to consider air-core inductor devices, since the high frequency behavior of many magnetic core materials is relatively poor. The characteristics of integrated inductors with non-magnetic core can be determined solely by the coil geometry and location. In addition to accurate determination of the inductance of air-core inductors, it is also important at high frequency to assess parasitic effects such as stray capacitance and conductor skin effect in order to understand the effective operation of the device. Much research has been done to analyze the stray capacitance of integrated inductors in high frequency applications.[25] The stray capacitance of the inductor determines the self-resonant frequency (SRF) and the Q-factor. Various aspects of the inductor ge-

(a) (b)

Fig. 15. SEM pictures of integrated high Q solenoid type inductor (a) with an alumina core and copper coils (from University of California at Berkeley), (b) with an air core and electroplated thick copper coils suspended on an alumina substrate (from Georgia Institute of Technology).

(a) (b)

Fig. 16. SEM pictures of integrated high Q spiral inductors using (a) plastic deformation magnetic assembly on a silicon (from University of university of Illinois), (b) electroplated thick copper coils suspended on a silicon substrate (from LG Electronics).

ometry, such as the cross-sectional area of the conductor line, the line spacing, and the core size, must be considered so that the inductor has the desired performance after fabrication. The resistivity of the metal conductor should be low to obtain a high Q-factor at high frequency applications. Since high frequency current flows mostly along the surface area of a conductor, the skin effect causes the winding resistance to increase as the frequency is increased. Conductor lines of thickness much in excess of several skin depths at the frequency of interest will not reduce significantly the conductor resistance or increase the inductor Q-factor. High quality factor means that the inductor has the desirable properties of low dissipation and favorable frequency characteristics when utilized in filter circuits.

3.4. *Thin film bulk acoustic resonators*

Recent development of wireless communication technology demands the miniaturized on-chip RF band pass filters and duplexer filters with high performance characteristics, small volume, and light weight. However, the conventional methods using lumped elements, dielectric resonators, or surface acoustic wave resonators have serious difficulties in fabrication and integration with other electronics.[26] A thin film bulk acoustic resonator (TFBAR)[27] is a promising candidate for fabricating integrated RF filters, RF duplexers, and voltage controlled oscillators (VCO's), because the TFBAR can be easily scaled down and integrated with other electronics. It has high quality factor Q in the wide spectral range from 0.1 GHz to 10 GHz,[28] and better power handling characteristics compared to the SAW's, especially at high frequencies where the pitch of the interdigitated structures must be reduced.

The TFBAR can be distinguished as a membrane type resonator and a solidly mounted resonator (SMR) based on the fabricated structure configurations. The membrane type device is a resonator fabricated on top of supported nitride, silicon, or oxide membrane formed by etching the substrate selectively, while the SMR is a resonator fabricated on top of multiple reflectors formed on the substrate. The reflector comprised of alternating high and low mechanical impedance materials to give a large reflection and to reduce the substrate impedance to near zero. In SMR type, energy loss to the substrate is avoided by acoustically isolating the substrate from the piezoelectric resonator materials using impedance transformations obtained through proper selection of the number and thickness of layers separating the two media. A vibration or resonance is created by expansion and contraction of the piezoelectric layer, when an alternating electrical potential is applied across the TFBAR with a structure made of metal-piezoelectric material-metal, as shown in Fig. 17. AlN and ZnO is widely used as a piezoelectric material for fabricating RF TFBAR's due to their high acoustic velocity, low dielectric constant, low acoustic loss, and high acoustic coupling factor at high frequency. As depicted in Eq. (2), a resonant frequency, f_r can approximately be determined by the thickness of piezoelectric material.

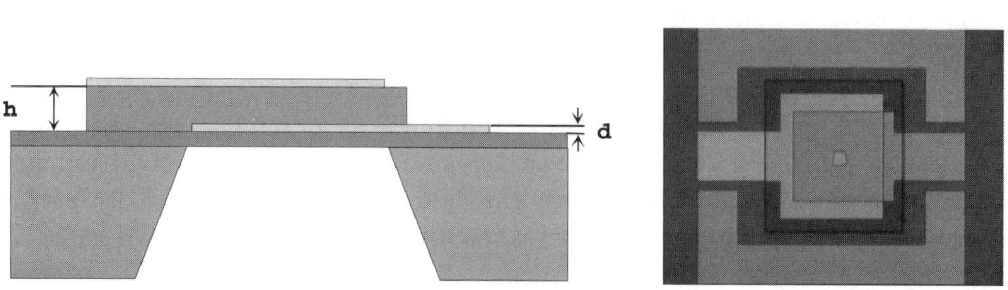

(a) (b)

Fig. 17. (a) A cross-sectional view of bulk micromachined membrane type thin film bulk acoustic resonator (TFBAR), (b) SEM photo of the fabricated TFBAR (from LG Electronics).

$$2f_r = n\left(\frac{h}{v_p} + \frac{d}{v_m}\right) \tag{2}$$

where h is the thickness of piezoelectric layer, d is a thickness of metal electrode, V_p is a propagation velocity of acoustic wave in piezoelectric material, and V_m is a propagation velocity of acoustic wave in metal electrode. With all of their advantages, there are still several important drawbacks such as the need for trimming and tuning when these resonators are utilized to implement small percent bandwidth filters. Table 3 shows the recent research history of TFBAR filters. As shown in Table 3, membrane type resonator structure and aluminum nitride piezoelectric material are most actively investigated.

Table 3. Recent history of TFBAR filter research.

Parameters	Westing House	Motorola	TFR	Agilent	RSC	LG Electronics
Resonator type	membrane	membrane	SMR	membrane	SMR	membrane
Piezoelectric materials	ZnO	ZnO	AlN	AlN	ZnO	AlN/ZnO
Filter type	ladder	lattice	ladder	ladder	ladder	ladder

4. Summary

The recent progress in optical and RF MEMS for advanced telecommunication applications was reviewed in this paper. The optical MEMS components reviewed include micromachined optical switches and silicon optical benches that are in the early stages of commercialization. The micromachined optical switches have advantages such as high scalability, low insertion loss, and low-level dependency on wavelength and polarization. The use of silicon optical benches promise significant cost reduction as well as better manufacturability in the precise alignment of photonic devices in the optical modules.

The RF MEMS devices reviewed include micromechanical switches, tunable capacitors, integrated high Q inductors, and thin film bulk acoustic resonators which will become core components for constructing miniaturized transceivers operating at multi-band and/or multi-mode. The micromechanical switches, the first devices studied in the RF MEMS field, are widely being developed due to their low resistive loss, negligible power consumption, good isolation, and high power handling capability compared to the semiconductor switches. The solid-state varactors suffer from small tuning ratio, poor device linearity, excessive resistive loss, low Q caused by large series resistance, and low self-resonant frequency due to the large parasitic effects from the silicon substrate. Thus, MEMS based high performance tunable capacitors are being realized to overcome the shortcomings of these solid-state varactors. These MEMS tunable capacitors tune their capacitance by adjusting device physical dimensions such as varying the gap between the top and bottom electrodes or overlapping capacitive area of the interdigitated comb structures. They can be utilized in various applications including harmonic frequency generators, low noise parametric amplifiers, matching networks, and frequency controllers such as voltage controlled oscillators. Large suspended integrated high Q microinductors are realized by using the MEMS technology for constructing miniaturized one chip filters, oscillators, and low-power converters. The TFBAR can be realized for fabricating integrated RF filters, RF duplexers, and voltage controlled oscillators (VCO's) with high performance characteristics, small size/volume, and light weight, since it can be easily scaled down and integrated with other electronics. A miniaturized single chip RF transceiver, one of the key elements in the wireless communication system, can be realized by fabricating RF MEMS components with good performance characteristics and integrating MMIC's or RFIC's with these newly developed components.

Acknowledgments

The authors would like to thank research members of Micro System group and MEMS product group of LG Electronics Institute of Technology for their contributions.

References

1. http://france.lucent.fr/presse/dossiers/pdf/Lambdarouter.pdf.
2. Y. Yee, J. U. Bu, M. Ha, J. Choi, H. Oh, S. Lee, and H. Nam, "Fabrication and characterization of a PZT actuated micromirror with two-axis rotational motion for free-space optics", *Proc. 14th IEEE Int. Micro Electro Mechanical Systems Conf. (MEMS 2001)*, Interlaken, Switzerland, January 2001, pp. 317–320.
3. L. Y. Lin, E. L. Goldstein, and R. W. Tkach, "Free-space micromachined optical switches for optical switching", *IEEE J. Selected Topics Quantum Electron.* **5**, 1 (1999) 4–9.
4. T.-S. Kim, S.-S. Lee, Y. Yee, J. U. Bu, H.-H. Oh, C.-G. Park, and M.-H. Ha, "Electrostatic micromirror with built-in large air-gap for wide range of rotational actuation", *Proc. 2001 IEEE/LEOS Int. Conf. Optical MEMS*, Okinawa, Japan, September 2001, pp. 99–100.
5. L. Y. Lin and E. L. Goldstein, "Optical-layer networking: Opportunities for and progress in light wave micromachines", *2000 Optical Fiber Communication Conf.*, Baltimore, USA, pp. 152–163.
6. J. Fouquet, "Compact optical cross-connect based on total internal reflection in a fluid-containing planar light wave circuit", *2000 Optical Fiber Communication Conf.*, Baltimore, USA, TuM1 (invited).
7. A. Gopinath and J. B. Rankin, "GaAs FET RF switches", *IEEE Trans. Electron. Devices* **32** (1985) 1272–1278.
8. Y. Ota, M. Sakakura, K. Fujimoto, S. Yamamoto, and H. Fujimoto, "High isolation and low insertion loss switch IC using GaAs MESFET's", *IEEE Trans. Microwave Theory. Technol.* **43** (1995) 2175–2177.
9. K. W. Kobayashi, A. K. Oki, D. K. Umemoto, S. K. Z. Claxton, and D. C. Streit, "Monolithic GaAs HBT p-I-n diode variable gain amplifiers, attenuators, and switches", *IEEE Trans. Microwave Theor. Techniq.* **41** (1993) 2295–2302.

10. K. E. Petersen, "Micromechanical membrane switches on silicon", *IBM J. Res. Dev.* **23** (1979) 376–385.
11. M.-A. Grétillat, F. Grétillat, and N. F. de Rooij, "Micromechanical relay with electrostatic actuation and metallic contacts", *J. Micromech. Microeng.* **9** (1999) 324–331.
12. J. J. Yao and M. F. Chang, "A surface micromachined miniature switch for telecommunications applications with signal frequencies from DC up to 4 GHz", *Tech. Digest, 8th Int. Conf. Solid-State Sensors and Actuators*, 1995, pp. 384–387.
13. D. Hyman *et al.*, "Surface-micromachined RF MEMS switches on GaAs substrates", *Int. J. RF Microwave CAE* **9** (1999) 348–361.
14. D. Hyman *et al.*, "GaAs-compatible surface-micromachined RF MEMS switches", *Electron. Lett.* **35** (1999) 224–226.
15. J. B. Muldavin and G. M. Rebeiz, "30 GHz tuned MEMS switches", *Tech. Digest, IEEE Microwave Theory and Techniques Symp.*, 1999, pp. 1511–1514.
16. C. Goldsmith, T.-H. Lin, B. Powers, W.-R. Wu, and B. Norvell, "Micromechanical membrane switches for microwave applications", *Tech. Digest, IEEE Microwave Theory and Techniques Symp.*, 1995, pp. 91–94.
17. Z. J. Yao, S. Chen, S. Eshelman, D. Denniston, and C. Goldsmith, "Micromachined low-loss microwave switches", *IEEE J. MEMS* **8** (1999) 129–134.
18. J. Y. Park, G. H. Kim, K. W. Cheong, and J. U. Bu, "Electroplated RF MEMS capacitive switches", *IEEE 13th Int. MEMS Conf.*, Miyajaki, Japan, January 2000, pp. 639–644.
19. D. J. Young and B. E. Boser, "A micromachined variable capacitor for monolithic low-noise VCOs", *Tech. Digest, Solid State Sensor and Actuator Workshop*, 1996, pp. 86–89.
20. A. Dec and K. Suyama, "Micromachined varactor with wide tuning range", *Electron. Lett.* **33** (1997) 922–924.
21. K. F. Harsh, W. Zhang, V. M. Bright, and Y. C. Lee, "Flip-chip assembly for Si-based RF MEMS", *Proc. IEEE, 12th Ann. Int. Workshop Micro Electro Mechanical Syst.*, 1999, pp. 273–278.
22. J. Y. Park, Y. J. Yee, H. J. Nam, and J. U. Bu, "Micromachined RF MEMS tunable capacitors using piezoelectric actuators", *Tech. Digest, IEEE Microwave Theory and Techniques Symp.*, May 2001, pp. 2111–2115.
23. J. J. Yao, S. Park, and J. DeNatale, "High tuning-ratio MEMS-based tunable capacitors for RF communications applications", *Tech. Digest, Solid State Sensor and Actuator Workshop*, 1998, pp. 124–127.
24. J. Y. Park and M. G. Allen, "Micromachined high Q inductors for high frequency applications", *Micromachining & Microfabrication '98 Conf.*, Santa Clara, CA, *SPIE* **3514** (1998) 218–228.
25. A. Massarini, M. Kazimierczuk, and G. Grandi, "Lumped parameter models for single- and multiple-layer inductors", *27th Ann. IEEE Power Electron. Specialists Conf.* Vol. 1, 1996, pp. 295–301.
26. R. B. Strokes and J. D. Crawfold, "X-band thin film acoustic filters on GaAs", *IEEE Trans. Microwave Theory Tech.* **41** (1993) 1075–1080.
27. S. H. Kim, J. S. Lee, H. C. Choi, and Y. H. Lee, "The fabrication of thin-film bulk acoustic wave resonators employing a ZnO/Si composite diaphragm structure using porous silicon layer etching", *IEEE Electron. Device Lett.* **20**, 3 (1999) 113–115.
28. K. M. Lakin, G. R. Kline, and K. T. McCarron, "High Q microwave acoustic resonators and filters", *Proc. 1993 IEEE MTT-S International Microwave Symp.*, Vol. 5, 1993, pp. 1517–1520.

10. L. E. Petersen, "Micromechanical membrane switches on silicon", IBM J. Res. Dev. 23(1979) 376–385.
11. M. A. Gretillat, F. Gueissaz, and N. F. de Rooij, "Micromechanical relay with electrostatic actuation and metallic contacts", J. Micromech. Microeng. 9 (1999) 324–331.
12. J. J. Yao and M. F. Chang, "A surface micromachined miniature switch for telecommunications applications with signal frequencies from DC up to 4 GHz", Tech. Digest 8th Int. Conf. Solid-State Sensors and Actuators, 1995, pp. 384–387.
13. D. Hyman et al., "Surface-micromachined RF MEMS switches on GaAs substrates", Int. J. RF Microw. CAE 9 (1999) 348–361.
14. D. Hyman et al., "GaAs-compatible surface-micromachined RF MEMS switch", Electron. Lett. 35 (1999) 224–226.
15. P. M. Zavracky and [illegible] [illegible] IEEE Microwave Theory and Techniques Symp., 1998, pp. 1141–1144.
16. [illegible]
17. [illegible]
18. [illegible]
19. [illegible]
20. [illegible]

International Journal of High Speed Electronics and Systems, Vol. 12, No. 2 (2002) 235–265

LOW TEMPERATURE PHYSICS AT ROOM TEMPERATURE IN WATER: CHARGE INVERSION IN CHEMICAL AND BIOLOGICAL SYSTEMS*

A. YU. GROSBERG, T. T. NGUYEN, B. I. SHKLOVSKII
Department of Physics, University of Minnesota
116 Church Street Southeast, Minneapolis, Minnesota 55455

We review recent advances in the physics of strongly interacting charged systems functioning in water at room temperature. We concentrate on the phenomena which go beyond the framework of mean field theories, whether linear Debye-Hückel or non-linear Poisson-Boltzmann. We place major emphasis on charge inversion - a counterintuitive phenomenon in which a strongly charged particle, called macroion, binds so many counterions that its net charge changes sign. We discuss the universal theory of charge inversion based on the idea of a strongly correlated liquid of adsorbed counterions, similar to a Wigner crystal. This theory has a vast array of applications, particularly in biology and chemistry; for example, the DNA double helix in the presence of positive multivalent ions (e.g., polycations) acquires a net positive charge and drifts as a positive particle in electric field. This simplifies DNA uptake by the cell as needed for gene therapy, because the cell membrane is negatively charged. We discuss also the analogies of charge inversion in other fields of physics.

1. Introduction

Molecular biological machinery functions in water at around room temperature. For a physicist, this very limited temperature range contrasts unfavorably to the richness of low temperature physics, where one can change temperature and scan vastly different energy scales in an orderly manner. In this Colloquium, we review the recently developed understanding of highly charged molecular systems in which Coulomb interactions are so strong that we are effectively in the realm of low temperature physics.

To be specific, imagine a problem in which one big ion, called a *macroion*, is screened by much smaller but still multivalent ions with a large charge Ze each (e is the proton charge); for brevity, we call them Z-ions. A variety of macroions are of importance in chemistry and biology, ranging from the charged surface of mica or charged solid particles, to charged lipid membranes, colloids, DNA, actin, and even to cells and viruses. Multivalent metal ions, charged micelles, dendrimers, short or long polyelectrolytes including DNA - to name but a few - can play the role of the screening Z-ions.

The central idea of this Colloquium is that of *correlations*: due to strong interactions with the macroion surface and with each other, screening Z-ions do not

*This paper is a reprint, with minor changes, of the paper "Rev. Mod. Phys., **74**, 329 (2002)" by the same authors.

position themselves randomly in three-dimensional space, but form a strongly correlated liquid on the surface of the macroion. Moreover, in terms of short range order this liquid is reminiscent of a Wigner crystal, as the cartoon in Fig. 1 depicts. Because of its central importance we shall use special abbreviation SCL to denote a strongly correlated liquid of adsorbed Z-ions.

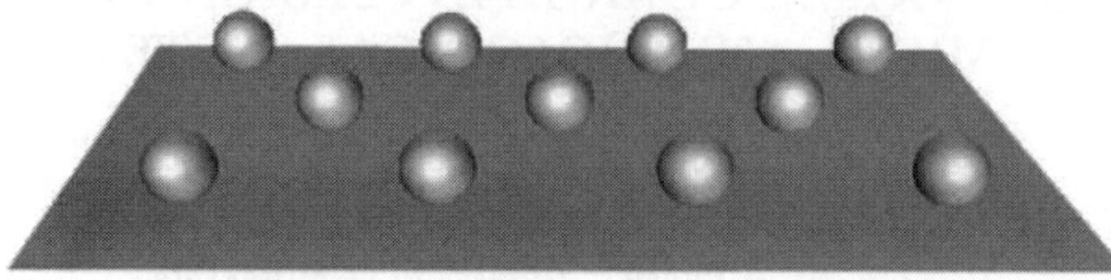

Fig. 1. Strongly correlated liquid (SCL) - almost a Wigner crystal - of Z-ions on the oppositely charged macroion surface. The figure is characteristic in showing the degree to which we are willing to ignore the microscopic details.

Depending on the system geometry and other circumstances, correlations between screening ions may appear in many different ways. To create some simple images in the reader's mind, it is useful to begin with a few examples. One example is that in Fig. 1, which may be the surface of, e.g., a latex particle screened by some compact ions. With a modest leap of the imagination, we can envision the same picture to represent the surface of a DNA double helix screened by multivalent counterions, such as spermine with $Z = 4$ [3]. Here, we imagine DNA as a thick and long cylinder, diameter 2 nm and charge $-e$ per 1.7 nm along the cylinder, or, in other words, with a huge negative surface charge density $-0.9\ e/\mathrm{nm}^2$. We study correlations between point-like Z-ions in Secs. 3 - 5. One obvious problem with the model Fig. 1 is that it ignores the discreteness of charges, both in the macroion and in the Z-ions; chemists, for instance, hardly ever accept such continuous models. In Sec. 11, we address the electrostatic correlations in the systems with discrete charges.

As another example, consider DNA molecules which screen the positive surface charge of a colloid particle. Obviously, DNA chains play the role of Z-ions. For very short DNA pieces we are back again to Fig. 1, but longer DNA are spaghetti-like, and in this situation, correlations mean parallel arrangement, as we see in Fig. 2. Theoretical treatment of this problem in Sec. 6 makes a simplifying assumption that it may be modeled as a system of parallel rods.

Electrostatic correlations are strong not only for artificial systems involving DNA, but also for the DNA in the cell [2]. In particular, to organize DNA in chromatin (in eukariotic cells), nature uses proteins having large positive charges - histones. Higher levels of chromatin hierarchical structure can be disassembled, with some of the histones released, by an elevated concentration of salt. The resulting most stable lower level structure is a bead-on-a-string necklace, as shown in the Fig. 3. It is called 10 nm fiber, because the beads, called nucleosomes, are about that big. Each nucleosome consists of a core particle, called an octamer, with a total charge of about $+170e$. As shown in the inset in Fig. 3, the octamer is encircled

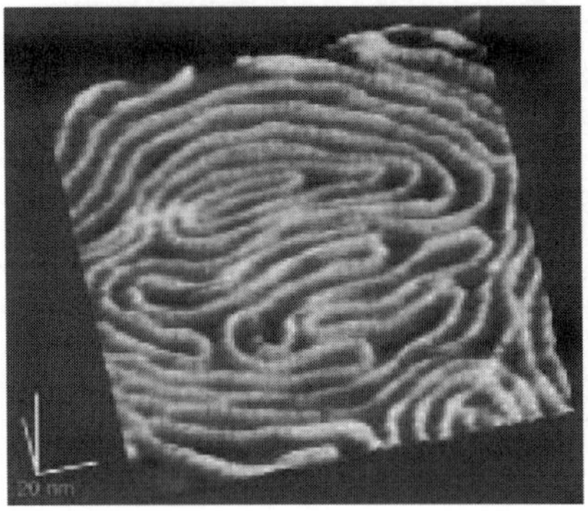

Fig. 2. DNA adsorbed on a positively charged surface, as seen in an atomic force microscopy image [61].

1.8 times by a DNA having a charge $-292e$. The electrostatic interaction energy between the DNA molecule and the histone octamer dwarfs the bending energy of the DNA molecule. Strikingly, a simple theory discussed in Sec. 7 yields a similar structure, Fig. 4, based on purely electrostatic correlations in a simple model.

Correlated screening has also many useful practical applications. The first to be mentioned is the technology of dressing of DNA with polycations[34,36], positively charged star-like polymers called dendrimers[98,37,17], or liposomes[85] in order to produce positive complexes with DNA. This facilitates gene delivery through a negative cell surface membrane [19]. There is also the idea to manufacture nano-wires by attaching positive silver or gold colloids to DNA[5,39].

We view these examples as both important and convincing enough to engage in the study of electrostatic correlations between strongly charged Z-ions. Strong correlations manifest themselves in a number of ways and alter dramatically the whole picture of screening. In the familiar Debye theory, screening somewhat reduces the effective value of charge as seen from a finite distance in the outside world. With strongly correlated ions, over-screening becomes possible, in which case the shielded macroion charge is seen from outside as having the opposite sign. This counterintuitive phenomenon is called *charge inversion*.

The first experimental study related to charge inversion was reported a long time ago by De Jong[32]. More recently, charge inverted complexes of polyelectrolytes were directly observed in electrophoresis experiments (see review articles by Kabanov and Kabanov[34,36] and references therein). It is now understood that charge inversion is a generic phenomenon and it is expected in all systems in which strongly charged ions participate in screening. It turns out to be a natural manifestation of correlations between screening ions. In recent years, the phenomenon of charge inversion has attracted significant attention of theorists [16,102,81,95,55,78,31,92,63,64,103,65,66,56,69,70,71,72,73,10,13,1,83,96].

Another equally interesting manifestation of correlations is the possibility of attraction between like charged macroions mediated by Z-ions[90,27,48,59]. This has

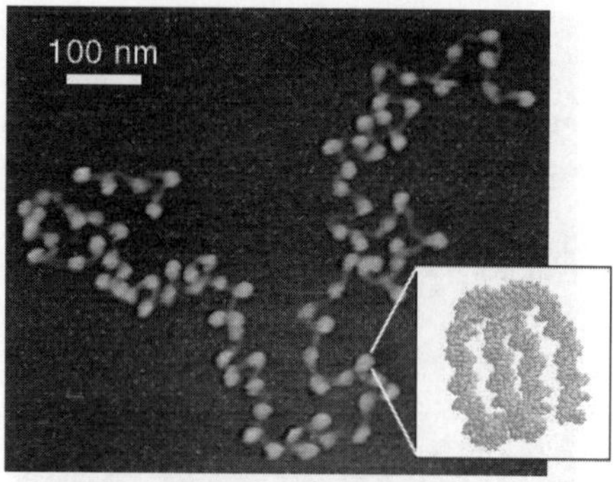

Fig. 3. Electron microscopy image of 10 nm chromatin fiber [93]. The beads are nucleosomes. The structure of a nucleosome is known to about 0.2 nm resolution [52], and the inset shows how the DNA double helix is bent in the nucleosome. While the overall shape of the fiber is perhaps due to the sample preparation procedure, the array of nucleosomes on the DNA is close to periodic, similar in this respect to the theoretical model shown in Fig. 4.

Fig. 4. Self-assembled complex of a negative polyelectrolyte molecule and many positive spheres in a necklace-like structure. On the surface of a sphere, neighboring polyelectrolyte turns are correlated similar to the rods in Fig. 8 below. On a larger scale, charged spheres repel each other and form one-dimensional Wigner crystal along the polyelectrolyte molecule.

implications in the large field of self-assembly of charged biological objects, ranging from RNA[104,77] to virus heads [21]. In Sec. 9, we show how this attraction of like charges competes with repulsion due to inverted charges, which combine to induce reentrant condensation of DNA or colloids out of solutions.

Although the subject matter of our Colloquium belongs to chemical and biological physics, it has remarkably many deep physical analogies in other branches of physics, from atomic physics to quantum Hall effect; we discuss this in Sec. 12.

In what follows, we re-examine the entire concept of screening to include correlations. Although this requires stepping beyond the mean field approximation, we shall start with some historical remarks which also serve to define the notations.

2. Historical remarks: Mean field theories

The history of our subject is almost one century long. While we are unable in this brief Colloquium to discuss it in any depth, we shall mention three key steps.

In the first step, Gouy[26] and independently Chapman[9] examined the double layer at an electrode surface, an intensely disputed subject at the time. Following Gouy and Chapman, let us consider a massive insulating macroion. To be specific, assume that its charge is negative, with surface density $-\sigma$. Assume further that the only counterions in the system are those dissociated from the surface. Since a macroion is large, the problem of the counterion distribution near the surface is one-dimensional, with both counterion concentration $N(x)$ and electrostatic potential $\phi(x)$ depending on the distance x from the surface. Gouy and Chapman addressed this problem by solving the following equation:

$$\Delta\phi = -\frac{4\pi}{\epsilon} N_s e Z \exp\left[-eZ\frac{\phi(x)-\phi_s}{k_B T}\right] . \tag{1}$$

Here $\epsilon \approx 80$ is the dielectric constant of water, ϕ_s and N_s are the potential and counterion concentration at the surface. Formula (1) follows from the Poisson equation for the potential $\phi(x)$ and the assumption that the ions determining the charge density are Boltzmann-distributed in the same potential. Because of this self-consistency assumption, this equation describes a mean field approximation. It is universally called the Poisson-Boltzmann equation. One boundary condition $d\phi/dx|_{x=0} = 4\pi\sigma/\epsilon$ follows from the fact that the field vanishes on the other side of the boundary (as there is no electrolyte and there are no charges). The second condition is that the concentration $N(x)$ must be normalized to σ/Ze ions per unit area. The exact solution of the thus formulated Gouy-Chapman problem reads

$$N(x) = \frac{k_B T\epsilon/2\pi e^2 Z^2}{(x+\lambda)^2} , \tag{2}$$

where λ is called Gouy-Chapman length, which is equal to

$$\lambda = k_B T\epsilon/2\pi\sigma Z e . \tag{3}$$

The interpretation of Eq. (3) is interesting. If a Z-ion were to be alone next to the charged plane, it would be confined by the surface field $2\pi\sigma/\epsilon$ to such a "height" λ

that its energy change $2\pi\sigma Ze\lambda/\epsilon$ is about k_BT - which leads to the correct answer Eq. (3). Other Z-ions cancel the field inside the macroion but double the field in the electrolyte at the macroion surface (see above the boundary condition at $x = 0$). On the other hand, every particular ion at every moment is higher than, roughly, half the other ions, whereupon it finds itself in a partially screened field. Exact solution of the Poisson-Boltzmann equation indicates that these two factors cancel each other, yielding Eq. (3).

The next step of the screening story, well known to every physicist, is the theory of Debye and Hückel[11] initially developed for electrolytes - the overall neutral mixture of mobile ions of both signs, and now widely used in plasma and solid state physics. Debye and Hückel linearized the Poisson-Boltzmann equation (1) (generalized by introducing the sum over ion species on the right-hand side). Of course, linearization can be done if the potential is not too strong anywhere in the system, which is often the case if charges involved are small enough. Debye-Hückel screening leads to exponential decay of the potential around a point-like charge Q:

$$\phi(r) = \frac{Q}{\epsilon r} e^{-r/r_s} , \tag{4}$$

where the Debye-Hückel radius is given by

$$r_s = \left(\frac{k_BT}{8\pi N_1 e^2} \right)^{1/2} , \tag{5}$$

and N_1 is the concentration of monovalent salt.

Less well known among physicists is the fact that Debye-Hückel theory ignited a heated debate: Bjerrum[4] commented that $\int \exp\left[e^2/rk_BT\right] r^2 dr$ diverges at $r \to 0$ and, therefore, point-like charged particles are expected to associate in neutral pairs. The discussion led to the realization of the important role of short range repelling forces. In order to prevent association of monovalent ions and formation of Bjerrum pairs, the repulsion forces should take over at the distance which is not much smaller than the so-called Bjerrum length

$$\ell_B = e^2/\epsilon k_BT , \tag{6}$$

which is about 0.7 nm in water at room temperature. (This is why hydration layer of a few water molecules around each ion is essential to stabilize dissociated ions).

The last step we mention here is relatively recent, it has to do with non-linear screening of cylindrical charges, such as the DNA double helix[74,53,76]. Consider a cylinder charged to the linear density $-\eta$. Since the potential is logarithmic, its competition with entropy is quite peculiar. Indeed, releasing counterions to some distance r requires energy $(2eZ\eta/\epsilon)\ln(r/a)$ (a being the cylinder radius), while the corresponding entropy gain is $k_BT\ln(\pi r^2/\pi a^2)$. Therefore, counterions are released only as long as $\eta < \eta_Z$, where

$$\eta_Z = k_BT\epsilon/eZ . \tag{7}$$

When a cylinder is charged in excess of $-\eta_Z$, some of the ions remain *Onsager-Manning condensed* on the cylinder, so that its effective net charge is equal to $-\eta_Z$. To emphasize the importance of this subject, let us mention that the DNA double helix has a bare charge density of about $-4.2\eta_1$, where $\eta_1 = \eta_Z|_{Z=1}$ (see Frank-Kamenetskii et al[20] for further DNA applications). Onsager-Manning condensation was more accurately justified by Zimm and Le Bret[107]. These authors addressed non-linear Poisson-Boltzmann equation in cylindrical geometry (refining earlier works - see the collection of papers[38]).

Gouy-Chapman, Debye-Hückel, and Onsager-Manning theories are all of mean field type, based on the Poisson-Boltzmann equation. This approach works well when screening charges are small, $Z = 1$. In the case of strongly charged macroions and Z-ions with $Z \gg 1$, however, correlations are important and mean field theory fails. Hence it is necessary to step beyond this approximation. This is precisely the subject matter of the present Colloquium. Charge inversion in this context should be viewed as the most obvious manifestation of the failure of the mean field approximation.

3. Strongly correlated liquid of multivalent ions

To begin with, let us explain why the Gouy-Chapman solution (2) fails at large Z. Apart from λ (3), there is a second length scale in the problem due to the discreteness of charge. It is associated with the distance between ions in the lateral direction, along the plane. As long as the system as a whole is neutral, the two-dimensional concentration of Z-ions is $n = \sigma/Ze$, and the surface area per ion can be characterized by a radius R such that $\pi R^2 = 1/n$ (see Fig. 5). Thus $R = (\pi n)^{-1/2} = (Ze/\pi\sigma)^{1/2}$, and hence (cf. Eq. (3)

$$\frac{R}{\lambda} = 2\Gamma \, , \quad \Gamma = \frac{Z^2 e^2/\epsilon R}{k_B T} \, . \tag{8}$$

Here Γ is the Coulomb coupling constant, or the inverse dimensionless temperature measured in the units of a typical interaction energy between Z-ions. A system of monovalent ions, $Z = 1$, is weakly coupled, $\Gamma \sim 1$, this is why classical mean field theory applies. By contrast, a system in which Z-ions have large Z is strongly coupled, and we see that R becomes larger than λ. For example, at $Z = 3$ and $\sigma = 1.0\ e/\mathrm{nm}^2$ we get $\Gamma = 6.4$, $\lambda \simeq 0.1$ nm and $R \simeq 1.0$ nm. Clearly, mean field treatment along the lines of Poisson-Boltzmann theory fails in this situation, since Z-ions do not affect each other when they are at distances smaller than R from the plane. It is worth emphasizing once again that it is not only the ***linearized*** Debye-Hückel theory that fails in the strong coupling regime, but so does also the non-linear Poisson-Boltzmann theory. It is the mean field approximation that is inapplicable because of correlations between discreet charges.

An alternative theory appropriate for the regime $\Gamma \gg 1$ was suggested by Perel and Shklovskii[81]. The main idea of this theory is that at $\Gamma \gg 1$ the screening atmosphere is narrowly confined at the surface (see Fig. 1), and it should be approximated as a two-dimensional SCL.

A two-dimensional liquid of classical charged particles on a neutralizing background, the so-called one component plasma, is well understood [100]. At zero temperature, it acquires the minimal energy state of a Wigner crystal, shown in the Fig. 5, in which the correlation energy per ion and the chemical potential are given by

$$\varepsilon(n) \simeq -1.11 Z^2 e^2 / R\epsilon = -1.96 n^{1/2} Z^2 e^2/\epsilon \,, \tag{9}$$

$$\mu_{WC} = \frac{\partial[n\varepsilon(n)]}{\partial n} = \frac{3}{2}\varepsilon(n) = -1.65 Z^2 e^2/\epsilon R \,. \tag{10}$$

We interpret R here as the radius of a Wigner-Seitz cell approximated by a disc (see Fig. 5).

At non-zero temperature, the chemical potential of one component plasma can be written as $\mu = \mu_{id} + \mu_{WC} + \delta\mu$. Here μ_{id} is the chemical potential of an ideal gas at the same concentration. Accordingly, $\mu_{WC} + \delta\mu$ part is entirely due to correlations. Furthermore, it turns out that $\delta\mu$, which is the thermal correction, is negligible for $\Gamma \gg 1$. Although a Wigner *crystal*, in terms of long range order, melts at $\Gamma \approx 130$, the value of $\delta\mu$ is controlled by short range order and remains negligible as long as $\Gamma \gg 1$. It is in this sense that a SCL of Z-ions is similar to a Wigner crystal.

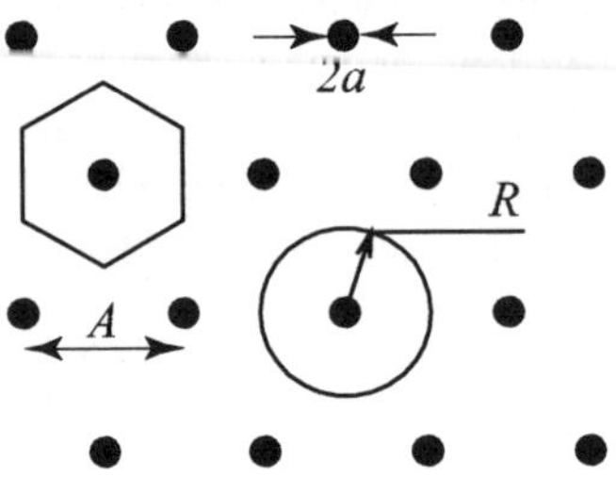

Fig. 5. A Wigner crystal of positive Z-ions on a uniform background of negative surface charge. A hexagonal Wigner-Seitz cell and its simplified version as a disk with radius R are shown.

Thus, the correlation part of the chemical potential can be approximated by μ_{WC} (10), which is negative and large: $-\mu_{WC}/k_BT = 1.65\Gamma \gg 1$. The physics of a large and negative μ_{WC} can be understood as follows. Pretend for a moment that the insulating macroion is replaced by a neutral metallic particle. In this case, each Z-ion creates an image charge of opposite sign inside the metal. The energy of attraction to the image is $U(x) = -(Ze)^2/4\epsilon x$, where x is the distance to the surface. This energy is minimal when the Z-ion is placed next to the surface, at a distance equal to its radius a; therefore, the Z-ion sticks to the surface. With this idea in mind, consider bringing a new Z-ion to the insulating macroion surface already covered by an adsorbed layer of Z-ions (Fig. 6). This layer behaves like a metal surface in the sense that the new Z-ion repels adsorbed ones, creating a correlation hole. In other words, it creates a negative image. Because of the discreteness of charges, the adsorbed layer is a good metal at length scales above R only. Accordingly, attraction to the image gets saturated at $x \sim R$. This is why the chemical potential of an ion in a Wigner crystal scales as $\mu_{WC} \sim -(Ze)^2/\epsilon R$. Eq. (10) specifies the numerical coefficient in this expression.

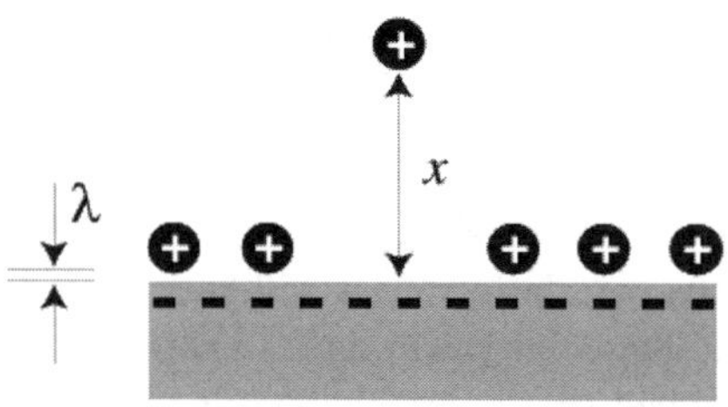

Fig. 6. The origin of attraction of a new positive Z-ion to the already neutralized surface. Z-ions are shown by solid circles. The new Z-ion creates its negative correlation hole.

Using the concept of images, we can now understand the distribution of Z-ions, $N(x)$, near the surface. To do this, let us extract one Z-ion from the SCL and move it along the x axis. As long as $x \ll R$, its correlation hole does not change, and, therefore, the Z-ion is attracted to the surface by the uniform electric field $E = 2\pi\sigma/\epsilon$; other Z-ions do not affect this attraction in any way. Therefore, $N(x) = N_s \exp(-x/\lambda)$ for $x \ll R$. Here $N_s \simeq n/\lambda$ is the three-dimensional concentration of Z-ions close to the surface plane (Eq. (1)). For $x \gg R$, the correlation hole acts as a point-like image charge, the corresponding interaction energy being $-Z^2e^2/4\epsilon x$. At $x = Z^2e^2/4\epsilon k_BT = R_0\Gamma/4 = \lambda\Gamma^2/2$, the interaction with the image charge drops to about k_BT, i.e., negligible, and the Z-ion concentration becomes

$$N_0 = N_s \exp\left(-\frac{|\mu_{WC}|}{k_BT}\right) = N_s \exp\left(-\frac{1.65Z^2e^2}{\epsilon R k_BT}\right) . \tag{11}$$

We shall further comment on the physical meaning of N_0 after Eq. (12). Note that the correction term $-Z^2e^2/4\epsilon x$ to Z-ion energy, which is important in the interval $R \ll x \ll \lambda\Gamma^2/2$, is similar to the "image" correction to the work function of a metal [46].

The dramatic difference between the exponential decay of $N(x)$ and the Gouy-Chapman $1/(\lambda + x)^2$-law (2) is due to correlation effects. Moreira and Netz[58] re-derived these results in a more formal way and confirmed them by Monte-Carlo simulations. Recently, Moreira and Netz[60] also showed that discreteness of the surface charge neglected above leads to the lateral pinning of Z-ions. This brings Z-ions even somewhat closer to the surface. General direction of this effect can be understood from the limit (although unrealistic for strongly charged surface) when the distance of closest approach between a discrete surface charge and a Z-ion is so small that they form isolated Bjerrum pairs (see (6)).

At larger x, correlations and interactions with image charges are unimportant, and the Poisson-Boltzmann equation applies. In this region $N(x)$ varies so smoothly that $N(x) = N_0$ provides an effective boundary condition for the Poisson-Boltzmann equation [95]. At this stage we must remember that in a real physical situation there is always some concentration of Z-ions, N, in the surrounding solution. It can be either larger or smaller than N_0. In the latter case, the surface is overcharged, as we show in Sec. 4 below.

4. Correlation-induced charge inversion

Let us return once again to the physical argument illustrated by Fig. 6. It explains why an extra Z-ion may be attracted to the macroion surface despite the fact that it is already neutralized by the previously adsorbed Z-ions. What happens if another Z-ion approaches? Clearly, correlation effect will keep providing an attractive force for this and subsequent Z-ions, but it will have to compete with the repulsive force which is simply due to the fact that macroion has already too many Z-ions adsorbed and the whole complex is, therefore, positively charged. Thus, the question is this: what is the *equilibrium* amount of (over)charge?

One useful way to think about it is to realize that correlation mechanism provides voltage to drive overcharging, but the actual amount of (over)charge depends on both voltage and capacitance. Since the latter depends strongly on the geometry, we will have to explore several cases - spherical macroion, cylindrical, etc.

Another, equivalent, view, involves the comparison of chemical potentials of adsorbed Z-ions and Z-ions in the bulk solution. This approach immediately suggests that in equilibrium the total charge, Q^*, depends on the concentration, N, of Z-ions in the surrounding bulk solution. Here Q^* is the net charge of the entire complex which includes bare charge of the macroion, $-Q < 0$, and the proper (determined by the equilibrium condition) number of adsorbed Z-ions with charge $Ze > 0$ each.

Let us see now how we can implement the condition of equal chemical potentials for the case of a spherical macroion, with radius r. As regards adsorbed Z-ions, we argue[†]that their chemical potential is only different from the one-component plasma expression (10) by the energy of the Z-ion in the potential $\psi(0) = Q^*/\epsilon r$ created by the net charge Q^* . Therefore, equilibrium condition reads $\mu_{id}+\mu_{WC}+Ze\psi(0) = \mu_b$, where μ_b is the bulk chemical potential. To determine Q^* from here, we first note that $\mu_{id} - \mu_b = k_BT\ln(N_s/N)$; we further express μ_{WC} in terms of N_0 (see Eq. (11)), and finally obtain

$$Q^* = \frac{\epsilon r}{Ze} k_BT \ln(N/N_0) \ . \tag{12}$$

Clearly, the net charge Q^* is indeed positive when $N > N_0$, i.e. it has the sign

---

[†]The argument goes as follows. Spherical surface of the macroion has bare surface charge density $-\sigma = -Q/4\pi r^2$. Let us pretend to place there, along with real charge $-\sigma$, also two imaginary spheres, with uniform charge densities $\sigma^* = Q^*/4\pi r^2$ and $-\sigma^*$. Their total charge is zero, so they have no effect. However, we can now think that the Z-ions are adsorbed on the sphere with the charge density $-\sigma - \sigma^*$, and with this sphere they form a ***neutral*** SCL, quite like that considered in Sec. 3. The remaining sphere has the charge density σ^* and creates spherically symmetric field with potential on the surface $\psi(0)$. We emphasize that the macroscopic net charge σ^* does not interact with the one component plasma, because the potential $\psi(0)$ is constant along the surface, while one component plasma is neutral.

opposite to the bare charge Q.

The result (12) sheds also light on the meaning of the quantity N_0 defined above, in Eq. (11): this is the concentration of Z-ions in the surrounding bulk solution at which macroion is exactly neutralized by the adsorbed Z-ions. This concentration is very small because $|\mu_{WC}|/k_BT \gg 1$. For example, $N_0 = 0.3$ mM and 0.8 μM for $Z = 3$ and 4 respectively (1 M $\approx 6 \times 10^{20}\text{cm}^{-3}$). Therefore, it is easy to increase charge inversion by increasing N. How far does it go? At large enough N, translational entropy terms $\mu_b - \mu_{id}$ become negligible compared to μ_{WC}, yielding

$$Q^*/\epsilon r = \psi(0) = |\mu_{WC}|/Ze \,. \tag{13}$$

Expressing R and $|\mu_{WC}|$ through Q and Z with the help of Eq. (10) (and remembering that $\sigma^* = -\sigma + Zen$), Shklovskii[95] arrived at the prediction for the *maximal* inverted charge for the spherical macroion which can be achieved by increasing concentration of $Z:1$ salt:

$$Q^* = 0.83\sqrt{QZe}. \tag{14}$$

This charge is much larger than Ze, but still is smaller than Q because of limitations imposed by the large charging energy. For example, for $Q = 100e$, $Z = 4$, we get $Q^* = 17e$. Eq. (14) was recently confirmed by numerical simulations[56,96]. Further increase of charge inversion beyond the level dictated by Eq. (14) is achievable with the help of a monovalent salt (see Sec 5).

Applied in cylindrical geometry, similar arguments lead to a revision of the conventional Onsager-Manning condensation theory (Sec. 2) when dealing with multivalent Z-ions. Consider a cylinder with a negative linear charge density $-\eta$ and assume that $\eta > \eta_Z$. Mean field Onsager-Manning theory (7) predicts $\eta^* = -\eta_Z$. By contrast, Perel and Shklovskii[81] showed that a correlation-induced negative chemical potential μ_{WC} results in

$$\eta^* = -\eta_Z \frac{\ln(N_0/N)}{\ln\left(4/\pi Z^6 N \ell_B^3\right)} \,, \tag{15}$$

where ℓ_B is the Bjerrum length (6). This result reproduces the Onsager-Manning one (7) only at extremely small values of N, which are unrealistic at $Z \geq 3$. On the other hand, at $N = N_0$ the net charge flips sign, resulting in charge inversion at $N > N_0$ (which is absent in Onsager-Manning theory). At large enough N, the inverted charge density η^* can reach $k_BT\epsilon/e = \eta_1$.

5. Enhancement of charge inversion by monovalent salt

Most water solutions, particularly biological ones, contain significant amounts of monovalent salt, such as NaCl. Correlations between these monovalent ions are negligible, and, therefore, their only role is to provide Debye-Hückel screening with a decay length r_s (5). This screening makes charge inversion substantially stronger. Indeed, screening by a monovalent salt diminishes the charging energy of the macroion much stronger than the correlation energy of Z-ions. Furthermore, in a sufficient concentration of salt, the macroion is screened at the distance smaller than its size. Then, the macroion can be thought of as an over-screened surface, with inverted charge Q^* proportional to the surface area. In this sense, overall shape of the macroion is irrelevant, at least to a first approximation. Therefore, we consider here a simpler case: screening of a planar macroion surface with a negative surface charge density $-\sigma$ by a solution with a concentration N of $Z:1$ salt and a large concentration N_1 of a monovalent salt.

Nguyen et al[65] calculated analytically the dependence of the charge inversion ratio, σ^*/σ, on r_s, in two limiting cases $r_s \gg R_0$ and $r_s \ll R_0$, where $R_0 = (\pi\sigma/Ze)^{-1/2}$ is the radius of a Wigner-Seitz cell at the neutral point $n = \sigma/Ze$. At $r_s \gg R_0$ calculation starts from Eq. (13). The electrostatic potential of a plane with the charge density σ^* screened at the distance r_s reads $\psi(0) = 4\pi\sigma^* r_s$. At $r_s \gg R_0$ screening by monovalent ions does not change Eq. (10) substantially so that we still can use it in Eq. (13) which now describes charging of a plane capacitor by voltage $|\mu_{WC}|/Ze$. This gives

$$\sigma^*/\sigma = 0.41(R_0/r_s) \ll 1 \qquad (r_s \gg R_0)\ . \tag{16}$$

Thus, at $r_s \gg R_0$ inverted charge density grows with decreasing r_s.

Now we switch to the case of strong screening by monovalent salt. To begin with, let us assume that screening is already so strong that $r_s \ll R_0$, but energy of SCL is still much greater than k_BT per Z-ion. In this regime, free energy consists of Debye-Hückel screened nearest neighbor repulsion energies of Z-ions and the attraction energy of Z-ions to the charged surface:

$$F = (3nZ^2e^2/\epsilon A)\exp(-A/r_s) - 4\pi\sigma r_s Zen/\epsilon\ , \tag{17}$$

where $A = (2/\sqrt{3})^{1/2} n^{-1/2}$ is the lattice constant of the hexagonal Wigner crystal (Fig. 5). Minimizing F with respect to n one arrives at

$$\frac{\sigma^*}{\sigma} = \frac{\pi}{2\sqrt{3}}\left(\frac{R_0}{r_s\ln(R_0/r_s)}\right)^2 \qquad (r_s \ll R_0)\ . \tag{18}$$

Thus σ^*/σ grows with decreasing r_s and can become larger than 100%. At $r_s \sim R_0$, Eq. (16) and Eq. (18) match each other. As we see, σ^* continues to grow with decreasing r_s. This is because the repulsion between Z-ions becomes weaker, so that it is easier to pack more Z-ions on the surface. Of course, when r_s decreases even further, binding energy of Z-ions becomes small, SCL dissolves, and charge inversion disappears.

The results above are in a good agreement with simulations. Terao and Nakayama[99] reported the results of a Monte Carlo simulation for the system consisting of a macroion with charge $Q = 20e$ surrounded by 20 monovalent counterions and 1500 ions of $2:1$ electrolyte. Tanaka and Grosberg[96] performed molecular dynamics simulations with a spherical macroion of the charge $Q = -28e$, spherical Z-ions ($2 \le Z \le 7$), and up to 500 monovalent ions; the system is neutral overall, which determines the number of Z-ions to be between 180 and 52. Simulations confirmed the strong adsorption of the overcharging amount of Z-ions on the surface of macroion. On a more quantitative level, Tanaka and Grosberg[96] examined the dependence of the inverted charge on the ionic strength, I, and found the crossover between $Q^* \propto \sqrt{I} \propto 1/r_s$ and $Q^* \propto I \propto 1/r_s^2$, consistent with Eqs. (16) and (18). Tanaka and Grosberg[96] attempted also to maximize the charge inversion ratio Q^*/Q. In agreement with the theoretical views presented above, the growth of charge inversion is capped when correlations between Z-ions are suppressed. We have mentioned eventual reduction of correlations for large monovalent salt concentration, or small r_s, when the SCL evaporates. If one tries to increase Z instead of lowering r_s, then, at large Z, correlations become suppressed because monovalent ions condense on Z-ions, forming Z'-ions with smaller net charge Z'. This effect is clearly seen in Fig.7 (see Nguyen et al[66] for the conditions under which this phenomenon is and is not important). Nevertheless, Q^*/Q up to about 150% is easily observed.

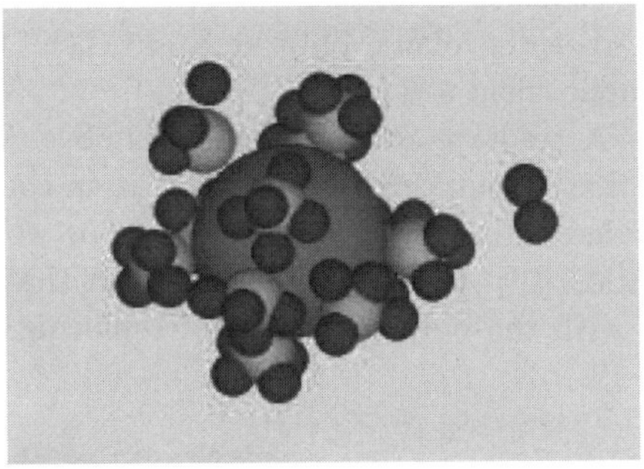

Fig. 7. The snapshot of the system simulated by Tanaka and Grosberg[96]. The spherical macroion is grey, Z-ions are white, and monovalent negative ions are black. Bare charge of the spherical macroion is $Q = -28e$, $Z = 7$, and $\Gamma \approx 30$. There are 52 Z-ions and 336 monovalent ions in the simulation domain. What is clearly seen is the formation of Z'-ions: because Z is so large, Z-ions adsorb monovalent ions. This reduces correlations between Z-ions and restricts charge inversion. Nevertheless, bare charge of the complex shown in the figure is $+16e$.

6. Screening of a charged plane by polyelectrolytes

A practically important class of Z-ions are charged polymers, i.e., polyelectrolytes. Let us start with a rigid polyelectrolyte and discuss charge inversion caused by adsorption of long rod-like Z-ions. For example, the moderately long (up to about 50 nm, or about 150 base pairs) DNA double helix can be well approximated as a rod. Actin is another example of an even more rigid polyelectrolyte. Apart from the uninteresting regime of extremely small macroion surface charge density (in which case the elongated shape of molecules is irrelevant, rendering our previous results applicable), charged rods adsorbed at the surface tend to be parallel to each other due to the strong lateral repulsion. In other words, there is the short range order of a one-dimensional Wigner crystal with lattice constant A in the direction perpendicular to the rods (Fig. 8).

To make the signs consistent with the case of DNA, we assume that the polyelectrolyte charge is negative and equal to $-\eta$ per unit length, while the macroion surface is a plane with positive charge density σ. We assume also that there is a certain concentration of monovalent salt, N_1, in the solution, corresponding to the Debye screening radius (5). To begin with, let us assume that charge density of the rods, $-\eta$, is below the Onsager-Manning threshold, Eq. (7), and let us apply the Debye-Hückel approximation to describe the screening of the charged surface by the monovalent salt. We can then directly minimize the free energy of the one-dimensional crystal of negative rods on the positive surface written similarly to Eq. (17). Then the competition between the attraction of the rods to the surface and the repulsion of the neighboring rods results in the negative net surface charge density $-\sigma^*$ similar to Eq. (18)[63,65]:

$$\frac{\sigma^*}{\sigma} = \frac{\eta/\sigma r_s}{\ln(\eta/\sigma r_s)} \; , \; r_s \ll A_0 \; . \tag{19}$$

Here the applicability condition involves $A_0 = \eta/\sigma$, which is the distance between rods when they neutralize the plane; only at $r_s \ll A_0$ is the overcharged plane linearly screened by monovalent salt.

Speaking about DNA, we have already discussed in Sec. 2 that the DNA charge density, $-\eta$, is such that about three quarters of it is compensated by positive Onsager-Manning-condensed monovalent ions. In other words, the net charge of DNA in the bulk solution is $\eta^* = -\eta_1$ (7). It turns out that at $r_s \ll A_0$ the result (19) applies to DNA with the only correction of replacing $-\eta$ with $\eta^* = -\eta_1 \equiv -k_B T\epsilon/e$ [65].

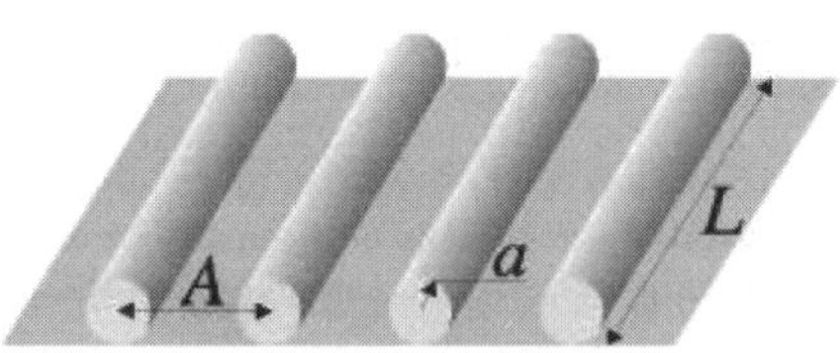

Fig. 8. Rod-like negative Z-ions, adsorbed on a positive uniformly charged plane.

Thus the inversion ratio grows with decreasing r_s as in the case of spherical Z-ions. At small enough r_s and σ, the inversion ratio can reach 200% before DNA molecules are released from the surface. It is larger than for spherical ions, because in this case, due to the large length of the DNA helix, the correlation energy remains large and the Wigner crystal-like short range order is preserved at smaller values of σr_s. Nguyen et al[65] called this phenomenon "giant charge inversion."

Let us switch now to the opposite extreme of weak screening by a monovalent salt, $r_s \gg A_0$. In this case, screening of the overcharged plane by monovalent salt becomes strongly nonlinear, with the Gouy-Chapman screening length $\lambda^* = \epsilon k_B T/(2\pi e \sigma^*)$ much smaller than r_s. Furthermore, some of the positive monovalent ions Onsager-Manning condensed on DNA are released from it upon adsorption, as the plane repels them[78,21]. As a result the absolute value of the net linear charge density of each adsorbed DNA, η^*, becomes larger than η_1. To determine σ^* and η^*, Nguyen et al[65] considered two equilibrium conditions, dealing with chemical potentials of rods and small ions, respectively. As a result, they arrived at the following beautiful formulae valid at $r_s \gg A_0$:

$$\frac{\sigma^*}{\sigma} = \frac{\eta_1}{2\pi a \sigma} \exp\left(-\sqrt{\ln\frac{r_s}{a}\ln\frac{A_0}{2\pi a}}\right) , \tag{20}$$

$$\eta^* = \eta_1 \sqrt{\frac{\ln(r_s/a)}{\ln(A_0/2\pi a)}} , \quad r_s \gg A_0 . \tag{21}$$

At $r_s \simeq A_0/2\pi$ we get $\eta^* \simeq \eta_1$, $\lambda^* \simeq r_s$ and $\sigma^*/\sigma \simeq \eta_1/(2\pi r_s \sigma)$ so that Eq. (20) crosses over smoothly to the strong screening result of Eq. (19).

So far in this Section we assumed a rod-like polyelectrolyte. Let us now discuss how chain flexibility affects charge inversion. We argue that the results for rod-like Z-ions are remarkably robust. To begin with, consider a polyelectrolyte having several charged groups per each persistence length. We argue that our results remain valid as long as adsorption energy *per one persistence length* is large compared to $k_B T$. Indeed, under this condition even flexible polyelectrolyte chains lay flat on the surface, in which case they are ordered in Wigner crystal-like SCL and, therefore, behave similarly to rods.

Dobrynin et al[13] addressed the opposite extreme, namely, weakly charged polyelectrolytes, with so small a fraction of charged monomers, f, that a link between two neighboring charges is already a flexible polymer; in other words, the distance between charges is larger than the persistence length. It was discovered by de Gennes et al[22] that a weakly charged polyelectrolyte chain in a bulk solution consists of electrostatic blobs. Inside each blob polymer is only marginally perturbed by Coulomb interactions, while chain of blobs is fully stretched, rod-like. Dobrynin et al[13] have found that this blob structure remains valid for the adsorbed chains, which form SCL of effective rods of blobs. Speaking of charge inversion, this means that Eq. (19) remains valid for the weakly charged chains, provided η is replaced with a linear charge density of the string of blobs $\eta^* = (f\epsilon e k_B T/l^2)^{1/3}$, where l is the chain persistence length, and r_s is larger than a blob size.

With increasing σ, adsorbed rods, either real or blob ones, start to touch each other leading to multilayer adsorption. It is only in this regime that the real and blob rods behave differently, as we discuss in Sec. 8.

7. Polyelectrolytes wrapping around charged particles

As we have mentioned in the Introduction, one of the practically important situations is that of a long charged polymer forming complexes with oppositely charged particles. Motivated by the problem of nucleosome charge inversion, Mateescu et al[55], Park et al[78], Sens and Gurovich[92], and Netz and Joanny[64] considered the complex of a positive sphere with charge q and a negative polyelectrolyte, such as a DNA double helix, which has to make some $n_t > 1$ turns around the sphere to neutralize it (Fig. 9). These authors predicted a substantial charge inversion: more of the polyelectrolyte is wound than is necessary to neutralize a sphere. Furthermore, Mateescu et al[55] found that a tightly coiled polyelectrolyte conformation becomes unstable when the chain length exceeds a certain threshold, and then an almost straight tail abruptly stretches out (Fig. 9).

Nguyen and Shklovskii[69] emphasized the role of correlations in this case of charge inversion. Indeed, neighboring turns repel each other and form an almost equidistant solenoid, which locally resembles SCL. The tail of the polyelectrolyte repels the already adsorbed part of the polyelectrolyte and creates a correlation hole, which attracts the tail back to the surface (compare Fig. 6). As a result, the net charge of the sphere with wrapped polyelectrolyte, q^*, is negative. It is shown that at $r_s \to \infty$ the charge inversion ratio scales as $|q^*|/q \sim 1/n_t$. On the other hand, at small enough r_s it can exceed 100%.

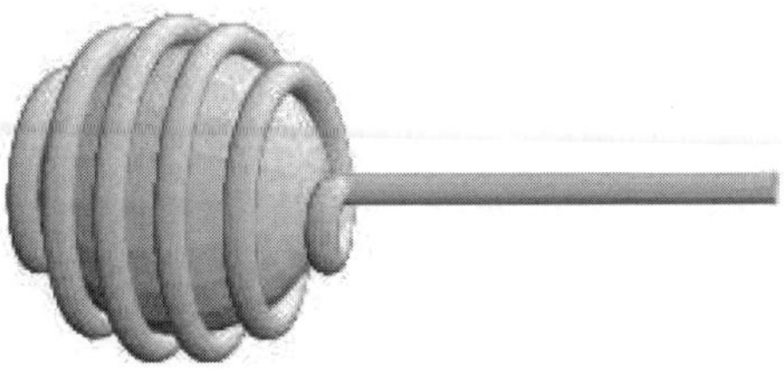

Fig. 9. A polyelectrolyte molecule winding around a spherical macroion. Due to the Coulomb repulsion, neighboring turns, which play the role of Z-ions, are strongly correlated.

Even more interesting is the system in which charged polymer is so long that it forms complexes with many oppositely charged particles (see Fig. 4). Examples include micelles [103], globular proteins[33,105], colloids[5,39], or dendrimers[37,17], and, last but not least, histone octamers forming 10 nm chromatin fiber with DNA (Fig. 3). To be specific, we remain with the signs consistent with the DNA case and consider a long negative polymer chain in the solution of positive spheres. If the concentration of spheres is large, it is favorable to adsorb many spheres on the polyelectrolyte chain. As a result, each sphere is under-screened by polyelectrolyte and has positive net charge. Then, adsorbed spheres repel each other and the complex forms a periodic necklace (see Fig. 4). This necklace is, in fact, a one-dimensional Wigner crystal, or SCL, of spheres, which serve as Z-ions . Indeed, since the segment of the polyelectrolyte wound around one sphere interacts almost exclusively with this sphere, it plays the role of the Wigner-Seitz cell. Because of correlations, spheres bind to polyelectrolyte in such a large number that the net charge of the polyelectrolyte molecule becomes positive [70]. In this case, the charge inversion ratio scales as $Q^*/Q \sim n_t^{1/4}$ in the absence of monovalent salt, where $-Q$

and Q^* are the bare and net charges of the polyelectrolyte molecule, respectively. This means that charge inversion may be larger than 100%. As we discussed in Sec. 5, charge inversion can be further enhanced by a monovalent salt, in which case $Q^*/Q \sim n_t$. We shall return to the complexes of a charged chain with spheres in Sec. 9.

8. Multilayer adsorption

So far we have not considered the possibility that Z-ions fully cover the macroion surface. This may sometimes happen, particularly when a macroion is very strongly charged, or Z-ions are large. Suppose, for instance, that each Z-ion has a hard core of radius a. In this case, the effect of excluded volume of the hard cores of the Z-ions adds positive contributions to the surface pressure and chemical potential of a SCL (10) that are proportional to k_BT and that diverge at the full coverage. Close to the full layer this term compensates and then over-compensates the negative Coulomb term μ_{WC}, so that charge inversion disappears. Indeed, a full layer is incompressible (see Fig. 10b), and, unlike a partially filled layer (see Fig. 6 or Fig. 10a), it does not allow for the creation of an image-like correlation hole.

At even larger macroion charge, the second layer starts to form, launching a new wave of charge inversion. In the beginning, charge inversion is small because all the attraction of a new Z-ion approaching the surface is provided by a weak interaction with an inflated image in the emerging second layer, where once again $A \gg a$ (Fig. 10c). Continuing, Nguyen and Shklovskii[71] arrived at the prediction of an oscillating inverted charge Q^* as a function of Q (see Fig. 10), where charge inversion vanishes every time the top layer of Z-ions is full.

Another way to look at this phenomenon is to examine a metallic electrode screened by Z-ions, when the potential of the electrode is controlled instead of its charge. In this case, oscillations of charge inversion and compressibility lead to oscillations of capacitance of this electrode with the number of adsorbed layers of Z-ions. This is similar to oscillations of compressibility and magneto-capacitance in the quantum Hall effect, which are related to consecutive filling of Landau levels[14,43,15]. In this sense, we deal with a classical analog of the quantum Hall effect.

To conclude this section, let us return to the adsorption of weakly charged polyelectrolytes that we discussed briefly in the Sec. 7. Dobrynin et al[13] have shown that parallel chains of adsorbed blobs start touching each other above the same threshold surface charge density $\sigma_e = ef/l^2$, which corresponds to the onset of squashing blobs on the surface. As a result, these authors arrived at the conclusion that if weakly-charged chains are adsorbed on the surface with $\sigma > \sigma_e$, they form a polymer liquid. In this liquid, correlations and image formation are only due to the uppermost layer, with the thickness about that of an unperturbed blob. There are no oscillations of inverted charge; instead, charge inversion saturates at about one layer of blobs and remains unchanged afterwards.

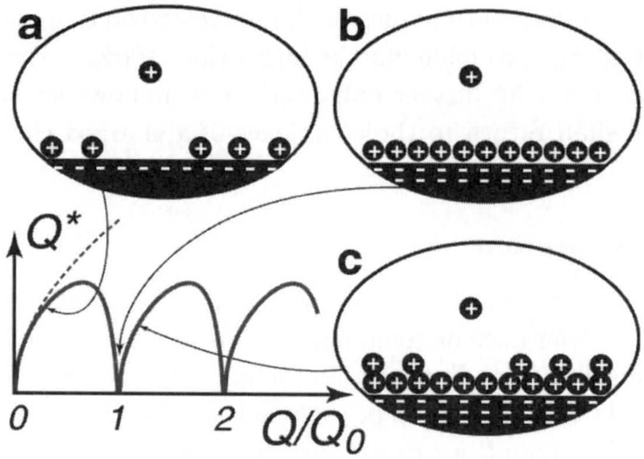

Fig. 10. Inverted charge Q^* as a function of the absolute value Q of the bare charge; Q_0 is the charge of one full layer of Z-ions. The dashed line corresponds to the case of Z-ions with vanishing radius, Eq. (14). (a) The first layer is not full, as in Fig. 6. An approaching new ion creates a correlation hole and is attracted to it. (b) The layer is full, there is no place for a correlation hole. (c) More than one layer is full. A correlation hole exists in the top layer only.

9. Correlation-induced attraction of like charges

The idea of a single screened macroion is a useful one in a theoretical context, but in practice it is rarely true that there is only one macroion. Typically, there is a certain concentration of them, so that interactions between them can be important. Let us start with the simplest question: consider two macroions, and suppose the concentration of Z-ions in solution is equal to N_0 (see Eq. (11)), such that each macroion forms a neutral complex with Z-ions. How do these two neutral complexes interact? It turns out that they attract each other at short distances and, therefore, tend to coagulate. In other words, two macroions of the same charge may attract each other because of the presence of Z-ions. In general, this attraction of like charges is as interesting a manifestation of correlations as charge inversion, even though our present Colloquium emphasizes the phenomenon of charge inversion. Nevertheless, we must discuss attraction at least briefly, in order to prepare the ground for the subsequent discussion of experiments (Sec. 10).

Medium-induced attraction of like charges is nothing new in physics, with Cooper pairs of electrons being the most prominent example. In the context of molecular physics, the most popular explanation of attraction is in terms of *salt bridges*: a divalent ion, such as Mg^{2+}, can form ionic bonds with two groups with charges of -1 each, in effect connecting them together. This idea is indeed adequate if we have, say, two macroion surfaces with regularly placed charges of -1, and there are ions of charge $+2$ between them. However appealing, the bridge concept becomes increasingly fuzzy when Z-ions have charges of 3 or higher and when charges in the macroion are not positioned regularly.

Motivated by experimental observations of DNA condensation (see Sec. 10), there was a significant effort by theorists to try to explain attractive forces by going beyond the bridge model. For simplicity, and following the majority of works, let us

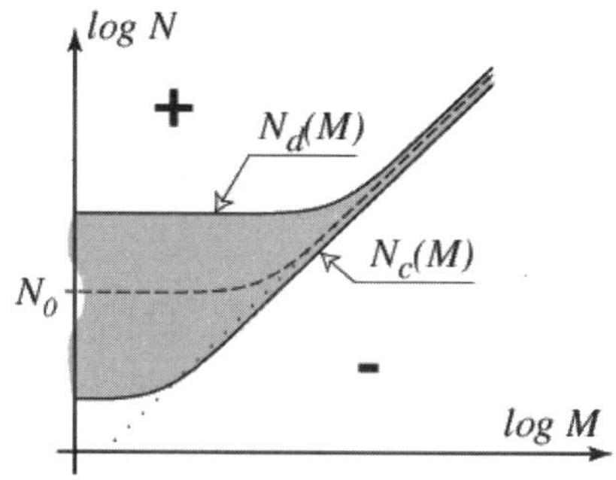

Fig. 11. Generic phase diagram of reentrant condensation and charge inversion in terms of macroion concentration M and Z-ion concentration N [72]. Isoelectric composition is shown by the dotted line. The dashed "neutrality line" corresponds to neutral complexes in the dilute phase. The segregation region is shaded. Minus and plus indicate the signs of complexes of DNA with Z-ions.

consider two planar macroion surfaces with some Z-ions between them. Of course, Poisson-Boltzmann theory predicts pure repulsion for such a system. However, attraction was observed in several computer experiments[28,40,27,49,59]. On the theory side, several groups attempted to go beyond the mean field approximation. An important observation is that due to dynamic fluctuations of counterions, there is an attractive component (similar to Van der Waals interactions), but at effectively high temperature, or small Γ (see Eq. for the definition of (8)), the Poisson-Boltzmann repulsion still dominates and the force remains mainly repulsive[75,47,29,82,23,24].

On the other hand, attraction of like charges dominates at effectively low temperatures, when $\Gamma \gg 1$, and the idea of spatial correlations between Z-ions, which is the central idea of this Colloquium, sheds light on the nature of this attraction. Indeed, for extremely large Γ, we deal with two Wigner crystals on the two opposing plates; they gain energy when approach each other by properly positioning themselves in the lateral direction. This was shown by Rouzina and Bloomfield[90] (see also Gronbech-Jensen et al[27]; Levin et al[48]; Moreira and Netz[59]). Furthermore, Gronbech-Jensen et al[27] and Shklovskii[94] pointed out that the long range order of a Wigner crystal is not important for this attractive force. As in the case of charge inversion, what is important is correlation and short range order. As we know, Z-ions form a SCL on the macroion surface as soon as Γ becomes large. Imagine now bringing two planar surfaces, along with their respective SCL, very close to each other. What we have now is essentially two copies of Fig. 5 superimposed on top of one another, with Z-ions confined in between. Clearly, two SCL merge, lowering the energy per Z-ion from $\varepsilon(n)$ to $\varepsilon(2n) < \varepsilon(n)$ (see Eq. (9)). Physically, every Z-ion in the merged SCL is sandwiched between two macroion surfaces, and its Wigner-Seitz cell can be approximated as a pair of discs, one on each surface (see again Fig 5). The charge of the cell must still be $-Ze$, but since there are two surfaces, the radius of the cell is reduced by the factor $1/\sqrt{2}$, leading to the energy gain. In some sense, this theory returns us to the idea of bridges, albeit on a completely new level, with each Z-ion bridging between two sides of its Wigner-Seitz cell, which can include many surface charges.

These arguments hold, at least qualitatively, not only for plates, but also for macroions of other shapes, including DNA double helices. To be specific, consider two DNA double helices. When the concentration of Z-ions is equal to N_0, each DNA is neutralized by Z-ions, and the two neutral complexes attract each other at short distances. What happens if the concentration of Z-ions is higher or lower than N_0? In this case, correlation-induced attraction, which is short-ranged, competes with Coulomb repulsion, which is much longer-ranged. Note that the Coulomb repulsion force is present both at $N < N_0$, when the DNAs are partially screened by Z-ions and negative, and at $N > N_0$, when they are overcharged and positive.

What are the implications of this competition between attraction and repulsion? They are summarized in Fig. 11 which shows a phase diagram of the solution with number concentrations of macroions, M, and Z-ions, N (along with the neutralizing amount of monovalent ions and salt.) The major feature of the phase diagram is the segregation region, which is the shaded area in the Fig. 11. As the figure indicates, the generic scenario is that of *reentrant condensation.* DNA molecules stay in solution and remain negative at $N < N_c(M)$, forming undercharged complexes with Z-ions. At some concentration of Z-ions, $N = N_c(M)$, repulsion loses to the correlation attraction and a condensed phase of DNA is formed, coexisting with a dilute phase. The condensed phase for DNA represents a (nematic) bundle of helices, it exists in the interval $N_c(M) < N < N_d(M)$. Finally, at $N = N_d(M)$, repulsion overcomes the correlation attraction, and the DNA molecules dissolve and form positive (overcharged) complexes with Z-ions.

Inside the coexistence region, there is a neutrality line, on which the equilibrium dilute phase consists of *neutral* complexes. At very small DNA concentrations, the neutrality condition corresponds to the concentration N_0 of Z-ions (Eq. 11). To see what happens at larger DNA concentration, consider increasing M, starting from the overcharged complexes, well above segregation on the phase diagram in Fig. 11. When M grows, the solution runs out of Z-ions when it approaches the "isoelectric line" $-\eta LM + ZeN = 0$, with $-\eta < 0$ and L being the DNA linear charge density and length, respectively. Near this line, the charge of complexes flips sign. Thus, the neutrality line crosses over from $N = N_0$ to the isoelectric line. The border lines $N_c(M)$ and $N_d(M)$ follow a similar pattern. Although not plotted in Fig. 11, at extremely small values of M these two lines join together at a critical point, and for smaller M only intramolecular condensation of DNA (the coil-globule transition) is possible if the DNA molecule is long enough.

We have considered the phase diagram in Fig. 11 for a solution of DNA chains with small Z-ions. As a matter of fact, the diagram is qualitatively quite general [72]. For instance, it applies to a solution of DNA with large positively charged particles. In Sec. 7, we considered the case of a small DNA concentration, M, and a large concentration of spheres, N, which corresponds to the region above the coexistence region on the phase diagram in Fig. 11. We found that complexes have the form of necklaces, as shown in Fig. 4, and that they are overcharged, i.e., contain more spheres than necessary to neutralize DNA molecule. Large spheres are so strongly bound to DNA that the concentration N_0 for them is extremely small and any real experiment deals with the narrow upper-right part of the diagram. Suppose now that there are relatively few spheres in the solution, so we are below the neutrality line. In this situation, chains make an overcharging number of turns around each sphere. This is energetically favorable due to the repulsive correlations between subsequent turns on a sphere surface. The inverted net charge of each sphere is about as large as for the case of a single sphere discussed in Sec. 7. Furthermore, the inverted charge of spheres determines their distribution along the chain of polyelectrolyte. Negative spheres repel each other and, therefore, the complex once again has the periodic beads-on-a-string structure, Fig. 4, which resembles the 10 nm chromatin fiber. In the narrow vicinity of the neutrality line, even a small correlation attraction between touching spheres is sufficient to drive aggregation (or coil-globule collapse for long chains) of DNA with spheres.

10. Experimental evidence of charge inversion

How does the theory of correlated screening compare with experiment? For the purposes of this Colloquium, we restrict ourselves to a qualitative comparison only.

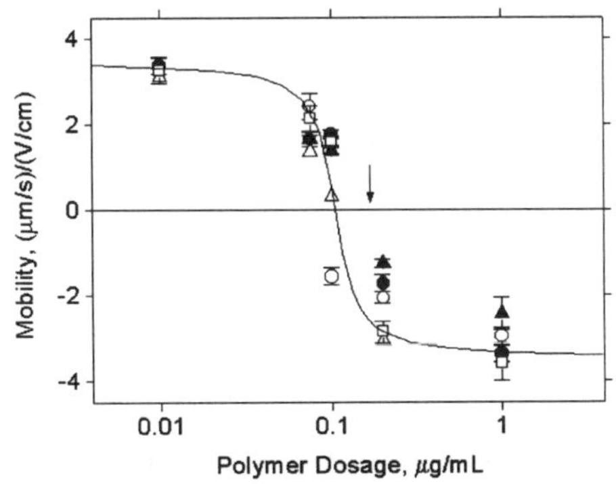

Fig. 12. Mobility of positive latex particles (macroions) in the presence of 0.005 M (moles per liter) concentration of NaCl as a function of polymer mass concentration. Polymers (single-stranded DNA chains) play the role of Z-ions. Different symbols correspond to DNA of the following lengths (in monomers): ▲ - 8, ○ - 10, △ - 40, • - 80, □ - 1400. The line is drawn to guide the eye. The arrow indicates the isoelectric point, the polymer mass concentration of 0.17 μg ml^{-1} at which DNA charge neutralizes the latex particles.

First observations of charge inversion were reported a long time ago by De Jong (1949). He was able to measure electrophoresis of macroscopic aggregates in the phase segregation region of the phase diagram Fig 11, and observed the reverse of their mobility upon crossing the isoelectric line. More recently, Kabanov and his co-workers[34,35,36] examined mixtures of positive and negative polymers and directly observed the interpolyelectrolyte complexes in which a larger polymer of one sign (playing the role of a macroion) was seen to bind an overcharging amount of smaller polymers of the opposite sign (playing the role of Z-ions). The effect was directly seen due to the reversal of the electrophoretic mobility of complexes. Furthermore, Wang et al[103] observed similar reversal for a mixture of polyelectrolytes (macroions) and micelles (Z-ions). Gotting et al[25] found the reversed mobility for the nanoparticles (macroions) and short single stranded DNA (elongated Z-ions). Walker and Grant[101] demonstrated this phenomenon for 120 nm latex particles (macroions) with single stranded DNA (Z-ions) ranging from 8 to 1400 nucleotides; their data are presented in Fig. 12. Kabanov et al[37] and Evans et al[17] observed the reversed electrophoretic mobility for DNA with the dendrimers as Z-ions.

An interesting observation, apparent from Fig. 12, is that the data for different DNA lengths collapse onto a single master curve upon re-scaling (in which mobility is plotted against mass concentration of DNA instead of number concentration, N). We note in passing that this observation can be rationalized by the following argument. Isoelectric point is obviously determined by the number of charged groups per unit volume, but this quantity is insensitive to the overall length of DNA and is simply proportional to the mass concentration.

All of the above mentioned experimental works rely on the reverse of electrophoretic mobility as an indication of charge inversion. Indeed, this is conceptually the most straightforward approach. It is valid because Z-ions are strongly bound to the macroion, with energies larger than k_BT, and move together with it. On the other hand, monovalent ions screening the net charge Q^* are attracted to macroion with energy much smaller than k_BT and move in electric field in the opposite direction. For these reasons, it is the net charge Q^* that determines both magnitude and sign of the observed electrophoretic mobility. This remains correct also in the case when monovalent ions adsorb on Z-ions, effectively reducing them to Z'-ions with $Z' < Z$, as in Fig. 7. In all cases, the net charge includes all ions bound with energies in excess of k_BT. In a recent molecular dynamics simulation, Tanaka and Grosberg[97] have directly examined the mobility of charge inverted macroion complexes similar to the one shown in Fig. 7. They confirmed that adsorbed Z-ions drift together with the macroion in a weak electric field and that the sign of the net charge Q^* determines the direction of electrophoresis.

It is also worth saying that the interpretation of electrophoretic experiments on charge inversion is not affected by the recent discoveries of Long et al[50,51]. These authors noted that electrophoretic mobility, under some circumstances, may not be entirely determined by the charge. For instance, even an overall neutral object may move in an electric field provided that it has strong asymmetry of charge distribution. A simplest example consists of two balls of different radiuses, rigidly connected by a thin rod, and having opposite charges. The effect is due to the fact that external electric field acts not only on macroion itself, but also on surrounding co- and counterions, causing the latter to flow and to exert viscous friction drug forces on the macroion. Strong geometrical asymmetry of positive and negative charges leaves these drug forces unbalanced. In this colloquium, we discuss only macroions in which the bare charge is uniformly distributed on the surface. Then, the charge of Z-ions is practically uniform, too (apart from the small length scales of the order of R, the distance between neighboring adsorbed Z-ions). For such macroions, reversal of electrophoretic mobility does indeed indicate the inversion of charge.

For a more detailed comparison with experiments, we should remember that charge reversal is expected to be accompanied by coagulation, as discussed above in Sec. 9. Whether in equilibrium or not quite in equilibrium, these large complexes should scatter light strongly. There are many experiments reporting such observations.

Let us begin with DNA. It has been known for some time that at some critical concentration of Z-ions, N_c, DNA abruptly condenses into large bundles [3]. Recently it was discovered that at a much larger critical concentration, N_d, the bundles dissolve back[91,79,80,87,88]. Specifically, for the spermine ions ($Z = 4$), it was found experimentally that $N_c = 0.025$ mM and $N_d = 150$ mM. If one interprets these data in the framework of the theory of Nguyen et al[67], the experimental values of N_c and N_d imply that, for spermine, $N_0 = 3.2$ mM and the binding energy of two helices per one spermine ion is $u = 0.3k_BT$. The last value agrees with the one obtained by a different method [89].

Let us now discuss some other systems. Wang et al[103] studied complex formation in mixture of micelles and oppositely-charged polyelectrolytes. In this experiment, the total charge of micelles was controlled by changing the concentration of the cationic lipid in the solution. In agreement with the above theory, measurements of dynamic light scattering and turbidity (coefficient of light scattering) show that complexes condense in bundles and the solution coacervates in the vicinity of the point where mobility crosses over between two almost constant values, positive and negative.

For the complexes of latex particles with DNA of various lengths, examined by Walker and Grant[101], equilibrium conditions were not found, but a significant rate of aggregation of latex particles was observed in the same narrow range of DNA concentrations where the mobility flips the sign.

There is a large body of interesting experimental[86,41] and theoretical[30,7] work on phase diagrams and overcharging of lamellar cationic lipid-DNA self-assembled complexes. These solutions always seem to have aggregates due to hydrophobic attraction of lipids. A phase diagram of this kind has been sketched by Rädler[86].

Another large group of works with nice (and technologically useful) examples of overcharging deals with the sequential adsorption of multiple layers of polyelectrolytes of alternating sign (see Decher[12] and references therein). Say, a positive surface is first treated with negative polymers, gets overcharged and becomes negative, then it is treated with positive polymers, gets overcharged etc. This procedure works reliably with up to many tens of layers. Theoretical interpretation of this technique was discussed by Castelnovo and Joanny[8]. It falls beyond the framework of the present Colloquium as it involves certain kinetic considerations while we deal in this Colloquium only with equilibrium phenomena.

11. Correlations "in a sheep's skin"

We mentioned correlations so many times in this Colloquium, that the reader may want to ask: Are there alternative, correlation-independent, electrostatic mechanisms leading to charge inversion? Our answer is no, and we argue that the correlation-based mechanism is the universal one. This fact notwithstanding, we should say that correlations may show up in a number of ways, sometimes masked like a wolf in a sheep's skin.

To understand better the role of correlations, let us first consider the case of no correlations. Namely, suppose, we have a set of randomly positioned point-like charges, equal numbers of $+e$ and $-e$. It is easy to establish that the averaged interaction energy in such system is exactly zero, where average is taken over independent random positions of all charges. Similarly, averaged electric field in the system is also zero. Both these statements remain correct also for the one component plasma, in which point-like charges of one sign are randomly positioned on the smeared uniform background of the opposite sign.

The essence of screening is that charges in plasma are *not* positioned randomly. Correlations happen because ions reconfigure themselves non-randomly to gain some energy. This so-called correlation energy is well known in plasma physics [44]. It is negative, meaning that correlated configuration is lower in energy, or more thermodynamically favorable, than random configuration.

These simple facts allow us to understand the underlying role of correlations in one of the theories suggested in the literature to explain charge inversion. This theory, put forward by Park et al[78], views monovalent *counterion release* from Z-ions as the driving force behind charge inversion. We argue that while counterion release accompanies, it is itself driven by correlations. As we did already more than once, let us imagine that DNA molecules, along with their Onsager-Manning-condensed small ions, are being adsorbed on the macroion one at a time (possibly releasing some of their counterions). Let us further consider the moment when the neutralization condition has just been achieved. Pretend now that DNA rods (Z-ions) are distributed randomly on the surface, uncorrelated in both positions and orientations. In this case, next arriving DNA molecule feels no average field, so that it has no reason to release its counterions. The situation is completely different if DNA molecules are correlated on the surface (see Figs. 2 or 8), where locally each molecule is surrounded by a correlation hole - the positive stripe of the background charge (the Wigner-Seitz cell). The corresponding field, or the positive potential of the Wigner-Seitz cell, causes the release of counterions from DNA not only at the neutrality point, but even if the surface overall is overcharged (see the solution of this problem given by Eqs. (20), (21)). These qualitative arguments can be formulated also more quantitatively [68]. Thus, correlation hole, or adjustment of DNA molecules to each other, or image charge, or correlations (all synonyms!) is the driving force for *both* counterion release *and* charge inversion; under a sheep's skin of counterions release, there is a wolf's face of correlations.

Let us now discuss another approach which we call ***metallization***. It was pioneered by Mateescu et al[55], who considered complexation of a polyelectrolyte with a sphere, and by Joanny[31], who examined adsorption of flexible polymers on a charged plane. Metallization theory considers adsorbed Z-ions as a continuous medium similar to a metal, while still treating the bulk solution as consisting of discrete charges.

We argue that metallization theory is in fact an overestimate of correlation effects. Indeed, as we illustrated with Fig. 6, SCL behaves as a metal on the length scale above R, the distance between neighboring adsorbed Z-ions. It behaves as a metal in the sense that it responds to the approaching new ion by forming an image. Clearly, the smeared continuum is also a metal in this sense, and even a better metal - it is a metal on all length scales. Another way to view that same physics is to note that correlation suppresses electric field of every Z-ion beyond certain distance of order R, while for the smeared continuum the field is suppressed everywhere and, therefore, the effect is overestimated.

The latter view suggests also another fruitful interpretation of metallization approach, which is in terms of the self-energy of Z-ions. Here, we resort to the terminology in which self-energy of an ion is identified with the energy of the electric field outside certain cutoff length, such as an ion size. (We are not interested in the electric field on the smaller length scales: although energy of this field is infinite, it does not change upon any processes considered here, such as adsorption of Z-ions.) Using this language, the metallization theory becomes physically transparent. Indeed, in this theory adsorption of Z-ions is energetically favorable because they do have self-energy while in the bulk solution and loose it completely upon adsorption. Once again, real screening, or correlations between Z-ions, correspond to suppression of a part of self-energy, corresponding to the energy of the electric field beyond the distance of order R. Energy of the field between the ion size and R is the amount of overestimation by the metallization theory. By contrast, the Poisson-Boltzmann approximation fails to describe charge inversion precisely because it smears Z-ions everywhere and neglects their self-energies.

Representation of correlations in terms of self-energy allows us to address one more practically important problem, namely, the discreteness of charges[73]. Indeed, instead of uniformly charged surface, like Fig. 1, it would be closer to reality (and to a chemist's heart) to think of a macroion as having some charged groups with unitary charge each. For instance, negative $(-e)$ charges of DNA are positioned on an external spiral rim of the double helix, at the distance $A = 0.67$ nm from each other along the rim.

To be specific, let us consider a macroion which is a regular lattice of charges $-e$ ("unfolded DNA strand") and Z-ion having also a linear array of charges $+e$ (a short polyelectrolyte). Importantly, Z-ions always have some degree of flexibility; for instance, in the case $Z = 3$, as shown in Fig. 13, we can imagine that they freely bend in the middle. For simplicity we assume that the distance between

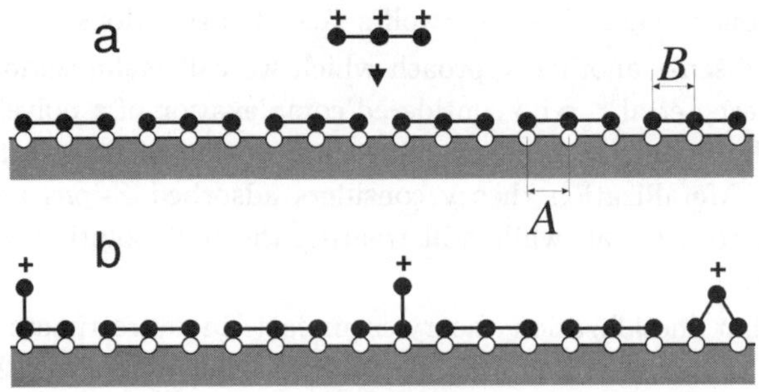

Fig. 13. Schematic representation of charge fractionalization. a) One strand of negative charges of DNA (empty balls) is completely neutralized by positive Z-ions with $Z = 3$, their charged groups are shown by black balls. A new Z-ion is approaching. b) The new Z-ion is "digested." Its charge is split in $+e$ charges of Z defects, tails and archs.

charges in the Z-ion, B, matches exactly that in the macroion, A: $A = B$. Finally, we assume that the Z-ion charges and the macroion charges can approach each other to the minimal distance much smaller than A. Then Z-ions can attach to the macroion locally compensating each charge and, therefore, achieving complete neutralization, as we show in Fig. 13a. The neutralization is so perfect that it is difficult to imagine how another Z-ion can be attached. Fig. 13 explains why this happens. Similarly to Fig. 6, suppose that the macroion is already neutralized by Z-ions and a new Z-ion comes. Then it turns out energetically favorable to disturb order among the neutralizing Z-ions by Z defects in Z independent places, thus opening a room for a new Z-ion. In each defect, one charge of Z-ion is detached from the corresponding macroion charge, forming positive tail or arch above the surface and leaving negative vacancy on the macroion. Then, shifting Z-ions along the macroion, Z vacancies can join together and form a large vacancy capable to accommodate an entire new Z-ion. A net result is that Z disconnected charges $+e$ appear on top of completely neutralized macroion (Fig. 13b), or, in other words, the charge of Z-ion is fractionalized. To avoid misunderstanding, we emphasize that none of the chemical bonds is really cut.

Fractionalization effectively eliminates self-energy of the free Z-ion. Indeed, self-energy of the Z-ion is simply the energy of repulsion between Z positive constituent charged groups in the extended conformation which the Z-ion assumes in the solution. In the fractionalized state, charges are far apart and practically do not interact, so that self-energy is gained. These results are easily generalized for the case $B < A$, when Z-ions have larger linear charge density than macroion. For instance, if $B = A/2$, then Z-ions repelling each other form a "Wigner crystal" on top of the lattice of macroion charges where they alternate with vacant places (similarly to Figs. 1, 5, 6, and 8).

Is fractionalization a correlation-independent mechanism of charge inversion? Of course, not: this phenomenon is solely due to the correlated distribution of Z-ions, which avoid each other at the macroion. Fractionalization is yet another mask under which correlations may show up.

12. Charge inversion in a broader physics context

In conclusion, we would like to show that charge inversion studied in this Colloquium has many physical analogies. There are other "charge inverted" systems in physics. Let us start from the hydrogen atom. It is known that it can bind a second electron, forming the negative ion H^- with an ionization energy of approximately $0.05Ry$ [54]. We can consider this effect as the inversion of proton charge. Attraction of the second electron to the neutral atom is due to the Coulomb correlation between electrons: the first electron avoids the second one, spending more time on the opposite side of the proton. In other terms one can say that binding is related to polarization of the neutral core.

Negative ions - nuclei overcharged by electrons - exist also for larger atoms. Mean field Thomas-Fermi or Hartree theories fail to explain negative ions [45]. One must include exchange and Coulomb correlation holes to arrive at a satisfactory theory explaining bound state and the nonzero ionization energy of a negative ion[54]. The Thomas-Fermi theory of an atom is an analog of the Poisson-Boltzmann theory of electrolytes. It is not surprising that both fail to explain charge inverted states.

Similar considerations apply also for a macroscopic metallic particle. Electrons in such a particle have a negative (compare to vacuum) chemical potential or, in other words, positive work function. The work function is known to vanish in the Thomas-Fermi or Hartree approximations [46].

The energy $|\mu_{WC}|$ plays the same role for Z-ions on the insulating macroion surface as the ionization energy of a negative ion or the work function of metallic particle for electrons. Similarly to electrons, charge inversion of a charged insulating macroion by Z-ions can not be obtained in the mean field Poisson-Boltzmann approximation. Only correlations of Z-ions on the surface of the macroion can lead to charge inversion.

Let us now return to the Onsager-Manning condensation[53], Sec. 2. Kosterlitz and Thouless[42] discovered a similar threshold phenomenon for generation of free vortexes in two dimensional superfluids or superconductors. They noticed that due to the logarithmic form of attractive interaction, two vortexes of opposite sign decouple only above some critical temperature, T_{KT}. Later, Kosterlitz-Thouless theory was applied to unbinding of dislocations and disclinations in the theory of defect-induced melting of two-dimensional crystals[62,106].

In the Kosterlitz-Thouless theory one can identify analog of the short-range correlation contribution to the chemical potential of Z-ions (which we called $|\mu_{WC}|$). This is the energy of creation of the two vortex cores. Similarly to $|\mu_{WC}|$, this energy provides additional binding of vortexes and strongly reduces the concentration of free vortexes at $T > T_{KT}$ [57]. In contrast to the Kosterlitz-Thouless theory, the short range contribution $|\mu_{WC}|$ was only recently introduced [81].

Another physical analogy already mentioned in Sec. 8 is the integer quantum Hall effect. We should add that charge fractionalization, as illustrated by Fig. 13, is an analog of the fractional quantum Hall effect [84]. Finally, Fig. 13 bears also analogy with electron charge fractionalization in the polyacetilene [6].

13. Conclusions and outlook

To conclude, we have discussed the physical picture of screening for the case of strongly interacting ions. This case appears to have been overlooked for many decades, since Debye. Its theoretical study, motivated by the experiments mainly on gene delivery and chromatin structure, revealed new interesting physical insights. Specifically, correlations between screening ions lead to such counterintuitive phenomena as charge inversion, reverse electrophoretic mobility, attraction of like charged molecules or colloids, etc. The physical theory of these phenomena is aesthetically attractive as it presents many parallels with other areas of physics, ranging from quantum Hall effect to atomic physics and metals. The potential applications are many, in both chemical and biological realms. We can mention here all sorts of manipulations with DNA, both for biological purposes and for using DNA as an assembly tool for non-biological nanotechnology. They all require understanding of electrostatic properties of DNA chains. Many diseases have to do with mis-assemblies of charged proteins, such as actin; we need to understand better the assembly of such objects. Food, cosmetic, paper, and waste water treatment industries, are all about charged colloids, and the list of applications is easy to continue. In brief, this theory is one of the busy junctions where physics meets chemistry and biology.

Acknowledgements

We enjoyed collaboration with V. I. Perel, I. Rouzina, and M. Tanaka. We are grateful to V. Bloomfield, E. Braun, R. Bruinsma, P. Chaikin, A. Dobrynin, P. Dubin, M. Dyakonov, W. Gelbart, S. Girvin, W. Halley, C. Holm, J.-F. Joanny, A. Kabanov, V. Kabanov, A. Khokhlov, R. Kjellander, K. Kremer, A. Koulakov, V. Lobaskin, F. Livolant, D. Long, G. Manning, R. Netz, P. Pincus, R. Podgornik, E. Raspaud, M. Rubinstein, J.-L. Sikorav, U. Sivan, M. Voloshin, and J. Widom for useful discussions. We thank Zhifeng Shao for the permission to use Figs. 2 and 3 and S. Grant for the permission to use Fig. 12. T. T. N. and B. I. S. are supported by NSF DMR-9985785.

References

1. Andelman, D., and J.-F. Joanny, 2000, cond-mat/0011072.
2. Aberts, B., D.Bray, J.Lewis, M.Raff, K.Roberts, J.D.Watson, 1994, *Molecular Biology of the Cell,* Galland Publishing, New York.
3. Bloomfield, V. A., 1996, Current Opinion in Structural Biol., **6**, 334; 1998, Biopolymers, **44**, 269; Ch. 7 at http://biosci.umn.edu/biophys/BTOL/supramol.html
4. Bjerrum, N. J., 2000, Zs. f. Phys. Chem., **119**, 145.
5. Braun, E., Y. Eichen, U. Sivan, G. Ben-Yoseph, 1998, Nature, **391**, n. 6669, 775.
6. Brazovskii, S., N. Kirova, 1991, JETP Lett. **33**, 4.

7. Bruinsma, R., 1998, Europ. Phys. J. **B4**, 75.
8. Castelnovo, M., and J.-F. Joanny, 2000, Langmuir **16**, 7524.
9. Chapman, D., 1913, Philos. Mag. Sixth Ser. **25**, 475.
10. Chodanowski, P., and S. Stoll, 2001, Macromolecules **34**, 2320.
11. Debye, P., and E. Hückel, 1923, Phys. Zeitsch. **24**, 185.
12. Decher, G., 1997, Science **277**, 1232.
13. Dobrynin, A. V., A. Deshkovski, and M. Rubinstein, 2001, Macromolecules **34**, 3421.
14. Efros, A. L., 1988, Solid State Comm. **65**, 1281.
15. Eisenstein, J. P., L. N. Pfeifer, and K. W. West, 1992, Phys. Rev. Lett. **68**, 674.
16. Ennis, J., S. Marcelja, and R. Kjellander, 1996, Electrochim. Acta **41**, 2115.
17. Evans, H. M., A. Ahmad, T. Pfohl, A. Martin, C. R. Safinya, 2001, Bull. APS **46**, 391.
18. Fang, Y., and J. Yang, 1997, J. Phys. Chem. **B101**, 441.
19. Felgner, P.L., 1997, Scientific American **276**, 102.
20. Frank-Kamenetskii, M. D., V. V. Anshelevich, and A. V. Lukashin, 1987, Soviet Physics Uspekhi **30**, 317.
21. Gelbart, W., R. Bruinsma, P. Pincus, and A. Parsegian, 2000, Physics Today **53**, 38.
22. De Gennes, P.-G., P. Pincus, R. M. Velasco, F. Brochard, 1976, J. de Physique (Paris), **37**, 1461.
23. Golestanian, R., M. Kardar, and T. B. Liverpool, 1999, Phys. Rev. Lett. **82**, 4456.
24. Golestanian, R., and M. Kardar, 1999, Rev. Mod. Phys. **71**, 1233.
25. Götting, N., H. Fritz, M. Maier, J. von Stamm, T. Schoofs, and E. Bayer, 1999, Colloid Polym. Sci. **277**, 145.
26. Gouy, G., 1910, J. Phys. Radium **9**, 457.
27. Gronbech-Jensen, N., R. J. Mashl, R. F. Bruinsma, and W. M. Gelbart, 1997, Phys. Rev. Lett. **78**, 2477.
28. Guldbrand, L. G., Bo Jonsson, H. Innerstrom, and P. Linse, 1984, J. Chem. Phys. **80**, 2221.
29. Ha, B.-Y., and A. J. Liu, 1998, Phys. Rev. Lett. **81**, 1011; 2000, "Physical Chemistry of Polyelectrolytes," ed. T. Radeva, Marcel Dekker (New York); cond-mat/0003162.
30. Harries, D., S. May, W. M. Gelbart, A. Ben-Shaul, 1998, Biophys. J. **75**, 159.
31. Joanny, J.-F., 1999, Europ. J. Phys. B **9**, 117.
32. M. G. Bungenberg de Jong, 1949, in "Colloid Science", edited by H. R. Kruyt (Elsevier, New York)
33. Kabanov, V. A., V. P. Evdakov, M. I. Mustafaev, and A. D. Antipina, 1976, Molekulyarnaya Biologia **11** 52.
34. Kabanov, A. V., and V. A. Kabanov, 1995, Bioconjug. Chem. **6**, 7.
35. Kabanov, V. A., A. A. Yaroslavov, S. A. Sukhisvili, 1996, J. of Control Release **39**, 173.
36. Kabanov, A. V., and V. A. Kabanov, 1998, Advanced Drug Delivery Reviews **30**, 49.
37. Kabanov, V. A., V. G. Sergeyev, O. A. Pyshkina, A. A. Zinchenko, A. B. Zezin, J. G. H. Joosten, J. Brackman, and K. Yoshikawa, 2000, Macromolecules **33**, 9587.
38. Katzir-Katchalsky, A., 1971, Biophysics and Other Topics: Selected Papers, Academic Press, Orlando.
39. Keren, K., Y. Soen, G. Ben Yoseph, R. Yechieli, E. Braun, U. Sivan, and Y. Talmon, 2001, preprint.
40. Kjellander, R., and S. Marcelja, 1985, Chem. Phys. Lett. **114**, 124(E).
41. Koltover, I., T. Salditt, and C. R. Safinya, 1999, Biophys. J. **77**, 915.
42. Kosterlitz, J. M., and D. J. Thouless, 1972, J. Phys C **5**, L124; 1973, J. Phys C **6**, 1181.
43. Kravchenko, S. V., D. A. Rinberg, S. G. Semenchinsky, and V. Pudalov, 1990, Phys. Rev. B **42**, 3741.
44. Landau, L. D., and E. M. Lifshitz, 1977, ***Statistical Physics, Part 1*** (Pergamon Press, Oxford), Ch. VII.

45. Landau, L. D., and E. M. Lifshitz, 1980, *Quantum Mechanics (nonrelativistic theory)* (Pergamon Press, Oxford), Ch. X.
46. Lang, N. D., 1973, *Solid state physics*, Edited by H. Ehrenreich, F. Seitz, and D. Turnbull, (Academic Press, New York), Vol. 28.
47. Lau, A. W. C., P. A. Pincus, 1998, Phys. Rev. Lett **81**, 1338.
48. Levin, Y., J. J. Arenzon, and J. F. Stilck, 1999, Phys. Rev. Lett. **83**, 2680.
49. Linse, P., and V. Lobaskin, 1998, Phys. Rev. Lett., **83**, 4208.
50. Long, D., J.-L. Viovy, and A. Ajdari, 1996, Phys. Rev. Lett. **76**, 3858.
51. Long, D., A. V. Dobrynin, M. Rubinstein, and A. Ajdari, 1998, J. Chem. Phys. **108**, 1234.
52. Luger, K., A. Mader, R. Richmond, D. Sargent, T. Richmond, 1997, Nature, **389**, 251. For pictures visit, e.g. http://info.bio.cmu.edu/Courses/03438/Nsome/1AOI.htm
53. Manning, G.S., 1969, J. Chem. Phys. **51**, 924.
54. Massey, H.S.W., 1938, *Negative Ions*, Cambridge University Press.
55. Mateescu, E. M., C. Jepperseni, and P. Pincus, 1999, Europhys. Lett. **46**, 454.
56. Messina, R., C. Holm, and K. Kremer, 2000, Phys. Rev. Lett. **85**, 872; 2000, Europhys. Lett. **51**, 461.
57. Minnhagen, P., 1987, Rev. Mod. Phys **59**, 1001.
58. Moreira, A. G., R. R. Netz, 2000a, Europhys. Lett. **52**, 705.
59. Moreira, A. G., R. R. Netz, 2000b, cond-mat/0009377.
60. Moreira, A. G., R. R. Netz, 2001 Europhys. Lett. (in press).
61. Mou, J., D. M. Czajkowsky, Y. Zhang, and Z. Shao, 1995, FEBS Letters **371**, 279.
62. Nelson, D. R., and B. I. Halperin, 1979, Phys. Rev. B **19**, 2457.
63. Netz, R. R., and J.-F. Joanny, 1999a, Macromolecules, **32**, 9013.
64. Netz, R. R., J.-F. Joanny, 1999b, Macromolecules, **32**, 9026.
65. Nguyen, T. T., A. Yu. Grosberg, and B. I. Shklovskii, 2000a, Phys. Rev. Lett. **85**, 1568.
66. Nguyen, T. T., A. Yu. Grosberg, and B. I. Shklovskii, 2000b, J. Chem. Phys. **113**, 1110.
67. Nguyen, T. T., I. Rouzina, B. I. Shklovskii, 2000, J. Chem. Phys. **112**, 2562.
68. Nguyen, T. T., A. Yu. Grosberg, and B. I. Shklovskii, 2001, in *Electrostatic Effects in Soft Matter and Biophysics*, Proceedings of the NATO Advanced Study Institute, Les Houches, France, 1-13 October 2000, C. Holm, P. Kekicheff, R. Podgornik eds, NATO SCIENCE SERIES: II: Mathematics, Physics and Chemistry, Volume 46, Kluwer Academic Publishers, Dordrecht.
69. Nguyen, T. T., and B. I. Shklovskii, 2001a, Physica A **293**, 324.
70. Nguyen, T. T., and B. I. Shklovskii, 2001b, J. Chem. Phys. **114**, 5905.
71. Nguyen, T. T., and B. I. Shklovskii, 2001c, Phys. Rev. E **64**, 041407.
72. Nguyen, T. T., and B. I. Shklovskii, 2001d, J. Chem. Phys. **115**, 7298.
73. Nguyen, T. T., and B. I. Shklovskii, 2001e, Phys. Rev. ; cond-mat/0109002
74. Onsager, L., 1967, private communication to G. Manning.
75. Oosawa, F., 1968, Biopolymers **6**, 134.
76. Oosawa, F., 1971, Polyelectrolytes, Marcel Dekker, NY.
77. Pan, J., D. Thirumalai, S. Woodson, 1999, Proc. Natl. Acad. Sci., **96**, 6149.
78. Park, S. Y., R. F. Bruinsma, and W. M. Gelbart, 1999, Europhys. Lett. **46**, 493.
79. Pelta, J., D. Durand, J. Doucet, and F. Livolant, 1996, Biophys. J. **71**, 48.
80. Pelta, J., F. Livolant, and J.-L. Sikorav, 1996, J. Biol. Chem. **271**, 5656.
81. Perel, V. I., and B. I. Shklovskii, 1999, Physica A **274**, 446.
82. Podgornik, R., and V. A. Parsegian, 1998, Phys. Rev. Lett. **80**, 1560.
83. Potemkin, I. I., K. B. Zeldovich, A. R. Khokhlov, 2000, Polymer Science, Ser. C, **42**, 154.

84. Prange, R. E., and S. M. Girvin, eds., 1990, *The Quantum Hall Effect* (Springer, Heidelberg), 2nd edition.
85. Rädler, J. O., I. Koltover, T. Saddit, and C. R. Safinya, 1997, Science, **275**, 810.
86. Rädler, J. O., 2000, in *Electrostatics effects in Biophysics and soft matter*, C. Holm, P. Kekicheff, R. Podgornik eds, Kluver.
87. Raspaud, E., M. Olvera de la Cruz, J.-L. Sikorav, and F. Livolant, 1998, Biophys. J. **74**, 381.
88. Raspaud, E., I. Chaperon, A. Leforestier, and F. Livolant, 1999, Byophys. J. **77**, 1547.
89. Rau, D. C., and A. V. Parsegian, 1992, Biophys. J. **61**, 246.
90. Rouzina, I., and V. A. Bloomfield, 1996, J. Phys. Chem. **100**, 9977.
91. Saminathan, M., T. Antony, A. Shirahata, L. Sigal, T. Thomas, and T. J. Thomas, 1999, Biochemistry **38**, 3821.
92. Sens, P., and E. Gurovitch, 1999, Phys. Rev. Lett. **82**, 339.
93. Shao, Z., 1999, News in Physiological Sciences, **14**, 142.
94. Shklovskii, B. I., 1999a, Phys. Rev. Lett. **82**, 3268.
95. Shklovskii, B. I., 1999b, Phys. Rev. E **60**, 5802.
96. Tanaka, M., and A. Yu. Grosberg, 2001, J. Chem. Phys. **115**, 567.
97. Tanaka, M., A. Yu. Grosberg, 2001, cond-mat/0106561.
98. Tang, M. X., C. T. Rederman, F. C. Szoka JR, 1996, Bioconjugate Chem. **7**, 703.
99. Terao, T., and T. Nakayama, 2001, Phys. Rev. E **63**, 041401.
100. Totsuji, H., 1978, Phys. Rev. A **17**, 399.
101. Walker, H. W., and S. B. Grant, 1996, Colloids and Surfaces A, **119**, 229.
102. Wallin, T., P. Linse, 1996, J. Phys. Chem. **100**, 17873; 1997 *ibid* **101**, 5506.
103. Wang, Y., K. Kimura, Q. Huang, P. L. Dubin, and W. Jaeger, 1999, Macromolecules **32**, 7128.
104. Woodson, S., 2000, Nature Structural Biology, **7**, 349.
105. Xia, J., and P. L. Dubin, 1994, *Macromolecular complexes in Chemistry and Biology* P. L. Dubin et al. eds. (Springer-Verlag, Berlin).
106. Young, A., 1979, Phys. Rev. B **19**, 1855.
107. Zimm, B., and M. Le Bret, 1983, J. Biomol. Struct. and Dyn. **1**, 461.

International Journal of High Speed Electronics and Systems, Vol. 12, No. 2 (2002) 267–293

MOSFET AND FRONT-END PROCESS INTEGRATION: SCALING TRENDS, CHALLENGES, AND POTENTIAL SOLUTIONS THROUGH THE END OF THE ROADMAP

PETER M. ZEITZOFF, JAMES A. HUTCHBY*, and HOWARD R. HUFF

International SEMATECH Inc., 2706 Montopolis Drive, Austin, TX 78741
**Semiconductor Research Corp., P.O. Box 12053, Durham, NC 27709*

e-mail: peter.zeitzoff@sematech.org

The development of advanced MOSFETs for future IC technology generations is discussed from the perspective of the 2001 International Technology Roadmap for Semiconductors (ITRS). Starting from overall chip circuit requirements, MOSFET and front-end process integration technology requirements and scaling trends are discussed, as well as some of the key challenges and potential solutions. These include the use of high-k gate dielectrics, metal-gate electrodes, and perhaps the use of non-classical devices such as double-gate MOSFETs in the later stages of the ITRS.

1. Introduction

By following Moore's Law for over thirty-five years, the IC industry has rapidly and consistently scaled the design rules, increased the chip and wafer size, and improved the design of devices and circuits[1,2]. As a result, chip speed and functional density have increased exponentially with time while chip power dissipation and cost per function have decreased exponentially with time[3]. In particular, the number of memory bits per chip has quadrupled every three to four years, while the speed of microprocessors has more than doubled every three years, based on the increase from about 2 MHz for the Intel® 8080 in the mid-1970's to well over 1 GHz for current leading-edge chips[4]. The design rules have been scaled from about 8 μm in 1972 to the 130 nm (0.13 μm) DRAM half pitch (which is defined as half the minimum metal or polysilicon pitch for dense features) of today's leading-edge technology. This is a reduction factor of about 0.87 per year, or $(1/\sqrt{2}) \sim 0.7$ in a time interval of between two and three years, where the $(1/\sqrt{2})$ factor is the traditional reduction in design rules between successive technology generations. However, because certain key material, process, and MOSFET limits are being approached, there are increasing problems in continuing to scale at this rate.

In the International Technology Roadmap for Semiconductors (ITRS)[5], the progression of the IC technology generations is projected over the next 15 years. The goal is to aid the IC industry and its supporting infrastructure (especially the material and equipment vendors and university and national laboratories researchers) in continuing to follow Moore's Law during that period. One key change in the most recent ITRS is a major acceleration in the scaling of the physical gate length (L_g) of the MOSFETs, where L_g is the final, etched length at the bottom of the gate electrode. The L_g projections for high-performance transistors from both the 1999[6] and 2001[7] ITRS are plotted in Figure 1. For any given year, the 2001 projection is reduced from the 1999 projection by an amount ranging from about 35% to 50%. Alternatively, the 2001 projections lead the 1999 projections by four to six years. This accelerated scaling is important because L_g is

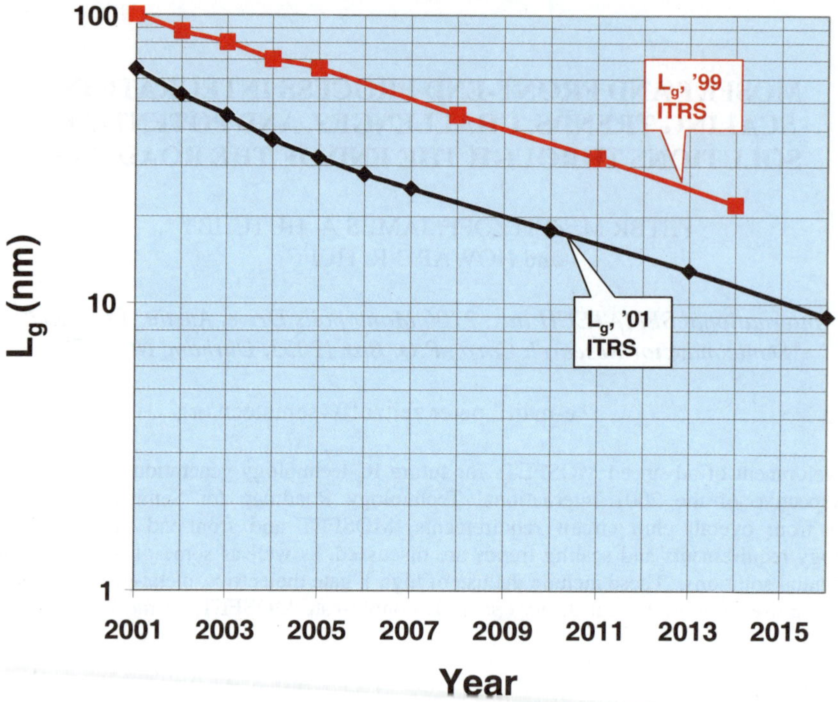

Figure 1. Projections of physical gate length (L_g) for high-performance logic: 2001 vs. 1999 ITRS.[6,7]

a key parameter driving both the overall MOSFET scaling and performance. The accelerated scaling is driven by the IC industry's need to maximize the chip speed, and it reflects actual trends in the industry since 1999, which are projected to continue.

The overall chip requirements for speed, power dissipation, functional density, cost, etc., drive the ITRS projections for MOSFETs. For logic chips there are two main application categories: high-performance and low-power logic. Since the overall chip requirements for each of these application areas are quite distinct, the scaling goals are different for each of them. High-performance logic is mainly used for high-end desktop and server applications (such as for microprocessor unit *(*MPU*)* chips), where the main goal is to maximize chip speed. Consequently, the key MOSFET design goal is to maximize transistor speed, and the tradeoff is relatively high MOSFET leakage current. In contrast, low-power logic is used mainly in mobile systems, where the over-riding goal is to preserve battery life by minimizing the chip power dissipation, particularly the static power dissipation. Hence, the device design is aimed at minimizing the MOSFET leakage current, and the tradeoff is reduced MOSFET speed.

Due to the rapid scaling projected in the 2001 ITRS, the IC industry faces a number of difficult MOSFET scaling challenges in the foreseeable future. These include excessive gate leakage current and boron penetration from the P+ polysilicon gate electrode as the gate dielectric thickness is scaled, as well as an increasing impact of polysilicon depletion in the gate electrode. Also, with MOSFET scaling, it will become increasingly difficult to get simultaneously shallow junction depth, low sheet resistance, low contact resistance, and low junction leakage as required for the source/drain *(S/D)* extension and contact regions. In the long term, improved electron and hole mobility in the inversion layer may be required to meet the transistor performance goals. Furthermore, the classical planar bulk MOSFET may eventually be inadequate, especially as a result of excessive short-channel and quantum effects as well as excessive statistical variability in dopant placement for very small transistors[8]. As a result, non-classical MOSFETs, such as ultra-thin body double-gate devices, may be used.

The discussion in this paper is limited to logic technology, since many of the key technology issues are driven by the logic technology. For both high-performance and low-power logic technologies, the 2001 ITRS scaling approach and results are presented, as well as key front-end process challenges and potential solutions. Finally, possible solutions for mobility improvement and the use of non-classical transistors for very small MOSFETs are discussed. Along similar lines, a recent publication[9] by one of the authors of this paper contains a discussion of circuit and MOSFET scaling and of front-end process challenges and potential solutions.

2. Circuit and MOSFET Scaling Approach and Results

Table 1 gives projections from the 2001 ITRS[10] of some key parameters for high-performance logic. The overall chip circuit requirements are in the last three rows. As is typical of the ITRS, the projections are given for each of the near-term years, from 2001 through 2007, while for the long-term years beyond 2007, projections are given only for every third year. The DRAM half pitch is a key parameter characterizing the leading-edge technology node in production for each year, and this half pitch is projected to be reduced by a factor of $(1/\sqrt{2}) \sim 0.7$ every three years, consistent with the long-term trends noted in the Introduction. In all cases the projected L_g is no more than half the DRAM half pitch. The local clock frequency and the number of transistors per chip are both increasing rapidly, becoming almost 29 GHz and 8.85 billion transistors per chip, respectively, at the end of the roadmap in 2016. In contrast, the allowable maximum power dissipation per chip (which is limited by system level cooling, test constraints, and the ability of the package to dissipate heat) is increasing slowly.

		Near-term years							Long-term years		
Year		*2001*	*2002*	*2003*	*2004*	*2005*	*2006*	*2007*	*2010*	*2013*	*2016*
DRAM Half Pitch	*nm*	130	115	100	90	80	70	65	45	32	22
MPU Physical Gate Length, Lg	*nm*	65	53	45	37	32	28	25	18	13	9
V_{dd}	*V*	1.2	1.2	1.1	1	0.9	0.9	0.8	0.6	0.5	0.4
Maximum on-chip local clock frequency	*GHz*	1.7	2.3	3.1	4.0	5.2	5.6	6.7	11.5	19.3	28.8
Allowable maximum power dissipation, with heatsink	*W*	130	140	150	160	170	180	190	218	215	288
Number of transistors per chip	*Millions of transistors*	276	348	439	553	697	878	1106	2212	4424	8848

Table 1. Key parameters for high-performance logic, data from 2001 ITRS.[10]

The chip power dissipation and clock frequency scaling requirements in Table 1 drive the scaling of key MOSFET parameters. The total chip power dissipation, P, is the sum of the dynamic power dissipation, P_d, and the static power dissipation, P_{st}. The former is:

$$P_d = C_a V_{dd}^2 f_c \quad (1)$$

where C_a is the total chip active capacitance (i.e., the total capacitance being switched during an average clock cycle), V_{dd} is the power supply voltage, and f_c is the clock frequency. For succeeding technology generations, C_a and f_c increase sharply, so V_{dd} must be reduced to keep P_d within tolerable limits. This is particularly critical for high-performance logic, where P_d is typically considerably larger than P_{st}[11]. Reducing V_{dd} with scaling is also beneficial for controlling short channel effects, improving reliability, and for general device scaling. In contrast to P_d, P_{st} is dependent on the transistor leakage current:

$$P_{st} = N_{off}\, W\, I_{off}\, V_{dd} \qquad (2)$$

where N_{off} is the average number of transistors that do not switch during a clock cycle (it generally is a substantial fraction of the total number of transistors per chip, listed in Table 1), W is the average transistor width, and I_{off} is the total transistor leakage current (in units of amperes per micron of transistor width). Since N_{off} generally increases with scaling, while W and V_{dd} generally decrease, it is difficult to generalize about the behavior of the product, $(N_{off}\, W\, V_{dd})$, with scaling. Nevertheless, any increase of I_{off} with scaling is clearly constrained by the need to keep P_{st} within tolerable limits.

The 2001 ITRS utilizes the transistor intrinsic switching frequency, f_i, as the figure of merit for transistor speed. f_i is defined as the reciprocal of the transistor intrinsic delay, τ_i:

$$\tau_i = 1/f_i = (C_g\, V_{dd})/\, I_{on} \qquad (3)$$

where C_g is the gate capacitance per micron of transistor width for a MOSFET of gate length, L_g, and I_{on} is the transistor saturation drive current (in units of amperes per micron of transistor width). τ_i can be understood as approximately the time required for the MOSFET to charge or discharge the gate of another identical MOSFET through a potential difference of V_{dd} (in this approximation, the S/D junction capacitance of the driving MOSFET is ignored). Alternatively, utilizing the explanation in Taur and Ning[12], τ_i is approximately the switching delay of an inverter for a capacitive load (C_L) equal to the gate capacitance of one transistor, (i.e., $C_L = W\, C_g$, units are farads). As seen from Eq. (3), f_i is dependent only on the transistor parameters and characteristics, and to maximize f_i, I_{on} must be maximized. f_i is a good figure of merit for the transistor intrinsic speed in dense logic circuits, where the interconnect wire lengths are short and C_L is dominated by the gate capacitance of the transistors in the fanout path. In contrast, for logic circuits with long interconnect wires, C_L is dominated by the wiring capacitance. As a result, the switching frequency is a function of both the driving transistor(s) layout and characteristics and the interconnect layout and characteristics. The interconnect dominated case is beyond the scope of this paper, but transistors optimized to maximize f_i for dense logic should also be reasonably optimal for interconnect dominated logic.

In the 2001 ITRS, the following approach was used to scale the MOSFETs. Simplified models were developed to capture the essentials of the impact of such key MOSFET parameters as V_{dd}, gate dielectric equivalent oxide thickness *(EOT)*, physical gate length (L_g), etc., on the important transistor electrical output characteristics such as I_{off}, I_{on}, etc. These models were then embedded into a commercial spreadsheet. Key output characteristics calculated by the spreadsheet included the MOSFET intrinsic delay, $\tau_i = (C_g\, V_{dd})/I_{on}$, and $f_i = 1/\tau_i$. In the expression, C_g includes the ideal gate capacitance per micron of width (defined as $[\varepsilon_{ox} L_g / EOT]$, where ε_{ox} is the dielectric constant of silicon dioxide) plus both parasitic gate overlap and fringing capacitance per micron transistor width. To determine the projected MOSFET parameter values for the tables, a target was set for either the MOSFET leakage current or f_i. Then an initial set of MOSFET parameters was chosen based on scaling rules, engineering judgment, and physical device principles. Using the spreadsheet, the input parameters were iteratively varied until the target was met, and the final value of the parameters was then used for the projected MOSFET parameter values in the tables.

For high-performance logic, an average 17% per year rate of increase in f_i was targeted, which matches the historic rate of device performance improvement. Using the iterative approach described above, the projections shown in Table 2 were obtained,

		Near Term							Long Term		
Calendar Year		**2001**	**2002**	**2003**	**2004**	**2005**	**2006**	**2007**	**2010**	**2013**	**2016**
DRAM half pitch	**nm**	**130**	**115**	**100**	**90**	**80**	**70**	**65**	**45**	**32**	**22**
Physical Gate Length, L_g	nm	65	53	45	37	32	28	25	18	13	9
Equivalent oxide thickness, T_{ox}	nm	1.3-1.6	1.2-1.5	1.1-1.6	0.9-1.4	0.8-1.3	0.7-1.2	0.6-1.1	0.5-0.8	0.4-0.6	0.4-0.5
Nominal Power Supply Voltage (Vdd)	V	1.2	1.1	1.0	1.0	0.9	0.9	0.7	0.6	0.5	0.4
Nominal High-Performance NMOS Sub-Threshold Current, $I_{sd,leak}$ (@25C)	uA/um	0.01	0.03	0.07	0.1	0.3	0.7	1	3	7	10
Nominal NMOSFET saturation drive current, I_{on}	uA/um	900	900	900	900	900	900	900	1200	1500	1500
Required Percent Current-Drive "Mobility/Transconductance Improvement"		0%	0%	0%	0%	0%	0%	0%	30%	70%	100%
Parasitic Series S/D resistance, $R_{sd,series}$	ohm-um	190	180	180	180	180	170	140	110	90	80
Parasitic Source/Drain Resistance (Rsd) Percent of Ideal Channel Resistance (Vdd/Idd)		16%	16%	17%	18%	19%	19%	20%	25%	30%	35%
NMOSFET intrinsic transistor switching frequency, $f_i = 1/t_i$	GHz	610	740	890	1000	1200	1300	1500	2600	4400	6500
Energy per (W/Lgate=3) Device Switching Transition (Cgate*(3*Lgate)*V^2)	fJ/Device	0.347	0.212	0.137	0.099	0.065	0.052	0.032	0.015	0.007	0.002
Static Power Dissipation Per (W/Lgate=3) Device	Watts/Device	5.6E-09	6.7E-09	1.0E-08	1.1E-08	2.6E-08	5.3E-08	5.3E-08	9.7E-08	1.4E-07	1.1E-07

Table 2. Technology requirement projections for high-performance logic, data from 2001 ITRS.[13]

where the data is from the 2001 ITRS[13]. The colors of the cells in the table have the following meaning: white indicates that manufacturable solutions are well understood, yellow indicates that manufacturable solutions are somewhat understood and are being worked out, while red indicates that manufacturable solutions are not known and further research is required. Note that *EOT* (which is denoted by "T_{ox}" in the table) is projected to scale down rapidly along with L_g. From the previous discussion, I_{on} must stay relatively high to maximize f_i. Hence, the projected NMOSFET I_{on} stays constant at 900 μA/μm until the 65 nm technology node in 2007, and then it increases somewhat in the later years. (The PMOSFET I_{on} is 40 – 50% of the NMOSFET I_{on} because of the lower mobility of holes.) V_{dd} is decreasing with device scaling, for control of power dissipation and for reliability, as discussed above. Since I_{on} is strongly dependent on the overdrive, ($V_{dd} - V_t$), where V_t is the effective threshold voltage, V_t must be reduced along with V_{dd} to keep I_{on} at the specified values. However, $I_{sd,leak}$ increases rapidly as V_t decreases[14], where $I_{sd,leak}$ is the source-drain subthreshold leakage current of the MOSFET, and is a

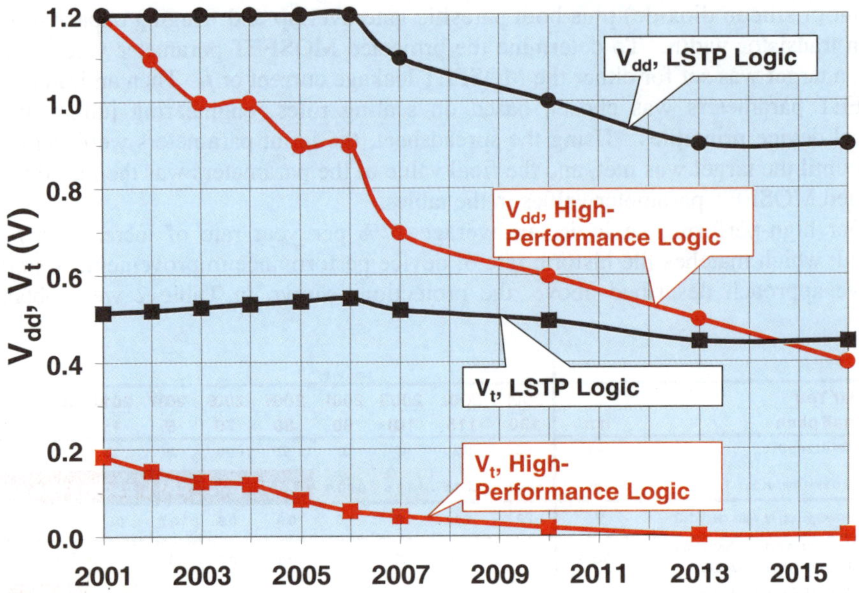

Figure 2. Scaling of V_{dd} and V_t for high-performance and LSTP logic, data from 2001 ITRS.

component of I_{off}. The scaling of V_{dd} and V_t is shown in Figure 2. Both V_{dd} and V_t scale with succeeding years, and V_t becomes very close to zero for the long-term years, beyond 2007. Hence, as shown in the table, $I_{sd,leak}$ increases rapidly with succeeding years, and becomes particularly large, at greater than 1μA/μm, for the long-term years. As a result of the increasing $I_{sd,leak}$, the static power dissipation per device increases with device scaling (see the last row of Table 2) despite the reduced V_{dd} and device dimensions with scaling. This increased static power dissipation has an impact, which will be discussed later, on the approaches to controlling chip power dissipation. Finally, $I_{sd,leak}$ is only one of three components of the total transistor leakage current, I_{off}. The other components are the gate leakage current and the S/D junction leakage current, including band-to-band tunneling. An important restriction imposed in the 2001 ITRS is that each of these other components of I_{off} must be less than $I_{sd,leak}$.

For high-performance chips, the rapid increase in $I_{sd,leak}$ (and hence in I_{off}) with scaling must be dealt with to keep the chip static power dissipation within tolerable limits. This is illustrated graphically in Figure 3, where the relative total allowable chip power dissipation from Table 1 is plotted, as well as the relative chip dynamic and static power dissipation with scaling. The dynamic and static dissipation were calculated using Equations (1) and (2), respectively, with the parameters scaled per Tables 1 and 2. In this figure, it is assumed that all transistors on the chip are the high-performance, low V_t, high leakage MOSFETs discussed in the previous paragraph and characterized in Table 2. The dynamic power dissipation scales up at about the same rate as the total allowable chip power dissipation, with an increase of about three times by 2016, while the static power dissipation scales up much more rapidly, with an increase of over three orders of magnitude by 2016. Clearly, the rate of increase of the static power dissipation is so large that, unless measures are taken to reduce it, the total allowable chip power dissipation limit will be exceeded within several technology nodes. An approach that is

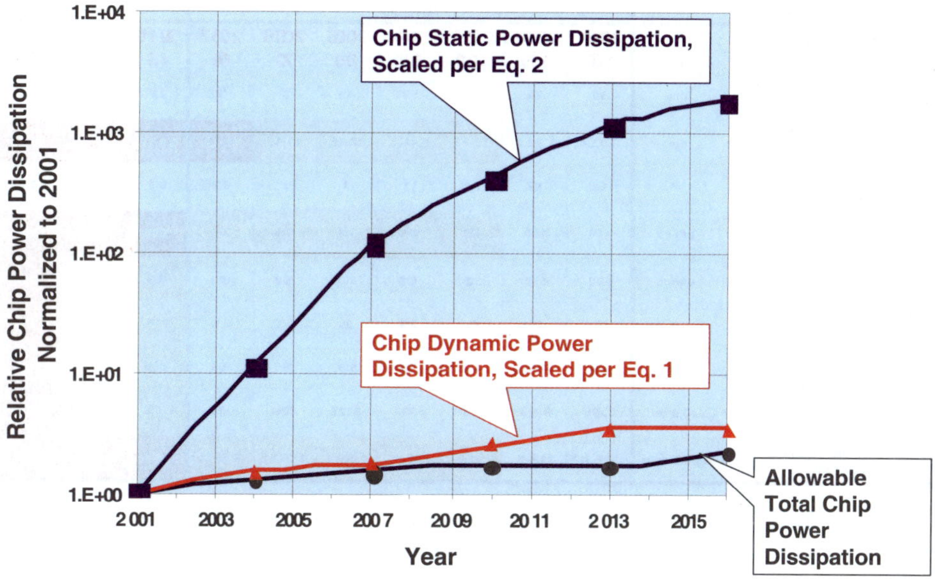

Figure 3. Relative chip power dissipation normalized to 2001 versus calendar year. Data from 2001 ITRS is utilized.

becoming common is to fabricate multiple transistor types on the chip in addition to the high-performance, low V_t device described above. The other MOSFET(s) have higher V_t and larger *EOT* to reduce I_{off}[15,16,17], but they also have lower I_{on} and lower f_i than the high-performance devices. The high-performance MOSFETs are used only in critical paths or in circuits which are constantly switching, while the lower leakage devices are used everywhere else. This approach can significantly reduce the chip static power dissipation without seriously degrading chip performance, while also increasing the chip flexibility for diverse, system-on-a-chip (SOC) functions. Other techniques utilized to curtail static power dissipation include well-biasing, use of electrically or dynamically adjustable V_t devices, and circuit/architectural techniques such as pass gates to cut off access to power/ground rails to power down circuit blocks[18]. Hence, a realistic picture of scaled high-performance ICs is that the static power dissipation will be controlled by utilizing the high-performance, high leakage MOSFETs sparingly, only when needed for performance, and by utilizing sophisticated device and IC design and architectural techniques. The ITRS technology requirements table (see Table 2) only characterizes the high-performance transistor because this transistor is the technology driver.

For low-power logic, which is focused on mobile applications, the ITRS targets the transistor leakage current to ensure reasonable battery life for mobile applications. Two types of low-power logic are considered: low operating power *(LOP)* and low standby power *(LSTP)*. LOP chips are used for higher performance mobile applications such as notebook computers, where the battery capacity is relatively large. In contrast, LSTP chips are used for consumer type applications such as cellular telephones, where the performance requirements and the battery capacity are both lower than for LOP chips. The leakage current targets are thus most stringent for LSTP. In the 2001 ITRS, as

		Near Term							Long Term		
Calendar Year		**2001**	**2002**	**2003**	**2004**	**2005**	**2006**	**2007**	**2010**	**2013**	**2016**
Technology Node	**nm**	**130**	**115**	**100**	**90**	**80**	**70**	**65**	**45**	**32**	**22**
Physical Gate Length Low-Operating Power (LOP)	nm	90	80	65	53	45	37	32	22	16	11
Equivalent Physical Oxide Thickness for Low-Operating Power LOP Tox (EOT)	nm	2.0-2.4	1.8-2.2	1.6-2.0	1.4-1.8	1.2-1.6	1.1-1.5	1.0-1.4	0.8-1.2	0.7-1.1	0.6-1.0
Nominal Low-Operating Power LOP Power Supply Voltage (Vdd)	V	1.2	1.2	1.1	1.1	1	1	0.9	0.8	0.7	0.6
Nominal Low-Operating Power LOP NMOS Sub-Threshold Current (@25C)	pA/um	100	100	100	300	300	300	700	1000	3000	10000
Nominal Low-Operating Power LOP NMOS Saturation Current Drive (Idd) (@Vdd, @25C)	uA/um	500	500	500	600	500	500	600	600	700	900
Required Percent Current-Drive "Mobility/Transconductance Improvement"	%	0%	0%	0%	0%	0%	0%	0%	10%	30%	70%
NMOSFET intrinsic transistor switching frequency, f = 1/	GHz	327	340	412	544	529	590	750	1005	1577	2820
Energy per (W/Lgate=3) Device Switching Transition (Cgate*(3*Lgate)*V^2)	fJ/Device	0.496	0.424	0.260	0.193	0.128	0.094	0.069	0.032	0.015	0.006
Static Power Dissipation Per (W/Lgate=3) Device	Watts/Device	3.2E-11	2.9E-11	2.1E-11	5.2E-11	4.1E-11	3.3E-11	6.0E-11	5.3E-11	1.0E-10	2.0E-10

Table 3. Technology requirement projections for low operating power (LOP) logic , data from 2001 ITRS.[19]

shown in Tables 3 and 4,[19] the LOP logic target for $I_{sd,leak}$ (and using the same reasoning as with the high performance logic, the maximum gate leakage and S/D junction leakage current) starts at 100 pA/μm in 2001 and rises steadily in succeeding years. For LSTP logic, the $I_{sd,leak}$ target starts at 1 pA/μm in 2001 and stays at that level until 2007, after which it rises slowly. L_g and *EOT* scale down with succeeding years, but the actual values of L_g lag behind those used in high- performance logic by two years (i.e., L_g is 65 nm for high performance logic in 2001, but it is 65 nm for LOP and LSTP logic in 2003).

		Near Term							Long Term		
Calendar Year		**2001**	**2002**	**2003**	**2004**	**2005**	**2006**	**2007**	**2010**	**2013**	**2016**
Technology Node		**130nm**			**90nm**			**65nm**	**45nm**	**32nm**	**22nm**
Physical Gate Length Low-Standby Power (LSTP)	nm	90	80	65	53	45	37	32	22	16	11
Equivalent Physical Oxide Thickness for Low-Standby Power LSTP Tox (EOT)	nm	2.4-2.8	2.2-2.6	2.0-2.4	1.8-2.2	1.6-2.0	1.4-1.8	1.2-1.6	0.9-1.3	0.8-1.2	0.7-1.1
Nominal Low-Standby Power LSTP Power Supply Voltage (Vdd)	V	1.2	1.2	1.2	1.2	1.2	1.2	1.1	1	0.9	0.9
Nominal Low-Standby Power LSTP NMOS Sub-Threshold Current (@25C)	pA/um	1	1	1	1	1	1	1	3	7	10
Nominal Low-Standby Power LSTP NMOS Saturation Current Drive (Idd) (@Vdd, @25C)	uA/um	300	300	300	300	400	400	400	500	600	700
Required Percent Current-Drive "Mobility/Transconductance Improvement"	%	0%	0%	0%	0%	0%	0%	0%	10%	30%	50%
NMOSFET intrinsic transistor switching frequency, f = 1/	GHz	217	227	253	280	398	432	443	700	1096	1522
Energy per (W/Lgate=3) Device Switching Transition (Cgate*(3*Lgate)*V^2)	fJ/Device	0.448	0.381	0.277	0.204	0.163	0.123	0.095	0.047	0.024	0.014
Static Power Dissipation Per (W/Lgate=3) Device	Watts/Device	3.2E-13	2.9E-13	2.3E-13	1.9E-13	1.6E-13	1.3E-13	1.1E-13	2.0E-13	3.0E-13	3.0E-13

Table 4. Technology requirement projections for low standby power (LSTP) logic, data from 2001 ITRS.[19]

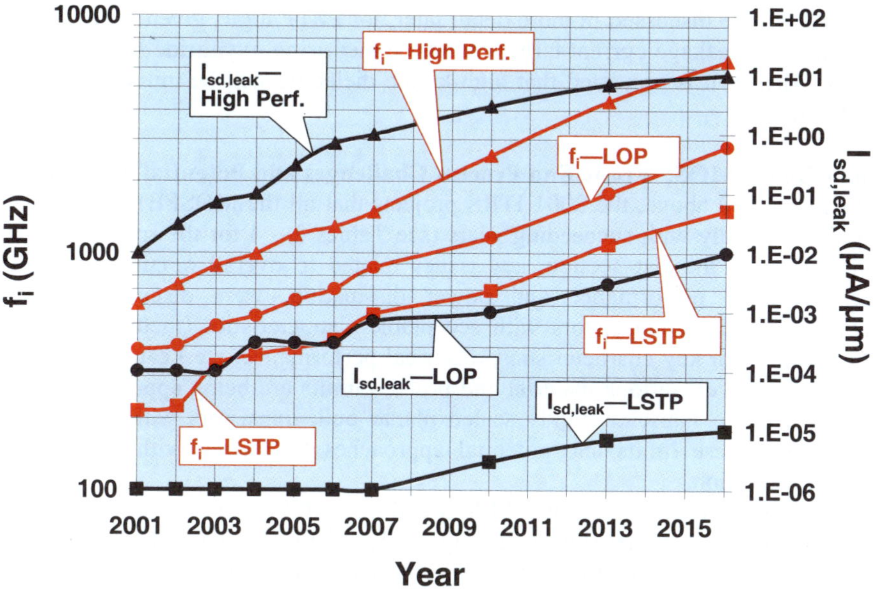

Figure 4. Projections of f_i and $I_{sd,leak}$ for high-performance and low-power logic, data from 2001 ITRS. (Figure is from Zeitzoff et al.[9], © 2002 IEEE.)

Tables 3 and 4 show the results of using the iterative approach described previously to meet the $I_{sd,leak}$ targets. The average rate of increase in f_i of about 14% per year for both LOP and LSTP is smaller than the 17% per year rate obtained with high-performance logic. $I_{sd,leak}$ and f_i are plotted versus calendar year for high-performance, LOP, and LSTP logic in Figure 4. As expected, $I_{sd,leak}$ and f_i are lowest for LSTP, intermediate for LOP, and highest for high-performance logic, with $I_{sd,leak}$ for high-performance logic ranging from four to six orders of magnitude larger than for LSTP. Figure 2 shows the scaling of V_{dd} and V_t for LSTP logic, as well as for high–performance logic. One key issue is the slow scaling of V_{dd} for the LSTP transistors. This is a result of the slow scaling of V_t to meet the low and relatively constant targets for $I_{sd,leak}$. V_{dd} follows V_t in scaling slowly for two reasons: first, for reasonable device performance I_{on} must be sustained, and hence the overdrive, (V_{dd}–V_t), must remain relatively large. Secondly, for adequate circuit switching noise margins, V_{dd} must be larger than at least $2 \times V_t$. (The curves of V_{dd} and V_t versus year for LOP logic are not shown in the figure, but they are similar to those for LSTP, with slightly smaller values.) Dynamic power dissipation per transistor can be characterized by the energy per device switching transition (see Tables 2 - 4, next to last row). This energy is proportional to $(V_{dd})^2$, and is highest for LSTP, intermediate for LOP, and lowest for high-performance logic. However, the static power dissipation per device (see last row of the tables) follows the $I_{sd,leak}$ values, and is largest for high performance, intermediate for LOP, and smallest for LSTP logic devices. Another important issue is caused by the sharp increase in the lateral electric field, $E_{lat} \sim V_{dd}/L_g$, with device scaling. This increase will result in difficulty in controlling short channel effects and possibly in significant reliability problems in the long-term years for low-power logic, particularly LSTP logic. A final critical issue is with gate leakage current. As will be discussed in more detail later, for LSTP logic, given the *EOT* and V_{dd} in 2005, the gate leakage current target cannot be met using oxynitride because of direct tunneling. Hence, it is projected that high-k gate dielectric will be initially required for LSTP ICs in 2005.

3. Planar Bulk CMOS: Front-End Process Challenges and Potential Solutions

As mentioned above, the 2001 ITRS projects that all the MOSFET dimensions are scaling down rapidly with succeeding years (see Tables 2 – 4 for the specific scaling of L_g and *EOT*). In the last section, key issues related to overall circuit and MOSFET scaling, including performance and power dissipation, were discussed under the assumption that scaled transistors with acceptable characteristics (including reliability, statistical control of key characteristics, electrical performance, etc.) can be successfully fabricated. However, serious material and process limits are being approached that will make it difficult to fabricate highly scaled planar bulk transistors with all the desired properties, and these limits and potential approaches to dealing with them form the subject of this section.

A simplified cross section of a PMOS and an NMOS transistor typical of current leading-edge technology is shown in Figure 5, where L_g, gate dielectric physical thickness, T_d, and the junction depths *(x_j)* of the deep source/drain *(S/D)* and S/D extension are defined. (Traditionally, the gate dielectric is silicon dioxide, also called oxide.) Important features illustrated here include: (a) shallow trench isolation *(STI),* (b) deposited, anisotropically etched spacers, and (c) dual (N+ for NMOS, P+ for PMOS) polysilicon ("*poly*") gate electrodes. In addition, "*salicide*"(self-aligned silicide) for the poly gate electrodes and the deep S/D's (which are also known as the *S/D contacts*) are shown, as well as the shallow S/D extensions.

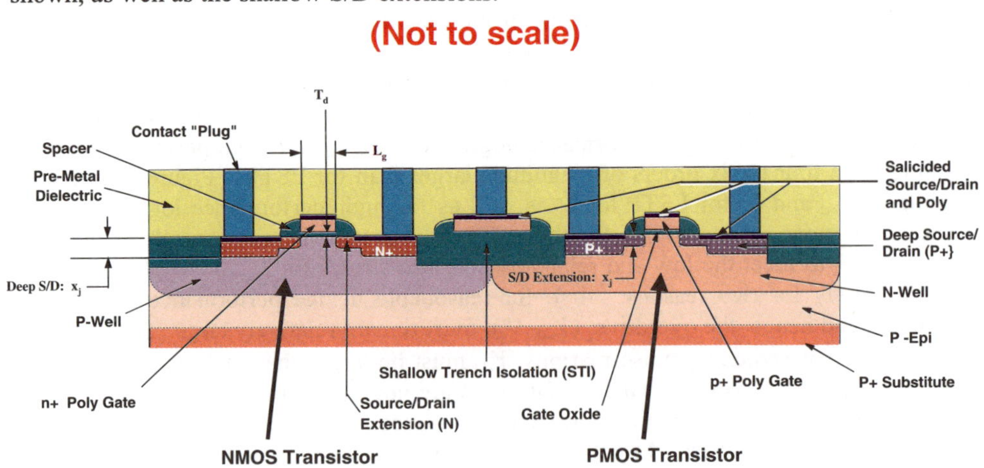

Figure 5. Simplified cross section of a typical PMOSFET and NMOSFET.
(Figure is from Zeitzoff et al. [9], © 2002 IEEE.)

For all the logic technologies, several problems are associated with the rapid scaling of *EOT* (see Tables 2 – 4 for the specific values). The first problem, boron penetration, is the uncontrolled diffusion of boron from the heavily doped P+ poly gate of the PMOSFETs through the gate oxide and into the MOSFET channel. The result is an uncontrollable shift in V_t in a positive direction. Adding a relatively small amount of nitrogen to the gate oxide to suppress the boron diffusion through the oxide is generally effective in dealing with this problem,[20,21] and is the standard technique used in current leading edge technologies. The other major problem is the rapid increase in gate leakage current with scaling. This is due to direct tunneling, which increases exponentially with decreasing thickness for thin oxides. As noted before, the gate leakage current, I_g (units are amperes/micron of width), must be less than the S/D subthreshold leakage current, $I_{sd,leak}$, which is projected for the various types of logic chip in Tables 2 – 4. One approach to reducing I_g is to use a gate dielectric with a higher relative dielectric constant, K, than the 3.9 value of thermal silicon dioxide (SiO_2). For a film of such a dielectric with thickness, T_d, the equivalent oxide thickness, EOT, is:

$$EOT = T_d\,(3.9/K) \tag{4}$$

To first order, a transistor with such a high-k gate dielectric has the same gate capacitance and hence the same electrical performance as an otherwise identical transistor with a thermal silicon dioxide film of thickness EOT. With K larger than 3.9, T_d is larger than EOT. Since direct tunneling decreases exponentially with T_d, it is expected that I_g will be smaller for the high-k gate dielectric compared to the thermal oxide gate dielectric. However, another important factor impacting tunneling is the barrier height between the conduction band of silicon and that of the high-k material. Tunneling increases sharply as the barrier height decreases, and as the K of different materials increases, the energy gap and the barrier height tend to decrease[22]. Hence, in evaluating candidate high-k materials, barrier height should be examined as well as K, and I_g should be experimentally determined to see whether the material is advantageous for use as a gate

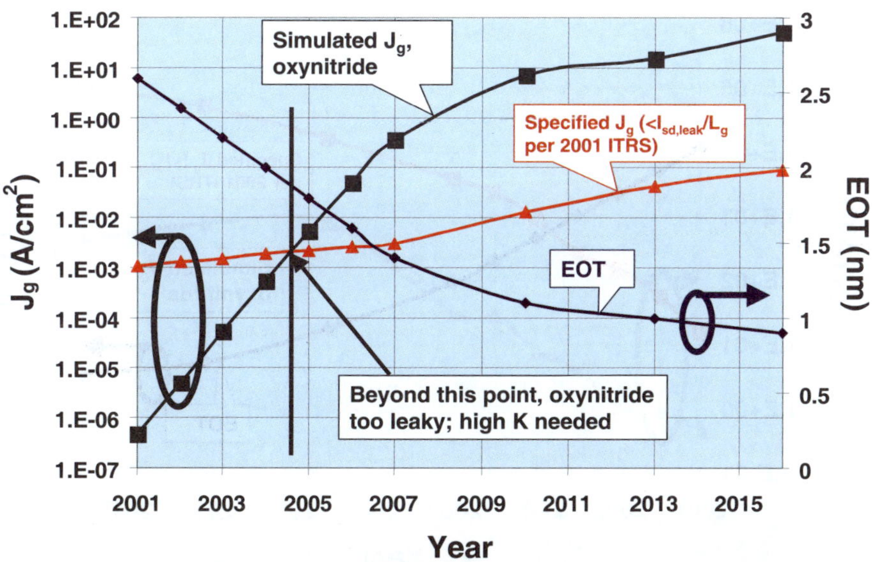

Figure 6. 2001 ITRS projections versus simulations of gate leakage current density for low standby power (LSTP) logic. (Figure is from Zeitzoff et al. [9], © 2002 IEEE.)

dielectric. Extending current approaches using lightly nitrogen doped silicon dioxide (called oxynitride), heavily nitrogen doped oxynitride or oxynitride/silicon nitride stacks have been shown to have a $K \sim 6.5$ and I_g reduced by about two orders of magnitude relative to thermal oxide with the same *EOT*[23,24,25]. In Figure 6, the 2001 ITRS projections for *EOT* and gate leakage current density, $J_g = I_g/(L_g)$, are plotted for LSTP logic, which has the most stringent targets for gate leakage. Also plotted are simulations of the expected J_g using heavily nitrogen doped oxynitride, as calculated in the 2001 ITRS[26]. By 2005, J_g due to direct tunneling through the oxynitride is well beyond the projection in the 2001 ITRS. (In Table 4, which shows the LSTP Logic Technology Requirements in the 2001 ITRS, this inability to meet the targeted gate leakage is reflected in the red color for 2005 and beyond in the row, "Equivalent Physical Oxide Thickness for Low-Standby Power LSTP Tox (EOT).") Hence, the introduction of high-k gate dielectric (with *K* larger than the ~6.5 value of heavily nitrogen doped oxynitride) is expected to be driven by LSTP logic in 2005, with a projected $EOT \sim 1.8$ nm and $I_g = 1$ pA/µm[27]. Some of the high-k materials currently being considered are hafnium oxide, aluminum oxide, their silicates, and quarternary combinations[28]). Active research and development programs are being carried out globally on high-k gate dielectrics and reported on at the VLSI Symposia on Technology (for example, in the 2001 Symposium see the papers by Choi et al.[29], Lee et al.[30] and others), the IEDM's (for example, in the 2001 IEDM, see the papers by Kim et al.[31], Onishi et al.[32] and others), and other conferences and workshops.

In contrast to LSTP logic, for LOP and especially for high-performance logic, the allowed gate leakage current is relatively large and increases significantly with scaling. For high-performance logic, Figure 7 has plots of the 2001 ITRS projections for gate

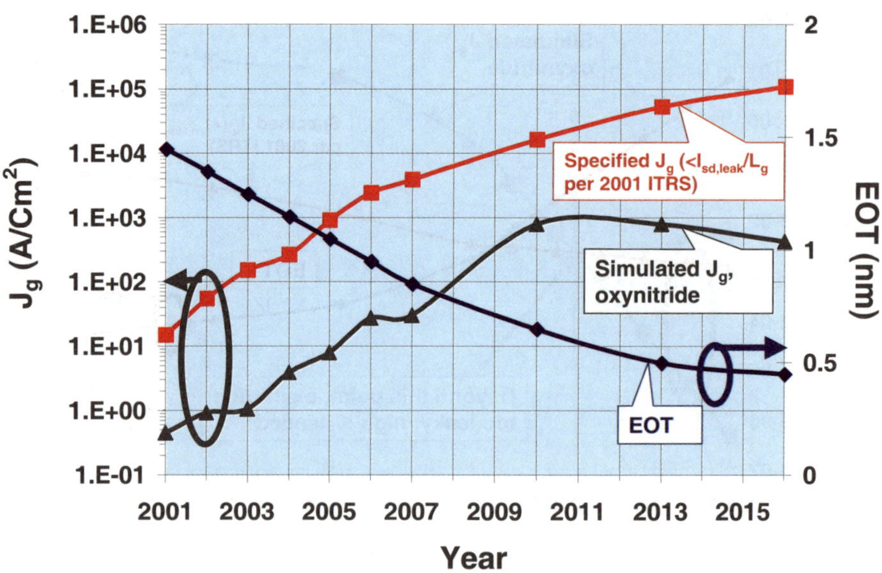

Figure 7. 2001 ITRS projections versus simulations of gate leakage current density for high-performance logic.

leakage current density, J_g, as well as simulations from the ITRS of the expected J_g for heavily nitrogen doped oxynitride[33]. From the figure, heavily nitrogen doped oxynitride or oxynitride/silicon nitride stacks appear to be adequate to meet the ITRS gate leakage projections through the end of the Roadmap in 2016; the same appears to be true for LOP logic. However, there are serious questions about oxynitride's reliability for the thin films and large gate tunneling currents projected in the ITRS, particularly for high-performance logic[34]. Also, achieving acceptable levels of thickness control for these thin films will become extremely difficult, particularly since I_g increases by ~one-half order of magnitude with a decrease of 0.1 nm in *EOT*[35]. (In Tables 2 and 3, which show the High-Performance and LOP Logic Technology Requirements in the 2001 ITRS, these reliability and process control issues are reflected in the yellow and red colors in the row for Equivalent Oxide Thickness.) Finally, it may be useful to reduce the gate leakage current below the targets in the ITRS to help reduce the chip static power dissipation. Hence, a likely scenario is that high-k gate dielectric will be implemented for high-performance and LOP logic after several years of learning on LSTP chips[36], in 2007 or beyond.

Another major front-end issue is polysilicon depletion in the gate electrode. When a MOSFET is turned on, a depletion region of thickness W_d forms in the polysilicon adjacent to the polysilicon-dielectric interface. Consequently, the centroid of the charge is not at the interface, but at a distance $x_{c,poly}$ from the interface, where $x_{c,poly}$ is proportional to W_d (in particular, $x_{c,poly} = W_d/2$ for uniform doping.). The effective electrical thickness of the gate dielectric, $EOT_{elec} = EOT + \Delta_{poly}$[37], where $\Delta_{poly} = (x_{c,poly}K_{ox}/K_{si}) = (x_{c,poly}/3)$. In this expression, $K_{ox} = 3.9$ is the relative dielectric constant of oxide and $K_{si} = 11.9$ is the relative dielectric constant of silicon[38]. According to Wilk et al.[39], Δ_{poly} can be as much as 0.4 nm. For a MOSFET with a given gate voltage applied, the increase in EOT_{elec} due to the non-zero Δ_{poly} results in a reduction in the inversion layer charge and hence in I_{on} compared to the case where there is no polysilicon depletion (i.e., where $EOT_{elec} = EOT$). The impact of polysilicon depletion becomes more severe in succeeding technology generations because of *EOT* scaling. Increased doping of the polysilicon can ameliorate this effect since W_d decreases with doping. However, this solution will eventually become inadequate because of the limited solubility of the dopants, particularly boron, even if poly-SiGe, which has higher solid solubility for boron, is used[40]. The projections in the 2001 ITRS regarding the required polysilicon active doping level are shown in Table 5 in the row labeled "Active poly doping for 25% depletion allowance (cm-3)"[41]. From the red coloring in the row, polysilicon will become inadequate beyond the 65 nm technology node in 2007, when the required activated doping levels approach 2×10^{20} cm^{-3}. An alternate solution, metal gate electrodes, is being pursued because metal gates have virtually no depletion, no boron penetration, and the gate electrodes have very low sheet resistance. However, CMOS optimization requires different work functions for PMOS and NMOS devices (~5.0 and ~4.1 eV, respectively[42]) to set the desired symmetric threshold voltages of 0.2 to 0.5 V for NMOS and –0.2 to –0.5 V for PMOS. For poly gates, this is done with P+ doping of the gate electrode for PMOSFETs and N+ doping of the gate electrodes for NMOSFETs. For metal gates, different metals with the required different work functions could be used, at the cost of difficult process integration problems (particularly including the thermal stability of the metal electrode) and increased chip processing complexity and cost. For NMOSFETs, TaSiN[43], W/TiN[44], and other combinations are being investigated, while TaN[45], RuO_2[46], TiAlN[47], and others are being investigated for PMOSFETs. An

Year of Production	2001	2002	2003	2004	2005	2006	2007	2010	2013	2016
DRAM ½ Pitch (nm)	130	115	100	90	80	70	65	45	32	22
Active poly doping for 25% depletion allowance (cm^{-3})	9.2E19	9.2E19	1.14E20	1.50E20	1.66E20	1.66E20	1.87E20	1.80E+20	2.50E+20	2.99E+20
Maximum drain extension sheet resistance (PMOS) (/sq)	400	460	550	660	770	830	760	830	940	1210
Drain extension Xj (nm)	27-45	22-36	19-31	15-25	13-22	12-19	10-17	7–12	5–9	4–6
Extension lateral abruptness (nm/decade)	7.2	5.8	5	4.1	3.5	3.1	2.8	2	1.4	1
Contact silicide sheet Rs (/sq)	4.2	5.1	6.1	7.4	8.5	9.7	10.9	15.2	21	30.3
Maximum silicon consumption (nm)	23–46	19–38	16–32	13–26	11–23	10–20	9–18	6–13	5–9	3–6
Contact Xj (nm)	48–95	39–78	33–66	27–45	24–47	21–42	18–37	13–26	10–19	7–13
Contact maximum resistivity ($-cm^2$)	4.10E-07	3.20E-07	2.70E-07	2.10E-07	1.80E-07	1.60E-07	1.10E-07	6.40E-08	3.80E-08	2.40E-08
Channel concentration for $W_{depletion} < 1/4 L_{eff}$ (cm^{-3})	4.0E18	6.0E18	8.0E18	1.1E19	1.4E19	1.6E19	2.3E19	5.00E+19	1.30E+20	5.00E+20
Uniform channel concentration (cm^{-3})*, for* V_t*=0.4 [V]*	0.8–1.5 E18	0.8–1.5 E18	1.5–2.5 E18	1.5–2.5 E18	1.5–2.5 E18	2.0–4.0 E18	2.5–5.0 E18	5.0–9.0 E18	0.9–1.8 E19	1.5–3.0 E19

Table 5. Key front end process projections, data from 2001 ITRS.[41]

alternate approach that promises to reduce the added process complexity and cost is to use one material system for the metal electrodes, and to vary the work function for PMOS and NMOS by varying the alloy composition or by doping. Zhong et al.[48] describe using a RuTa binary alloy in which they varied the alloy composition to obtain work functions appropriate for PMOSFETs (Ta < 20%) and for NMOSFETs (40% < Ta < 54%). Q. Lu et al.[49] used (110)-Mo for the PMOSFET metal electrode, and they doped the Mo with nitrogen to shift the work function toward the NMOSFET required value.

Another important effect is due to the displacement away from the silicon-dielectric interface of the inversion charge centroid in the silicon substrate. Following the same reasoning as in the case of polysilicon depletion, this displacement of the inversion charge gives rise to an increase, Δ_{inv}, in EOT_{elec}. Taking account of quantum mechanical effects, Δ_{inv} ~ 0.3 – 0.6 nm[50,51] when the transistor is fully turned on. It should be noted that polysilicon depletion and quantum effects in the substrate limit the effectiveness of gate dielectric scaling. $\Delta_{poly} + \Delta_{inv}$ ~ 0.7 nm or more, and since $EOT_{elec} = EOT + \Delta_{poly} + \Delta_{inv}$, it becomes very difficult to scale EOT_{elec} much below 1.0 nm, no matter how much the *EOT* of the dielectric is reduced. Furthermore, Δ_{inv} cannot be significantly reduced, so even if metal gate electrodes are used to make Δ_{poly} ~ zero, the ability to scale EOT_{elec} is still limited by Δ_{inv}[52].

The series S/D parasitic resistance, $R_{sd,series}$, has increasingly stringent limits with scaling (see the "Parasitic Series S/D resistance, *Rsd,series*" Row in Table 2 for the specific numbers). These requirements are critical, because high series parasitic resistance degrades the saturation current drive, I_{on}. $R_{sd,series}$ is made up of components from both the S/D extension regions and the S/D contact regions[53] (see Figure 5).

For the S/D extension regions, very shallow junctions are required with scaling, mainly to minimize short channel effects[54,55,56]. However, the shallow profiles tend to lead to high sheet resistance, R_s, which increases $R_{sd,series}$[57]. This tradeoff between x_j and R_s (see Table 5 for ITRS 2001projections for both x_j and R_s) will become increasingly difficult with scaling. Another requirement is that the S/D extension lateral doping profile in the vicinity of the extension-channel junction must become increasingly abrupt with scaling to minimize $R_{sd,series}$ and short channel effects[58,59,60] (see Table 5 for the 2001 ITRS projections of the S/D extension lateral abruptness). The highest dose implant consistent with meeting the junction depth target will be used to minimize the series resistance. For the PMOSFET, where boron is the poly dopant, it is expected to be especially difficult to meet the junction depth and lateral abruptness requirements with scaling. Since boron is a light atom, it has a large implantation projected range, it diffuses rapidly, and it is strongly affected by transient enhanced diffusion (TED)[61]. Ultra-low energy implants (0.5 keV or below at the 90 nm technology generation) are expected to be utilized to reduce the projected range of the boron as-implanted profile. In addition, rapid thermal processing (RTP) will be used extensively to minimize the x_j and to improve the lateral abruptness, since RTP is effective at reducing both the thermal budget and TED. Particularly with the implementation of "spike annealing", which has very high ramp rates, these enhanced RTP solutions will probably suffice through the 90 nm node in 2004. However, beyond the 90 nm node, novel doping schemes such as atomic layer epitaxy, plasma doping, or co-implantation[62] and novel annealing schemes such as laser thermal annealing, microwave annealing, or low temperature annealing[63,64] will eventually be required[65]. The ultimate solution may be deposited layers[66,67].

For the deep S/D regions (also known as contact S/D's), low energy implantation followed by RTP for anneal is the current mainstream process[68]. After the RTP anneal, self-aligned silicide (salicide) is formed selectively on all deep S/D regions and the polysilicon electrodes (i.e., on all exposed silicon) through deposition of the metallic component of the silicide (i.e., titanium for titanium silicide or cobalt for cobalt silicide) followed by low-temperature RTP. During the low temperature RTP, the metallic component combines with silicon to form silicide, and during that process a portion of the silicon is consumed by converting it to silicide. To lower the R_s of the silicide, a thicker silicide layer must be formed, which results in more silicon being consumed. If the consumed silicon depth gets too large, excessive junction leakage results because the silicide-silicon interface gets too close to the p-n junction. This leakage is caused by junction depletion region encroachment upon the silicide-silicon interface, which is quite rough and hence has many defects that can serve as leakage current generation centers[69]. Table 5 shows 2001 ITRS projections of the silicide R_s, the silicon consumption, and the S/D contact junction depth[70]. To maintain junction leakage within tolerable limits, the silicon consumption depth is limited to half of the S/D contact junction depth. For previous mainstream process flows, titanium salicide was used, but cobalt salicide, which has a more favorable tradeoff between silicon consumption and low R_s[71], is becoming predominant at the 130 nm node. Eventually, cobalt salicide may be replaced by nickel salicide, which has an even more favorable tradeoff between silicon consumption and low R_s[72,73]. These approaches appear to be adequate through the 65 nm technology generation in 2007. Beyond that, novel schemes such as elevated S/D may be needed, possibly including SiGe or Ge materials[74]. A final issue is the specific contact resistance (also known as the contact resistivity) between the silicide and the deep S/D, which is an important component of $R_{sd,series}$[75]. According to the projections in the 2001 ITRS (see Table 5), the contact resistance must decrease sharply with scaling to meet the overall requirements on $R_{sd,series}$. Starting with the 65 nm technology generation in 2007, where the specific contact resistance is projected to be a low 1.1 x 10^{-7} ohm-cm^2, there is no known manufacturable solution for meeting these requirements. Potential solutions being examined include selectively deposited silicide and selectively deposited metals[76], as well as selectively deposited, heavily doped Si-Ge contacted by tungsten, which results in as low as ~2 x 10^{-8} ohm-cm^2 specific contact resistance for P+ material[77]. Finally, dual low barrier silicide films (platinum silicide for PMOSFETs and erbium silicide for NMOSFETs, where platinum silicide has a Schottky barrier height, ϕ_{b0p}, of 0.24 V to P+ silicon and erbium silicide has a Schottky barrier height, ϕ_{b0n}, of 0.32 V to N+ silicon) show promise of reducing the specific contact resistance to acceptable levels for 20 nm and below MOSFETs[78,79].

Figure 8 shows the Logic portion of the Logic and Memory Potential Solutions table from the 2001 ITRS[80]. The timing of many of the potential solutions discussed here can be seen in this table.

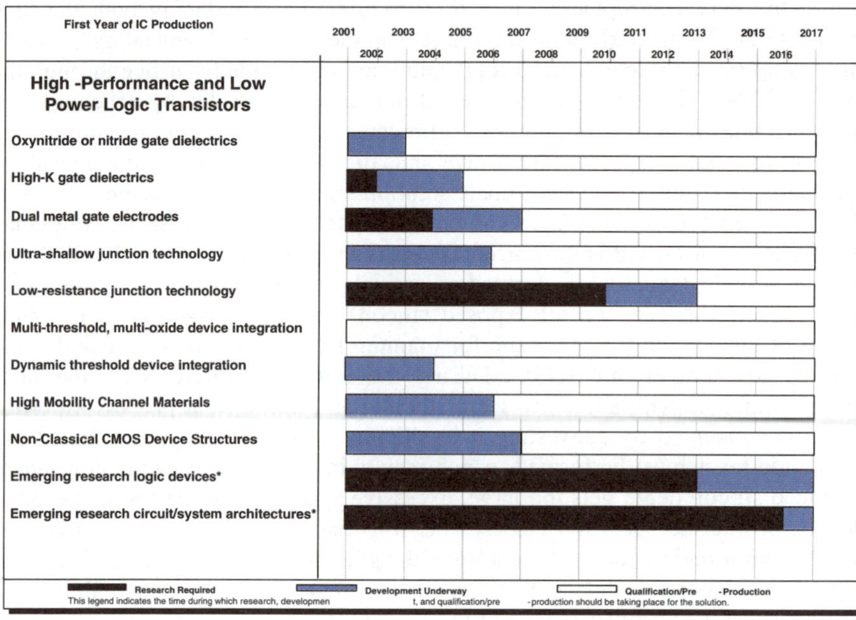

Figure 8. Logic potential solutions, from ITRS 2001 PIDS section[80]

4. Non-Classical CMOS

As CMOS technology is scaled to the 65 nm node and beyond, planar bulk CMOS technology will face increasing difficulties in meeting all of the transistor requirements (high I_{on}, low I_{off}, reasonable short channel effects, good noise margins, adequate statistical control, etc.), even with the utilization of high-k gate dielectric, metal electrodes, novel annealing schemes, and the other material and processing type potential solutions discussed previously. These problems are particularly acute with high-performance logic, which scales more rapidly than low-power logic. Key difficulties include controlling short channel effects adequately, the impact of the high channel doping required for such small devices, the increasing impact of quantum effects and statistical variation, and others. As a solution, non-classical CMOS devices such as enhanced mobility strained silicon/silicon-germanium channels, ultrathin body SOI, and various types of double gate MOSFET may be utilized[81,82]. Furthermore, use of these non-classical CMOS devices may facilitate a slowdown in the scaling of some of the parameters such as *EOT*, thus allowing a delay in the implementation into manufacturing of some of the solutions such as metal electrodes and high-k gate dielectric for high-performance logic.

The need for enhanced mobility channels is indicated in Table 2 in the row labeled "Required Percent Current-Drive Mobility/Transconductance Improvement" for high-performance logic. This is the improvement in the effective mobility beyond that of standard silicon needed to meet the targets for I_{on} and f_i. In the long-term years, this required improvement increases with scaling. To a lesser extent, the same need for enhanced mobility holds true for LOP and LSTP logic, as shown in Tables 3 and 4. An approach to realizing this enhanced mobility is to utilize a relaxed silicon-germanium *(SiGe)* layer on a silicon substrate with a surface layer of strained silicon on top. The silicon surface layer is the channel of the MOSFETs, and this channel is under strain due to its having a smaller lattice constant than the underlying relaxed SiGe layer. For NMOSFETs, tensile strain splits the degeneracy in the silicon conduction band so that the minima with lower effective mass are lower in energy. These lower mass minima are therefore preferentially populated with carriers when the device is turned on and an inversion layer forms. As a result, since mobility = $\mu = q\tau/m^*$, where q = *electronic charge*, τ = *mean free time against scattering, and* m^* *is the effective mass*[83], the effective mobility of those carriers is increased[84,85,86]. For PMOSFETs, either tensile or compressive strain lifts the degeneracy in the valence band and reduces the effective mass of the bands, thereby increasing the hole effective mobility[87,88]. In addition, for both NMOSFETs and PMOSFETs, interband scattering is reduced, which also contributes to increasing the effective mobility[89,90]. The details of the SiGe layers, including the Ge concentration, the layer thickness, the layer doping, and any grading, are different for the NMOSFET and PMOSFET to separately optimize the mobility enhancement for each. Cross sections of a PMOSFET and an NMOSFET with strained silicon channels are shown in Figure 9[91,92]. Rim et al.[93] utilized strained silicon on relaxed SiGe for

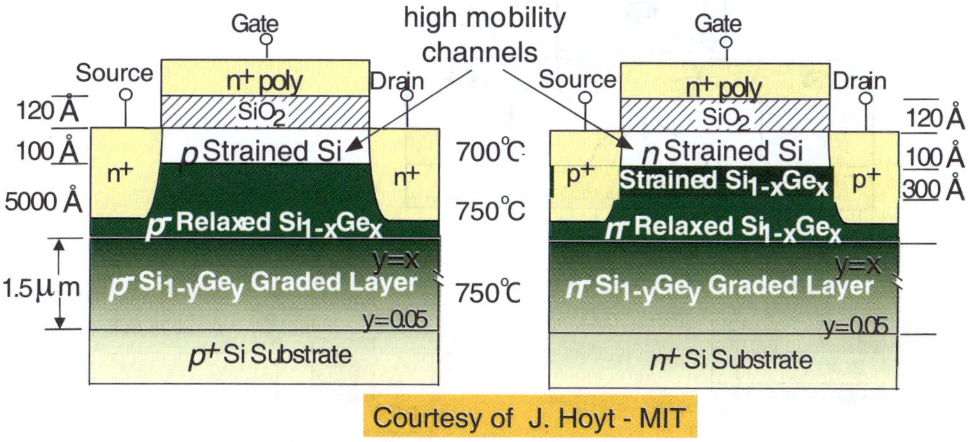

Figure 9. Cross sections of PMOS and NMOS Devices with Strained Si enhanced mobility channel[91,92]

NMOSFETs, and found that the effective mobility was increased by a factor of about 1.8 times compared to unstrained silicon control samples for values of the vertical effective electric field, E_{eff}, up to 1 MV/cm. An improvement by a factor of 1.8 in bulk PMOSFETs[94] and a factor of 1.45 in SOI PMOSFETs[95] has been found. An interesting approach was taken by J. Alieu et al.[96]. They used one type of multiple SiGe well for both NMOSFETs and PMOSFETs to form strained PMOS and NMOS channels. They realized a PMOSFET transconductance gain of 50% and an NMOSFET transconductance gain of 10% compared to unstrained Si. Use of enhanced mobility channels is a promising technique, but numerous issues need to be addressed, including such process integration issues as how to deposit the different SiGe layers needed for the NMOS and PMOS devices and the thermal stability of the SiGe and strained silicon layers during subsequent IC processing operations.

Aside from the enhanced mobility issues discussed above, planar bulk CMOS transistors will face additional difficult issues as they are scaled to the 65 nm technology generation (in 2007) and beyond. This is reflected by the predominantly red color of the rows for 2007 and beyond in Table 2 (from PIDS high-performance logic) and in Table 5 (from Front End Processes). Perhaps the most difficult issue in scaling planar bulk CMOS devices to the 65 nm technology generation (L_g = 25 nm) and especially beyond is achieving adequate control of short channel effects (SCEs). The difficulty is caused by electrostatic coupling between the charge in the source and drain regions and depletion charge in the MOSFET channel. This is illustrated schematically in the bulk MOSFET of Figure 10, where field lines from some of the charges in the source and drain regions terminate on charges in the channel depletion region, as shown schematically by the red lines connecting the S/D and the depletion charges. The field lines from the rest of the depletion charge (not drawn) will originate from the charge in the gate electrode. The

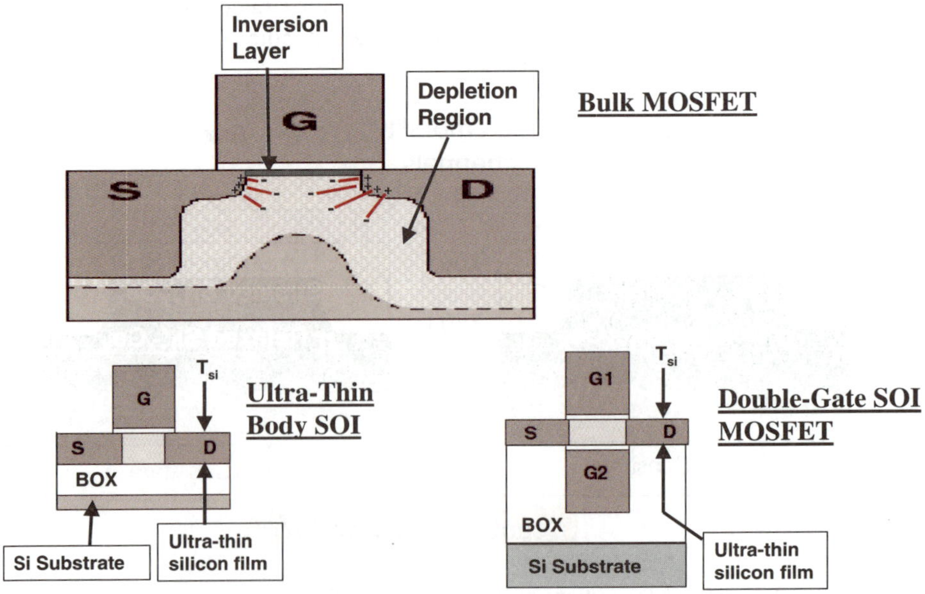

Figure 10. Schematic cross sections of planar bulk, UTB SOI, and DG SOI MOSFET

SCEs are caused by the increasing proportion of the channel depletion charge that is electrostatically coupled to charge in the source and drain as the gate length becomes shorter. As a result, SCEs such as threshold voltage rolloff, drain induced barrier lowering *(DIBL)*, and punchthrough (for very short gate length) will become more predominant as the gate length is scaled down.[97] The techniques to control SCEs include:

a) scaling the vertical device dimensions such as *EOT*, S/D extension x_j, channel depletion region width (W_{ch}) (specifically, x_j and W_{ch} must be significantly less than L_g)[98]
b) super steep retrograde *(SSR)* profiles[99,100] (vertical profiles relatively lightly doped near the silicon/dielectric interface with a peak in the interior)
c) halo implant profiles[101,102,103], where heavy, laterally assymetric doping is implanted in the channel near the source and drain.

The extremely sharp, asymmetric profiles needed for SSR and halo implant profiles at the 65 nm technology node and beyond will be very difficult to successfully fabricate. The vertical dimensions will also be difficult to scale, as indicated in Tables 2 and 5.

A major issue is the impact of high doping in the channel. Scaled technologies will require high channel doping (well over 10^{18} cm^{-3}), both to set the threshold voltage to the desired levels ~ 0.2 to 0.5 V (NMOS) and ~ -0.2 to -0.5 V (PMOS) and to keep W_{ch} significantly less than L_g to adequately control short channel effects. The last two rows of Table 5 contain projections from the 2001 ITRS concerning the required channel concentration levels to keep V_t and W_{ch} (referred to as "Wdepletion" in the table) at their desired values with scaling, assuming a uniform channel concentration. At the 65 nm technology node, ~3 x 10^{18} cm^{-3} doping is required to meet the V_t requirement, and ~2 x 10^{19} cm^{-3} doping to meet the W_{ch} requirement, and these increase with further scaling. Doping levels of 3 x 10^{18} cm^{-3} and higher lead to reductions in the effective mobility in the channel[104], and to possible problems with band-to-band leakage tunneling between the transistor drain and body, thereby increasing I_{off}[105106]. Another important issue is the impact of statistical variation on the control of the transistor characteristics. With scaling, the total number of dopant atoms, N_T, in the depletion region of a MOSFET's channel is significantly reduced; for example, N_T ~ 650 atoms at the 130 nm technology node, where L_g = 65 nm, $W = 3\ L_g$, the channel doping ~2 x 10^{18} cm^{-3} (see Table 5), and the depletion region width ~ 25 nm. On the other hand, for the 22 nm technology node, where L_g = 9 nm, the same type of analysis yields N_T ~ 60 atoms. N_T fluctuates stochastically for the different transistors[107] with a standard deviation, $\Delta N_T = N_T^{0.5}$. Hence, $(\Delta N_T/N_T) = 1/N_T^{0.5}$, which is about 0.04 (4%) for the 130 nm node, but is a quite large 0.13 (13%) for the 22 nm node. This increase in the percent fluctuation in the number of dopant atoms in the depletion region of the MOSFETs can give rise to increases in the statistical fluctuations of V_t, which is undesirable, particularly because these fluctuations in V_t cannot be reduced by improved process control of L_g, EOT, etc. Following Taur and Ning[108] and using the numbers quoted above, the standard deviation of V_t (denoted ΔV_t) due to the fluctuations in N_T is ~ 15mV for the 130 nm node and ~ 60 mV for the 22 nm node. At the 22 nm node, the full $3\Delta V_t$ of 180 mV is much larger than the 50 mV total statistical variation allowed for 10 nm gate length MOSFETs in Ren et al.[109]. Alternatively, the maximum variation in $I_{sd,leak}$ can be estimated using a typical 90 mV/decade subthreshold swing and allowing a full $+3\Delta Vt$. The result is an unacceptably large two orders of magnitude difference between the mean and the maximum value of $I_{sd,leak}$.

Active research and development is being carried out on ultra-thin body SOI (*UTB SOI*) and various types of double gate *(DG) MOSFETs* as a means to deal with many of the above problems for devices with L_g of 25 nm (characteristic of the 65 nm technology node in 2007) and below. UTB SOI and DG SOI MOSFETs are shown schematically in Figure 10. The UTB SOI has a very thin silicon film on top of a relatively thick buried oxide ("*BOX*"), which itself is on top of the silicon substrate. The thin film contains the active area of the MOSFET, and its thickness, T_{si}, is optimally $<L_g/3$,[110] and in all cases, it must be sufficiently small that the depletion region extends throughout the silicon thin film when the device is turned on. Such a device is called a fully depleted SOI *(FDSOI) MOSFET*, and it has a number of advantages. V_t is controlled by the work function of the gate electrode, rather than by the substrate doping. The substrate doping is relatively light compared to bulk planar MOSFETs, which will ameliorate the problems due to heavy channel doping discussed above. Furthermore, since V_t is not controlled by the channel doping, the stochastic variation of N_T will not be reflected in the statistical variation of V_t. Another advantage of the UTB SOI MOSFET is that the parasitic junction capacitance is significantly reduced from that of the bulk MOSFET because the bottom of the deep S/D is directly on top of the thick BOX, which is inherently low capacitance. However, there are several serious shortcomings with the UTB SOI device. One is the difficulty of establishing acceptable control of SCEs in highly scaled MOSFETs. This difficulty is due to the electrostatic coupling between charge in the source and drain and charge in the silicon body's depletion region. This is similar to the electrostatic coupling for planar bulk MOSFETs discussed above, except that the thick BOX strengthens the coupling and thus worsens the problem of controlling SCEs for highly scaled UTB SOI devices. Another difficulty is with the very small values of T_{si} that are needed: as mentioned above, optimally T_{si} should be less than $L_g/3$ to ameliorate the SCEs. For the 65 nm technology node or beyond, this amounts to silicon film thicknesses of 8 nm or less. Fabricating such thin layers of silicon with adequate process control is a significant issue. However, encouraging data on UTB SOI devices with L_g = 50 nm and T_{si} < 30 nm has been reported[111]. The transistor electrical characteristics are generally good, with PMOS I_{on} of 0.65 mA/μm and I_{off} = 9 nA/μm, subthreshold slope, S, of 75 mV/decade, and reduced DIBL < 50 mV/V. Y. K. Choi et al.[112] reported on L_g = 30nm, T_{si} = 4 nm PMOSFET devices with a selective Ge elevated S/D. They found generally good transistor electrical characteristics: S~ 75 mv/decade, DIBL < 50 mV/V, and I_{on} ~ 250 μA/μm for V_{dd} = 1 V. They also found that the SCEs got significantly worse if T_{si} was increased to ~ 8 nm; for example, S increased from about 75 to almost 100 mV/V.

A simplified cross section of a DG SOI MOSFET is shown in Figure 10. This is an UTB SOI MOSFET with a bottom gate in the BOX. The bottom gate is fully self-aligned with the top gate, and the equivalent oxide thickness of the gate dielectric for the bottom gate is the same as that of the top gate. Typically, the DG SOI MOSFET is a fully depleted device, and for optimal characteristics, $T_{si} \leq L_g/2$[113]. The DG MOSFET has the same advantages as the UTB SOI MOSFET, but in addition, it has an enhanced immunity to SCEs for highly scaled MOSFETs. This is because the majority of the charge in the source and drain regions is electrostatically coupled to charge on the two gate electrodes, rather than to depletion charge in the transistor channel region. For similar reasons, the subthreshold swing, S, can be close to the ideal value of 60 mV/decade. From the point of view of device design and electrical characteristics, the DG MOSFET is thus an optimal structure for highly scaled MOSFETs down to L_g ~ 10 nm[114,115,116,117]. However, series S/D parasitic resistance is of particular concern, because of the difficulty in making a low resistance S/D extension and a low resistance contact with the very thin silicon film. Heavy doping and a very abrupt lateral doping profile for the S/D extension as well as elevated S/D are required to deal with this issue[118,119]. Reports of such devices are in Neudeck et al.[120] and Guarini et al.[121] Guarini et al. used ~20 nm thick silicon film to fabricate MOSFETs with *Lg* as small as 50 nm, and found that the subthreshold slope, S, for both NMOSFETs and PMOSFETs is very close to the ideal 60 mV/decade if the back gate and front gate bias is equal. They also report that both the electron and hole effective mobilities match the universal curves from Takagi[122], and they demonstrate working logic circuits.

There are several other types of DG MOSFET that have been discussed in the literature: these include the vertical MOSFET and the FinFET. These have the same basic structure and the same optimal electrical characteristics as the DG SOI device, but the details of the layout and structure are different. A simplified cross-section of the vertical MOSFET is shown in Figure 11. This device is built above the silicon substrate

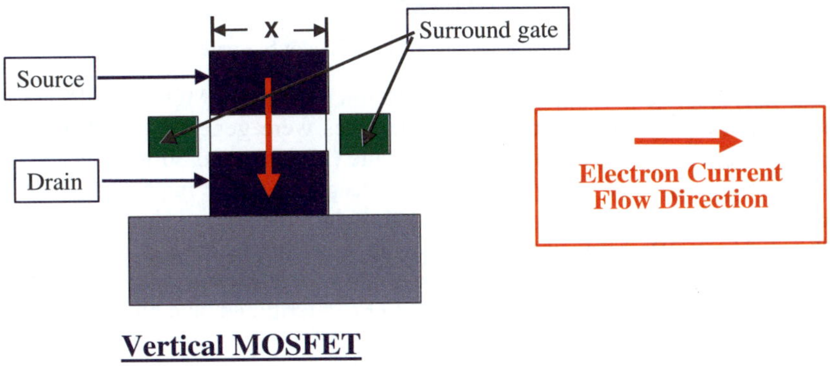

Figure 11. Simplified cross section of vertical MOSFET

using epitaxial growth techniques, and the polysilicon gate surrounds the device. The current flow direction is vertical, as shown, and the flow is at the silicon-oxide interface on either side of the device, so this functions like a double gate device with the two gates on either side, as shown in the figure. The width, *X*, is the analog of the film thickness, T_{si}, for the DG SOI. A number of reports on experimental vertical MOSFETs have appeared in the literature[123,124,125,126]. The smallest L_g achieved so far is about 50 nm, but the devices are not fully depleted because *X* is not small enough[127]. The electrical characteristics are generally acceptable, with NMOSFET I_{on} ~ 1000 μA/μm, and high-k gate dielectrics using HfO_2 or Al_2O_3 have been successfully integrated. Schultz et al. report on fully depleted MOSFETs with 70 nm gate length and *X* ~ 50 nm[128]. The SCEs are well under control, with negligibly small DIBL. The other type of DG MOSFET is the FinFET[129,130,131,132], schematically illustrated in Figure 12. In this device, very thin vertical fins (shown yellow in the figure) are formed in SOI thin films. Poly gates are formed surrounding these fins, and the current flow is from source to drain in the figure, with a MOSFET inversion layer along each vertical edge of the fins. The horizontal width of the fin in the direction perpendicular to the current flow direction is the equivalent of T_{si} in the DG SOI MOSFET. An advantage of the FinFET is that the process flow and process tools are relatively compatible with current techniques. Recent

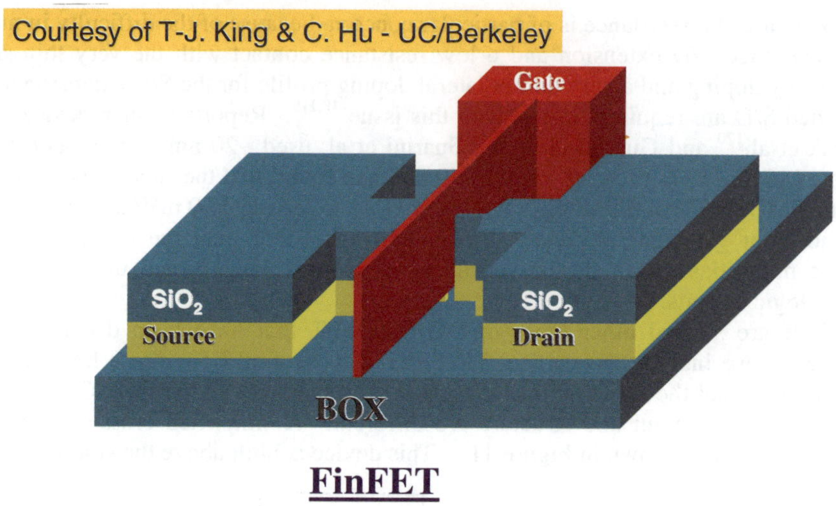

Figure 12. Simplified cross section of FinFET double-gate MOSFET

results for fully depleted, 20 nm gate length, 10 nm fin width PMOS and NMOS devices have been reported[133]. The electrical characteristics were generally good; for example, the subthreshold swing was around 70 mV/decade (NMOS) and 80 mV/decade (PMOS) for L_g down to 20 nm, and the Vt rolloff was acceptable. The impact of fin width was explored, and it was found that an increase in fin width (to 26 nm) significantly degraded the SCEs such as DIBL and subthreshold swing for short L_g devices and Vt rolloff. A main issue for all types of DG MOSFETs is how to fabricate these devices: a cost-effective, manufacturable process flow is still to be determined, and any process flow that emerges is likely to be complicated and difficult.

The IC industry is pursuing highly scaled, 25 nm or less gate-length devices (at the 65 nm technology node and beyond) along two parallel paths: planar bulk MOSFETs and non-classical MOSFETs such as UTB SOI or DG MOSFETs. Bulk planar MOSFETs are being aggressively scaled, and a recent paper reports on data from 15 nm gate-length devices with 0.8 nm EOT gate dielectric using a nitride/oxynitride stack[134]. The electrical results are reasonable but not optimal, with NMOS I_{on} = 615 μA/μm and PMOS I_{on} = 265 μA/μm, I_{off} < 500 nA/μm, and V_{dd} = 0.8 V. Very fast gate delays (CV/I) of 0.29 ps for NMOSFET and 0.68 ps for PMOSFET were obtained. In particular, the SCEs are a major issue, with DIBL, for example, well over 100 mV/V. In summary, both planar bulk CMOS and non-classical CMOS are candidates for the technology nodes toward the end of the Roadmap. Given the difficulties of scaling planar bulk CMOS into that regime, however, non-classical CMOS (particularly double gate) is preferred from a device point of view, but difficult challenges, particularly in processing, manufacturability, and fabrication cost, remain to be solved for non-classical CMOS. The ultimate CMOS device with L_g of 10 nm or less is likely to be a DG MOSFET with advanced materials and processes such as high-k gate dielectric, metal gate electrodes, elevated source/drains, and advanced annealing.

5. Conclusions

IC technology is continuing to scale according to Moore's Law, with the overall chip circuit requirements driving the MOSFET device and process integration requirements and optimal choices. In the 2001 ITRS the driver for the high performance logic is maximizing MOSFET intrinsic speed, while the driver for low power logic is minimizing MOSFET leakage current. In the front end, high-k gate dielectric is an important issue, and will apparently be needed initially for low-power logic in 2005, with subsequent implementation for high-performance logic. Other important front-end issues include polysilicon depletion in the gate electrode, which will probably eventually drive the industry to metal gates, and series source/drain parasitic resistance, which will probably drive the industry to various new solutions in source/drain processing. Also, the adequacy of planar bulk CMOS devices to meet the MOSFET requirements beyond 2007 appears questionable, and the industry is exploring alternatives in non-classical CMOS structures to deal with this issue.

References

[1] G. E. Moore, "Lithography and the Future of Moore's Law," *Proceedings of VIIIth Optical/Microlithography Conference,* SPIE Vol. **2440**, pp. 2 – 17, February 1995.

[2] Semiconductor Industry Association (SIA), *International Roadmap for Semiconductors 2001 edition*, Austin, TX: International SEMATECH, 2001, Executive Summary Chapter. (This is available for printing and viewing from the internet, with the following URL: http://public.itrs.net.)

[3] H. R. Huff et al., "Sub-100 nm Gate Stack/Ultrashallow Junction Integration Challenges," *ECS PV* **2001-09**, 263-297 (2001). [Originally presented at the ECS *RTA and Other Short Time Processing Technologies II* Symposium, March 2001.]

[4] P. K. Vasudev and P. M. Zeitzoff, "ULSI Semiconductor Technology Trends and Challenges in the Coming Decade," *IEEE Circuits and Devices Magazine,* **14**, p. 19 (March 1998).

[5] Semiconductor Industry Association (SIA), *International Roadmap for Semiconductors 2001 edition*, Austin, TX: International SEMATECH, 2001. (This is available for printing and viewing from the internet, with the following URL: http://public.itrs.net.)

[6] Semiconductor Industry Association (SIA), *International Roadmap for Semiconductors 2001 edition*, Austin, TX: International SEMATECH, 1999. (This is available for printing and viewing from the internet, with the following URL: http://public.itrs.net.)

[7] SIA, ITRS 2001, *op. cit.*, Executive Summary section.

[8] J. D. Meindl et al., "Limits on Silicon Nanoelectronics for Terascale Integration," *Science*, Vol. **293**, pp. 2044 – 2049 (14 Sept. 2001).

[9] P. M. Zeitzoff and J. E. Chung, "Weighing in on Logic Scaling Trends," *IEEE Circuits and Devices Magazine*, Vol. **18**, pp. 18 – 27, (March 2002).

[10] SIA, ITRS 2001, *op. cit.*, Executive Summary Chapter.

[11] S. Thompson et al., "MOS Scaling: Transistor Challenges for the 21st Century," 3rd Quarter 1998 on Intel's® website. URL is http://www.intel.com/technology/itj/q31998/articles/art_3.htm.

[12] Y. Taur and T. H. Ning, *Fundamentals of Modern VLSI Devices,* Cambridge University Press, New York, 1998, pp. 228-230.

[13] SIA, ITRS 2001, *op. cit.*, Process Integration, Devices, and Structures (PIDS) chapter, High-Performance Logic Technology Requirements Tables.

[14] Y. Taur and T. Ning, *op. cit.*, pp. 173 – 175.

[15] S. Thompson et al., "An Enhanced 130 nm Generation Logic Technology Featuring 60 nm Transistors Optimized for High Performance and Low Power at 0.7 – 1.4 V," *IEDM Tech. Digest*, pp. 257 – 260, Dec. 2001.

[16] S. F. Huang et al., "High Performance 50 nm CMOS Devices for Microprocessor and Embedded Processor Core Applications," *IEDM Tech. Digest*, pp. 237 - 240, Dec. 2001.

[17] S. Parihar, "A High Density 0.10μm CMOS Technology Using Low K Dielectric and Copper Interconnect," *IEDM Tech. Digest*, pp.249 - 252, Dec. 2001.

[18] SIA, ITRS 2001, *op. cit.*, PIDS and Design chapters.
[19] SIA, ITRS 2001, *op. cit.*, PIDS chapter, LSTP and LOP Logic Technology Requirements Tables.
[20] B.Maitl et al., "High Performance 20 Å NO Oxynitride for Gate Deielectric in Deep Sub-Quarter Microcon CMOS Technology," *IEDM Tech. Digest*, pp. 651- 654, Dec. 1997.
[21] B. Yu et al., "Scaling Toward 35nm Gate Length CMOS," *Symposium on VLSI Technology, Digest of Tech. Papers,* pp. 9 – 10, June 2001.
[22] G. D. Wilk et al., "High-k Gate Dielectrics: Current Status and Material Properties Considerations," *Journal of Applied Physics*, **89**, pp. 5243 – 5275 (May 2001).
[23] X. Guo and T. P. Ma, "Tunneling Leakage Current in Oxynitride: Dependence on Oxygen/Nitrogen Content," IEEE Elec. Dev. Letters, EDL-**19**, pp. 207-209 (June 1998).
[24] B. Yu et al., *op. cit.*
[25] A. Karamcheti et al., "Siilicon Oxynitride Films as a Segue to the High-K Era," *Semiconductor FABTECH Magazine*, **12th ed.**, pp. 207 – 214 (June 2000).
[26] SIA, ITRS 2001, *op. cit.*, Front End Processes chapter.
[27] SIA, ITRS 2001, *op. cit.*, PIDS chapter, Memory and Logic Potential Solutions section.
[28] H. R. Huff et al., "Integration of High-k Gate Stack Systems into Planar CMOS Process Flows," *Proceedings of International Workshop on Gate Insulators (IWGI)*, Tokyo, Japan, November 2001.
[29] R. Choi et al., "High-Quality Ultra-thin HfO2 Gate Dielectric MOSFETs with TaN Electrode and Nitridation Surface Preparation," *Symposium on VLSI Technology, Digest of Tech. Papers,* June 2001.
[30] S. J. Lee et al., "Performance and Reliability of Ultra Thin CVD HfO_2 Gate Dielectrics with Dual Poly-Si Gate Electrodes," *Symposium on VLSI Technology, Digest of Tech. Papers,* June 2001.
[31] Y. Kim et al., "Conventional n-channel MOSFET devices using single layer HfO_2 and ZrO_2 as high-k gate dielectrics with polysilicon gate electrode," *IEDM Tech. Digest,* pp. 455-458, Dec. 2001.
[32] K. Onishi et al., "Reliability Characteristics, Including NBTI, of Polysilicon Gate HfO_2 MOSFET's," *Tech. Digest of 2001 IEDM,* pp. 659-662, Dec. 2001.
[33] SIA, ITRS 2001, *op. cit.*, Front End Processes chapter.
[34] G. D. Wilk et al., *op. cit.*
[35] H. R. Huff et al., "Sub-100 nm Gate Stack/Ultrashallow Junction Integration Challenges," *op. cit.*
[36] SIA, ITRS 2001, *op. cit.*, PIDS chapter.
[37] Y.Taur and T. Ning, op. cit., pp. 75 – 78.
[38] S. M. Sze, *Physics and Technology of Semiconductor Devices, 2nd ed.*, Wiley and Sons, New York, 1981, Appendixes H and I.
[39] G. D. Wilk et al., *op. cit.*
[40] S. Gannavaram et al., "Ultra-Shallow P^+-N Junctions for 35-70 nm CMOS Using Selectively Deposited Very Heavily Boron-Doped Silicon-Germanium Films, *Rapid Thermal Processing and Other Short-Time Processing Technologies I,* (F. Roozeboom et al., eds.) ECS PV 00-9, pp. 73 – 82 (2000).
[41] SIA, ITRS 2001, *op. cit.*, Front End Processes chapter, Thermal and Thin Film, Doping, and Etching Technology Requirements Tables.
[42] H. Zhong et al., "Properties of Ru-Ta Alloys as Gate Electrodes for NMOS and PMOS Silicon Devices," *IEDM Tech. Digest,* pp. 467-470, Dec. 2001.
[43] Y. Suh et al., "Electrical Characteristics of TaSixNy Gate Electrodes for Dual Gate Si_CMOS Devices," *VLSI Symposium Dig. Of Tech. Papers,* pp.47-48, June 2001.
[44] D. Park et al., "Novel Damage-Free Direct Metal Gate Process Using Atomic Layer Deposition," *VLSI Symposium Dig. Of Tech. Papers,* pp. 65-66, June 2001.
[45] Y. H. Kim et al., "High Quality CVD TaN Gate Electrode for Sub-100nm MOS Devices," *IEDM Tech. Digest,* pp. 667-670, Dec. 2001.
[46] H. Zhong et al., "Characterization of RuO2 electrode on Zr silicate and ZrO_2 dielectrics," *Appl. Phys. Lett.*, **78,** pp. 1134-1136 (2001).

[47] D. Park et al., "Robust Ternary Metal Gate Electrodes for Dual Gate CMOS Devices," *IEDM Tech. Digest,* pp. 671-674, Dec. 2001.
[48] H. Zhong et al., *op. cit.,* 2001 IEDM.
[49] Q. Lu et al., "Metal Gate work Function Adjustment for Future CMOS Technology," *VLSI Symposium Dig. Of Tech. Papers,* pp. 45-46, June 2001.
[50] G. D. Wilk et al., *op. cit.*
[51] Y. Taur and T. Ning, *op. cit.*, pp. 194 – 200.
[52] G. D. Wilk et al., *op. cit*
[53] K. W. Ng and W. T. Lynch, ""The impact of intrinsic series resistance on MOSFET scaling", *IEEE Transactions on Electron Devices,* ED-**34**, p.502 (March 1987).
[54] Y. Taur et al., *op. cit.*, pp. 164 - 173.
[55] S. Thompson et al., "MOS Scaling: …," *op. cit.*
[56] L. Larson and H. Huff, "ITRS front-end process roadmap," *Solid State Technology*, pp. 32-34, January 2002.
[57] K. W. Ng and W. T. Lynch, *op. cit.*
[58] Y. Taur and T. Ning, *op. cit.*, pp. 214 – 219.
[59] S. Thompson et al., "MOS Scaling: …", *op. cit.*
[60] K. W. Ng and W. T. Lynch, *op. cit.*
[61] S. W. Crowder et al., "The Effect of Source/Drain Processing on the Reverse Short Channel Effect of Deep Sub-Micron Bulk and SOI NMOSFETs," *IEDM Tech. Digest*, pp. 427 - 430, Dec. 1995.
[62] H-J. Li et al., "Boron Diffusion and Activation in the Presence of Other Species," *IEDM Tech. Digest*, pp. 515 - 518, Dec. 2000.
[63] K. Tsuji et al., "High Performance 50-nm Physical Gate Length pMOSFETs by using Low Temperature Activation by Re-Crystallization Scheme", *VLSI Symposium Dig Of Tech. Papers*, pp. 9 - 10, June 1999.
[64] L. Larson and H. Huff, *op. cit.*
[65] SIA, ITRS 2001, *op. cit.*, Front End Processes Chapter.
[66] L. Larson and H. Huff, *op. cit.*
[67] H. Huff, G. Brown, et al., *op. cit.*
[68] SIA, ITRS 2001, *op. cit.*, Front End Processes chapter.
[69] W. P. Maszara, "Very thin $CoSi_2$ films by Co sputtering", *Appl. Phys. Lett.*, **62**, 961 (March 1993).
[70] SIA, ITRS 2001, *op. cit.*, Front End Processes chapter, Thermal and thin Film, Doping, and Etching Technology Requirements Tables.
[71] W. P. Maszara, *op. cit.*
[72] S. Inaba et al., "High Performance 35 nm Gate Length CMOS with NO Oxynitride Gate Dielectric and Ni SALICIDE," *IEDM Tech. Digest*, pp. 641 - 644, Dec. 2001.
[73] B. Yu et al., "15 nm Gate Length Planar CMOS Transistor," *IEDM Tech. Digest*, pp. 937- 939, Dec. 2001.
[74] SIA, ITRS 2001, *op. cit.*, Front End Processes Chapter.
[75] S. D. Kim et al., "Advanced Model and Analysis for Series Resistance in Sub-100nm CMOS Including Poly Depletion and Overlap Doping Gradient Effect," *IEDM Tech. Digest*, pp. 723 – 726, Dec. 2000.
[76] SIA, ITRS 2001, *op. cit.*, Front End Processes chapter, Figure 30: Doping Potential Solutions.
[77] S. Gannavaram, "Low Temperature (≤800°C) Recessed Junction Selective Silicon-Germanium Source/Drain Technology for sub-70 nm CMOS," *IEDM Tech. Digest*, pp. 437- 440, Dec. 2000.
[78] J. Kedzierski et al., "Complementary silicide source/drain thin-body MOSFETs for the 20nm gate length regime," *IEDM Tech. Digest*, pp. 57- 60, Dec. 2000.
[79] S. D. Kim et al., *op. cit.*
[80] SIA, ITRS 2001, *op. cit.*, PIDS chapter, Logic and Memory Potential Solutions table.
[81] SIA, ITRS 2001, *op. cit.*, PIDS chapter, Emerging Research Devices section.

[82] J. Hutchby et al., "Extending the Road Beyond CMOS," *IEEE Circuits and Devices Magazine*, March 2002.
[83] Y. Taur and T. Ning, *op. cit.,* Appendix 4.
[84] T. Mizuno et al., "Advanced SOI-MOSFETs with Strained-Si Channel for High Speed CMOS Electron/Hole Mobilty Enhancement," *VLSI Symposium Dig Of Tech. Papers*, pp.210 - 211, June 2000.
[85] Y. Taur and T. Ning, *op. cit.,* pp. 285 – 286.
[86] M. V. Fischetti and S. D. Laux, "Band Structure, Deformation Potentials, and Carrier Mobility in Strained Si, Ge, and SiGe Alloys, *Electron. Lett.*, **17**, pp. 503 – 504 (1996).
[87] Y. Taur et al., *op. cit.,* pp. 285 – 286.
[88] M. V. Fischetti and S. D. Laux, *op. cit.*
[89] K. Rim et al., "Strained Si NMOSFETs for High Performance CMOS Technology," *IEDM Tech. Digest*, pp. 59 - 60, *VLSI Symposium Dig Of Tech. Papers*, pp. 59 - 60, June 2001.
[90] K. Rim et al., "Enhanced Hole Mobilities in Surface-channel Strained-Si *p*-MOSFETs," *IEDM Tech. Digest*, pp. 517 – 520, Dec. 1995.
[91] J. Welser et al., "Strain Dependence of the Performance Enhancement in Strained-Si *n*-MOSFETs," *IEDM Tech. Digest*,pp. 373 – 376, Dec. 1994.
[92] K. Rim et al., 1995 IEDM, *op. cit.*
[93] K. Rim et al., "Transconductance Enhancement in Deep Submicron Strained-Si n-MOSFETs," *IEDM Tech. Digest*, pp. 707 – 710, Dec. 1998.
[94] K. Rim et al., 1995 IEDM, *op. cit.*
[95] T. Mizuno et al., *op. cit.*
[96] J. Alieu et al., "Multiple SiGe Well:a New Channel Architecture for Improving Both NMOS and PMOS Performances, *VLSI Symposium Dig Of Tech. Papers*, pp.130 - 131, June 2000.
[97] Y. Taur and T. Ning, *op. cit.,* pp. 139 – 149.
[98] SIA, ITRS 2001, *op. cit.*, Front End Processes chapter.
[99] S. Song et al., "Design of Sub-100nm CMOSFETs: Gate Dielectrics and Channel Engineering," *VLSI Symposium Dig Of Tech. Papers*, pp.190 - 191, June 2000.
[100] Y. Taur and T. Ning, *op. cit.,* pp. 189 – 194.
[101] H. Wakabayashi et al., "45-nm Gate Length CMOS Technology and Beyond Using Steep Halo," *IEDM Tech. Digest*, pp. 49 - 52, Dec. 2000.
[102] Y. Taur, "25 nm CMOS Considerations," *IEDM Tech. Digest*, pp. 789 - 792, Dec. 1998.
[103] Y. Taur and T. Ning, "CMOS Devices Below 0.1 μm: How High Will Performance Go?" *IEDM Tech. Digest*, pp. 215 - 218, Dec. 1997.
[104] S. Takagi et al, "On Universality of Inversion-Layer Mobility in Si MOSFET's: Part I—Effects of Substrate Impurity Concentration," *IEEE Trans. on Elec. Devices*, **41,** 2357 – 2362 (December 1994).
[105] L. Chang et al., "Gate Length Scaling and Threshold Voltage Control of Double-Gate MOSFETs," *IEDM Tech. Digest*, pp. 719 - 722, Dec. 2000.
[106] T. Schultz et al., 50-nm Vertical Sidewall transistors With High Channel Doping Concentrations," *IEDM Tech. Digest*, pp.61 - 64, Dec. 2000.
[107] Y. Taur and T. Ning, *op. cit.,* pp. 200 – 202.
[108] Y. Taur and T. Ning, *op. cit.,* pp. 200 – 202, Eq. 4.64.
[109] Z. Ren et al., "Examination of Design and Manufacturing Issues in a 10 nm Double Gate MOSFET Usng Nonequilibrium Green's Function Simulation," *IEDM Tech. Digest*, pp. 107 - 110, Dec. 2001.
[110] Y. K. Choi et al., "30 nm Ultra-Thin-Body SOI MOSFET with Selectively Deposited Ge Raised S/D," 58th Device Research Conf. (DRC), pp. 23 – 24.
[111] R. Chau et al., "A 50nm Depleted-Substrate CMOS Transistor (DST)," *IEDM Tech. Digest*, pp. 621 - 624, Dec. 2001.
[112] Y-K. Choi et al., "Ultra-Thin Body PMOSFETs with Selectively Deposited Ge Source/Drain," *VLSI Symposium Dig Of Tech. Papers*, pp.19 - 20, June 2001.

[113] Z. Ren et al., *op. cit.*
[114] D. T. Frank et al., "Monte Carlo Simulation of a 30 nm Dual-Gate MOSFET: How Short Can Si go?", *IEDM Tech. Digest*, pp. 553 - 556, Dec. 1992.
[115] Z. Ren et al., *op. cit.*
[116] L. Chang et al., *op. cit.*
[117] D. Hisamoto, "FD/DG-SOI MOSFET—a Viable Approach to Overcoming the Device Scaling Limit," *IEDM Tech. Digest*, pp.429 - 432, Dec. 2001.
[118] Z. Ren et al., *op. cit.*
[119] D. Hisamoto, *op. cit.*
[120] G. W. Neudeck et al, "Novel Silicon Epitaxy for Advanced MOSFET Devices," *IEDM Tech. Digest*, pp. 169 - 172, Dec. 2000.
[121] K. W. Guarini et al., "Triple-Self-Aligned, Planar Double-Gate MOSFETs: Devices and Circuits, *IEDM Tech. Digest*, pp. 425 - 428, Dec. 2001.
[122] S. Takagi et al., "On Universality of Inversion-Layer Mobility in n- and p-Channel MOSFETs, *IEDM Tech. Digest*, pp. 398 – 401, Dec. 1988.
[123] J. M. Hergenrother et al., "50nm Vertical Replacement Gate (VRG) nMOSFETs with ALD HfO_2 and Al_2O_3 Gate Dielectrics, *IEDM Tech. Digest*, pp.51 - 54, Dec. 2001.
[124] S-H. Oh et al., "50 nm Vertical Replacement-Gate pMOSFETs," *IEDM Tech. Digest*, pp.65 - 68, Dec. 2000.
[125] T. Schultz et al., *op. cit.*
[126] J. M. Hergenrother et al., "The Vertical Replacement Gate (VRG) MOSFET: a 50-nm Vertical MOSFET with Lithography-Independent Gate Length," *IEDM Tech. Digest*, pp.75 - 78, Dec.1999.
[127] J. M. Hergenrother et al., *2001 IEDM, op. cit.*
[128] T. Schultz et al., *op. cit.*
[129] Y-K Choi et al., "Sub-20nm CMOS FinFET Technologies," *IEDM Tech. Digest*, pp. 421 - 424, Dec.2001.
[130] J. Kedzierski et al., "High-Performance Symmetric-Gate and CMOS-Compatible Vt Assymetric-Gate FinFET Devices," *IEDM Tech. Digest*, pp. 437 - 440, Dec.2001.
[131] D. Hisamoto et al., "FinFET—A Self-Aligned Double-Gate MOSFET Scalable to 20 nm," *IEEE Trans. on Electron Devices*, **47**, pp. 2320 – 2325 (2000).
[132] X. Huang et al., "Sub 50-nm FinFET: PMOS," *IEDM Tech. Digest*, pp. 67 - 70, Dec.1999.
[133] Y-K Choi et al., *2001 IEDM, op. cit.*
[134] B. Yu et al., *2001 IEDM, op. cit.* [134] G. E. Moore, "Lithography and the Future of Moore's Law," *Proceedings of VIIIth Optical/Microlithography Conference,* SPIE Vol. **2440**, pp. 2 – 17, February 1995.

International Journal of High Speed Electronics and Systems, Vol. 12, No. 2 (2002) 295–304

ISSUES, ACHIEVEMENTS AND CHALLENGES TOWARDS INTEGRATION OF HIGH-k DIELECTRICS

M. CAYMAX, S. DE GENDT, W. VANDERVORST, M. HEYNS
H. BENDER, R. CARTER, T. CONARD, R. DEGRAEVE, G. GROESENEKEN, S. KUBICEK, G. LUJAN, L. PANTISANO, J. PETRY, E. ROHR, S. VAN ELSHOCHT, C. ZHAO, E. CARTIER[*], J.CHEN[*], V.COSNIER[*], S.E. JANG[*], V. KAUSHIK[*], A. KERBER[*], J. KLUTH[*], S. LIN[*], W. TSAI[*], E. YOUNG[*] AND Y. MANABE[+]

IMEC vzw, Kapeldreef 75, B-3001 Leuven, Belgium
[*]International Sematech residents at IMEC vzw
[+]Hitachi Semiconductor resident at IMEC vzw

Once the thickness of the gate dielectric layer in CMOS devices gets thinner than 1.2 nm, excessive gate leakage due to direct tunneling makes the use of alternative materials obligatory. Candidate high-k materials are metal oxides such as Al_2O_3, ZrO_2 and HfO_2 as well as their mixtures. Very promising results have been reported world-wide. Here, however, we show that there are a number of issues related to materials and electrical characteristics as well as to processing which are not always recognized and that necessitate more work to find solutions. Among these are problems with density, interface layer growth and island formation which are clearly related to the deposition process. Also thermal instabilities as well as interactions between the high-k material and poly-Si need attention. Further possible show-stoppers are electrical reliability issues and strongly reduced carrier mobility.

1 Introduction

Since the dawn of the integrated circuit in the early sixties, (C) MOS processing has relied on silica (SiO_2) as the gate dielectric. Feature size reduction and the ever demanding technology roadmaps have imposed scaling constraints on gate oxide thickness to the point where excessive tunneling currents make transistor design untenable; an alternative gate dielectric is needed.[1] Replacement of SiO_2 with a high dielectric constant material allows an increase in the physical oxide thickness, while maintaining a low equivalent oxide thickness (EOT) and low direct tunneling current.

Requirements to candidate dielectrics are: a high dielectric constant k, stable in contact with silicon at high temperature, a low density of defects in the bulk ($Q_{tot} < 5E^{10}$ cm^{-2}) and at the interface ($D_{it} < 5E^{10}$ cm^{-2}), an EOT of < 1.3nm, scalable to sub 1.0 nm with low leakage ((I_{leak} < 10^{-3}A/cm^2 at 25°C for VDD =1V), channel mobility equivalent to SiO_2.

Some promising materials are Al_2O_3, ZrO_2, HfO_2 or mixtures thereof (silicates, aluminates).[2] In this paper, we will show promising results obtained in the work at Imec. More importantly, however, we will discuss some of the challenges one faces when integrating high-k materials in CMOS processing; such as achieving a high quality Si/high-k interface, material stability and characterization, as well as intrinsic electrical and reliability issues.

2 Deposition of High-k Gate Stacks with ALCVD and MOCVD High-k Layers in Cluster Systems

From production point of view, there are two possible deposition techniques for the candidate high-k materials, *i.e.* Atomic Layer Chemical Vapor Deposition (ALCVD™) and Metal Organic CVD (MOCVD). At Imec, dedicated reactors are available for both techniques as part of larger cluster systems.

2.1 Why cluster systems ?

CVD reactors for high-k layer deposition can be either stand-alone or integrated into a cluster of reactors. This is helpful *e.g* when one wants to control the interface between the Si substrate and the high-k layer very carefully. The need for a very strict layer engineering becomes immediately clear if one considers as an example a high-k stack composed of ZrO_2 and Al_2O_3 (and potentially an additional interfacial SiO_2 layer as well). The corresponding EOT can be calculated as follows:

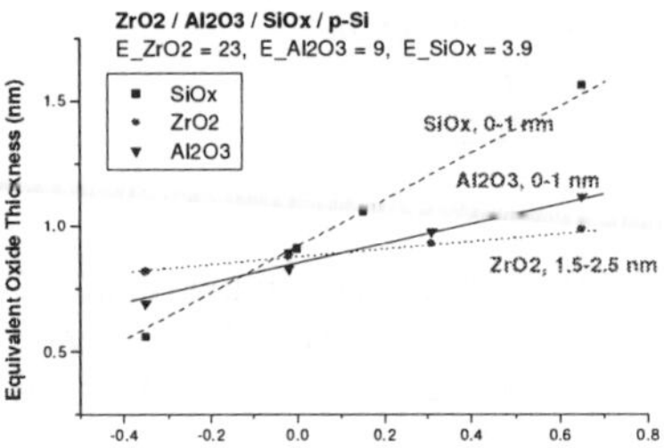

Figure 1 Impact on EOT of the nominal sub-layer thickness

$$EOT = \varepsilon_{SiOx}(t_{ZrO2}/\varepsilon_{ZrO2} + t_{Al2O3}/\varepsilon_{Al2O3}) + t_{SiOx}$$

whereby one normally assumes a dielectric constant for SiO_2, Al_2O_3 and ZrO_2 equal to 3.9, 9 and 23 respectively. If one targets 0.7 nm EOT, generally speaking this can be achieved by various combinations of material ranging from 0.7 nm SiO_2 to 4.1 nm ZrO_2. Due to the differences in dielectric constant, the sensitivity of the final EOT to a small thickness variation of the interfacial SiO_2 layer is extremely important, as is clear from Fig.1. So, in order to minimize EOT, the Si wafer native oxide needs to be removed before the high-k deposition, either by standard wet HF etching, or by means of a gas phase cleaning with a HF/Vapor chemistry. Additionally, it might be important to modify the Si surface before high-k deposition to obtain better interface properties (well-controlled interfacial SiO_x, lower trap density, better barrier properties...). Possible techniques include nitridation of the Si surface (remote plasma, thermal nitridation) perhaps in combination with a limited oxidation. Finally, deposition of a gate electrode (poly Si) layer is required, preferentially without exposing the wafer to the ambient air to avoid re-oxidation or other kinds of contamination. It can also be useful to anneal the high-k layer before poly-Si deposition in order to reduce its chemical reactivity. So, a high-k cluster typically can contain a gas-phase SiO_2 etching reactor, a high-k CVD chamber, a poly-Si reactor, and some reactor which allows to do all kinds of thermal annealing treatments in various gases (Rapid Thermal Processing RTP), plasma treatments, Rapid Thermal Oxidations (RTO) and so on.

2.2 ALCVD of high-k layers in the ASM Polygon™ P8200 cluster platform

The ASM Polygon™ P8200 cluster platform contains four reactor modules: an ALCVD™ reactor (with H_2O, Trimethylaluminum (TMA) and $ZrCl_4$), a poly-Si reactor, a nitride reactor and an HF/Vapor cleaning module. The poly (SiH_4, GeH_4, p and n type dopant gases) and nitride (SiH_4, NH_3, NF_3, O_2) modules are based on the ASM Epsilon™ epitaxial reactor, which is a single wafer, lamp heated horizontal CVD reactor with a rotating SiC coated graphite susceptor in a quartz chamber. The nitride module incorporates also a remote microwave plasma source (MRG). The HF/Vapor cleaning reactor consists of a closed loop reactor chamber and uses anhydrous HF and CH_3COOH vapor.

The ALCVD™ module is a Pulsar 2000 reactor, developed by ASM Microchemistry. It contains two liquid sources (for H2O and TMA) and one solid source for ZrCl4. It allows deposition of Al_2O_3, ZrO_2 and combinations of these, either as nanolaminates or as mixed oxides AlZrOx. The principle of ALCVD is as follows: the precursors for the different elements in the material to be deposited are introduced one after the other, separated each time by a pulse of some inert gas.[3,4] They react with the surface in a kind of layer-by-layer growth. So, for example, for the deposition of Al_2O_3, one could start with a pulse of TMA. This adsorbs on the surface, hereby releasing one ligand as follows:

$$\text{Si-OH} + \text{Al(CH}_3)_3 \rightarrow \text{Si-Al(CH}_3)_2 + \text{CH}_4.$$

The unreacted TMA as well as the reaction product CH_4 is purged out by means of a N2 purge pulse, followed by a pulse of H_2O. This reacts again with the remaining CH_3 ligands, converting them into OH groups and so preparing the surface for a next TMA pulse:

$$\text{Si-Al(CH}_3)_2 + 2\ \text{H}_2\text{O} \rightarrow \text{Si-Al(OH)}_2 + \text{CH}_4.$$

The combination of a TMA and a HYO pulse is called a cycle. Per cycle, one typically grows less than a full mono-layer of metal oxide. For Al_2O_3, it is typically about 0.1 nm, whereas for ZrO_2 this is about 0.05 nm. A full cycle takes about 2-5 seconds, so the deposition of a typical layer for gate applications takes a few minutes.

2.3 MOCVD in the AMAT Centura system

The MOCVD High-k cluster is an Applied Materials (AMAT) system build on the Thermal Processing Common Centura® (TPCC) platform. Attached to the platform are a RTPXEplus™ module with Remote Plasma Nitridation (RPN) capability, a Polygen™, and a high-k MOCVD chamber (based on the Tanox™ chamber). The RTPXEplus™/RPN chamber is capable of running Rapid Thermal Oxidation, In Situ Steam Generated oxidation, and RPN processes and contains two liquid source delivery systems aimed at the deposition of HfO_2 and $HfSiO_x$ (Hf silicates). The PolyGen module allows to deposit in-situ doped poly-Si films.

The metal precursors used in the high-k MOCVD chamber are Tetrakis Diethyl Amino Hafnium (TDEAH) and Tetrakis Dimethyl Amino Silicon (TDMAS). The principle of the deposition reaction is

Precursor + O_2 → Metal Oxide + CO_2 + H_2O +NO_x.

Typical growth conditions for HfO_2 are temperatures between 300 and 600 °C and a pressure of 3 torr. Growth rates for HfO_2 are about 3-4 nm/min.

3 Some Initial Results

Over the last few years, exciting results in the quest for the ultimate alternative gate material have been reported by several research groups all over the world.[2,5-9] A frequently used figure of merit is the combination of gate leakage and EOT as deduced from CV measurements on

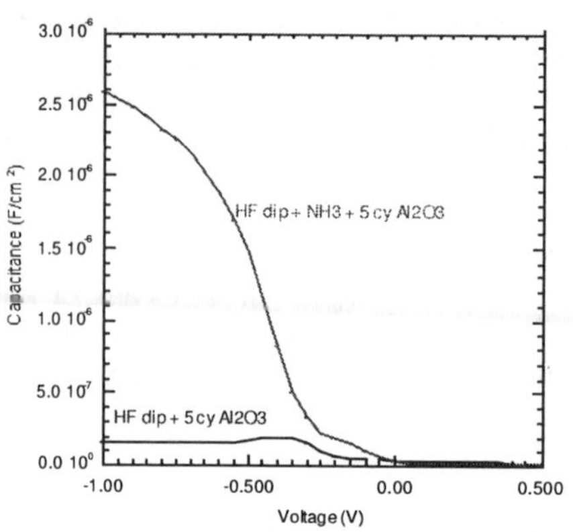

Figure 2 CV curve of high-k capacitor with and without NH_3 pre-anneal

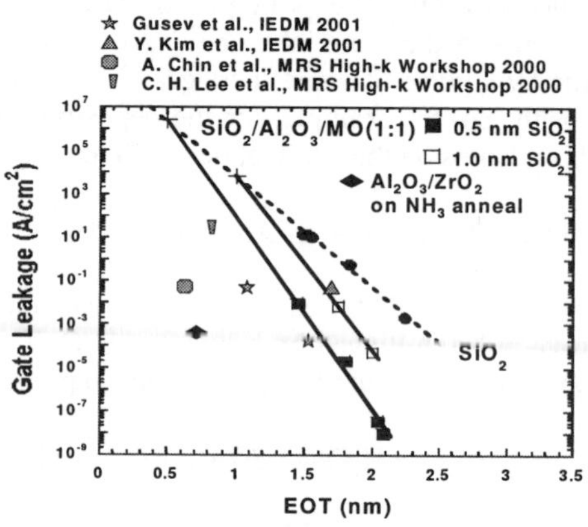

Figure 3 Comparison of NH_3-pre-annealed high-k capacitor vs. no-anneal and literature results

MOS capacitors. Encouraging low EOT values and leakage currents have been obtained for various high-k materials.[10-13]

At Imec, we investigated a wide range of process and material characteristics of both ALCVD and MOCVD based high-k layers, as well as electrical characteristics of MOS devices.[14-17] An interesting example of this work relates to the influence of the wafer surface preparation before high-k deposition. For this experiment, we deposited ZrO_2 layers of about 3 nm by means of ALCVD on top of Si wafers after a dip for 30 sec in a 2 % solution of HF to remove the native oxide. Alternatively, before high-k deposition the wafer was nitrided in

NH3 at 560 °C after the HF dip. The ZrO_2 deposition was preceded by 5 cycles of Al_2O_3, which results in approximately 0.02 nm of Al_2O_3 on an HF dipped wafer. As gate electrode, 100 nm of TiN was sputtered. MOS capacitors were defined by wet etch and contact lithography. Capacitor areas range from $5.0x10^{-6}$ to $1.0x10^{-3}$ cm^2. The gate stacks were annealed in dry N_2/H_2 for 30 min at 400 °C. The CV measurement in Fig. 2 shows how the capacitors on the HF dipped surface are completely leaky, whereas after the NH3 treatment, leakage is strongly reduced. Fitting the CV curve by means of the NCSU CVC program reveals an EOT of 0.77 nm.[19] The leakage through the gate stack with the NH_3 surface treatment at –1 volt is ~ $3x10^{-4}$ A/cm^2. This is about 6 orders of magnitude less than the stack with the HF pre-treatment. The ZrO_2/Al_2O_3 dielectric stack with the NH_3 pre-treatment is benchmarked with IMEC furnace grown oxides, high-k demo stacks of ZrO_2/Al_2O_3 and HfO_2/SiO_2 and some of the more promising results from literature in Fig. 3. It is clear, that this approach keeps some promise.

4 Issues Related to Materials Aspects of High-k Gate Stacks

Although promising results have been obtained, one should realize there are still a number of problems to be solved. We want to give a brief overview of some of these issues.

4.1 Deposition related problems: layer density, interfacial oxide and island formation

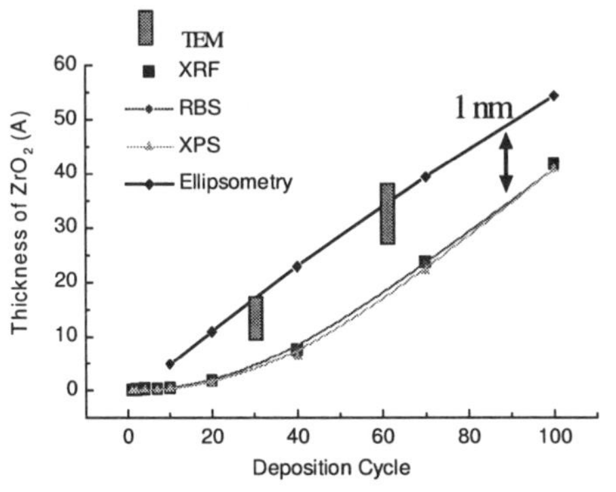

Figure 4 Growth curve of ZrO_2 on HF dipped Si surface

An important feature of the high-k deposition process is how the layer grows on a specific surface: we measured the layer thickness as a function of the number of growth cycles for ALCVD of Al_2O_3 and ZrO_2 on various surfaces, among which HF dipped Si and different kinds of thin SiO_2 layers.[16] We applied several techniques: spectroscopical ellipsometry (SE), Rutherford Backscattering (RBS), X-ray Photo-electron Spectroscopy (XPS), X-ray Fluorescence (XRD), Transmission Electron Microscopy (TEM). A summary is shown in Fig. 4 for ZrO_2 on HF dipped Si. An important difference is to be noted between SE and XRF, RBS and XPS. Ellipsometric analysis of very thin layers (< 5 nm) does not allow the simultaneous modeling of different layer characteristics (e.g. thickness, refractive index, roughness) in an accurate way due to the resulting strong correlation in the regression between these parameters. Therefore, the thickness determined in this way needs some precaution. The other techniques mainly determine an areal density (atoms/cm2), and a thickness can only be extracted by assuming a density such as e.g. normal bulk density. The only reliable absolute thickness determination comes from TEM. However, comparing TEM and the other techniques reveals a difference, which can only be explained by assuming that the film density is lower than normal bulk density with about 10-50 % depending on deposition conditions, and this both for ALCVD and for MOCVD films. This has been verified by means of X-ray Reflectivity on thicker layers.

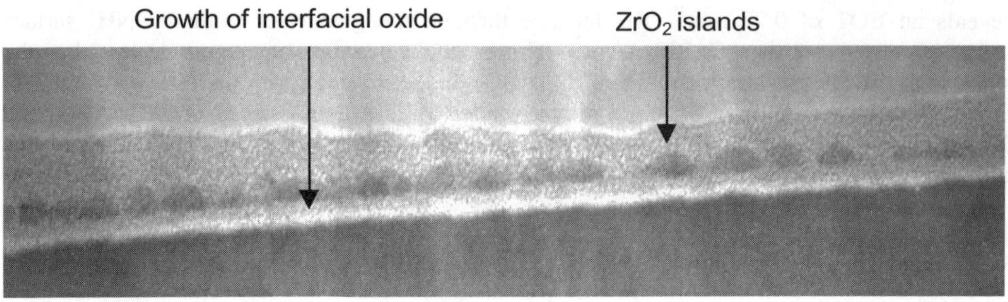

Figure 5 TEM cross-sectional image of a very thin ALCVD ZrO_2 film on HF dipped Si

A second feature is the strong growth incubation: it takes up to 20 cycles before the growth of ZrO_2 really takes off. This is due to the lack of active surface sites on the H passivated Si surface: as mentioned before, the preferred surface finish for ALCVD growth is OH groups. Although this by itself is not really an issue, the way how the material grows is a problem, as

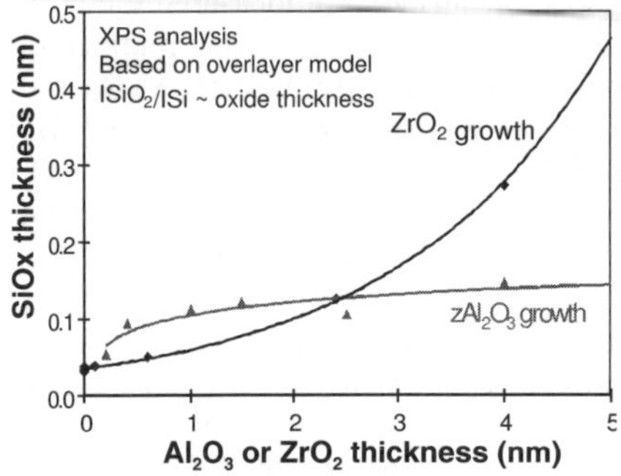

Figure 6 Growth of interfacial SiO_x layer during ALCVD (XPS result)

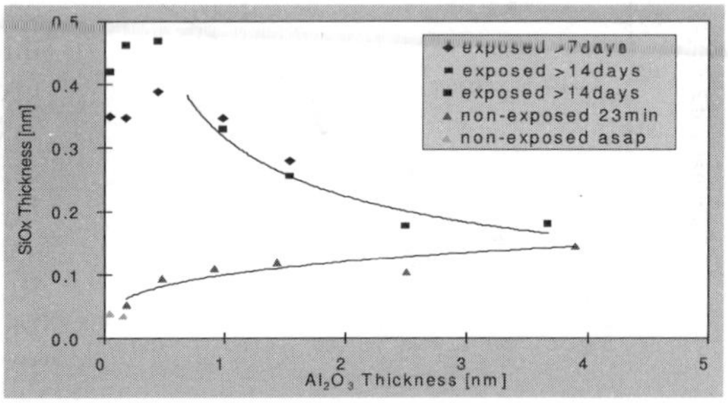

Figure 7 Interfacial SiO_x layer growth during air exposure of Al_2O_3 films

illustrated in Fig. 5: obviously the growth starts by formation of small nuclei of ZrO_2 on the Si surface. Growth takes place preferentially on these nuclei, such that islands of ZrO_2 are formed[18] which ultimately coalesce for a film thickness of more than 3 nm. This brings about two other problems. First, during film growth, the Si surface becomes oxidized due to the unhindered transport of oxidizing species through the holes which exist in the film, and secondly, it is very likely that some pinholes and other weak spots persist even after film coalescence, which will give problems with reliability of the devices (see further). By means of XPS, we have seen indeed the formation of interfacial SiO_x layers during the film growth, see Fig. 6. Moreover, also after taking out the wafers with a high-k layer from the cluster tool, this interfacial layer grows further upon air exposure, as illustrated in Fig. 7. It is obvious, that this behavior makes EOT control not straightforward. Moreover, although one might be able to produce a device which shows low leakage due to the formation of SiO_x in the pinholes, this will definitely result in a yield and reliability problem as these pinholes present weak spots which will break down at higher gate voltages.

4.2 Thermal stability of high-k layers

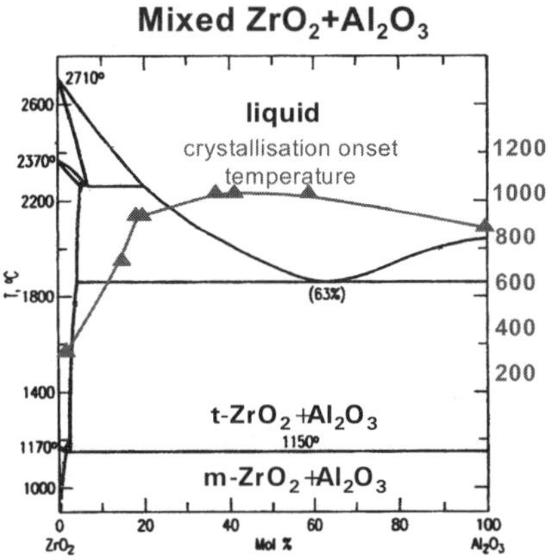

Figure 8 Phase diagram of Zr Aluminate and associated recrystallisation temperatures

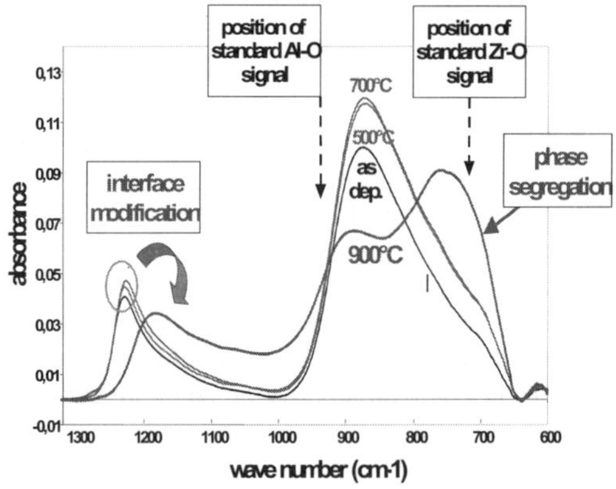

Figure 9 FTIR spectrum of Zr Aluminate layer before and after annealing

During subsequent processing, high-k layers must withstand thermal anneals at high temperatures up to 1000 °C (implant anneals) during which they should not re-crystallize. Single oxide layers of ZrO_2 and HfO_2 are known to re-crystallize at fairly low temperatures (500 – 600 °C), depending on layer thickness. As expected, mixtures of HfO_2 or ZrO_2 with Al_2O_3 are found to re-crystallize only at much higher temperature, up to 1000 °C see Fig. 8.[15] However, FTIR studies indicated a less desirable phenomenon, i.e. phase separation at temperatures of only 900 °C (Fig. 9). This could become a serious problem with respect to reliability of the completed device, see 5.1.

4.3 Interaction of poly-Si with the underlying high-k material

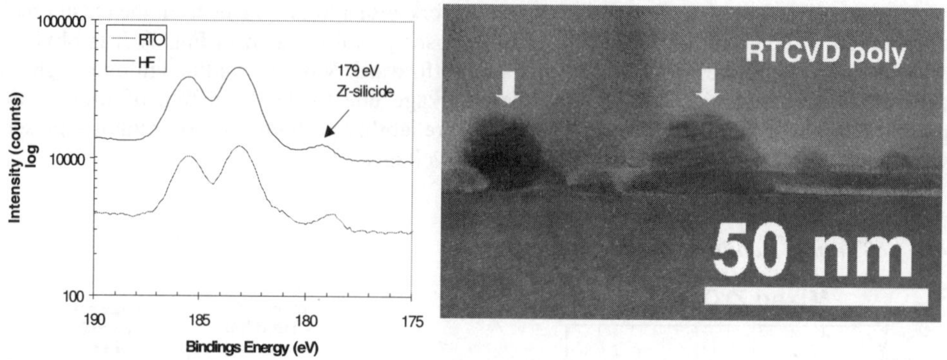

Figure 10 XPS and TEM study of poly-Si/High-k interaction on ZrO_2

On top of the high-k layer, an electrode of poly-Si needs to be grown. After a lot of initial difficulties, we managed to obtain optically smooth layers of in-situ doped poly-Si. However, cross-sectional TEM images revealed the formation of small and big nodules throughout the high-k layer, which were identified by XPS as Zr silicide material, see Fig. 10. I-V measurements on capacitors in similar layers indicated huge gate leakage. Although poly-Si growth on HfO_2 apparently presents no problems, and almost no silicide formation has been noticed in XPS, still important problems have been seen with broken capacitors on HfO_2 /poly-Si layers, probably also due to interaction between the two materials. Also this problem will require a lot of attention.

5 Electrical Behavior Related Issues

5.1 Reliability of high-k devices

In order to become accepted in CMOS production lines, high-k containing transistors should show reliable operation over many years without stress related break-down or other gate leakage problems. A quantitative evaluation of the reliability of the dielectric layer can be done by means

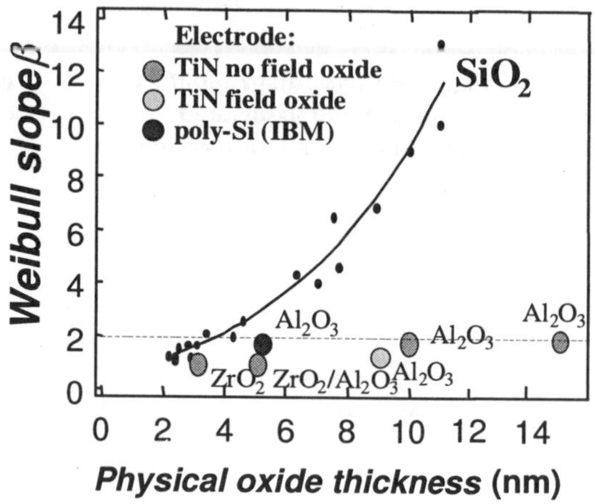

Figure 11 Reliability of SiO_2 and various high-k combinations

of time-to-breakdown measurements under constant voltage stress of area scaled capacitors. The statistical distribution of the time-to-breakdown is known as the Weibull distribution. The Weibull slope shows a strong dependence on layer thickness for thermal SiO_2. This can be understood by means of the percolation model: under electrical stress, a chain of defect related traps becomes formed which short-circuits anode and cathode. We have found that various high-k layers and layer combinations of Al_2O_3 and ZrO_2 show very low Weibull slopes as compared to SiO_2, see Fig. 11.[17] Moreover, there is no thickness dependence. These observations indicate, that an extrinsic breakdown mechanism is operational, probably due to process related weak spots in the gate stack.

5.2 *Carrier mobility in high-k devices*

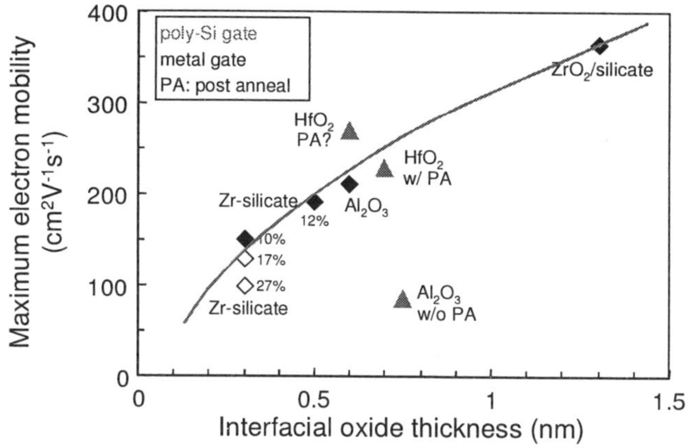

Figure 12 Influence of interfacial oxide layers on mobility

CV and I-V measurements on high-k capacitors are used by many authors to show the promising characteristics of those new materials. Recent measurements however of mobility on MOSFET transistors at Imec as well as in other groups have indicated that electron mobility underneath high-k layers quite often is strongly reduced as compared to SiO_2. A summary of recent data published at IEDM 2001 (Fig. 12) indicates that this mobility is strongly influenced by the presence and characteristics of an interfacial layer, either SiO_2 or silicate-like. According to an analysis by Fischetti et al., this mobility reduction is due to phonon scattering and thus, related to the specific material which is used.[20] The only apparent solution to this problem might be screening of the high-k material by some interface layer as suggested by Fig. 12. This, however, could become a show-stopper in view of the final aim of very low EOT values.

6 Conclusion

A world-wide research effort to find valuable candidate high-k materials and process flows to replace SiO_2 as gate dielectric has shown very promising results so far. However, it is important to realize that there are still several problem areas around, which are not always obvious. A lot of work still has to be spent on identifying and solving the materials and electrical characteristics related issues which possibly could act as real show-stoppers.

Acknowledgements

The authors are strongly indebted to the teams of ASM Belgium and AMAT working at Imec for their contribution in layer deposition. They also gratefully acknowledge the financial support by International Sematech and Hitachi in the frame of the Imec Industrial Affiliation Program on High-k Gate stacks, as well as by the Flemish IWT in the frame of the Medea+ Research Program Alad1n T302

References

1. M.L. Green, E.P. Gusev, R. Degrave and E. Garfunkel, J. Appl. Phys. 90 (2001) 2057
2. G.D. Wilk, R.M. Wallace and J.M. Anthony, J.Appl. Phys. 89 (2001) 5243
3. T. Suntola. *Handbook of thin Film Process Technology.* IOP Publishing Ltd 1995 ch B1.5 "Atomic Layer Epitaxy"
4. M. Tuominen, T. Kanniainen and S. Haukka, "High-k Oxides by Atomic Layer Chemical Vapour Deposition" presented at Rapid Thermal and Other Short-Time Processing Technologies, 197th Meeting of the Electrochemical Society, Toronto, May 14-18, 2000
5. W.J. Qi, R. Nieh, B.H. Lee, L. Kang, Y. Jeon. K. Onishi, T. Ngai, S. Banerjee and J. Lee, IEDM p. 145, 1999
6. A. Chin, Y.H. Wu, S.B. Chen, C.C. Liao and W.J. Chen, VLSI Symposium 2000.
7. M. Copel, M. Gribelyuk and E. Gusev, *Appl. Phys. Lett.* 76, 436-438, 2000
8. G. D. Wilk, R. M. Wallace and J. M. Anthony, J. Appl. Phys. 89, 5243 (2001)
9. B. Rayner, R. Therrien and G. Lucovsky, 2000 MRS Spring Meeting, April 24-28, San Francisco, California, USA
10. A. Chin et al., MRS Workshop on High-k Gate Dielectrics, June 1-2 2000.
11. C. H. Lee et al, MRS Workshop on High-k Gate Dielectrics, June 1-2 2000
12. E.P. Gusev et al. IEDM 2001
13. Y. Kim et al. IEDM 2001
14. O. Richard, H. Bender, M. Houssa, C. Zhao; Inst. Phys. Conf. Ser., in press (2001). Twelfth International Conference on Microscopy of Semiconducting Materials, Oxford, 25-29 March 2001
15. C. Zhao, O. Richard, H. Bender, M. Houssa, R. Carter, S. De Gendt and M. Heyns, E. Young , W. Tsai , G. Roebben and O. Van Der Biest, S.Haukka; European-MRS 2001, Strasboug, France, June 5-8, 2001
16. Bender H., T. Conard, H. Nohira, J. Petry, O. Richard, C. Zhao, B. Brijs, W. Besling, C. Detavernier, W. Vandervorst, M. Caymax, S. De Gendt, J. Chen, J. Kluth, W. Tsai and J.W. Maes; Invited paper at the International Workshop on Gate Insulator, Nov. 1-2 2001, Japan
17. R. J. Carter, E. Cartier, M., S. De Gendt, R. Degraeve, G. Groeseneken, M. Heyns, T. Kauerauf, A. Kerber, S. Kubicek, G. Lujan, L. Pantisano, W. Tsai, E. Young, Invited paper at the International Workshop on Gate Insulator, Nov. 1-2 2001, Japan
18. M. Copel, M. Gribelyuk and E. Gusev, *Appl. Phys. Lett.*, vol. 76, no. 4, p. 436, 2000
19. J.R. Hauser, CVC ©2000 NCSU software, version 5.0, Dept. Electr. Comput. Eng., North Carolina State University, Raleigh, NC
20. M. Fischetti , E. Cartier, D. Neumayer, JAP, Nov. 1, 2001

International Journal of High Speed Electronics and Systems, Vol. 12, No. 2 (2002) 305–314

MATERIALS FOR STRAINED SILICON DEVICES

P.M. MOONEY
IBM Research Division, T.J. Watson Research Center
PO Box 218, Yorktown Heights, NY 10549

Strained Si devices exhibit enhanced carrier mobility compared to that of standard Si CMOS devices of the same dimensions. Recent strained Si CMOS device results are reviewed. Materials issues related to the strained Si/relaxed SiGe heterostructures required for a strained Si CMOS technology are discussed.

1. Introduction

The amazing performance improvements in CMOS circuits during the last 40 years are primarily a result of reductions in the dimensions of the individual transistors by orders of magnitude enabled by advances in photolithography. The smaller device size permits a much higher device density that has, for example, resulted in logic chips with clock speeds greater than 1 GHz. As device dimensions approach <100 nm, scaling becomes increasingly difficult. Therefore, new materials that improve circuit performance cost are of interest. Examples of new materials that have been implemented recently are silicon-on-insulator (SOI) wafers, consisting of a thin Si layer separated from the bulk Si substrate by a thin layer of SiO_2 and Cu interconnections.

Improved performance due to increased carrier mobility has been reported for CMOS devices fabricated on a strained Si layer.[1-9] Theoretical calculations predict higher mobility for both electrons and holes in silicon under tensile strain.[10-14] Tensile strain removes the 6-fold degeneracy in the conduction band, lowering the energy of the two λ valleys having electrons with a lower in-plane effective mass with respect to the four λ valleys having electrons with a higher in-plane effective mass. This energy splitting reduces intervalley scattering and preferential occupation of the two λ valleys with lower in-plane effective mass electrons results in higher electron mobility. Tensile strain also removes the valence band degeneracy at the Gamma-point and shifts the spin-orbit band,[1-5] thus improving the hole mobility. Typically a Si layer under tensile strain is achieved by epitaxial growth of Si on a strain-relaxed SiGe buffer layer on a Si substrate. The relaxed SiGe buffer layer serves as a "virtual substrate" for these devices. The electron mobility increases with increasing Ge mole fraction in the SiGe buffer layer, saturating at about 20% Ge.[11] The hole mobility increases more slowly and saturates at about 35% Ge.[14] Therefore structures consisting of strained Si on a relaxed SiGe buffer layer containing 15-35% Ge are interesting for CMOS applications.

Fig. 1 shows a cross section of a SSCMOS device fabricated on a 200 mm wafer using a CMOS fabrication process that was modified to accommodate the SiGe buffer layer.[7] The electron mobility measured on large area, long channel devices on SiGe buffer layers having 15 and 20 % Ge is shown in Fig. 2. The mobility of the strained Si device is about 70% higher than the universal MOSFET mobility than the mobility of control Si devices and it increases with increasing Ge content in the buffer layer as expected.[7] The mobility enhancement, which persists at high vertical field in the range

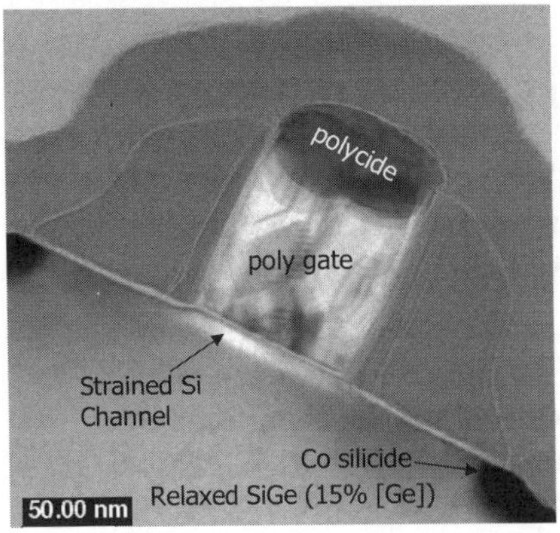

Fig. 1. Cross sectional TEM image of a fabricated strained Si MOSFET.[7]

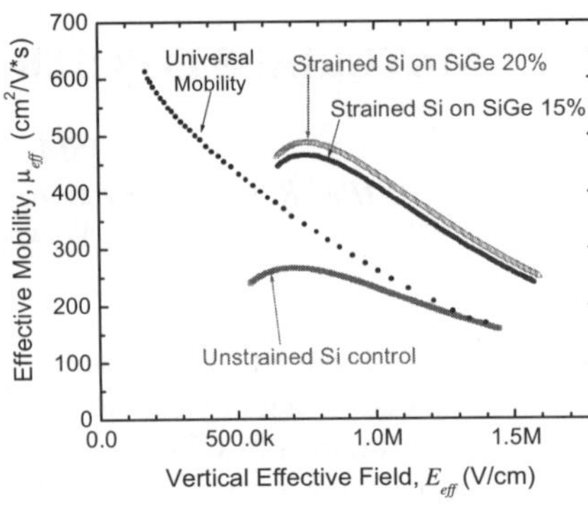

Fig. 2. Effective electron mobility of strained Si MOSFETs.[7]

required for future CMOS generations, cannot be explained entirely by the band structure modifications, indicting that the tensile strain may also reduce the so-called surface roughness scattering.[7,11] This mobility enhancement is very exciting. However, there are still many questions related to process integration and materials issues to be addressed. In this paper recent work on the strained Si/relaxed SiGe structures required for strained Si devices are discussed.

2. Strain-relaxed SiGe buffer layers on Si (001)

An essential feature of strained Si CMOS device structures is the strain-relaxed SiGe buffer layer or virtual substrate. The alloy composition and degree of strain relaxation of the SiGe buffer layer determines the magnitude of the strain in the pseudomorphic Si cap layer. Moreover, the defect density in the upper part of the buffer layer must be sufficiently low to achieve the necessary yield of devices with good characteristics. When the epitaxial strained layer exceeds a critical thickness, strain relaxation occurs in $Si_{1-x}Ge_x$/Si(001) having $x<0.5$ by the introduction of 60° misfit dislocations near the SiGe/Si interface. Since Si(001) substrates are essentially dislocation free, the critical thickness in this case is that needed for dislocation nucleation; the critical thickness decreases as the Ge mole fraction of the SiGe layer is increased and also depends on the growth temperature.[16-18] A relatively thick SiGe layer may remain fully strained (pseudomorphic) when grown at low temperature; however, strain relaxation occurs when a metastable structure is subsequently heated, e.g. during device fabrication.

Since dislocations must terminate at a free surface, each 60° misfit dislocation nucleated at the initial growth interface or in the epitaxial layer ends with two threading arms that go through the SiGe layer to the wafer surface. Relatively low threading dislocation densities have been achieved by means of compositional grading. The graded buffer typically consists of a compositionally graded SiGe layer followed by a uniform composition SiGe layer. The alloy composition may be linearly graded or step-wise graded. Threading dislocation densities in the range 10^5-10^8 cm^{-2} have been reported, with the lowest densities occurring for grading rates as slow as 10% Ge per micron of layer thickness.[19] Thus graded $Si_{1-x}Ge_x$ buffer layers for strained Si CMOS applications may be as thick as 2-4 ℓm. Because SiGe has a lower thermal conductivity than Si, self-heating effects are observed in strained Si CMOS devices fabricated on step-graded buffer layers.[7] Graded buffer layers also have relatively rough surfaces; a cross hatch surface pattern is created by the surface steps associated with the 60° misfit dislocations. The surface of the relaxed SiGe buffer layer can be planarized prior to the growth of the active device layers using chemical-mechanical polishing methods. However, the polishing step adds to the cost of fabrication, as does the growth of such a thick SiGe buffer layer.

For these reasons, alternative methods to achieve a much thinner relaxed SiGe buffer layer are under investigation.[20-24] One approach, using molecular beam epitaxy (MBE), employs a special low temperature (200 °C) growth step.[20] Another approach, also by MBE, uses plasma cleaning prior to growth.[21] Ion-implantation of hydrogen or helium has been used to create dislocation nucleation sources and thus enhance strain relaxation in thin SiGe/Si structures.[22-24] In this method, a pseudomorphic SiGe layer is first grown on Si(001) at low temperature. The SiGe layer must be thin enough that no strain relaxation occurs during the growth, but thick enough that it exceeds the critical thickness for relaxation at higher temperatures. Hydrogen or helium is then implanted at or below the SiGe/Si interface. After annealing, spherical defects (bubbles) were observed at or below the SiGe/Si interface as well as a dense network of misfit dislocations that relieve the strain in the SiGe layer.

We have investigated strain relaxation mechanisms in He-implanted SiGe/Si structures.[25] Depending on the He dose and implantation energy, i.e. the depth of the He atoms below the SiGe/Si interface, three different strain relaxation mechanisms were observed, as shown in Fig. 3. For some implant conditions, we observed that platelets are formed upon annealing (group 2 samples) rather than bubbles (groups 1 and 3) as reported previously. Dislocations nucleate at the platelets by a prismatic punching mechanism as seen in Fig. 4, and a dense regular misfit dislocation network is formed at the SiGe/Si interface as shown in Fig. 5. Dislocations that terminate at platelets or bubbles below the SiGe/Si interface do not have threading arms running through the SiGe layer. Therefore implanted/annealed buffer layers are expected to have low threading dislocation densities. Threading dislocation densities comparable to those found for graded buffer layers have been reported.[22-24]

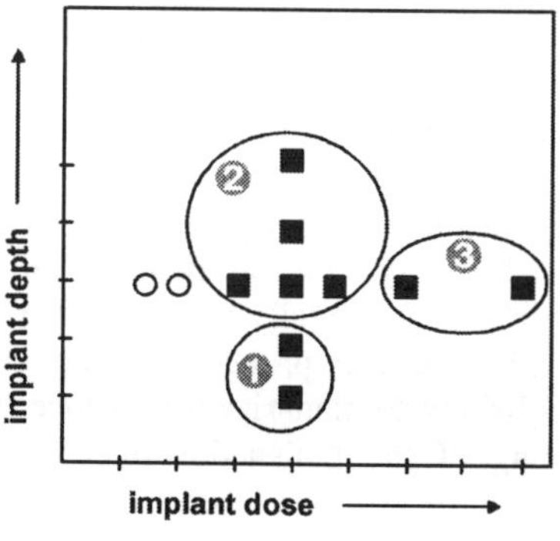

Fig. 3. Plot showing strain relaxation at various implant doses and depths below the SiGe/Si interface for both $Si_{0.85}Ge_{0.15}$ and $Si_{0.80}Ge_{0.20}$ samples: circles indicate layers that relaxed only slightly (<10%), squares indicate samples having substantial (30-80%) relaxation. The more relaxed samples are grouped according to the strain relaxation mechanism.[25]

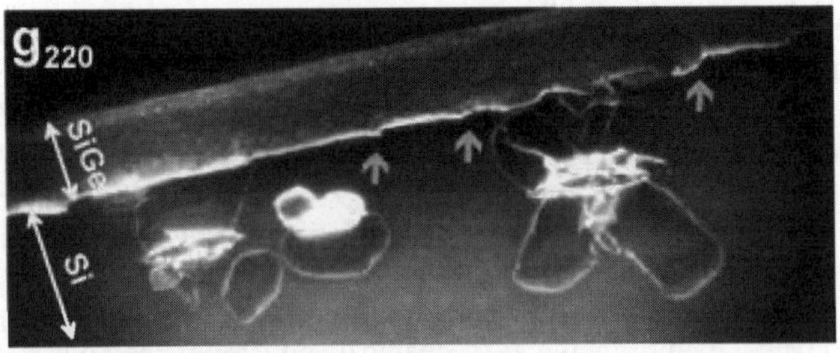

Fig. 4. Cross-sectional TEM micrograph (weak beam conditions) showing platelets below the SiGe/Si interface. Dislocation loops extend from the platelets in all eight <110> directions and glide to form misfit dislocation segments at the interface as indicated by arrows.[25]

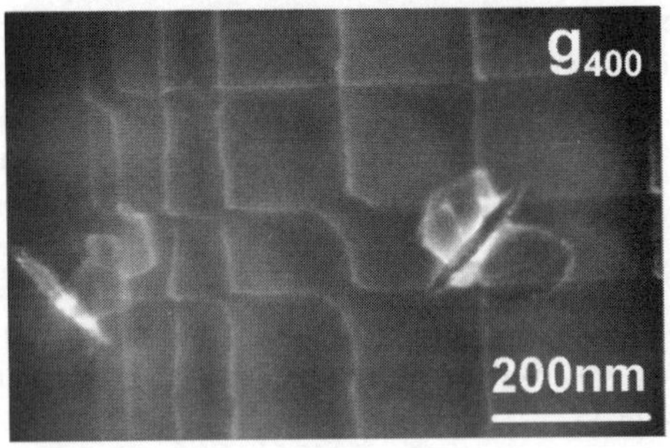

Fig. 5. Planar view TEM micrograph (weak beam conditions) showing platelets along the <100> directions below the SiGe/Si interface with a regular array of 60^0 misfit dislocations, having line directions parallel to the <110>-directions at the interface.[25]

3. SiGe-on-insulator substrates

The advantages of strained Si can be combined with the advantages of SOI substrates by fabricating SiGe-on-insulator (SGOI) substrates. Several different methods to fabricate SGOI using a relaxed SiGe buffer layer on Si(001) have been implemented. One group fabricated SGOI substrates by means of a SIMOX-like process, i.e. implantation of oxygen below the surface of the SiGe buffer layer and then annealing at high temperature (>1300 °C) to form a layer of SiO_2.[26,27] Apparently this method works only for SiGe buffer layers having low (~10%) Ge. Another approach uses dry oxidation at high temperature after deposition of a pseudomorphic SiGe layer with low Ge content.[28] During oxidation, Ge atoms are rejected from the oxide and condensed in the remaining SiGe layer, which is partially relaxed without introducing a significant number of dislocations. Raman spectroscopy shows that a strained Si cap layer grown on such a structure has about 1% tensile strain.

Wafer bonding and layer transfer is a very promising method to fabricate SGOI substrates with higher Ge content.[8,29,30] The fabrication steps are shown schematically in Fig. 6; and involve implantation of hydrogen into the relaxed SiGe buffer layer, then bonding the implanted wafer to a Si wafer with an SiO_2 layer on top, then annealing to split the SiGe layer and strengthen the bonded interface, and finally epitaxial growth of the strained Si device layers on the bonded SGOI substrate. Key process steps not shown explicitly in Fig. 6 include chemical-mechanical polishing to planarize the surface of the relaxed SiGe buffer layer and subsequent cleaning of the polished surface prior to bonding and also after layer splitting before growing the device layers. Temperature

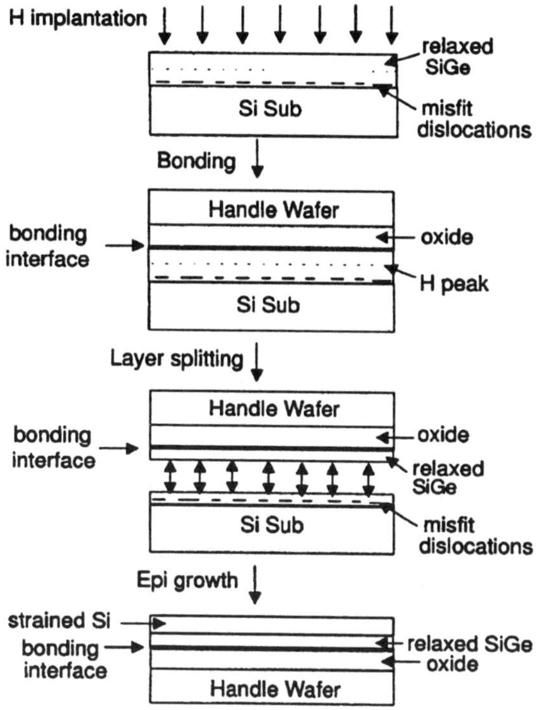

Fig. 6. Diagram showing fabrication steps for bonded SGOI substrates.[29]

dependent Hall effect measurements performed on modulation-doped structures grown simultaneously on a polished SiGe buffer layer and on a polished SGOI substrate having the same SiGe alloy composition show the presence of a two dimensional electron gas and identical electron mobility, ~2000 cm2/V-s at 300 K.[29] This indicates that the quality of the SiGe layer on the bonded SGOI wafer is the same as that of the relaxed SiGe buffer layer used to fabricate the SGOI wafer.

The effective electron mobility in strained Si MOSFETS fabricated on a bonded SGOI substrate having 15% Ge is shown in Fig. 7. An enhancement of about 50% is observed,[8] comparable to the results for strained Si MOSFETs on thick relaxed SiGe buffer layers on bulk Si substrates[7] Fig. 8 shows the effective hole mobility in MOSFETs fabricated on a bonded SGOI substrates having 25% Ge. This work extends earlier results[6] to higher effective vertical field. Hole mobility enhancements of 15-20% were found at low vertical field:[8] at high vertical field, however, there is negligible mobility enhancement, suggesting that SGOI wafers with higher Ge content are desirable.

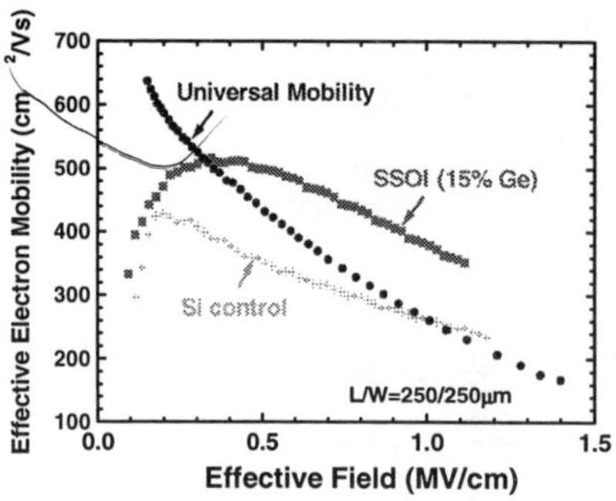

Fig. 7. Electron mobility enhancement in strained Si MOSFETs on a bonded SGOI substrate.[8]

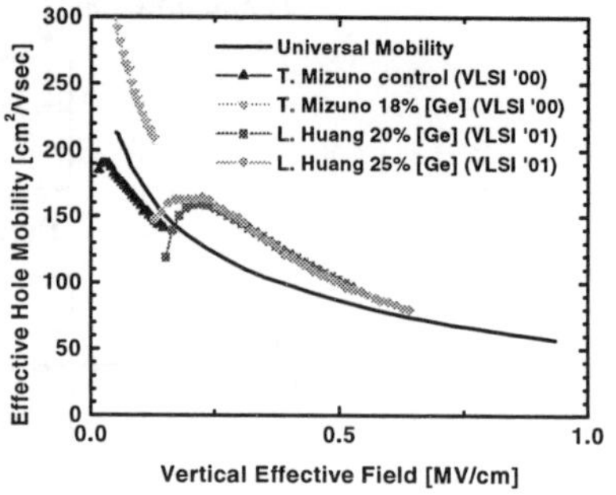

Fig. 8. Hole mobility enhancement in strained Si MOSFETs on a bonded SGOI substrate.[8]

4. Thermal stability of strained Si structures

Strained Si CMOS devices, implemented by the epitaxial growth of a 10-30 nm-thick strained Si layer on a relaxed $Si_{1-x}Ge_x$ buffer layer on Si(001), are typically grown by a variety of epitaxial methods at relatively low temperatures. However, device fabrication includes standard processes, e.g. the activation of ion-implanted dopant atoms and gate oxidation, requiring temperatures as high as 1000 °C. When threading dislocations are present in the substrate, an epitaxial strained layer that exceeds the critical thickness for dislocation glide will relax by the formation of misfit dislocations at the Si/SiGe interface.[31] The rate of misfit dislocation formation increases exponentially with temperature. Interdiffusion at the Si/SiGe interface also occurs at this temperature, resulting in a graded interface and effectively reducing the thickness of the strained Si layer. We have investigated these effects by annealing $Si/Si_{1-x}Ge_x$ structures, consisting of a strained Si cap layer of varying thickness on a strain-relaxed step-graded $Si_{1-x}Ge_x$ buffer layer of varying $Si_{1-x}Ge_x$ alloy concentration, at 1000 °C for 5, 30 and 300 seconds, times which far exceed those normally used for Si CMOS fabrication.[32] The goal of this work was to measure the magnitude of the strain relaxation of the Si layer and the interdiffusion that occurs at the Si/SiGe interface during annealing.

High-resolution x-ray diffraction (HRXRD) was used to determine both the thickness and strain of the Si cap layer. X-ray diffraction measurements were performed at beamline X20 at the National Synchrotron Light Source at Brookhaven National Laboratory using a triple-axis configuration as described previously.[33] The composition and strain of the uniform $Si_{1-x}Ge_x$ layer were determined from the 004 and 224 (grazing exit) reflections. The 004 reflection measures the crystal lattice spacing in the direction of the surface normal, i.e. the out-of-plane lattice spacing. The 224 reflection measures the spacing of lattice planes at an angle to the surface normal, i.e. a set of planes whose lattice spacing has both an out-of-plane and an in-plane component. The composition and the strain are calculated from the angular separation between the SiGe layer and the Si substrate peaks in the two measurements.[33]

Fig. 9 shows the 004 data for a 21 nm-thick Si cap layer on a $Si_{0.72}Ge_{0.28}$ buffer layer taken at two different regions, one with and one without the strained Si cap layer. Comparing the two scans, the features to the right of the Si substrate peak are from the strained Si cap layer. Although the Si layer thickness can be determined from the solid curve in Fig. 9, comparison with the simulation is more precise when the two scans are subtracted. The difference of the two scans in Fig. 9 is shown in Fig. 10, along with a simulated diffraction scan from RADS.[34] The thickness of the Si cap layer determines the spacing of the oscillations and the strain determines their position with respect to the Si substrate peak. The shift of the experimental data from the Si cap layer with respect to the Si substrate peak is clearly visible. With this method of analysis the strain in the Si cap layer can be determined to within 1 or 2%, depending on the sample.

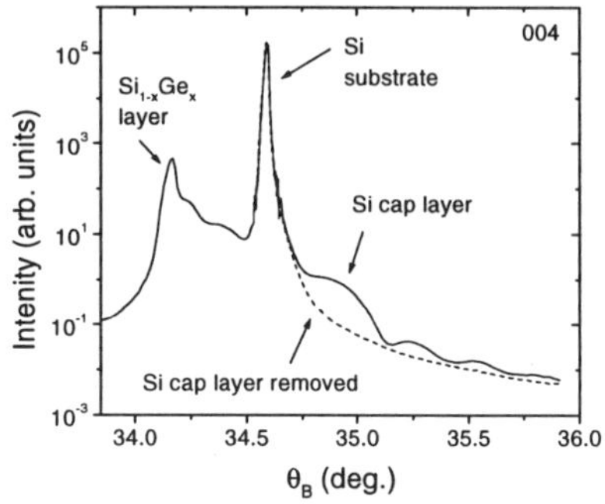

Fig. 9. 004 x-ray scan taken at regions with and without the 21nm-thick strained Si cap layer on $Si_{0.72}Ge_{0.28}$.[32]

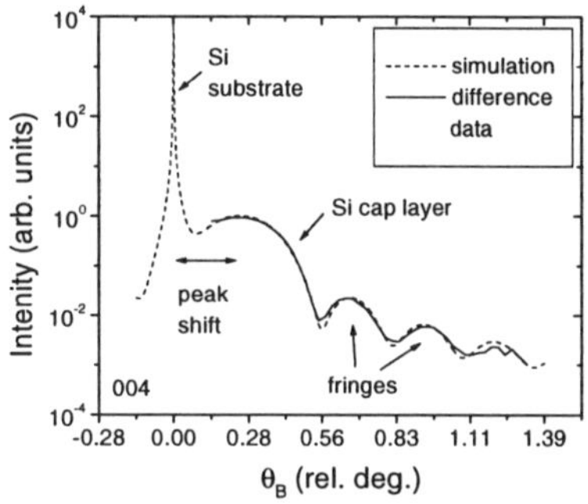

Fig. 10. The difference of the two x-ray scans shown in Fig. 8 with the simulated x-ray data.[32]

As was expected from previous Raman spectroscopy measurements on the same samples,[35] x-ray diffraction results show that the thickness of the Si cap layer decreases with the annealing time (see Fig. 11). The change in the Si cap layer thickness showed no dependence on the initial thickness of the Si cap layer or the alloy composition of the SiGe layer over the ranges investigated.[32] Since these samples were annealed in an inert atmosphere, the most likely explanation is that interdiffusion occurs at the $Si/Si_{1-x}Ge_x$ interface, effectively reducing the thickness of the pure Si layer. These changes in thickness are consistent with experimental values for the diffusion coefficients reported for Ge in Si.[36-37] Interdiffusion results in a graded interface rather than the abrupt one achieved during epitaxial growth at low temperature. This is an advantage for the pMOSFETs, since a graded interface reduces the probability of parallel hole conduction in the top of the relaxed $Si_{1-x}Ge_x$ buffer layer.[3]

Fig. 12 shows the time dependence of the strain relaxation for two samples. The Si cap layer is 0% relaxed when the in-plane lattice parameter matches that of the underlying $Si_{1-x}Ge_x$ layer and it is fully relaxed when the in-plane lattice parameter has the value of bulk Si. Even in the worst case, very little strain relaxation occurs (~10%), consistent with previous Raman scattering measurements on the same samples.[35] The trends observed are exactly as expected; greater relaxation occurs for higher Ge mole fraction and for thicker Si layers. The x-ray results agree well with the strain relaxation calculated from the misfit dislocation density determined from planar view TEM images.[32] TEM images also clearly show that the misfit segments terminate at threading dislocations in the upper part of the $Si_{1-x}Ge_x$ buffer layer, indicating that strain relaxation occurs by the glide of threading dislocations in the relaxed $Si_{1-x}Ge_x$ layer.

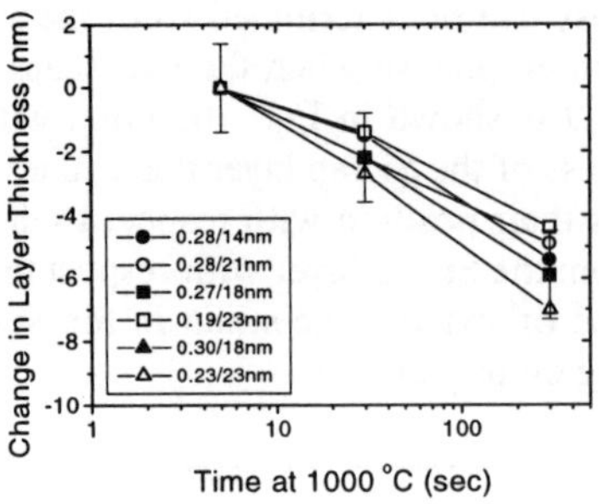

Fig. 11. Change in strained Si layer thickness with annealing time in samples having initial Si layer thickness and $Si_{1-x}Ge_x$ alloy Composition indicated.[32]

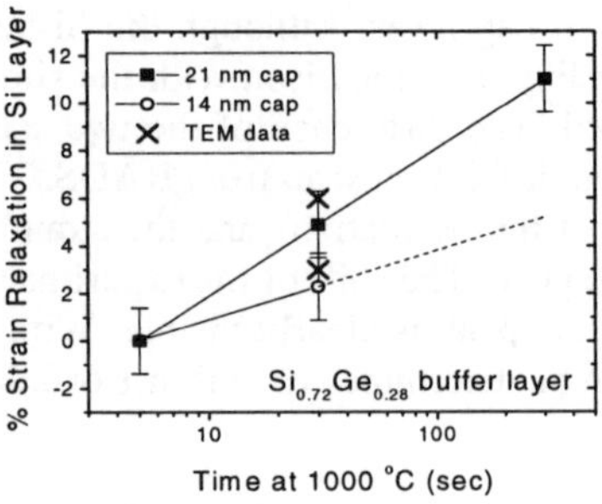

Fig. 12. % strain relaxation as a function of annealing time. Strain relaxation calculated from the misfit dislocation density from planar view TEM images is also plotted.[32]

5. Conclusions

Strained Si CMOS devices fabricated on 200 mm wafers using standard CMOS fabrication processes exhibit increased carrier mobility due primarily to the modified band structure in Si under tensile strain. These devices were fabricated in a strained Si layer grown epitaxially on a strain-relaxed SiGe layer that acts as a virtual substrate. The relaxed SiGe layer may also be the thin SiGe layer of an SGOI substrate. No matter how the virtual substrate is fabricated, the defect density in the relaxed SiGe buffer layer must be sufficiently low to achieve device yields required for the desired applications of strained Si CMOS technology, e.g., ULSI logic circuits. To avoid the formation of misfit dislocations at the strained Si/SiGe interface during device fabrication, the strained Si/SiGe heterostructures must be thermodynamically stable, i.e., the strained Si layer thickness must be less than the critical thickness for dislocation glide. If metastable heterostructures are used, device fabrication process conditions must be restricted to temperatures and times for which dislocation glide is negligible.

Acknowledgements

The contributions of many IBM colleagues to the work discussed in this paper are gratefully acknowledged. The work at NSLS is partially supported by the Dept. of Energy, Materials Science and Chemical Science Divisions.

References

1. J. Welser, J.L. Hoyt, S. Takagi and J.F. Gibbons, "Strain Dependence of the Performance Enhancement in Strained-Si n-MOSFETs", IEDM Tech Dig. (1994) 373-6.
2. D.K. Nayak, K. Goto, A. Yutani, J. Murota and Y. Shiraki, "High Mobility Strained Si pMOSFETs", IEEE Trans. Elect. Dev. **43** (1996) 1709-16.

3. K. Rim, J. Welser, J.L. Hoyt and J.F. Gibbons, "Enhanced Hole Mobilities in Surface Channel Strained Si p-MOSFETs", IEDM Tech Dig. (1995) 517-20.
4. K. Rim, J.L. Hoyt and J.F. Gibbons, "Transconductance Enhancement in Deep Submicron Strained-Si n-MOSFETs", IEDM Tech. Dig. (1998) 707-10.
5. T. Mizuno, S. Takagi, N. Sugiyama, J. Koga, T. Tezuka, K. Usuda, T. Hatakeyama, A. Kurobe and A. Toriumi, "High Performance Strained Si pMOSFETs on SiGe on Insulator Substrates Fabricated by SIMOX Technology", IEDM Tech. Dig. (1999) 934-6.
6. T. Mizuno, N. Sugiyama, H. Satake, S. Takagi, "Advanced SOI MOSFETs with Strained Si Channel for High Speed CMOS Electron/Hole Mobility Enhancement", Symposium on VLSI Technology Dig. (2000) 210-11.
7. K. Rim, S. Koester, M. Hargrove, J. Chu, P.M. Mooney, J. Ott, T. Kanarsky, P. Ronsheim, M. Ieong, A. Grill, H.-S.P. Wong, "Strained Si NMOSFETs for High Performance CMOS Technology" Symposium on VLSI Technology Dig. (2001) 59-60.
8. L.J. Huang, J.O. Chu, S.A. Goma, C.P. D'Emic, S.J. Koester, D.F. Canaperi, P.M. Mooney, S.A. Cordes, J.L. Speidell, R.M. Anderson, H.-S.P. Wong, "Carrier Mobility Enhancement in Strained Si-on-Insulator Fabricated by Wafer Bonding", Symposium on VLSI Technology Dig. (2001) 57-58.
9. N. Sugii, D. Hisamoto, K. Washio, N. Yokoyama and S. Kimura, "Enhanced Performance of Strained-Si MOSFETs on CMP SiGe Virtual Substrate", IEDM Tech. Dig. (2001) 737-40.
10. M.V. Fischetti and S.E. Laux, "Band Structure, Deformation Potentials, and Carrier Mobility in Strained Si, Ge, and SiGe Alloys", J. Appl. Phys. **80** (1996) 2234-52.
11. S. Takagi, J.L. Hoyt, J.J. Welser, and J.F. Gibbons, "Comparative Study of Phonon-Limited Mobility of Two-Dimensional Electrons in Strained and Unstrained Si Metal–Oxide–Semiconductor Field-Effect Transistors", J. Appl. Phys. **80** (1996) 1567-77.
12. J.B. Roldán, F. Gámiz, J.A. López-Villanueva, and J.E. Carceller, "A Monte Carlo Study on the Electron-Transport Properties of High-Performance Strained-Si on Relaxed $Si_{1-x}Ge_x$ Channel MOSFETs", J. Appl. Phys. **80** (1996) 5121-28.
13. G. Formicone, D. Vasileska and D.K. Ferry, "Transport in the Surface Channel of Strained Si on a Relaxed $Si_{1-x}Ge_x$ Substrate", Solid State Elec. **41** (1997) 879-85.
14. R. Oberhuber, G. Zandler and P. Vogel, "Subband Structure and Mobility of Two-Dimensional Holes in Strained Si/SiGe MOSFETs", Phys. Rev. B **58** (1998) 9941-8.
15. D.K. Nayak, J.C.S. Woo, J.S. Park, K.L. Wang and K.P. MacWilliams, "High-mobility p-channel metal-oxide-semiconductor field-effect transistor on strained Si", Appl. Phys. Lett. **62** (1993) 2853-5.
16. J.W. Matthews, S. Mader, and T.B. Light, "Accomodation of Misfit Across the Interface Bewteen Crystals of Semiconduction Elements or Compounds" J. Appl. Phys. 70 (1970) 3800-04; J.W. Matthews, A.E. Blakeslee and S. Mader, "Use of Misfit Strain to Remove Dislocations From Epitaxial Thin Films", Thin Solid Films **33** (1976) 253-66.
17. J.C. Bean, L.C. Feldman, A.T. Fiory, S. Nakahara and I.K. Robinson, "Ge_xSi_{1-x}/Si Strained-Layer Superlattice Grown by Molecular Beam Epitaxy", J. Vac. Sci. Technol. A **2** (1984) 436-440; R. Hull, "Metastable Strained Layer Configurations in the SiGe/Si System" in *Silicon Germanium and SiGe:Carbon,* EMS Datareviews Series No. 24, INSPEC, London (2002) 21-41.
18. S.R. Stiffler, J.H. Comfort, C.L. Stanis, D.L. Harame and E. de Fresart, "The Thermal Stability of SiGe Films Deposited by Ultra-High Chemical Vacuum Vapor Deposition", J. Appl. Phys. **70** (1991) 1416-20.
19. P.M. Mooney, "Strain Relaxation and Dislocations in SiGe/Si Structures", Materials Science and Engineering Reports, **R17** (1996) 105-146, and references therein.
20. M. Bauer, K. Lytovich, M. Oehme, E. Kasper, H.J. Herzog, F. Ernst, "Relaxed SiGe Buffers with Thickness below 0.1 Micometers", Thin Solid Films **369** (2000) 152-156.
21. J. Kuchenbecker, H. Kibbel, P. Muthsam, U Konig, "Thin SiGe Buffer Layer Growth by In Situ Low Energy Hydrogen Plasma Preparation", Thin Solid Films **389** (2001) 146-152.

22. D.M. Follstaedt, S.M. Meyers, S.R. Lee, "Cavity–Dislocation Interactions in Si–Ge and Implications for Heterostructure Relaxation", Appl. Phys. Lett. **69** (1996) 2059-61.
23. S. Mantl, B. Hollaender, R. Leidtke, S. Mesters, H.J. Herzog, H. Kibble, T. Hackbarth, "Strain Relaxation of Epitaxial Layers on Si(001) Improved by Hydrogen Implantation", Nucl. Instr. and Meth. Phys. Res. B **147** (1999) 29-34.
24. M. Luysberg, D. Kirch, H, Trinkhaus, B. Hollaender, S. Lenk, S. Mantl, H.J. Herzog, T. Hackbarth, P.F. Fichtner, "Relaxation of SiGe Buffer Layers on Si(100) Through Helium Ion Implantation", *Microscopy in Semiconducting Materials*, Institute of Physics Publishing, Oxford (in press).
25. S.H. Christiansen, P.M. Mooney, J.O. Chu and A. Grill, "Strain Relaxation Mechanisms in He^+-Implanted and Annealed SiGe Layers on Si(001) Substrates", in *Materials Issues in Novel Si-Based Technology,* Mat. Res. Soc. Symp. Proc. **686** (in press).
26. N. Sugiyama, T. Mizuno, S. Takagi, M. Koike, A. Kurobe, "Formation of Strained-Silicon Layer on Thin Relaxed-SiGe/SiO_2/Si Structure Using SIMOX Technology", Thin Solid Films **369** (2002) 199-202.
27. T. Mizuno, S. Takagi, N. Sugiyama, H. Satake, A. Kurobe, A. Toriumi, "Electron and Hole Mobility Enhancement in Strained Si MOSFETs on SiGe on Insulator Substrates Fabricated by SIMOX Technology", IEEE Elect. Device Lett. **21** (2000) 230-2.
28. T. Tezuka, N. Sugiyama, and S. Takagi, "Fabrication of Strained Si on an Ultrathin SiGe-on-Insulator Virtual Substrate with a High-Ge Fraction", Appl. Phys. Lett. **79** (2001) 1798-1800.
29. L.J. Huang, J.O. Chu, D.F. Canaperi, C.P.D'Emic, R.M. Anderson, S.J. Koester and H.-S. P. Wong, "SiGe-on-Insulator Prepared by Wafer Bonding and Layer Transfer for High-Performance Field-Effect Transistors", Appl. Phys. Lett. **78** (2001) 1267-9.
30. Zhi-Yuan Cheng, M.T Currie, C.W. Leitz, G. Taraschi, E.A. Fitzgerald, J.L. Hoyt and D.A. Antoniadas, "Electron Mobility Enhancement in Strained-Si n-MOSFETs Fabricated on SiGe-on-Insulator (SGOI) Substrates", IEEE Electron. Device Lett. **22** (2001) 321-3.
31. J.W. Matthews, "Defects Associated with the Accommodation of Misfit Between Crystals", J. Vac. Sci. Technol. **12** (1975) 126-33.
32. P.M. Mooney, S.J. Koester, J.A. Ott, J.L. Jordan-Sweet, J.O. Chu and K.K. Chan, "Thermal Stability of Strained Si on Relaxed $Si_{1-x}Ge_x$ Buffer Layers", in *Materials Issues in Novel Si-Based Technology,* Mat. Res. Soc. Symp. Proc. **686** (in press).
33. P.M. Mooney, J.L. Jordan-Sweet, G.B. Stephenson, F.K. LeGoues, and J.O. Chu, "High Resolution X-Ray Diffraction Measurements of Strain Relaxed SiGe/Si Structures", Advances in X-Ray Analysis **38** (1995) 181-93.
34. Rocking Curve Analysis by Dynamical Simulation, Bede Scientific, Inc.
35. S.J. Koester, K. Rim, J.O. Chu, P.M. Mooney, and J.A. Chu, "Effect of Thermal Processing on Strain Relaxation and Interdiffusion in Si/SiGe Heterostructures Studied Using Raman Spectroscopy", Appl. Phys. Lett. **79** (2001) 2148-50.
36. P. Baucaud, L. Wu, C. Guedji, H.F. Julien, I. Sahnes, Y. Campidelli and L. Garchery, "Photoluminescence and Intersubband Absorption Spectroscopy of Interdiffused Si/SiGe Quantum Wells", J. Appl. Phys. **80** (1996) 1414-21.
37. M. Griglione, T.J. Anderson, Y.M. Haddara, M.E. Law, K.S. Jones and A. van den Bogaard, "Diffusion of Ge in $Si_{1-x}Ge_x$/Si Single Quantum Wells in Inert and Oxidizing Ambients", J. Appl. Phys. **88** (2000) 1366-72.

International Journal of High Speed Electronics and Systems, Vol. 12, No. 2 (2002) 315–323

RECENT PROGRESS OF FERROELECTRIC MEMORIES

HIROSHI ISHIWARA

Frontier Collaborative Research Center, Tokyo Institute of Technology
4259 Nagatsuta, Midoriku, Yokohama 226-8503, Japan

Recent progress of ferroelectric random access memories (FeRAMs) is reviewed. First, novel ferroelectric materials, which are suitable for both low temperature crystallization and low voltage operation are introduced. Then, various cell structures in FeRAMs are discussed, in which particular attention is paid to non-destructive-readout-type cells such as a 1T-type cell composed of a single ferroelectric-gate field effect transistor. Finally, a novel 1T2C-type non-destructive-readout cell with good data retention characteristic is introduced and its basic operation is presented.

1. Introduction

As modern portable electronic devices such as mobile phones and notebook computers become more and more popular, there is confirmed increase in the demand for non-volatile memories. FeRAM (ferroelectric random access memory) is one of the most promising candidates for this application, because the power consumption of FeRAM is lowest among various semiconductor memories and because the operation speed is as fast as DRAM (dynamic RAM). However, there are still problems in high-density integration of FeRAM cells and no megabit-scaled memory has been commercialized yet. In this paper, current status and prospects of FeRAMs are discussed from viewpoints of materials, cell structures, and circuits. Particularly, operations of non-destructive-readout type cells such as a 1T-type cell composed of a ferroelectric-gate FET (field effect transistor) is precisely discussed.

2. Ferroelectric Materials

Typical ferroelectric materials for FeRAM are $PbZr_XTi_{1-X}O_3$ (PZT), SBT ($SrBi_2Ta_2O_9$), and $Bi_4Ti_3O_{12}$ (BIT). So far, many critical problems of these materials such as fatigue, retention, and imprint characteristics have substantially been solved. However, in order to fabricate a high-density memory or to integrate with an advanced logic circuit as an SOC (system-on-a-chip), it is necessary to decrease the crystallization temperature of a ferroelectric film so that the underlying CMOS circuit does not degrade during crystallization annealing. Furthermore, if the crystallization temperature decreases to below 450°C, integration of FeRAM and logic circuits becomes very simple, because ferroelectric capacitors can be formed on aluminum inter-connections at the last step of IC (integrated circuits) fabrication. It is also necessary to form a very thin ferroelectric film so that the circuit can be operated at a low voltage.

Recently, we succeeded in forming very thin ferroelectric films by adding sol-gel solution of Bi_2SiO_5 to those of PZT, SBT, and BIT and by spin-coating them on $Pt/Ti/SiO_2/Si$ structures[1]. The deposited films were pre-annealed at 400°C for 5 min in air and annealed for crystallization in O_2 atmosphere using an RTA (rapid thermal annealing) furnace. The annealing temperature ranged from 450°C to 650°C and the annealing time was typically 10 min. It was found from X-ray diffraction analysis that the crystallization temperature decreased by 150°C to 200°C compared to the original materials. That is, the crystallization temperatures were as low as 450°C, 550°C, and 500°C for PZT-, SBT-, and BIT-based materials, respectively. It was also found that the surface of these films was very smooth. Thus, we attempted to decrease the film thickness down to 13 nm.

Polarization vs. voltage characteristics of typical thin films are shown in Fig.1. It can be seen from the figure that the remnant polarization does not degrade even in a film with 13 nm in thickness and that the polarization of these films is saturated at an applied voltage of 0.5 V. It can also be seen that the hysteresis shapes of the films are almost the same as those of the original materials, even though the mole ratios of Bi_2SiO_5 to PZT, SBT, and BIT are as high as 0.1, 0.33, and 0.4, respectively. Comparison of the operation voltage of the ferroelectric capacitors with the technology roadmap of supply voltage in integrated circuits is shown in Fig.2. This figure suggests that these new materials play an important role in fabrication of future high-density FeRAMs.

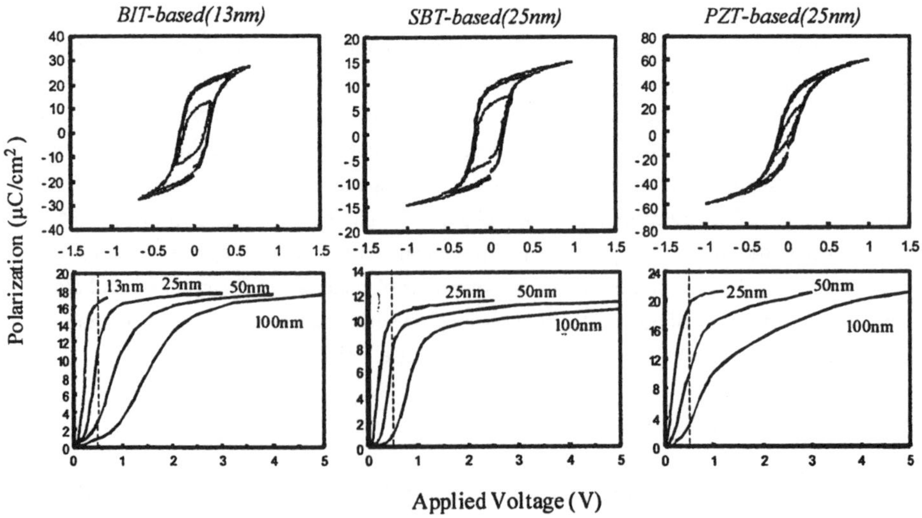

Fig. 1 P-V hystresis loops of thin ferroelectric films and the relation between the remnant polarization and applied voltage. The ramnant polarization values do not change even in the thinnest film in each group.

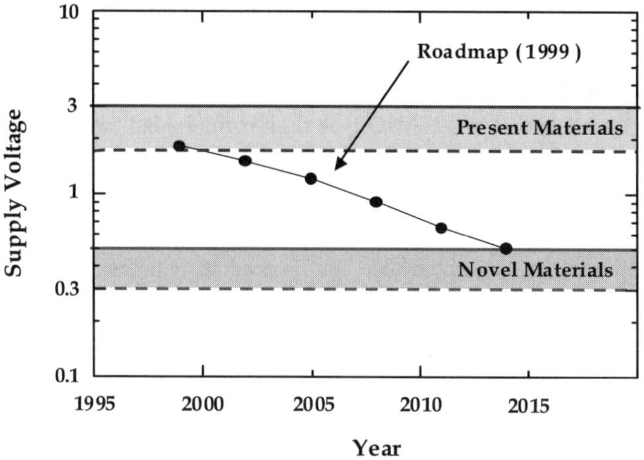

Fig. 2 Comparison of operation voltage of ferroelectric capacitors with the supply voltage of integrated circuits.

3. Cell Structures

A typical cell structure of FeRAM is the same as that of DRAM except for a plate line (PL), as shown in Fig3(a). In the "write" operation, the polarity of the ferroelectric capacitor is directed either upwards or downwards by applying a positive or negative pulse between the bit and plate lines and by switching-on the FET in the cell. In the "read" operation, a voltage pulse to align the polarization upwards, for example, is given to a selected cell and the response current of the capacitor is measured. In this operation, if the initial polarization of the ferroelectric film is downwards, the polarization reversal current flows as well as the displacement current does. Thus, the total switching current is larger than the non-switching current corresponding to the upward polarization of the film. Since the stored data are destroyed through the "read" operation, they have to be rewritten after each "read" operation.

In commercially available FeRAMs, a pair of the 1T1C cells is now used for storing a single bit, so that the operation margin of sense amplifiers becomes large. That is, a single cell is composed of 2 FETs and 2 capacitors (2T2C). Furthermore, in the present cell the ferroelectric capacitors are placed on a field oxide film and they are connected to FETs using Al interconnection. Thus, the cell size in the present FeRAM is fairly large. In order to achieve high-density integration, formation of a stacked structure of the FET and capacitor is necessary as well as use of a 1T1C cell. As a test chip, 4-Mbit FeRAM has been fabricated in Samsung Electronics using 0.6-μm design rule and 1T1C configuration[2].

A NAND-type cell called a chain FeRAM has also been proposed[3], which is schematically shown in Fig.3(b). In the "read" operation, the FET connected in parallel to the ferroelectric capacitor is off at a selected cell, keeping all other FETs on. Thus, the "read" voltage is applied only to the capacitor of the selected cell and the polarization reversal current which flows through other on-state FETs is detected between the bit and plate lines. This type of cell has an advantage that all capacitors are short-circuited when the supply voltage is off. The cell structure is similar to that of a NAND-type flash memory, but the actual cell size is not so small as that of flash memory, because of the contact hole areas existing in every spacing between the FETs.

A 1T-type cell shown in Fig.3(c) is promising in future high-density FeRAMs. It also has such a unique feature that the stored data can be readout non-destructively. The actual cell structure using an SOI (silicon-on-insulator) wafer has been proposed, as shown in Fig.3(d) [4]. In this figure, Si stripes with a lateral npn structure are placed on an insulating substrate and then covered with a uniform ferroelectric film, and metal stripes are placed on the film perpendicular to the Si stripes. Each Si stripe in the figure corresponds to parallel connection of ferroelectric-gate FETs and each metal stripe to a common gate electrode. In this structure, each FET represents a single memory cell and their packing density is expected to be very high, since no via-holes to the ferroelectric film exist in the array area.

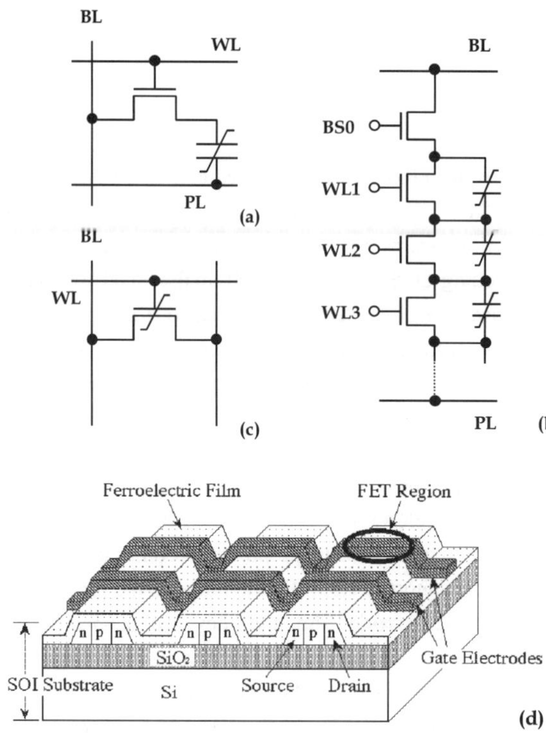

Fig. 3 Cell structures of FeRAMs. (a) NOR-type (usual 1T1C), (b) NAND-type (chain FeRAM), (c) AND-type (1T), and (d) Schematic of 1T-type cell array formed on an SOI structure.

4. Ferroelectric-gate FETs

As discussed in the previous section, a 1T-type cell composed of a single ferroelectric-gate FET seems to be promising in future FeRAMs. However, there is still a significant problem in the ferroelectric-gate FET that the data retention time is as short as several hours or a day. This problem arises from the gate structure with a dielectric buffer layer which is inserted between the ferroelectric film and Si substrate for obtaining good FET characteristics[5]. Thus, at present it is difficult to use an FET with the MFIS (metal-ferroelectric-insulator-semiconductor) gate structure as a general non-volatile memory to keep data for 10 years. Instead, applications to a latch circuit with temporary memory function seem to be more promising, because the stored data in this latch circuit are not destroyed even if the power supply of this circuit block is temporarily off in an MT (multiple threshold)-CMOS circuit. The new materials shown in Fig.1 seem to accelerate this application significantly, because there has been no material which can operate at such a low supply voltage that MT-CMOS circuits become important.

In order to improve the data retention characteristics of MFIS devices, comprehensive studies have been conducted. An interesting approach is to insert a floating gate electrode between the ferroelectric film and the dielectric buffer layer (MFMIS structure), and to optimize the area ratio between the MFM and MIS capacitors. Actually, such important knowledge has been obtained through this approach that a key factor to retain the data for a long time is the saturated polarization of the ferroelectric film[6]. However, the area of the optimized MIS capacitor is usually 10 times larger than that of the MFM capacitor, and thus this approach does not fit the request for high-density integratrion.

Considering both requests of the data retention and the packing density, we fabricated MFIS capacitors composed of a Si_3N_4 buffer layer and a c-axis-oriented $(Bi,La)_4Ti_3O_{12}$ (BLT) film[7]. The reasons why we chose these materials are as follows; (1) Si_3N_4 has a high dielectric constant and high resistivity against oxidation of the Si surface during crystallization annealing of the overlaid ferroelectric film, and (2) the remnant polarization of BLT is as small as 4 $\mu C/cm^2$ along a c-axis and it is expected that the polarization is well saturated before the Si_3N_4 buffer layer breaks down. In the experiment, a Si_3N_4 buffer layer was formed on a Si substrate by a direct nitridation method using atomic N radicals, and then a BLT film was deposited by a sol-gel spin-coating method.

Figure 4 shows X-ray diffraction patterns of BLT films crystallized at 800°C. As can be seen from the figure, an almost perfectly c-axis-oriented film was obtained, when the sample was pre-annealed at 400°C. On the other hand, as can be seen from a (117) peak located at an angle of 30°, a randomly oriented film was obtained for the pre-annealing at 500°C. In order to ascertain the effectiveness of a c-axis-oriented film, C-V characteristics were measured for the samples shown in Fig.4. The results are shown in Fig.5, in which good hysteretic characteristic with a counterclockwise direction can be obtained only in the c-axis-oriented film. These results clearly demonstrate the importance of optimization of the device structure as well as the buffer layer and ferroelectric materials.

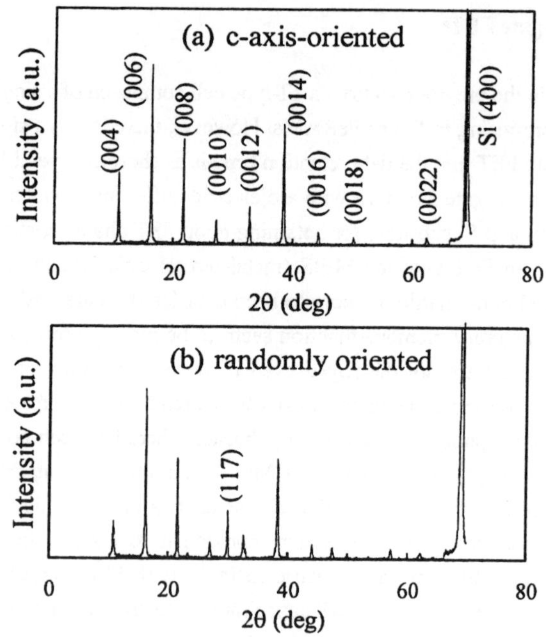

Fig. 4 X-ray diffraction pattern for 100-nm-thick BLT films crystallized at 800°C. Pre-annealing temperatures are 400°C (a) and 500°C (b).

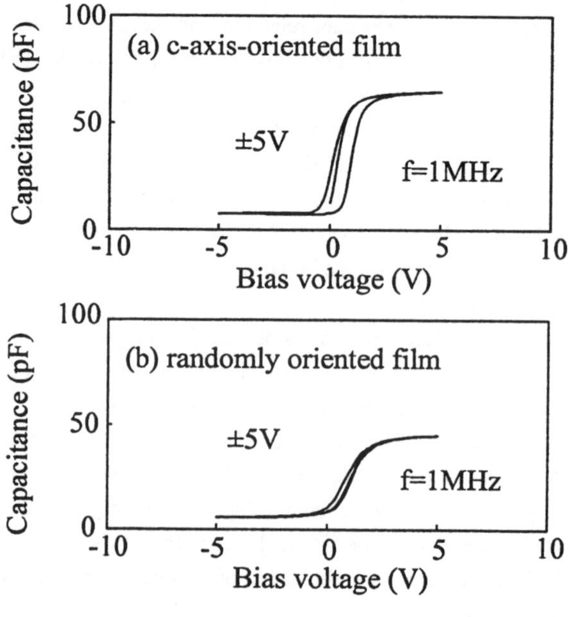

Fig. 5 C-V characteristics of Pt/BLT/Si_3N_4/n-Si diodes. (a) c-axis-oriented film and (b) randomly oriented film..

5. Non-destructive Readout Type 1T2C Memories

In order to keep the data retention time long in the FET-type memories, I proposed a novel cell structure, as shown in Fig.6, in which two ferroelectric capacitors with the same area are connected to the gate of a usual MOSFET[8]. The operation of this 1T2C-type memory cell has already been demonstrated using an SBT film and Pt top and bottom electrodes[9]. In order to write a datum in this cell, a positive or negative pulse is applied between the terminals A and B, so that the two ferroelectric films are polarized oppositely with respect to the gate electrode of MOSFET. In this case, the electric charges induced to the electrodes of both capacitors are canceled each other and no charge is induced to the gate electrode of FET. Under this condition, the floating gate potential, which is proportional to the induced charge on the gate, is zero and no depolarization field is generated in the ferroelectric film, when the terminals A and B are grounded. Thus, the data retention time in this cell is expected to be as long as the time of a usual 1T1C-type FeRAM.

For the "read" operation, positive voltage pulses are applied to the terminal B in Fig.6, keeping the terminal A open. In this operation, when the stored datum is "0", no polarization reversal occurs in the ferroelectric film and a little drain current flows through the MOSFET. To the contrary, when the stored datum is "1", the polarization is reversed and a large drain current flows. Figure 7 shows variation of the drain current of the FET by application of a "read" pulse of 2.5 V. It can be seen form the figure that the drain current increases only in the period when the "read" pulse is applied to the gate, and it almost comes back to the initial value when the pulse is removed. It can also be seen that the drain current during the "read" operation is very different between the data "1" and "0".

Concerning the non-destructive readout operation, it was found that variation of the drain currents for both "1" and "0" data was negligible even after the "read" operation was repeated 10^4 times. It was also found in the data retention measurement that the drain current did not change even if the stored data were readout after they had been retained for $6x10^4$ sec. The results are summarized in Fig.8. As can be seen from the figure, variation of the drain current is negligibly small when the two ferroelectric capacitors are polarized oppositely, while the drain current decreases much faster when the two capacitors are polarized in the same direction and the cell is operated as a single 1T-type cell. It is concluded from these results that the 1T2C-type

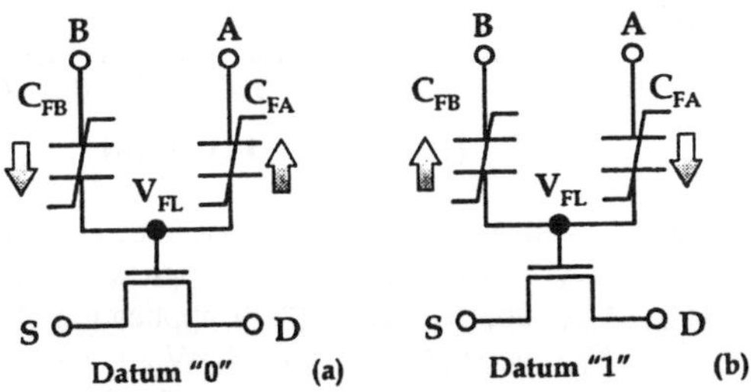

Fig. 6 1T2C-type ferroelectric memory cell.

cell is one of the most promising ferroelectric memories having both features of non-destructive data readout and excellent data retention characteristics.

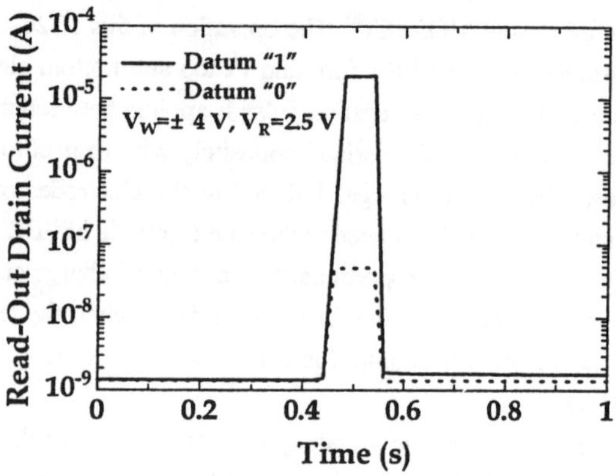

Fig. 7 "Read" operation of 1T2C-type cell.

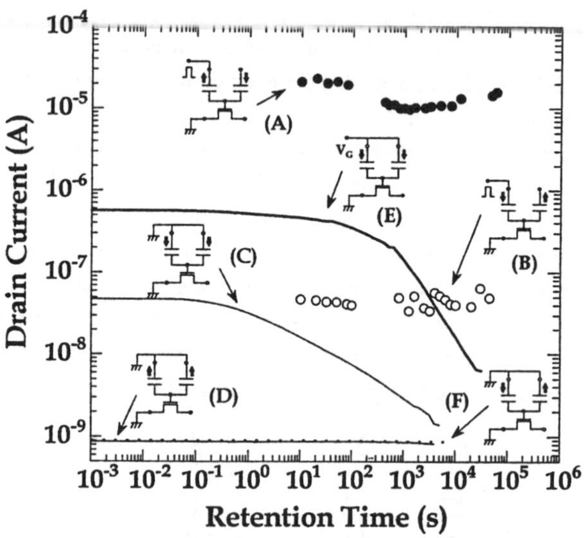

Fig. 8 Data retention characteristics of 1T2C cell.
(A) C_{FA} : down, C_{FB} : up ("1"), application of V_R ,
(B) C_{FA} : up, C_{FB} : down ("0"), application of V_R ,
(C) C_{FA} : down, C_{FB} : down, $V_A=V_B=0.5$ V.
(D) C_{FA} : down, C_{FB} : down, $V_A=V_B=0$ V.
(E) C_{FA} : up, C_{FB} : up, $V_A=V_B=0$ V.
(F) dashed line, C_{FA} : up, C_{FB} : down, $V_A=V_B=0$ V.

6. Summary

Current status and prospects of FeRAM were reviewed from viewpoints of materials, cell structures, and circuits. It was pointed out that improvement of the data retention characteristics was most important in practical use of FET-type ferroelectric memories and recent progress of this type of memories was mainly discussed. In the near future, studies on various kinds of FeRAMs are expected to become more and more popular, and in these studies the new ferroelectric materials which I discussed above are believed to become key materials for low voltage operation.

Acknowledgement

This work was performed under the auspices of the R&D Projects in Cooperation with Academic Institutions (Next-Generation Ferroelectric Memory) supported by NEDO (New Energy and Industrial Technology Development Organization in Japan) and managed by FED (R&D Association for Future Electron Devices).

References

1) T.Kijima and H.Ishiwara; "A novel solid solution exhibiting excellent ferroelectricity", Ext. Abstracts of 1st Intern. Meeting on FeRAM, Gotemba, (2001) 67-68
2) K.Kim; "Technology perspective for 1T/1C FRAM", Integrated Ferroelectrics, **25** (1999) 149 -167
3) D.Takashima and I.Kunishima; "High-density chain ferroelectric random access memory (Chain FRAM)", IEEE J. Solid State Circuits, **33** (1998) 787-792
4) H.Ishiwara, T.Shimamura, and E.Tokumitsu; "Proposal of a single-transistor-cell-type ferroelectric memory using an SOI structure and experimental study on the interference problem in the write operation", Jpn. J. Appl. Phys. **36** (1997) 1655-1658
5) H.Ishiwara; "Current status of fabrication and integration of ferroelectric-gate FETs", Mat. Res. Soc. Symp. Proc. **596** (2000) 427-436
6) S-M.Yoon, E.Tokumitsu, and H.Ishiwara: "Improvement of memory retention characteristics in ferroelectric neuron circuits using a $Pt/SrBi_2Ta_2O_9/Pt/Ti/SiO_2/Si$ structure field effect transistor as a synapse device", Jpn. J. Appl. Phys. **39** (2000) 2119-2124
7) T.Kijima, Y.Fujisaki and H.Ishiwara; "Fabrication and characterization of $Pt/(Bi,La)_4Ti_3O_{12}$ $/Si_3N_4/Si$ metal-ferroelectric-insulator-semiconductor structure for FET-type ferroelectric memory applications", Jpn. J. Appl. Phys. **40** (2001) 2977-2982
8) H.Ishiwara; "Recent progress of FET-type ferroelectric memories", Integrated Ferroelectrics, **34** (2001) 11-20
9) S-M.Yoon and H.Ishiwara: "Memory operations of 1T2C-type ferroelectric memory cell with excellent data retention characteristics", IEEE Trans. on Electron Devices, **48** (2001) 2002-2008

International Journal of High Speed Electronics and Systems, Vol. 12, No. 2 (2002) 325–332

System-on-Chip Integration

ROBERT R. DOERING

Texas Instruments, Inc.
P.O. Box 650311, MS 3730
Dallas, TX 75265, U.S.A.
e-mail: doering@ti.com

Numerous "signal-processing products" are now driving the semiconductor market for SOC solutions enabling real-time performance, low-cost, low-power, portability, etc. A primary limit on the types of electronic (or other) functions that will be integrated into future SOCs is cost of integration, which tends to grow non-linearly with process complexity and chip area. A near-continuum of System-on/in-X solutions is emerging between traditional System-on-Chip and System-on-Board. These approaches span the tradeoff between bandwidth and cost. For the foreseeable future, digital CMOS will continue to serve as a "host platform" for integrating a wide range of mechanical, optical, biological, and, perhaps, even "quantum" technologies.

1. Introduction

"System-on-Chip" (SOC) has become the catch phrase for integrated circuits incorporating diverse functionality. In this paper, we explore some of the drivers, trends, and potential limits of this approach to future electronics. In the near to mid term, we expect significant further progress in integrating logic, memory, analog, RF, and other electronic functions into individual ICs. In some cases, this will be through the co-fabrication of multiple device technologies. In other instances, it will be via the emulation of traditional analog and RF functions in the digital domain.[1] In part, this is being enabled by the increasing performance and density resulting from continued scaling of digital CMOS.[2] In the longer term, we anticipate the use of digital CMOS as a "host substrate" for integrating a wide range of mechanical, optical, biological, and, perhaps, "spintronic" or other "quantum" system technologies. In such hybrid systems, the digital CMOS will provide some of the traditional electronic (e.g., I/O) functions as well as a base volume manufacturing technology. Today's early MEMS and "gene chips" are forerunners of this class of SOCs. A core issue for SOC integration level is the tradeoff between performance and cost. For example: "what memory should be on-chip vs. off-chip?" Answering this question becomes even more complex with the development/improvement of "3D-integration" approaches such as "System-in-Package" (SIP) and "chip-bonding." Optimizing such performance-cost tradeoffs is a key element of electronic/system R&D strategy today and in the future.

2. SOC Drivers and Challenges

Figure 1 shows one of today's prototypical SOCs: the digital baseband chip in a cell phone. It employs CMOS "glue" logic to integrate the additional functions of digital signal processing, analog signal processing, central processing, read-only program-store memory, and data-store memory onto one chip rather than multiple chips. However, as illustrated in Fig. 2, there is a great deal of opportunity for further SOC integration in cell phones. At present, there are still about a dozen separate ICs, even more discrete transistors/diodes, and over two hundred passive electronic components. Later in this decade, we should be able to integrate everything but the power amplifier and a few of the large-value passives onto an SOC that will almost be a "single-chip cell phone."[1] One of the enabling elements of this program will be capitalizing on the increasing performance trend of CMOS to implement ADC at higher frequencies and perform what were previously analog functions in the digital domain. Note that even the power amplifier can be provided in currently available CMOS for short-range radio transmission at frequencies up to a few GHz, such as in the Bluetooth specification.

For implementation in a wide variety of end equipments, the general opportunities (as well as challenges!) for SOC integration include:

- Reduction of system cost (lowest end-equipment bill-of-materials) despite tendencies toward increasing process complexity and chip size
- Increase of bandwidth:
  - low RLC (e.g., shorter) interconnect
  - dense signal interconnects
- Dense, low-loss, power and ground distribution
- Reduction of operating and standby power

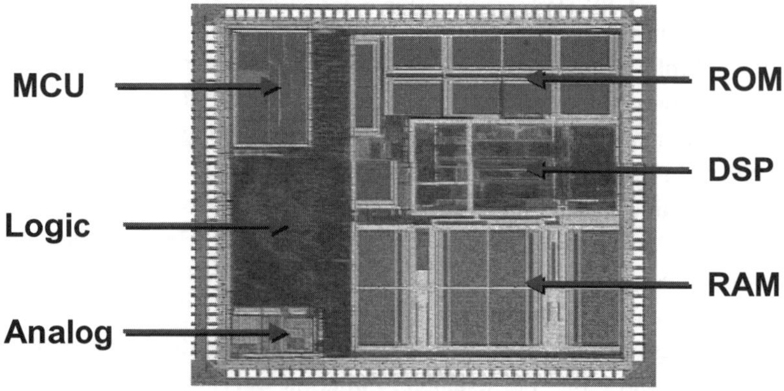

Figure 1. Cell-Phone Digital Baseband Chip

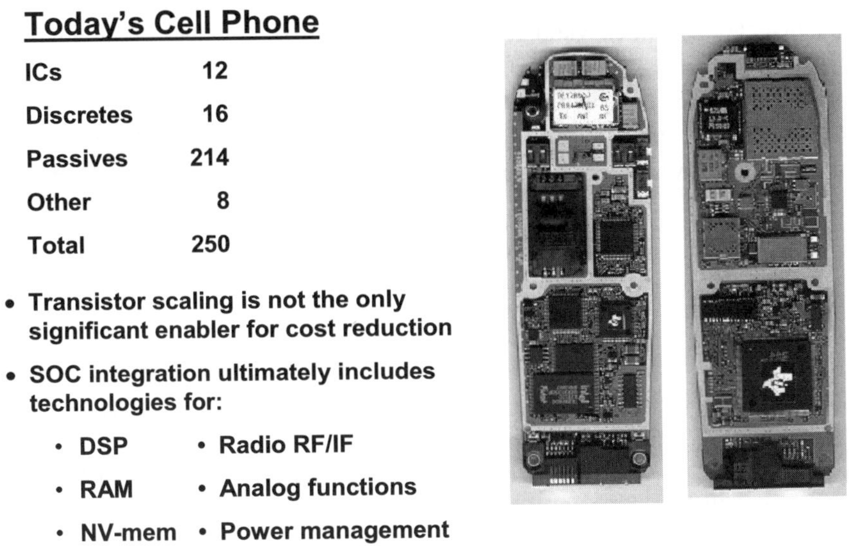

Figure 2. Opportunity for Integrating More Functions in Cell Phone SOCs

Of course, SOC solutions often have to contend with the additional challenges of supporting multiple operating voltages and insuring technology compatibility and reliability despite integrating diverse structures and materials.

In most cases, reduction of system cost is the principal driver for SOC implementation. Thus, we will discuss it in the next section.

3. SOC Cost Constraints

Figure 3 presents results from a relatively simple model of chip cost as a function of area and process complexity. Note that the number of lithography masks is used as an approximate representation of process complexity. Cost rises faster than linearly with either mask count or chip area, independently, due to their separate contributions to chip yield. For this figure, we have calculated the model with parameters appropriate for today's state-of-the-art CMOS technology: "130-nm node" on 300-mm-diameter silicon wafers.[2] For example, we have used a starting substrate cost of $500, a fab process cost of $140 per mask level, and a mask (reticle) cost of $30K per level. Of course, the mask cost is amortized over the total number of chips produced, as shown in Fig. 4 for a typical "baseline" 25-mask CMOS logic process. Note, that, for Fig. 3, we have assumed production of one million chips, which is approximately what is required these days to get mask cost under 10% of chip cost for a 25-mask process! In this model, we have also assumed that defect-limited yield is given by a negative-binomial distribution[3] with a defect density per mask level of 0.01/cm^2 and a defect cluster factor of 5. Costs associated with parametric yield loss, assembly and test, and circuit design have not been included in Figs. 3 and 4. Such "bare-chip" costs are an appropriate first step in evaluation of SOC vs. SIP vs. chip-set solutions, where the packaging costs must be added separately for each.

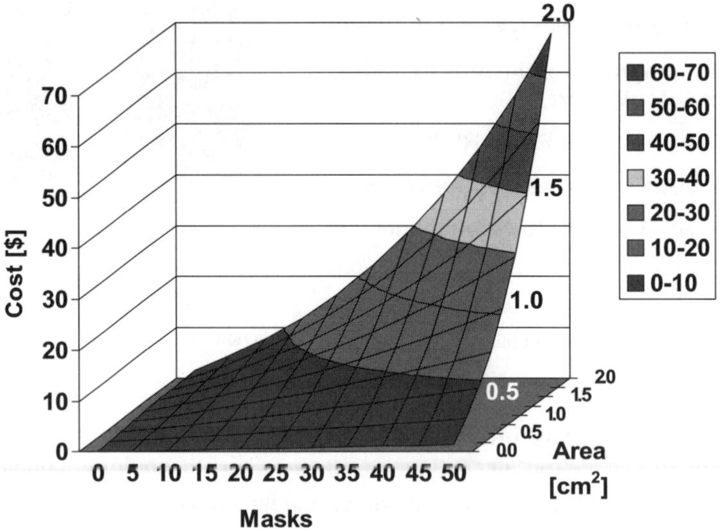

Figure 3. Chip Cost vs. Size and Process Complexity

A current hot example for illustrating the SOC cost issue is integration of large amounts of memory with logic. Suppose that we want to create an SOC with 5 million gates of (25-mask, 130-nm) logic and 128 Mbits of memory. At the 130-nm technology node, we presently get about 50 million transistors (12.5 million gates) per cm^2 of silicon. This results in a requirement of 0.4 cm^2 of silicon for 5 million gates of logic. Thus, from our model, the logic as a separate (bare) chip would cost $3.45.

The added area and cost for the memory obviously depends on the memory type. SRAM is the "simplest" memory option since it can be integrated with the logic without adding process complexity and its speed generally matches that of the logic (although there is a trade-off between speed and area). Figure 5 shows the trend for minimum SRAM cell area. At the 130-nm technology node, we have a 6-T SRAM cell area of approximately 2 μm^2 and, correspondingly, a density of 40-45 million bits/cm^2 at 80-90% memory array efficiency. Thus, adding the memory as SRAM would bring the total chip area to 3.24 cm^2 and the total cost to \$49.49!

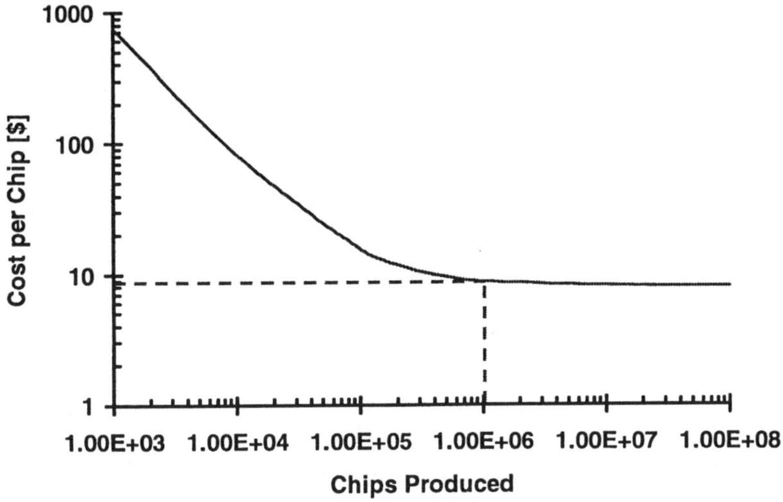

Figure 4. Amortization of Mask Cost

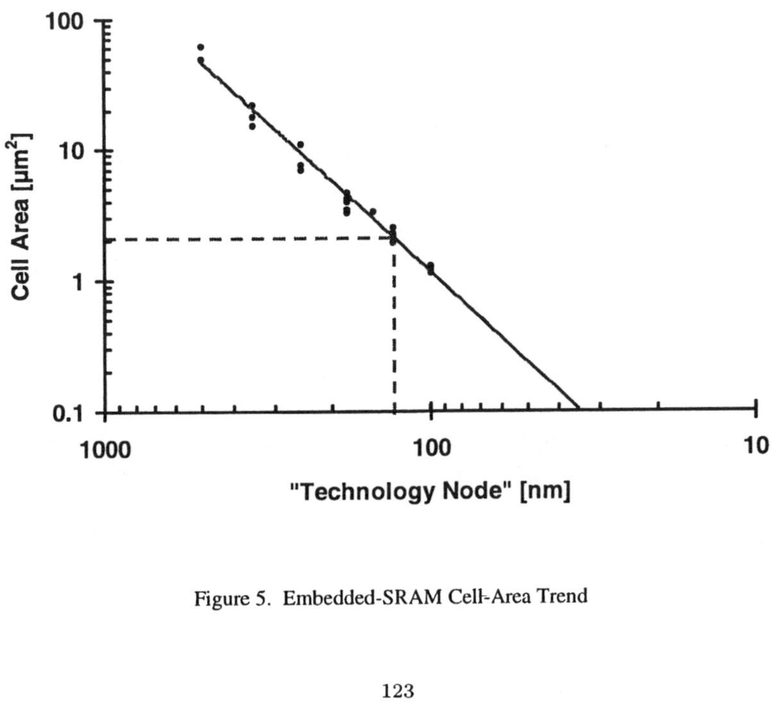

Figure 5. Embedded-SRAM Cell-Area Trend

Another option for the embedded memory would be 1-T eDRAM, which would require much less additional area. At the 130-nm node, there are 0.3 μm^2 1-T eDRAM cells. At 55% array efficiency, this would add only 0.7 cm^2, but at least 5 additional mask levels are required for the DRAM cell process complexity. In this case, the resulting 1.1-cm^2, 30-mask SOC would cost $12.25. However, note that stand-alone 128-Mbit DRAMs cost only about $2.40. Thus, it also makes sense to consider alternative forms of integration, such as System-in-Package. In this example, SIP would definitely be the best integration strategy as long as cost is indeed the major driver. The total cost of bare die for the logic and the DRAM plus assembly cost for SIP would be far less than the $12.25 (even before assembly) logic+eDRAM SOC. However, note that this assumes that all of the other requirements could be inexpensively incorporated. For example, depending on speed and architecture considerations, the DRAM might require a custom design and the SIP assembly might need to be one of the high-performance versions. Today, "flip-chip" attachment of the DRAM to the logic chip would give the maximum SIP performance. If the system application can be satisfied with less bandwidth between the logic and memory, a variety of less-expensive SIP assembly options are available, ranging down to simple side-by-side wire-bonding of the chips to a common interconnect substrate within the package. The bandwidth differences result directly from the number, placement, and length (i.e., RC) of inter-chip connections available in each SIP assembly technique. Until recently, SIP usage has been greatly suppressed by costs associated with "testing for known-good-die (KGD)." However, flip-chip assembly is now becoming widely used for high-performance single-chip packages, and that development is contributing solutions for KGD and "bonding over active area" that are also key enablers for SIP.

4. Future SOC Opportunities

As we look at the future of SOC, we envision many new types of functionality being integrated via an increasing diversity of techniques. In fact, the gap between SOC and "flip-SIP" will probably be bridged by other forms of "3D-integration" such as "chip bonding." One form of chip bonding currently undergoing demonstration allows interconnections between the "sub-chips from different starting substrates" to be as short as about 10 μm at pitches down to about 2 μm, compared to 50 μm and 125 μm, respectively, for the next generation of flip-SIP. Such technology offers the prospect of further blurring the distinction between SIP and SOC, and it should be scalable to sub-micron interconnect pitch. In fact, note that 3D interconnection of chips via either flip-SIP or bonding can often provide even shorter wires between circuit blocks (e.g., logic and memory) than can be achieved "in the plane" of a traditional SOC.

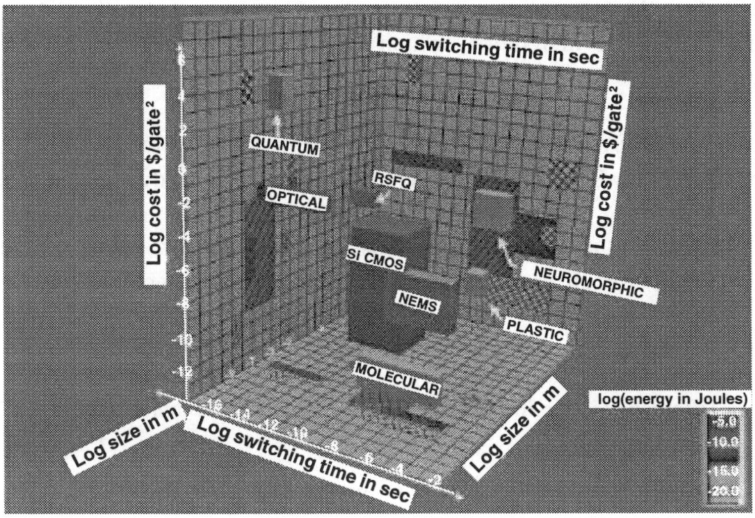

Figure 6. ITRS Comparison of Potential Future SOC Technologies

Perhaps the key questions with regards to the future of electronics are "how far can CMOS be scaled?" and "what, if anything, can succeed CMOS?" The 2001 International Technology Roadmap for Semiconductors (ITRS) targets 9-nm channel-length MOSFETs in 2016.[2] This is essentially the present consensus limit on the extent to which even "ultimate CMOS" (i.e., double-gate MOSFETs) can be scaled and still have useful on-off characteristics. Figure 6 shows part of the latest attempt by the ITRS Process Integration and Device Structures Working Group to answer the question of "what might come next?"[2] In this chart, they compare potential future logic technologies in terms of estimated cost, density, speed, and power usage. Note that none of the potential successors to CMOS are presently estimated to be capable of surpassing "ultimate CMOS" in all of these parameters simultaneously. In addition, most of them require CMOS for practical I/O. Thus, we believe that, at least initially, many such "emerging technologies" are most likely to be implemented as additional functions integrated into CMOS-based SOCs. Of those technologies shown in Fig. 6, RSFQ would be the most obvious exception, perhaps followed by some forms of "plastic" devices and quantum-computing devices.

5. Conclusions

Numerous "signal-processing" products are now driving the semiconductor market for SOC solutions enabling "real-time performance," low-cost, low-power, portability, etc.

A primary limit on the types of electronic (or other) functions that will be integrated into future SOCs is cost of integration, which tends to grow non-linearly with process complexity.

A near-continuum of System-on/in-X solutions is emerging between traditional System-on-Chip and System-on-Board that span the tradeoff between bandwidth and cost.

Digital CMOS will continue to serve as a "host platform" for integrating a wide range of mechanical, optical, biological, and, perhaps, "quantum" system technologies.

References

1. D. Buss, "Technology in the Internet Age," ISSCC Digest of Technical Papers, 2002, vol. 45, p. 18.

2. "The International Technology Roadmap for Semiconductors," 2001 edition, Semiconductor Industry Association, 2001.

3. R. Ross and N. Atchison, "Yield Modeling," in Handbook of Semiconductor Manufacturing Technology, Y. Nishi and R. Doering, eds., Marcel Dekker, New York, 2000, p. 851.

International Journal of High Speed Electronics and Systems, Vol. 12, No. 2 (2002) 333–342

ULTRA-THIN SOI TRANSISTORS FOR ULTIMATE CMOS TECHNOLOGY: FUNDAMENTAL PROPERTIES AND APPLICATION PERSPECTIVES

D. ESSENI[1], E. SANGIORGI[1], M. MASTRAPASQUA[2], C. FIEGNA[3]
G. K. CELLER[4], and L. SELMI[1]

[1] *DIEGM, University of Udine, Udine, 33100 Italy*
[2] *Agere Systems, 600 Mountain Avenue, Murray Hill, 07974 NJ, USA*
[3] *Dept of Engineering, Univ. of Ferrara, Ferrara, 44100 Italy*
[4] *SOITEC - USA*

In this paper we present experimental data of electron mobility in Ultra-Thin SOI MOSFETs operating in Single- and Double-Gate Mode. Mobility is measured for silicon thicknesses down to 5nm and for temperatures ranging from 225K to 375K. At large inversion densities (N_{inv}) ultra-thin SOI exhibit higher mobility than heavily doped bulk MOS and a small dependence of mobility on silicon thickness (T_{SI}). However, at small N_{inv} the mobility is clearly reduced for decreasing T_{SI}, possibly due to enhanced phonon scattering in the thin quantum well. Compared to Single-Gate, a modest but unambiguous mobility improvement is observed in Double-Gate mode, for silicon films around 10nm and small inversion densities. The mobility increase is more pronounced at low temperatures.

1 . Introduction

CMOS technology represents the backbone of today semiconductor industry and the enabler for the implementation of complex systems on single Chip. In order to fulfill demanding requirements in terms of performance/cost trade-offs, the ITRS Roadmap is pushing the scaling of the MOS transistor to physical channel lengths below 70nm for the year 2004-2005, despite the presence of several roadblocks that have no known solutions [1]. In particular, the scaling of the MOS transistor to physical channel lengths below 70nm, projected to year 2004-2005, requires channel doping concentrations larger than 10^{18} and S/D junctions shallower than approximately 30 nm to suppress Short Channel Effects (SCE). At such gate lengths, the statistical fluctuations of the doping in the channel introduce an appreciable dispersion of threshold voltages (V_T). Furthermore high channel doping results in increased source-drain junction capacitances and reduced effective mobility (μ_{eff}) [2] thus adversing the device performance for a given L_{Poly}. In addition, in order further control SCE, the ratio between the polysilicon gate-length L_{Poly} and the oxide thickness T_{ox} should not drop below approximately 50 [3], that results in an *equivalent* T_{ox} around 1.0-1.5nm for L_{Poly} below 70nm. Though high-K dielectrics could possibly replace SiO_2 as gate insulators in the future, the oxide scaling below an equivalent T_{ox} of approximately 1.3nm is presently an issue with no known solutions [1].

In this perspective, Fully-Depleted Silicon on Insulator (SOI) transistors realized on Ultra-Thin (UT) silicon films are probably the strongest candidates for the design of the ultimate MOSFET because, when the silicon thickness (T_{SI}) is reduced below 10nm, SOI transistors exhibit excellent scalability even with essentially undoped channels [4, 5]. Simulations have shown that the Double-Gate (DG) MOSFET on thin silicon layers is probably the most scalable design of the MOS transistor [6, 7]. In particular, UT-SOI MOSFETs can relax the ratio rement between L_{Poly} and equivalent T_{ox}; for example, for T_{SI}=10nm a gate length of 50nm seems feasible with T_{OX}=3nm [8]. Furthermore, undoped silicon films eliminate the issue of statistical doping fluctuations and result in lower effective fields [9] for a given inversion density with possible μ_{eff} improvements over bulk MOSFET. Moreover, T_{SI} inherently limits the depth of S/D junctions and the corresponding capacitances are greatly reduced.

Despite this remarkable applicative potential, the fundamental transport properties of UT-SOI transistors still lack an extensive, experimental investigation. In particular no experimental data have been presented for μ_{eff} of DG UT-SOI devices though possible improvements of μ_{eff} in DG versus Single-Gate (SG) SOI or conventional bulk transistors have been discussed in the literature [10, 11]. Since the DG configuration is an electrical condition that can be also obtained in conventional SOI transistors by using appropriate, asymmetrical front- and back-gate voltages (V_{FG} and V_{BG}, respectively), we used the UT-SOI devices presented in [12] to measure μ_{eff} in DG mode. Our results, obtained for silicon thicknesses (T_{SI}) down to 5nm and for temperatures between 225 and 375K, show that for the thinnest silicon films a modest but systematic μ_{eff} increase is observed in DG mode compared to SG.

2 . Experimental Procedure

The fabrication of the UT-SOI devices suitable for μ_{eff} measurements, and the determination of the silicon film thickness T_{SI} have been discussed in details in [12] and here are briefly summarized.

Low field effective mobility μ_{eff} is a key parameter for the current drivability of deep-submicron MOS transistors because a large μ_{eff} allows carrier velocity in the channel to exceed the bulk-silicon saturation value [13] thus approaching the ballistic transport regime [14]. The experimental determination of μ_{eff} is performed on *long channel* transistors and consists in the C-V based determination of the inversion density N_{inv} and of the *intrinsic conductance* $g_D^{(int)}$. Then mobility is simply calculated as

$$\mu_{eff} = (L/W) \cdot g_d^{(int)} / qN_{inv} \quad (1)$$

where L and W are device length and width, respectively.

However, when T_{SI} is very small and no elevated S/D structures are employed, the series resistance of the extremely shallow junctions can artificially reduce the device conductance, and the determination of the value of $g_D^{(int)}$ in UT-SOI becomes problematic. In order to circumvent this issue the μ_{eff} measurements presented here were obtained on the test structure illustrated in Fig. 1 that allows a determination of $g_D^{(int)}$ independent of series resistance [12]. The procedure to determine the T_{SI} values is reported in [12].

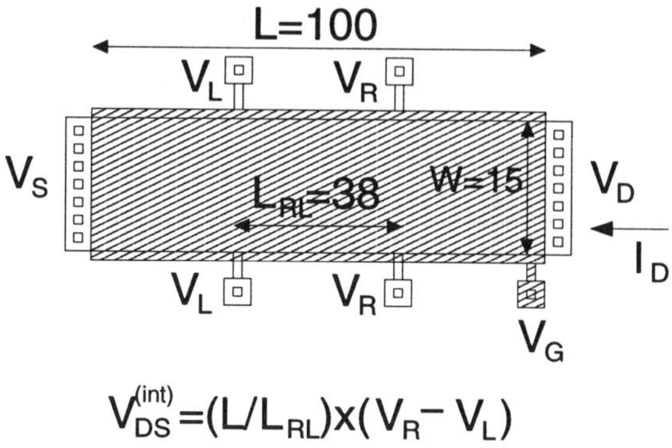

Figure 1: Schematic top-view of the UT-SOI MOSFETs used for μ_{eff} measurements. The voltage probes V_R and V_L allow to extract the *intrinsic* drain to source voltage $V_{DS}^{)int)}$ and conductance $g_D^{(int)}$. L, L_{RL} and W are in microns.

When the SOI transistor operates in DG mode, with generic V_{FG} and V_{BG} values, the determination of the inversion charge $Q_{inv}=qN_{inv}$ is not straightforward because it depends on both V_{FG} and V_{BG}. If we define $C_{FG}=[\partial Q_{inv}/\partial V_{FG}]$ as the front-gate capacitance, then Q_{inv} can be written as:

$$Q_{inv}(V_{FG}, V_{BG}) = Q_{inv}^{(0)}(V_{FG}^{(0)}, V_{BG}) + \int_{V_{FG}^{(0)}}^{V_{FG}} C_{FG}(V, V_{BG})dV \qquad (2)$$

where $V_{FG}^{(0)}$ and $Q_{inv}^{(0)}$ are generic, initial values which depend on V_{BG}.

Eq.2 states that, for a given V_{BG}, Q_{inv} can be obtained by integrating the corresponding C_{FG} versus V_{FG} characteristics, providing we start from an initial V_{FG} value such that $Q_{inv}^{(0)}(V_{FG}^{(0)}, V_{BG}) \simeq 0$. The C_{FG} characteristics for our devices are illustrated in Fig.2 for different V_{BG} and indicate that when V_{FG} is low enough it can dismiss the inversion layer induced at the back-interface by a large V_{BG} so that the condition $Q_{inv}^{(0)} \simeq 0$ can be achieved. Following the outlined procedure, for each V_{BG} we determined the $Q_{inv}(V_{FG}, V_{BG})$ value necessary for μ_{eff} calculation by integrating the corresponding C_{FG} characteristic as indicated in Fig.2.

As an example of Q_{inv} determination, Fig.3 reports the measured inversion charge as a function of V_{BG} for different values of V_{FG}. Because the data are taken in strong inversion, the slope of the Q_{inv} vs. V_{BG} curves represents the back oxide capacitance and provides an independent measure of the back oxide thickness T_{box}. The value obtained from the data on Fig.3 is 407nm compared to a nominal thickness of 400nm.

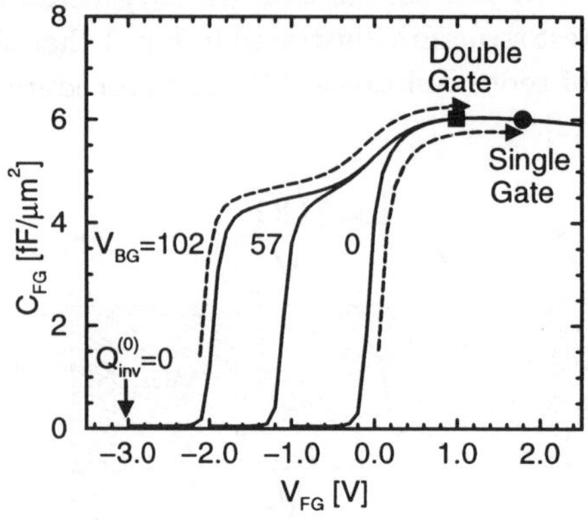

Figure 2: Measured C_{FG} versus V_{FG} characteristics for different back-gate voltages. The dashed arrows indicate the integrals performed to calculate Q_{inv} in either SG or DG operation mode.

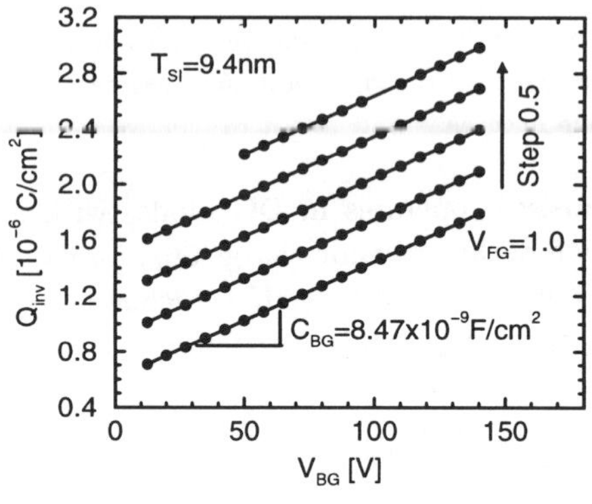

Figure 3: Measured Q_{inv} in DG mode as a function of V_{BG}, for different values of V_{FG}. In strong inversion the slope of each curve corresponds to the back gate capacitance.

Since in our samples the back-gate oxide is much thicker than the front-gate oxide (400nm versus 4.5nm) then in order to proceed with the mobility determination in DG mode it is necessary to identify, for each V_{FG}, the V_{BG} that induces the symmetric charge distribution in the silicon film. To this purpose we used a 1D Schrödinger-Poisson simulator that solves for the electrostatic configuration accounting for poly-depletion and for subband quantization in the silicon film [15].

Fig.4 shows potential energy and electron density obtained by Schrödinger-Poisson simulations, in symmetrical DG operation at fairly low $N_{inv} = 10^{12} cm^{-2}$.

The electron density exhibits a single peak at the center of the silicon film, a condition called "volume inversion". The value of N_{inv} indicates 1/2 of the total inversion density in the silicon layer.

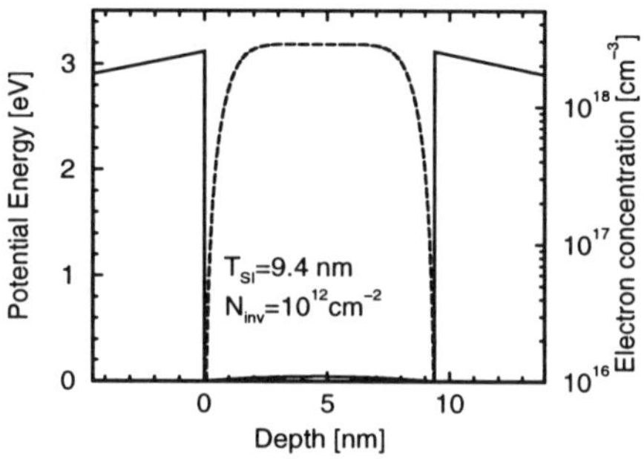

Figure 4: Potential energy (full line) and electron density (dashed line) in the vertical direction of an SOI structure operating in symetrical DG mode for T_{SI}=9.4nm and $N_{inv} = 10^{12} cm^{-2}$. N_{inv} indicates 1/2 of the total inversion density in the silicon layer.

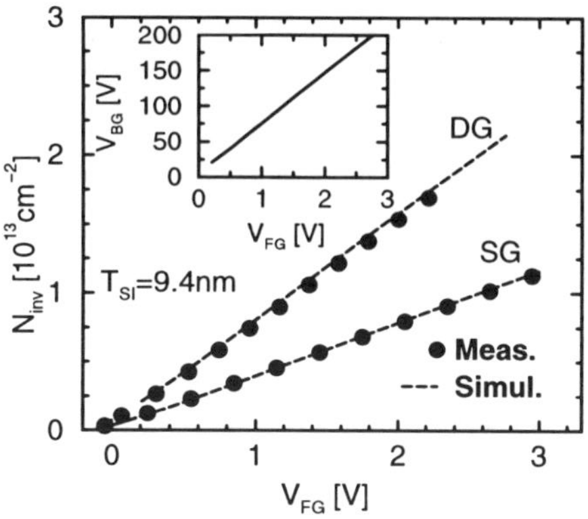

Figure 5: Comparison between measured and simulated inversion densities N_{inv} in either SG or symetrical DG mode for T_{SI}=9.4nm. In the inset it is shown the V_{BG} to V_{FG} relationship that realizes the symmetrical DG condition calculated by means of numerical simulations.

Finally the relationship between N_{inv} and V_{FG} is reported in Fig.5 for SG and symmetric DG. In the SG case V_{BG} is grounded, while in the DG case V_{BG} is a linear function of V_{FG} as illustrated in the inset of Fig.5. As can be seen the agreement between measurements and simulations is very good thus validating the whole procedure for the identification of the symmetrical DG condition and the experimental determination of the inversion charge.

3 .Double-Gate SOI Mobility

In the devices used for our study the same μ_{eff} values were found at the front and back interface when the transistor operates in SG mode and the inversion layer is induced at either one of the interfaces [12]. This result, due to the comparable quality of the front and back interfaces in our UNIBOND wafers, rules out spurious reductions of mobility in DG mode which may obscure the intrinsic physical picture.

Fig. 6 shows electron μ_{eff} obtained from our samples in symmetric DG conditions, for three different T_{SI} values, compared to previously published bulk MOSFETs data [2]. Mobility is plotted versus the inversion density N_{inv} because for current drivability it is interesting to compare μ_{eff} of different MOS structures at the same N_{inv}.

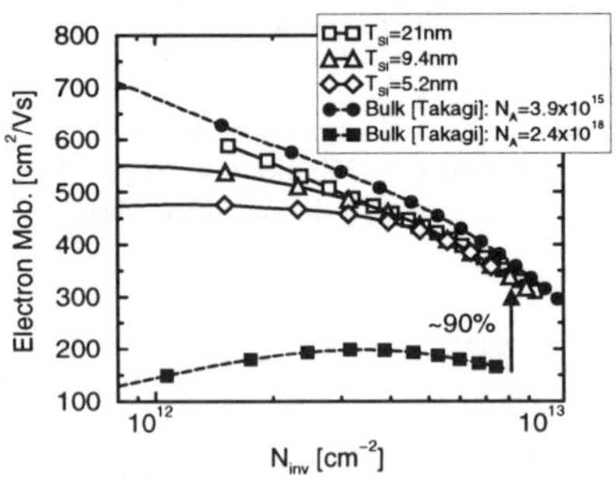

Figure 6: Measured room temperature electron mobility versus inversion charge density N_{inv} in symmetrical DG conditions and for different T_{SI}. A comparison with previously published data for bulk transistors [2] is illustrated.

At large N_{inv} the mobility of SOI devices is largely independent of T_{SI} and consistent with μ_{eff} of the lightly doped bulk MOSFETs because the silicon films of our devices have a very low boron concentration of approximately 10^{15}cm^{-3} [12, 2]. Since the heavily doped bulk transistor has a larger effective field than UT-SOI devices for a given N_{inv} due a large depletion charge [9], a remarkable μ_{eff} improvement can be obtained with SOI devices at the large N_{inv}.

At lower inversion densities, instead, μ_{eff} of SOI devices becomes more sensitive to T_{SI} and it is reduced with decreasing T_{SI}. Having ruled out possible non-idealities of the back-interface [12], a μ_{eff} dependence on T_{SI} is also expected because the carrier confinement in the *real space* is known to enhance phonon-scattering rate because of the corresponding *delocalization in the reciprocal, K space* [16]. At this regard we recall here that in [12] we measured the temperature dependence of μ_{eff} in SOI devices operating in SG mode. μ_{eff} exhibits an approximately $T^{-1.4}$ dependence that is close to the $T^{-1.7}$ dependence observed in bulk MOSFETs for phonon-limited mobility [2], suggesting that a modulation of the phonon-scattering rate could play a relevant role in the mobility dependence on T_{SI}.

Since we have evidence that the back-interface in our samples has a quality comparable to the front-interface [12], it is interesting to compare μ_{eff} in the SG and DG modes to verify if different electronic structures result in appreciably different μ_{eff}.

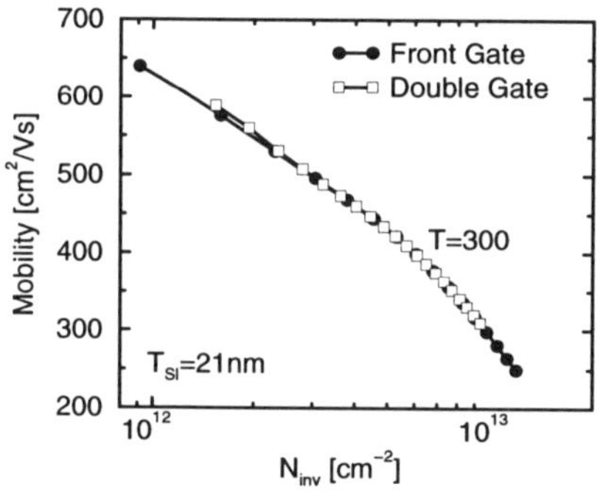

Figure 7: Measured electron mobility versus inversion charge density N_{inv} in either SG or DG mode for T_{SI}=21nm. Virtually the same mobility is found in the two operation modes. In DG mode N_{inv} indicates one half of the total inversion density.

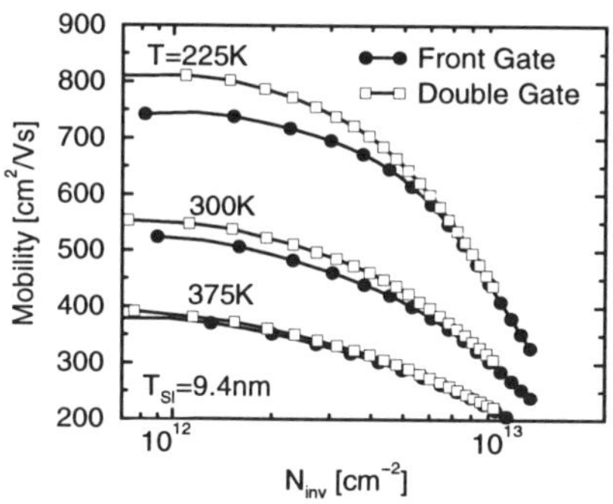

Figure 8: Measured electron mobility versus inversion charge density N_{inv} in either SG or DG mode for T_{SI}=9.4nm. A clear μ_{eff} improvement is observed in the DG mode at low N_{inv}. This effect is more pronounced for decreasing temperatures. In DG mode N_{inv} indicates one half of the total inversion density.

The results of such a comparison are illustrated in Figs.7,8 and 9 for different T_{SI}. N_{inv} in the DG mode is always one half of the entire inversion density. The general behavior of μ_{eff} is very similar in the two operating modes, including the silicon thickness dependence. However, while for $T_{SI} = 21nm$ the same mobility is observed in DG and SG mode, for the two thinnest silicon films and moderate inversion densities, instead, a modest but systematic and unambiguous increase of μ_{eff} in DG mode is observed which increases for decreasing temperatures. The results of Figs.8 and 9 indicate that mobility in SG and DG mode is virtually the same at large inversion densities where it is essentially limited by surface-roughness scattering suggesting that the improvement observed in DG mode at low N_{inv} could

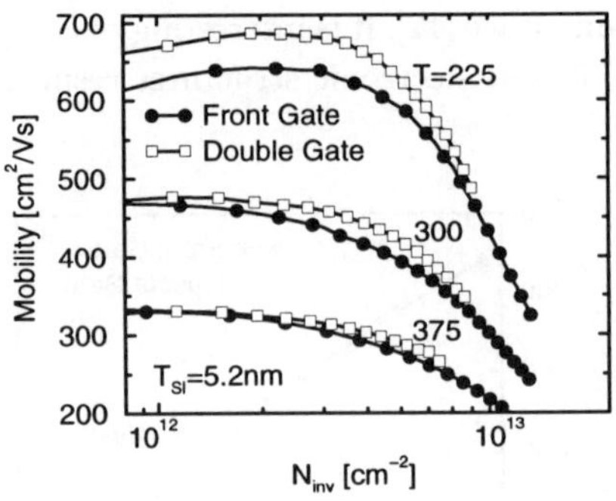

Figure 9: Measured electron mobility versus inversion charge density N_{inv} in either SG or DG mode for T_{SI}=5.2nm. A clear μ_{eff} improvement is observed in the DG mode at low N_{inv}. This effect is more pronounced for decreasing temperatures. In DG mode N_{inv} indicates one half of the total inversion density.

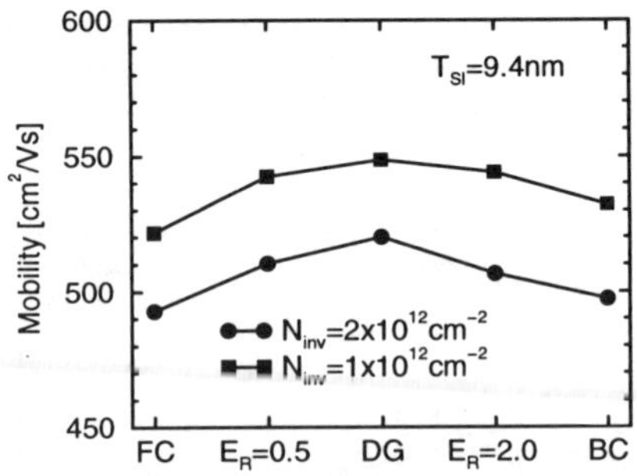

Figure 10: Measured mobility at two given inversion densities and different symmetry conditions. FC, BC, and DG refer to the front channel, back channel, and symmetric double gate conditions. E_R is the ratio of the surface field at the front and back-interface. N_{inv} is the total inversion density in the FC and BC case while it is one half of the total inversion density in the DG mode and for E_R=0.5 or 2.0.

be mostly caused by a variation of the phonon-scattering rate due to the different electronic structures of the SG and DG charge configurations that has been recently predicted by numerical simulations [17].

As a further confirmation that the change in the electronic structure between SG and DG operation mode is responsible of mobility modulation, Fig. 10 reports the measured mobility at two given inversion densities and different symmetry conditions, including front and back SG (FC, BC), symmetric DG, and two asymmetric DG conditions, characterized by the value of E_R, defined as the ratio of the two surface fields at the front and back-interface. N_{inv} is the total inversion density in the FC and BC case while it is one half of the total inversion density in the symmetric DG mode and for E_R=0.5 or 2.0. A maximum of μ_{eff} is observed in the symmetric DG condition.

4 Conclusions

In this paper we have presented an experimental characterization of low-field electron mobility μ_{eff} in ultra-thin SOI MOSFETs operating in symmetric DG conditions (two symmetric inversion layers at the two Si-SiO_2 interfaces) and compared it with single-gate (SG) mobility. Mobility was measured for silicon thickness down to 5nm and temperatures between 225 and 375K. The results show that at large inversion densities μ_{eff} is very similar to that of lightly doped bulk transistors in the entire T_{SI} range, resulting in a remarkable improvement for a given inversion density compared to heavily doped bulk devices. At relatively small N_{inv} a systematic mobility reduction with decreasing T_{SI} is observed, compared to undoped bulk MOSFETs, possibly due to increased phonon-scattering due to spacial confinement in the silicon layer.

For T_{SI} $\approx$20nm, μ_{eff} is essentially the same in the SG and DG conditions. For thinner silicon films and small N_{inv}, however, we observed a clear μ_{eff} improvement in DG mode, that increases for decreasing temperatures and exhibits a peak when the two channels are symmetric. The difference in the electronic structure between SG and DG mode is likely to be the main responsible of such behavior.

References

[1] *International Technology Roadmap for Semiconductors:1999.* Austin, TX, SEMATECH, 1999.

[2] S.Takagi, A.Toriumi, M.Iwase, and H.Tango, "On the Universality of Inversion-layer Mobilty in Si MOSFETs. Part I- Effect of Substrate Impurity Concentration," *IEEE Transaction on Electron Devices*, vol. 41, no. 12, pp. 2357–62, 1996.

[3] G. Baccarani, M. R. Wordeman, and R. H. Dennard, "Generalized Scaling Theory and its Application to a 1/4 Micrometer MOSFET Design," *IEEE Transaction on Electron Devices*, vol. ED-31, pp. 452–462, 1984.

[4] E. Suzuki, K. Ishii, S. Kanemaru, T. Maeda, T. Tsutsumi, T. Sekigawa, K. Nagai, and H. Hiroshima, "Highly Suppressed Short-Channel Effects in Ultrathin SOI n-MOSFET's," *IEEE Transaction on Electron Devices*, vol. 47, no. 2, pp. 354–359, 2000.

[5] M.Jurczak, T.Skotnicki, M. Paoli, B.Tormen, J.Martins, J.L. Regolini, D.Dutartre, P.Ribot, D.Lenoble, R.Pantel, and S.Monfray, "Silicon-on-Nothing (SON) - an Innovative Process for Advanced CMOS," *IEEE Transaction on Electron Devices*, vol. 47, no. 11, pp. 2179–2187, 2000.

[6] H.S.Wong, D. Frank, and P. Solomon, "Device Design Considerations for Double-Gate, Ground-Plane and Single-Gated Ultra-Thin SOI Mosfet's at the

25 nm Channel Length Generation," in *IEEE IEDM Technical Digest*, p. 407, 1998.

[7] K. Kim and J.G. Fossum, "Double-Gate CMOS: Symmetrical- versus Asymmetrical-Gate Devices ," *IEEE Transaction on Electron Devices*, vol. 48, no. 2, pp. 294–299, 2001.

[8] K.Suzuki, T.Tanaka, Y.Tosaka, H.Horie, and Y. Arimoto, "Scaling theory for double-gate soi mosfet's," *IEEE Transaction on Electron Devices*, vol. ED 40, p. 2326, 1993.

[9] A. G. Sabnis and J. T. Clemens, "Characterization of the electron mobility in the inverted ⟨100⟩ Si surface," in *IEEE IEDM Technical Digest*, pp. 18–21, 1979.

[10] F. Balestra, S.Cristoloveanu, M.Benachir, J.Brini, and T.Elewa, "Double-Gate Silicon-on-Insulator Transistor with Volume Inversion: A New Device with Greatly Enhanced Performance," *IEEE Electron Device Letters*, vol. EDL-8, pp. 410–412, 1987.

[11] S.Venkatesan, G.W.Neudeck, and R.F.Pierret, "Dual-Gate Operation and Volume Inversion in n-Channel SOI MOSFET's," *IEEE Electron Device Letters*, vol. 13, pp. 44–46, 1992.

[12] D. Esseni, M.Mastrapasqua, G.K.Celler, C.Fiegna, L.Selmi, and E.Sangiorgi, "Low Field Electron and Hole Mobility of SOI Transistors Fabricated on Ultra-Thin Silicon Films for Deep Sub-Micron Technology Application," *IEEE Transaction on Electron Devices*, vol. 48, no. 12, pp. 2842–2850, 2001.

[13] M. R. Pinto, E. Sangiorgi, and J. Bude, "Silicon MOS Transconductance Scaling into Overshoot Regime," *IEEE Electron Device Letters*, p. 375, 1993.

[14] F. Assad, Z. Ren, D. Vasileska, S. Datta, and M. Lundstrom, "On the Performance Limits for Si MOSFET's: A Theoretical Study," *IEEE Transaction on Electron Devices*, vol. 47, no. 2, pp. 232–240, 2000.

[15] C.Fiegna and A.Abramo, "Analysis of Quantum Effects in Nonuniformly Doped MOS Structures," *IEEE Transaction on Electron Devices*, vol. 45, no. 4, pp. 877–880, 1998.

[16] B. K. Ridley, "The electron-phonon interaction in quasi-two-dimensional semiconductor quantum-well," *Journ. of Physics C: Solid State Physics*, vol. 15, pp. 5899–5917, 1982.

[17] M.Shoji and S.Horiguchi, "Electronic structures and phonon-limited electron mobility of double-gate silicon-on-insulator Si inversion layers," *Journal of Applied Physics*, vol. 85, no. 5, pp. 2722–2731, 1999.

International Journal of High Speed Electronics and Systems, Vol. 12, No. 2 (2002) 343–351

FAR–FUTURE TRENDS IN SOI TECHNOLOGY: A GUESS

SORIN CRISTOLOVEANU
Institute of Microelectronics, Electromagnetism and Photonics (CNRS, INPG & UJF)
Centre for Projects in Advanced Microelectronics (CPMA)
ENSERG, B.P. 257, 38016 Grenoble Cedex 1, France

The horizon of the Silicon On Insulator (SOI) technology is explored based on the compilation of many expert viewpoints. An SOI panel was organized at WOFE'02 with the aim of 'guessing' the shape of SOI and its industrial place in 10 years from now. SOI is expected to gradually take the role of bulk-silicon CMOS, reach the final stages of the ITRS roadmap in microelectronics, and enter the nanoelectronics world. The debate exhaustively examined the strategic and critical aspects of SOI technology. Only a few issues were retained as serious challenges.

1. Introduction

Silicon On Insulator (SOI) technology has achieved significant progress and has been included in the International Technology Roadmap of Semiconductors (ITRS). After the pioneering involvement of IBM, other key microelectronics companies have announced the migration of IC's families (including microprocessors) from bulk-silicon into SOI. A typical example of strategic projection is shown in Figure 1. Bulk-Si CMOS technology was occupying the whole breathing space and, before the third millennium, there was no room left for SOI. The Moore's law cannot go on indefinitely and, already, we witness a tendency of inflection. At this point, SOI becomes extremely attractive because it offers a gain in performance of about one generation. In addition, SOI can continue on the 'MOS Avenue' until the very end, leaving bulk-Si on the sidewalk.

There is no doubt about the superior performance and scalability of SOI transistors, hence the immediate future of SOI is no longer a question, it is an affirmation. More interesting are the long-term prospects and the role of SOI components in the ultimate stages of the microelectronics. This was the topic of a technical debate organized during WOFE'02 conference.

In this paper, we reproduce the main interrogations together with the opinions expressed during the panel, by both the invited SOI experts and the top-level audience. For each issue, we indicate the general context, the median feeling, as well as the extreme positions adopted by several participants. The next section is dedicated to strategic considerations, whereas section 3 addresses more practical concerns related to the further development of SOI materials and devices.

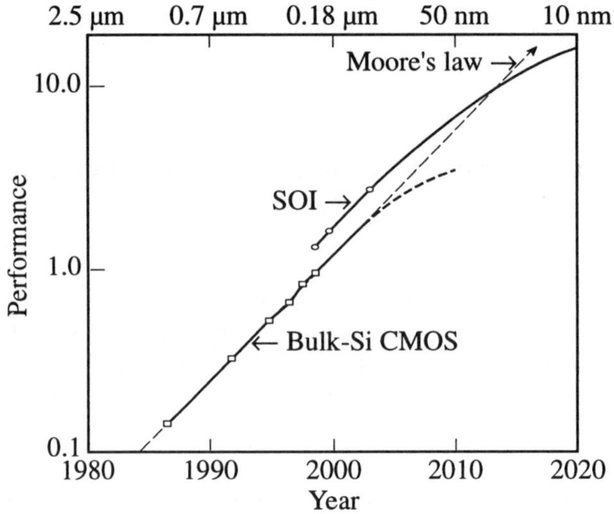

Fig. 1. Relative performance of bulk-silicon and SOI circuits in the past and future.

2. SOI Technology in 10 Years from Now

Will SOI be challenged by other Si-based technologies?

This is the fundamental question for existing and surviving. The general opinion is that bulk-Si will not be able to compete beyond a certain technology node. SOI will dominate, however future SOI might look different from the present SOI (100–200 nm silicon film over a 200–400 nm thick buried oxide). Possible competition can come from SiGe, but SiGe films will presumably be attached to an insulator, and this is still SOI.

How will SOI look like in 10 years?

Ultimate scaling requires ultra-thin films (< 30 nm) and buried oxides (< 100 nm). Achieving such structures is the current trend in SOI wafer manufacturing. On the other hand, enhanced functionality can be reached with new materials (Figure 2) obtained by replacing: (i) the silicon film with another semiconductor (XOI structure, where X stands for SiGe, SiC, III–V or II–VI compounds, etc), (ii) the buried oxide with a different high-K or low-K insulator (SOY structures), including air-gaps (SON), and (iii) the silicon substrate with glass, diamond, plastic, etc. 3–D stacks, sort of Si–SiO_2 'super-lattices' are suitable for more compact integration and special optical effects. It is admitted that all above structures are part of the SOI family. Most believe that ultra-thin SOI will dominate, but many other solutions will be available: SOY, XOI (mainly SiGe, also labelled SGOI), 3–D, etc.

Which gate length will allow SOI to definitely pass bulk-Si?

There was very little divergence on this point. 50 nm gate length seems indeed

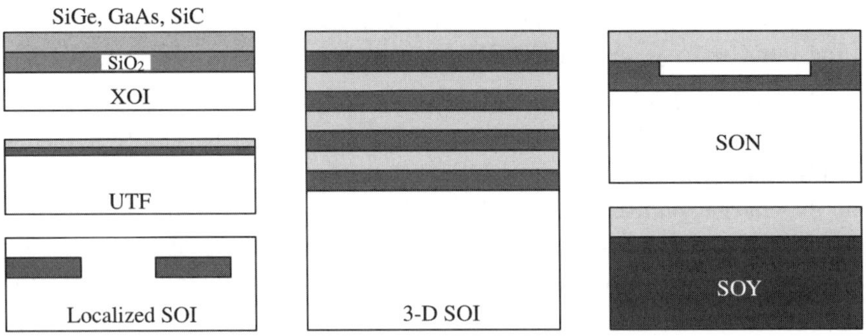

Fig. 2. Schematic configuration of future SOI-like structures.

to be the barrier beyond which the bulk-Si CMOS technology will not be able, or economically capable, to go. A few Si–addicted colleagues would like to see this barrier pushed a little further to about 35 nm.

What will be the minimum achievable channel length in SOI–MOSFETs?
In other words, how far will SOI be able to go from 50 nm down? A variety of opinions are offered due to the ambiguity of the question. It is admitted that, in the next 10 years, the industry will be capable to set a reliable 30 nm gate length process, whereas research laboratories will demonstrate sub-10 nm transistors. A 10 nm MOSFET is considered to be at the border of profitability. Yet, much smaller devices can be imagined. Electrostatic analysis indicates that short-channel effects can be kept under control when the film thickness is about 1/3 of the channel length. We extrapolate that the 'final' SOI–MOS transistor will be 1-nm-long, fabricated on one monolayer of Si on insulator. Recent simulations show that 2.5 nm long MOSFETs still exhibit MOS characteristics. However, most participants think that 5 nm is a more reasonable limit, imposed by direct tunneling between source and drain.

What will be the minimum feasible film thickness?
The general opinion is that 10–15 nm thick films will be utilized in about 10 years and, by the end of the roadmap, 3–5 nm thick films will be universally available. In this range, the transistor will enjoy special quantum and transport properties. Pessimists will be happy to stay above 8 nm thickness, whereas optimists regret to be blocked at half a nanometer!

Fully or partially depleted SOI MOSFETs?
Partially depleted integrated circuits are under development and optimization because they better correspond to the current capability of the SOI technology (thicker films, more similarity with bulk-Si CMOS, etc). In less than 10 years, fully depleted MOSFETs (first with single-gate and then with double-gate configuration) will be mainstream. It is noted that the acronym FD is threatened as companies will at-

tempt to claim, for pure marketing reasons, the paternity of a new technology, which is nothing but the 25-years old FD SOI.

What will be the limiting mechanism or issue?
The broad range of answers reflects the challenges faced by the SOI technology. From a physicist viewpoint, source-drain tunneling in sub-10 nm long MOSFETs and energy/charge quantization in ultra-thin quantum wells will dominate. Technologists worry about source-drain engineering, doping fluctuations, and series resistance problems, while managers are concerned about spread in characteristics and cost.

Is the double-gate MOSFET the ultimate, perfect transistor?
Of course, double-gate MOSFETs are unchallenged in terms of scalability, current drive, and transconductance. It is also noted that they are better adapted to high–K dielectrics, being able to lower the noise and the gate tunneling current, as illustrated in Figure 3. There are still question marks about the feasibility of a convincing double-gate process and the device topology: planar, vertical, or mixed (as in the Fin process).

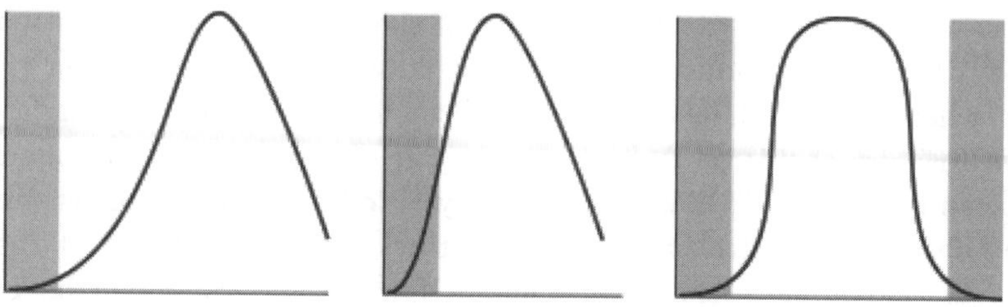

Fig. 3. Schematic distribution of the inversion charge in single-gate and double-gate MOSFETs with SiO_2 or higk-K dielectrics. In ultra-thin double-gate transistors the charge centroid is located in the middle of the film, further from the interface, which may improve the noise, the mobility, and the tunneling current.

Which will be the dominant SOI technology?
The opinion is divided between the two technologies expected to survive: bonding and layer transfer (as Unibond, Smart-Cut, and Eltran) or oxygen implantation (as SIMOX and ITOX).

How many SOI wafers per year?
The total number of SOI wafers (300 mm diameter) is estimated from 10 to 100 million. Whatever the figure, this is far more than the SOI production in 2001–2002 (1 million 200-mm-wafers).

What will be the main application of SOI technology?
SOI is envisioned for various domains: low-power/low-voltage, high performance, RF, MEMS, power, nano-electronics, optical devices and interconnects, or another

application ... unknown yet. In the far-end horizon, SOI appears as the most suitable structure for optics and for transition to nanoelectronics. On the practical side, the search for high performance will probably drive the development of the SOI technology in the next decade. However, LP/LV, RF, and System-On-Chip will stand as important segments of the market. Overall agreement is reached on the following statement: LP/LV will intrinsically come together with high performance and RF capabilities.

Are SOI–based quantum devices realistic?
This is an esoteric question as most quantum devices, which could be accommodated on SOI, have still to be imagined. There are just preliminary attempts to fabricate tunneling diodes, very narrow or very thin quantum wells, quantum dots and single-electron transistors. While basically in favor of quantum SOI devices, most participants think that the question is too premature and should be asked again in 10 years.

3. SOI–Related Concerns

When launched on the market place, an emerging technology as SOI brings many potential advantages but also generates a new set of problems. Most difficulties are related to the particular configuration of SOI, the material quality, and the transfer of expertise from bulk-Si to SOI. There are still several typical issues that SOI has to learn how to domesticate and take advantage of.

In this section, we do not focuss on immediate concerns that exist and are over-emphasized by SOI-reluctant engineers. These problems will be gradually solved when SOI will reach more industrial maturity. Instead, we attempt to anticipate those which will still be of concern in 10 years. Table 1 reproduces the items under debate as well as the median opinion. More details are given next.

Wafer quality and defects.
The quality of the silicon film and buried oxide has been continuously improved, resulting in a reduction of the defect density by orders of magnitude. Bonded SOI wafers have already reached the standard of bulk-Si material. This trend is expected to continue in 300 mm wafers and beyond, so that the quality will stop being a question mark. This holds true even for fully depleted transistors which are more sensitive to Si crystal defects.

Film thickness uniformity.
Thickness fluctuations are responsible for the spread of the threshold voltage in fully-depleted MOSFETs. The major effects are DIBL, fringing fields and interface coupling, rather than the fluctuation of the body charge. The present standard of ± 5 nm variation across the wafer has to be improved, which implies the development of new thinning techniques. This aspect is considered as a moderate challenge.

Integrity of thin buried insulators.
Bonded technologies offer a buried thermal oxide, whereas the BOX synthesized

Table 1: Summary of the debate on the critical aspects of SOI technology. The median opinion of the participants and the standard deviation is indicated. The rating was from 0 to 3: 0 = no problem, 1 = some concern, 2 = serious concern, 3 = show stopper.

Issues	Average	Deviation
Wafer quality & defects	0.5	0.6
Film thickness uniformity	1.2	0.6
Thin BOX integrity	0.6	0.5
Doping fluctuations	0.8	0.9
V_T control	1.8	0.5
Leakage currents	1.0	0.5
Series resistance	1.8	0.35
Design tools & libraries	0.7	0.5
Models & simulators	0.4	0.3
Difficult process modules	1.6	0.6
Available market space	0.2	0.3
Cost	0.6	0.7

by oxygen implants has a somewhat different micro-structure. Leakage problems, interface traps, and oxide defects will certainly be eliminated, even in ultra-thin insulators, by further optimization.

Doping fluctuations.
The control of the implant energy and doping profile is difficult in very thin films and affect the performance of partially depleted SOI MOSFETs. In principle, fully-depleted transistors do not require doping at all (except for V_T control) and are immune to fluctuations. One participant is concerned by doping fluctuations in the source and drain which may cause variations in series resistance.

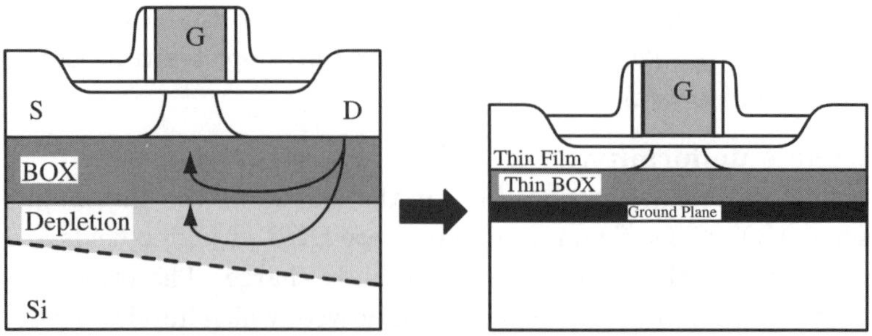

Fig. 4. Field penetration from drain to body (through the BOX and substrate) and optimized architecture using a ground plane.

Threshold voltage control.
Two different aspects are examined. First, the threshold voltage is governed by various short-channel effects, the magnitude of which depends on film thickness. Since film inhomogeneity results in fluctuations of short-channel effects, the threshold voltage and off-state leakage current will also fluctuate. A temporary solution consists in using a ground plane (Figure 4), before advanced double-gate MOSFETs will be achieved. A second problem, considered as serious, is the accurate tuning of the threshold voltage to proper values, which implies the development of new gate materials, with adequate work-functions. Another possible approach is to take advantage of back-gate biasing, in order to set the front-channel threshold by coupling effects.

Leakage currents.
For years, SOI has been advocated because of much lower junction leakage current (and capacitance) than in bulk Si. Also, the supplementary leakage current flowing through the buried oxide has radically been lowered. There are other sources of leakage, primarily through the thin gate insulator, but this is a consanguine illness in bulk-Si MOSFETs as well. The only special concern in SOI is the off-state leakage current whose control requires, besides the excellent subthreshold slope available in SOI MOSFETs, the proper (doping-free) control of the threshold voltage.

Series resistance.
Large series resistances can jeopardize the benefit brought about by the enhanced carrier mobility in fully-depleted films. Various solutions are under development: advanced silicidation (difficult in ultra-thin films), raised (thicker) source and drain, and electrically controlled junctions. The latter idea consists in using, for example, a longer back gate (or a second front gate) which overlaps the source and drain terminals; biasing the back gate induces an inversion channel in the body and, simultaneously, an accumulation layer in the terminals, so that their resistance is lowered (Figure 5). Unfortunately, the series resistance problem will be aggravated as the transistors will become shorter and thinner. It will continue to be a severe issue, presumably not a show stopper.

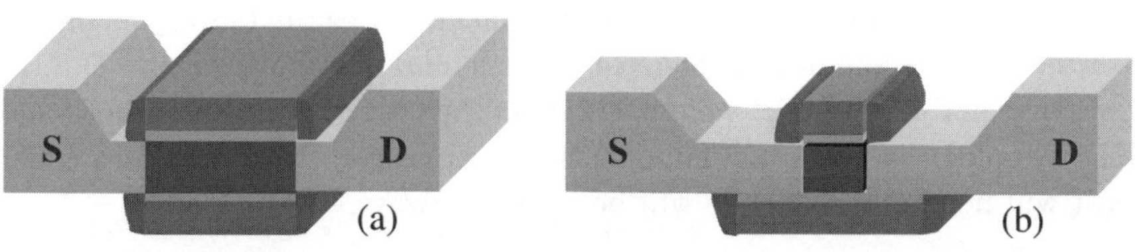

Fig. 5. Double-gate transistors with symmetric or length-asymmetric gates.

Availability of design tools and libraries.
The relative scarceness of SOI-dedicated libraries nowadays is viewed as one of the

weakest points. The companies, which already produce successful SOI circuits, have developed in-house compact models, tools, and libraries. Others are reluctant about moving to SOI just because of such necessary developments. However, the context is expected to change rapidly: libraries will enlarge with time and new companies are being created to provide SOI-design service to customers and chip manufacturers. There will be no design crisis in SOI!

SOI models and simulators.
The physics of conventional partially- or fully-depleted SOI devices is well covered and reliable physics-based models, validated by systematic measurements, have been published. Compact models are also proposed by several universities and companies. Most process and device simulators, utilized in the industry, do include SOI-dedicated modules. More modules are under development. There is still plenty of room for modelling in the domain of innovative SOI components: double-gate MOSFETs, tunneling and quantum devices, etc. This new generation of devices will require simulators clever enough to understand and solve simultaneously the 3–dimensional Schrödinger, Poisson, and non-stationary transport equations. Monte-Carlo codes will play an important role. Shaping the future landscape of modelling is nothing but a matter of routine.

Difficulty of process modules.
Full benefit of SOI circuit performance is not obtained by simply transposing the bulk-Si technology. Specific SOI modules need to be implemented or revisited. This is not an easy task, particularly for thin film structures. Gate and source/drain engineering stands as a difficult issue. This challenge is worth it because, *en revanche*, the nightmare of doping (heavy dose, profiling, etc) will be forgotten.

Availability of market space.
This is actually a fading problem, which was imposed by the hegemony of bulk-Si. Product families are now migrated from bulk to SOI, and new market segments are continuously opened. The participants are extremely optimistic about SOI chances of success.

Cost.
Adopting the SOI technology implies additional expenses related to the higher cost of SOI wafers and to the development of process modules and libraries. But, on one hand, wide availability of SOI wafers will drop the cost and, on the other hand, the bulk-Si technology is becoming exponentially expensive. It is therefore concluded that cost will not be an issue. It will be awful anyway ...

Other concerns.
No other critical issues could be detected or included in the above list. According to an interesting scenario, imagined by a participant and approved by many, SOI v progress a lot and, in 10 years, we will be facing problems that are not identifiable yet.

4. Conclusion

Even in the near past, SOI was object of reluctance and mixed feelings. The recent change in the context is effective and can be sensed in international meetings, where more and more SOI-dedicated sessions are included. A long-term progress in microelectronics cannot be imagined without SOI. The participants had an overall optimistic appreciation of SOI perspectives. A large set of possible problems and challenges have been examined and many question-marks have been erased.

Accurate predictions were formulated by recognized experts. In 10 years from now, we will enjoy verifying how wrong these predictions were.

Acknowledgements

The panel was composed of highly respected SOI connoisseurs: George Celler (SOITEC), Jerry Fossum (University of Florida), Enrico Sangiorgi (University of Udine), Christophe Tretz (AMD), and Alex Zaslavsky (Brown University). They all are deeply thanked for sharing, with enthusiasm, their expertise and visions. The debate was also made successful by the active contribution of a distinguished audience.

International Journal of High Speed Electronics and Systems, Vol. 12, No. 2 (2002) 353–364

NANOELECTRONICS: SOME CURRENT ASPECTS AND PROSPECTS

ROBERT HULL
University of Virginia, Department of Materials Science and Engineering, Charlottesville, Virginia 22920, USA
hull@virginia.edu

RICHARD MARTEL
IBM T. J. Watson Research Center, Yorktown Heights, NY, 10598, USA

J. M. XU
Division of Engineering, Brown University, Providence, RI 02912, USA

A brief summary is provided of selected current activities in the field of nanoelectronics, which is taken here to mean the fabrication and integration of active microelectronic components with feature dimensions of tens of nanometers or less. Particular emphasis is placed upon the classes of nanoelectronic devices that were discussed at the 2002 WOFE Conference.

Keywords: Carbon nanotubes; quantum cellular automata; DNA; molecular electronics; device scaling.

1. Evolution of Current Microelectronics Scaling

The relentless drive towards miniaturization in the silicon microelectronics industry has resulted in circuits with exponentially improved functionality over the past several decades. This has largely been driven by the reduction in the feature dimensions of individual devices within the circuit. Minimum dimensions in commercial production, are currently 0.13 μm and will decrease by a factor of two approximately every six years. This produces devices that are faster, use less power, and are cheaper on a cost-per-unit basis, resulting in the ever-increasing capacity and speed of commercial microelectronic circuits. However the trend of increasing miniaturization is approaching inevitable limits. The limit of the current metal oxide semiconductor field effect transistor (MOSFET) architecture is a matter of continuing debate, but it appears that it will be reached by about the 32 nm node (slated for 2013 on the International Semiconductor Technology Roadmap).[1] Beyond that, future scaling will be limited by control, metrology and statistical fluctuations of dopant concentration, availability of suitable gate dielectric materials that remain sufficiently immune to tunneling at the required ultra-thin layer dimensions, and by interconnect challenges, among many other factors. Thus, although the conventional MOSFET architecture has proven to be extraordinarily robust with respect to continuous scaling (and its demise prematurely predicted for the past ten years or so), it does appear that its practical limits will be reached early in the next decade. Hence the great interest in exploring new device and circuit architectures, both those that are evolutionary extensions of the Si MOSFET, and those that are based upon entirely new materials and concepts.

While the current microelectronics industry relies most heavily on silicon as its fundamental "building" material, there are of course extensive parallel efforts in compound semiconductor devices, for both future and current applications. In this summary, however, we will focus on silicon-based devices when discussing existing semiconductor architectures.

2. New Architectures for Silicon MOSFETS: Scaling Towards Ten Nanometers

Conventional Si MOSFET devices are limited (amongst other factors) by the fact that tunneling through SiO_2 limits gate dielectric thicknesses to about 1 nm. Ultra-high precision control of the Si/SiO_2 interfacial roughness/thickness uniformity, and source and drain junction depths have enabled fabrication of both p- and n-MOSFETS with gate lengths of order 50 nm, and effective channel lengths in the 20 nm range.[2] Scaling to the 32 nm node on the current SIA roadmap seems plausible, particularly if gate materials with higher dielectric constant (and that are compatible with the full spectrum of materials and processing requirements for the MOSFET manufacturing process) become available.

Extension to even smaller devices, with gate lengths of order 10 nm, will likely require the development of new MOSFET geometries. These include double gate structures,[3–5] that promise to allow further MOSFET device scaling by virtue of the electrostatic confinement of the channel afforded by the two gates, until source-to drain tunneling occurs at dimensions less than 10 nm. A double-gate device has a Si body thinner than one half the gate length, sandwiched by thin dielectric and gate stacks on both sides. In this configuration, the channel can also be lightly doped thus obviating the needs for high channel doping levels and complex doping profiles. Three broad implementations of such structures have been proposed. (1) The planar configuration mirrors the existing MOSFET and is thus compatible with conventional CMOS processing techniques.[3] (2) Vertical surround-gate devices, of either the etched pillar type or epitaxial trench-refill type.[4] (3) In the FinFET design,[5,6] conduction channels are on the vertical surfaces of a silicon fin etched from an SOI substrate, and a prong-shaped gate straddles the fin.

Parallel development of higher dielectric constant gate materials is another critical activity. As SiO_2 gate dielectric thicknesses decrease as the gate length scales, tunneling through the oxide becomes an increasingly significant factor. A typical leakage current density for a 1.5 nm thick film of thermal SiO_2 is of order 1 A/cm^2.

At an oxide thickness of 1.0 nm, this increases to about 100 A/cm^2. Further, detailed atomic-scale compositional analysis using high angle annular dark field imaging and electron energy loss spectroscopy in the transmission electron microscope has shown that a far less insulating sub-oxide SiO_x layer of 0.35 nm exists at the Si/SiO_2 interface,[7] effectively creating electrical shorts at oxide layer thicknesses less than 0.7 nm (there being two Si-oxide interfaces at the channel and at the gate electrode). Thus there is extensive activity at identifying higher dielectric constant gate materials (as higher ε enables capacitance of the dielectric to be maintained at a higher film thickness, thereby greatly reducing tunneling currents) that have the required material properties and that may be integrated into conventional MOSFET processing schemes.[8] Thus continued scaling of MOSFET like structures will require parallel advances in both materials development/processing, and device design/implementation.

3. Quantum Dot-Based Nanoelectronics

Quantum-sized semiconductor structures may be synthesized through a variety of approaches including self assembly from solution,[9,10] through epitaxial growth of lattice mismatched heterostructures,[11,12] and through approaches employing lithographic definition.[13,14] Such structures may be employed in a range of potential nanoelectronic architectures. For example, quantum cellular automata (QCA) rely upon the transmission of signals through bistable states of quantum dot quadruplets (Fig. 1). Direct serial fabrication of structures with sufficiently small dimensions to allow operation at practical temperatures is likely to be too challenging and slow for commercial application. For example, dot dimensions of a few tens of nm correspond to energy differences between the states of order of only 1 meV;[15] dot spacings are required to be of order ten nm or less to allow rapid tunneling rates and high device speeds. To date QCA cells have been implemented only with metallic quantum dots with maximum operating temperatures of a few hundred millikelvin and maximum operating speeds of a few hundred MHz.[16] However, there are enormously attractive features of such circuits: power-delay products are many orders of magnitude less than in conventional CMOS devices, packing densities orders of magnitudes higher and interconnect challenges are largely circumvented. Self assembly and ordering

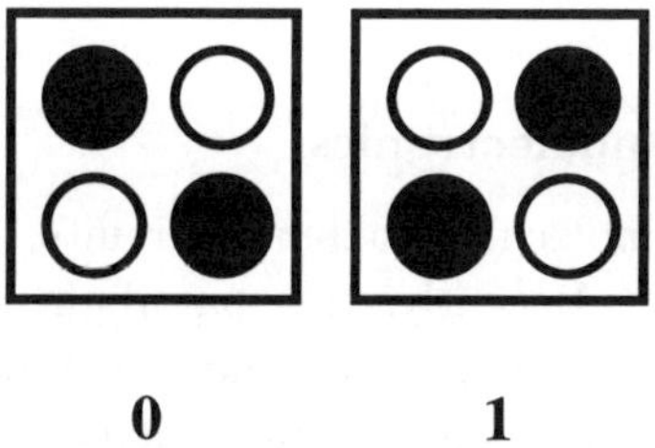

Fig. 1. The quantum cellular automata (QCA) concept: two bistable states exist for extra electronic charge (filled circles) in the quantum dot quadruplet.

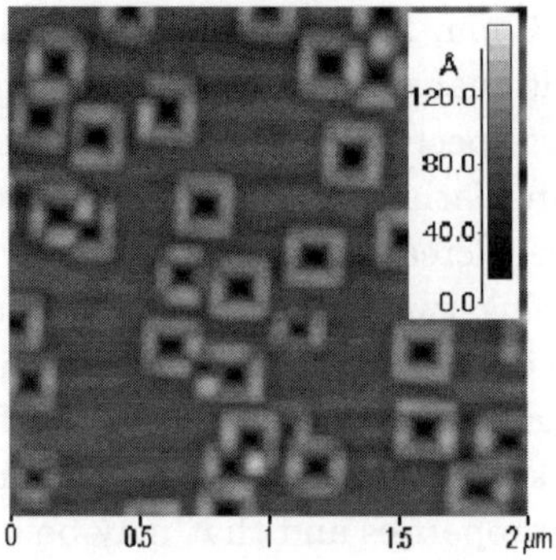

Fig. 2. AFM images of quadruplet clusters nucleating around stress-relieving pits formed during growth of $Ge_{0.3}Si_{0.7}$ alloys by Molecular Beam Epitaxy at 550°C and a growth rate of 0.09 nm/s. At this stage the mean film thickness is 30 nm.[17]

processes during strained semiconductor cluster nucleation by epitaxial growth offers one potential route for growth of QCA like structures of sufficiently reduced dimensions to enable operation at significantly higher temperatures and speeds than so far realized during operation of metallic QCA cells. For example, Fig. 2 illustrates the formation of quadruplet cluster geometries ("quantum fortresses") nucleating around pits formed during heteroepitaxial growth of $Ge_{0.3}Si_{0.7}$ alloys on Si(100) surfaces at a growth temperature of 550°C and a growth rate of 0.09 nm/s by Molecular Beam Epitaxy.[17]

Quantum dot/nanocrystalline architectures have also been demonstrated as the basis for ultra-high density memory structures,[18–20] and most typically comprise nanocrystals or single quantum dots embedded between a control oxide/polycrystalline gate and the source–drain conduction channel. Electrons are injected by tunneling into the nanocrystal floating gate, where trapped charges cause a threshold voltage shift of the device. Critical issues to address include fundamental understanding of how retention time depends upon the number of electrons and energy spacing in the dot and the role of defects (e.g., dislocations and grain boundaries) upon charge retention. Another very active area of nanoelectronic device research is the "single electron transistor" concept where single electrons are controllably transferred between nanoscale semiconductor islands separated by tunnel barriers (see Ref. 21 for a recent review).

4. Molecular-Based Nanoelectronics

The idea of using small molecules to build circuits, or "molecular electronics" started as early as 1974.[22] Recently, this paradigm attracted special attention because of the rapid advances in probing and understanding the conduction and switching behavior in single molecules.[23] In principle, the basic functions of novel molecular-based nanodevices can be probed and evaluated for electronic applications. Research programs in molecular electronics are based on a radical change in the vision of future computers, which has already attracted significant attention and investment.[24] Among the ideas that have already emerged, random assembly strategies (the "bottom-up" approach), the defect tolerant computer architectures,[25] and cross bar designs,[26,27] have been the source of much work. Little is known, however, about the operation of the basic unit, the molecular nanodevice. Among the new alternative structures and devices proposed, the molecular-scale switches or memory units in a cross bar geometry,[27–29] in a break junction,[30,31] or contacted with nanowires,[32] and the work on nanowire circuits by C. Lieber and co-workers,[33] are important to mention. In addition, carbon nanotube FETs[34] and very recent three-terminal molecular device structures[35] present characteristics that already resemble in many respects conventional MOSFETs.

The prospect of new technologies based on inexpensive molecular devices is attractive and radically novel. However, only simple demonstrations based on prototype devices have been made, and there is an immense technology gap to be cover by basic and applied research. In particular, the integration, the reliability issues, the feasibility, and even the basic physics of electron transport in most of the molecular nanostructures have not yet been resolved. In this part of the overview, we present recent results obtained with both carbon nanotubes and conjugated molecules with a special emphasis on three terminal device structures.

4.1. *Conduction through DNA*

The understanding of charge transport in molecules is at the core of the research in molecular electronics. Many theoretical studies have looked into the fundamentals of the conduction in single molecule and molecular assemblies[36] and several experiments have supplied conductance measurements across simple molecules using different innovative techniques.[28,37,38] Among the most surprising and interesting experimental results is the observation of electrical conduction through single DNA molecules or small bundles of DNA strands.

DNA is a well understood double-helical stacking of periodically spaced nucleotide base pairs. However, what appears to be perfectly periodic stacking of base pairs can in fact be intrinsically disordered. Each sequence of the four bases corresponding to a biologically different DNA is also a different series of electrical potentials for electron transport. Although the spacing between the stacked base pairs is constant, the location and frequency with which any one of the four bases appears in the sequences are not, and this can give rise to very different conductivities. In the extreme limit of all base pairs being the identical (e.g., the poly-G oligomer), one has an ideal periodic potential. Depending on the coupling strength between the adjacent potential wells, electron transport could in principle take place via band-like conduction or hopping. Such synthetic poly-G DNAs have indeed been shown to be semiconducting. In the opposite limit, namely complete sequence disorder, no band of the conventional kind can exist because of the absence of any translational symmetry (i.e., periodic potential). Here, hopping offers a natural alternative for electron transport. However, if the coupling between adjacent bases is strong, the density of states in a given range of k value can be sufficiently high; and as in amorphous semiconductors, delocalized (band-like) transport can take place. Adding to this complexity are the effects of the contacts, critical for all molecular scale transport measurement and becoming dominating in measurements of short molecules of low-conductivity because of the extended screening length.

Significant electrical conduction has been experimentally measured in DNA strands and this topic has rapidly become the subject of on-going active debate. It has been suggested that the aromatic bases form a stack of π-electron states that are delocalized along the length of the DNA molecule.[39–41] The π-states would offer an efficient way to transfer electrons across the molecule. DNA conductivities reported in the literature vary widely; reports range from metallic to semiconducting to insulating; semiconducting[42] and metallic[43,44] behaviors have been observed in native and modified DNA strands respectively. The very recent results by J. M. Xu and coworkers on a long nature DNA (i.e., with no apparent sequence ordering) are consistent with a band-gap of a few hundreds of meV.[45] Moreover, this work has also shown that metallic conduction can be obtained when the electronic structure of the native DNA is altered by inserting an divalent metal ion into each base pair replacing the imino proton. With this new degree of freedom of DNA engineering, one can now entertain ideas of developing molecular structures based on engineered DNA molecules with controllably altered physical properties, and exploiting such structures for molecular electronics and computing. Given the drastically increased conductivity, the engineered DNA (referred to as M-DNA[46]) also become less susceptible to the effects arising from the intrinsic base potential disordering.

Despite the inevitable complications arising from the studies of DNA's electronic properties, DNA remains the molecule of choice for molecular electronics. As the basic building block of life, DNA has been studied and researched for decades and is the best understood biomolecule. From its geometry to its chemical reactions with thousands of other compounds, there exists a wealth of knowledge on DNA. Furthermore, there exists a wealth of tools and expertise for the manipulation, synthesis and analysis of DNA. Equally important is the fact that DNA has the ability to self-assemble, self-repair and be reliable outside of a clean room in the changing conditions of biological systems. These attributes are perhaps even more important than the size consideration for molecular electronics. It is also likely that the self-organization and molecular recognition properties of DNA could be coupled with carbon nanotubes or other nanostructures to offer both conduction and assembly capabilities. As a matter of fact, the DNA engineering work in Xu's group was initiated by the need to reliably make contacts and assemble the Y-junction Carbon Nanotubes made in his lab in 1999.[47,48] More work is clearly needed so that we understand better its transport properties as a function of the sequence and alteration. However, DNA has clearly passed the initial tests of the conductivity.

4.2. *Self assembled monolayer (SAM) FET*

The most sophisticated molecular device to date appeared very recently in the form of a novel molecular field effect transistor (SAM FET) made by H. Schön and coworkers from Bell Laboratories.[35] Characteristics measured on a SAM FET made with conjugated molecules (e.g., 1,4-biphenyl dithiol) self-assembled on gold (SAM FET) are shown in Fig. 3. The SAM FETs show potentially ground breaking results because the observed current drive and transconductance are larger than the best semiconductor transistors. The source–drain separation of the SAM FET is assumed to be the length of the single molecule, i.e., ~ 1.5 nm, while the gate-to-channel distance is long (~ 30 nm). The aspect ratio of the device is therefore a real puzzle. The characteristics of the SAM FETs look disarmingly normal, however, for an ultra short device. There are concerns because the switching characteristics of the SAM FET are those of a long channel device, which appear to be in violation to the basic laws of electrostatics (i.e., superposition of electric fields).[49] Nevertheless, these results are extremely interesting and promising. Based on these SAM FET characteristics, it is likely that such a transistor, if reliably reproduced, can find its use in molecular-based electronics.

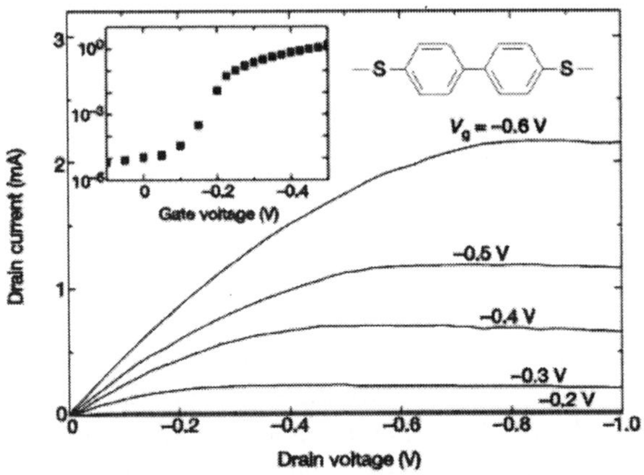

Fig. 3. Output and sub-threshold (inset) characteristics of a SAM FET.[35]

5. Carbon Nanotube Electronics

The electronic structure of single-wall carbon nanotubes presents unique properties for electronic applications (see Fig. 4). It can be either metallic or semiconducting depending on the arrangement of the carbon atoms.[34] Metallic nanotubes present interesting properties for future interconnect applications while semiconducting nanotubes can be used for making field-effect transistors.[50,51] The carbon nanotube FET (CNFET) exhibits the desired properties of a logical switch.[52] It functions much like a conventional semiconductor transistor, with some important differences. First, the carrier transport in single-wall carbon nanotubes is

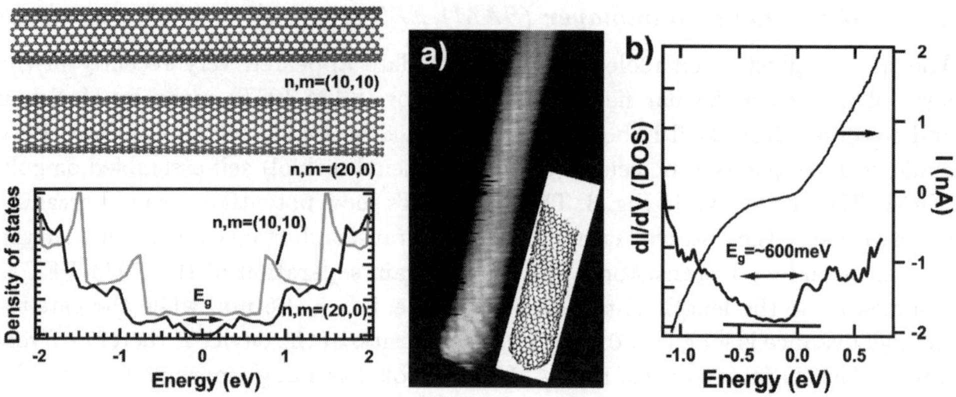

Fig. 4. Top left: model of the structure of a metallic and a semiconductor nanotubes. Bottom left: calculated density of states (DOS) for a $[n, m] = [20, 0]$ semiconducting and a $[10, 10]$ metallic single wall nanotube. (a) Scanning tunneling microscopy image and (b) spectroscopy (STS) of a semiconducting (chiral) carbon nanotube dispersed on a gold substrate. A band gap of around 600 meV is measured by comparing the resulting dI/dV as a function of tip voltage.

one-dimensional. The reduced dimensionality of this semiconductor greatly reduces electron scattering and these devices would likely operate in the ballistic regime. Second, all chemical bonds are satisfied, and stable. Therefore, a careful passivation of the interface between the channel and dielectric is not necessary, i.e., there is no equivalent of the silicon/silicon dioxide interface. Third, the Schottky contact is the active switching element in an intrinsic nanotube device.[53] This one-dimensional contact switching is bound to have important implications while scaling the device structure.

Early nanotube FETs were back-gated devices with the nanotube side-bonded to noble metal electrodes leading to large contact resistances.[50,51,54,55] Since then, significant improvements in the performance of CNFETs have been achieved. The best devices are now built with top-gate geometry and they resemble, in many respects, to conventional silicon CMOS devices.[56] An example of the characteristics of CNFETs is shown in Fig. 5. This device is a top-gated CNFET with titanium carbide contacts and the gate is an Al electrode separated by 15 nm gate oxide (SiO_2) (see Fig. 5). The relevant parameters extracted from the device characteristics show very large improvements compared to the early nanotube devices. This is most likely due to the scaling of the dimensions and the improvement of the device geometry. While scaling the device, we noted, however, that the scaling properties of the CNFETs are not conventional. They rather suggest that contact switching (as oppose to bulk switching) is involved. In this respect, the extrapolation of the measured performance to shorter channel devices, i.e., in the limits of scaling, is still unclear for those nanotube transistors. However, the performance of the recent devices placed in the structure of an array is expected to out perform silicon-based devices.[52,56]

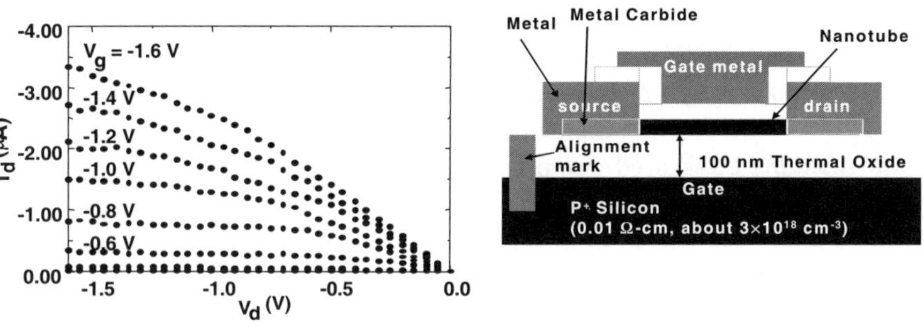

Fig. 5. Example of output characteristics of a top-gated CNFETs with aluminum electrodes and a schematic of the devices structure.[56]

5.1. *Complementary CNFETs, doping, and logic circuits*

Two different methods have been used to make circuits with nanotubes. One is based on *p*-type CNFETs connected by independent gates[57] and the other involves complementary nanotube transistors, i.e., both *p*- and *n*-type, integrated in the same structure.[58,59] The optimal way of forming such a molecular logic circuit involves complementary FETs. They consume less power, can achieve higher gain, are immune to fluctuations in the characteristics of the FET, and are easier to implement in integrated circuits. The main obstacle to make complementary nanotube logic is to make *n*-type transistors. Although the "as-grown" carbon nanotube are

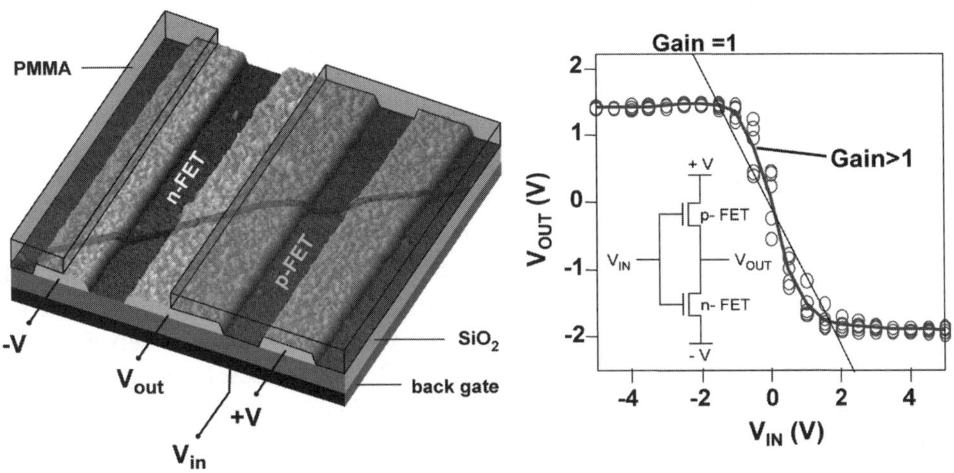

Fig. 6. (Left) AFM image showing the intra-molecular logic gate. A single nanotube bundle is positioned over the gold electrodes to produce two *p*-type CNFETs in series. The device is covered by PMMA and a window is opened by *e*-beam lithography to expose part of the nanotube. Potassium is then evaporated through this window to produce an *n*-CNTFET, while the other CNTFET remains *p*-type. (Right) Characteristics of the resulting intra-molecular voltage inverter. The thin straight line corresponds to an output/input gain of one. The corresponding circuit is shown in the inset [From Ref. 58].

undoped semiconductors, the resulting CNFETs are always *p*-type, i.e., hole conductors. *N*-type CNFETs can be made either by direct doping of the nanotubes with an electron donor,[60] or by a special treatment of the metal-nanotube Schottky contacts.[61,62] Either way, complementary CN circuits can be produced. In Fig. 6 we show a simple integrated circuit, a voltage inverter (NOT gate), integrated on a small bundle of carbon nanotubes.[58] The *p*- and *n*-type portions of the CNFET are defined along the length of an individual tube using spatially-resolved doping. This intra-molecular circuit is an example of the potential integration that can be achieved using nanotubes. Gain higher than one has been obtained.

The use of independent gates, such as the top-gate device in Fig. 5 or local bottom-gate structures as in Ref. 57, allow for the construction of more complex circuits. In principle, much higher gain is possible. However, the fabrication of arrays of nanotube devices is not yet possible. Making arrays of semiconducting nanotubes is the main challenge and will need major developments in both the synthesis and the assembly of nanotubes on a substrate.

6. Summary

Entire new classes of nanoscaled electronic devices are being explored, of which we have presented only a subset here. While the continuing miniaturization of conventional Si CMOS does march on, it appears highly likely that the fundamental MOSFET structure will need to be supplanted as devices scale below about the 32 nm node, either through re-design of the MOSFET device architecture, as described in Sec. 2, or through entirely new semiconductor device paradigms (Sec. 3), or through development of entirely new materials sets (Secs. 4 and 5). Continuing extensive research into this full set of alternatives can be expected over the coming decade.

Acknowledgments

Research described here in RH's group was supported by the National Science Foundation through DMR Grant #0075116 (Focused Research Group) and through the Materials Research Science and Engineering Center at UVa, "The Center for Nanoscopic Materials Design". The work described here in JX's group was supported by NSF, ONR, DARPA, AFOSR, CIAR, and Motorola, and performed by a number of most able former or current members of the lab (e.g., J. Li, C. Papadopoulos, A. Rakitin, and P. Aich).

Note Added in Proof

Since writing this paper, data in Ref. 35 by H. Schön *et al.* has been part of an investigation of scientific misconduct allegations (see http://www.lucent.com/new_events/researchreview.html). Section 4.2 of this paper should thus be understood to reflect understanding at the time of writing this article in early 2002.

References

1. International Technology Roadmap for Semiconductors, Semiconductor Industry Association, 2001, http://public.itrs.net/.
2. G. Timp, J. Bude, K. K. Bourdelle, J. Garno, A. Ghetti, H. Gossman, M. Green, G. Forsyth, Y. Kim, R. Kleiman, F. Klemens, A. Kornbilt, C. Lochstampor, W. Mansfield, S. Moccio, T. Sorch, D. M. Tennant, W. Timp, and R. Tung, *Proc. Int. Electron Device Meeting*, IEEE, Piscataway, NJ, 1999, p. 55.
3. J.-H. Lee, G. Traschi, A. Wei, T. A. Langdo, E. A. Fitzgerald, and D. A. Antoniiadis, *Proc. Int. Electron Device Meeting*, IEEE, Piscataway, NJ, 1999, p. 71.
4. R. Li, Y. Zhang, Y. Lu, D. S. Choi, M. Luo, and K. L. Wang, *Proc. Int. Electron Device Meeting*, IEEE, Piscataway, NJ, 2001, p. 63.
5. X. Huang, W.-C. Lee, C. Kuo, D. Hisamoto, L. Chang, J. Kedzierski, E. Anderson, H. Takeuchi, Y.-K. Choi, K. Asano, V. Subramanian, T.-J. King, J. Bokor, and C. Hu, *Proc. Int. Electron Device Meeting*, IEEE, Piscataway, NJ, 1999, p. 67.
6. D. Hisamoto, W.-C. Lee, J. Kedzierski, E. Anderson, H. Takeuchi, K. Asano, T.-J. King, J. Bokor, and C. Hu, *Proc. Int. Electron Device Meeting*, IEEE, Piscataway, NJ, 1998, p. 1032.
7. D. A. Muller, T. Sorch, S. Moccio, F. H. Baumann, K. Evans-Lutterodt, and G. Timp, *Nature* **399**, 758 (1999).
8. See special issue of MRS Bulletin on "Alternative gate dielectrics for microelectronics", **27**, 3 (2002).
9. N. Chestnoy, R. Hull, and L. E. Brus, *J. Chem. Phys.* **85**, 2237 (1986).
10. M. G. Bawendi, *Solid State Comm.* **107**, 709 (1998).
11. J. Tersoff, C. Teichert, and M. G. Lagally, *Phys. Rev. Lett.* **76**, 1675 (1996).
12. A. Madhukar, T. R. Ramachandran, A. Konkar, I. Mukhametzhanov, W. Yu, and P. Chen, *Appl. Surf. Sci.* **123–124**, 266 (1998).
13. L. P. Rokhinson, L. J. Guo, S. Y. Chou, and D. C. Tsui, *Superlattices & Microstructures* **28**, 413 (2000).
14. G. Jin, J. L. Liu, Y. H. Luo, and K. L. Wang, *Thin Solid Films* **380**, 169 (2000).
15. J. C. Lusth, C. B. Hanna, and J. C. Diaz-Velez, *Microelectronics J.* **32**, 81 (2001).
16. G. L. Snider, A. Orlov, R. K. Kummamuru, R. Ramasubramaniam, I. Amlani, G. H. Bernstein, C. S. Lent, J. L. Merz, and P. Wolfgang, *Proc. IEEE-NANO*, IEEE, Piscataway, NJ, 2001, p. 465.
17. J. L. Gray, R. Hull, and J. Floro, *Appl. Phys. Lett.* **81**, 2445 (2002).
18. K. Yano, T. Ishii, T. Hashimoto, T. Kobayashi, F. Murai, and K. Seki, *IEEE Trans. Electron. Devices* **41**, 1628 (1994).
19. S. Tiwari, F. Rana, H. Hanafi, A. Hartstein, A. F. Crabbe, and K. Chan, *Appl. Phys. Lett.* **68**, 1377 (1996).
20. L. Guo, E. Leobandung, and S. Y. Chou, *Science* **275**, 649 (1997).
21. K. K. Likharev, *Proc. IEEE* **87**, 606 (1999).
22. A. Aviram and M. A. Ratner, *Chem. Phys. Lett.* **29**, 277 (1974).
23. C. Joachim, J. K. Gimzewski, and A. Aviram, *Nature* **408**, 541 (2001).
24. K. K. Kwok and J. C. Ellenbogen, *Materials Today* **5**, 28 (2002) and references cited therein.
25. J. R. Heath, P. J. Kuekes, G. S. Snider, and R. S. Williams, *Science* **280**, 1716 (1998).
26. C. P. Collier, E. W. Wong, M. Belohradsky, F. M. Raymo, J. F. Stoddart, P. J. Kuekes, R. S. Williams, and J. R. Heath, *Science* **285**, 5426 (1999).
27. C. P. Collier, G. Mattersteig, E. W. Wong, Y. Luo, K. Berverly, J. Sampaio, F. M. Raymo, J. F. Stoddart, and J. R. Heath, *Science* **289**, 1172 (2000).
28. M. A. Reed, J. Chen, A. M. Rawlett, D. W. Price, and J. M. Tour, *Appl. Phys. Lett.* **78**, 3735 (2001).

29. J. Chen, M. A. Reed, A. M. Rawlett, and J. M. Tour, *Science* **286**, 1550 (1999).
30. M. A. Reed, C. Zhou, C. J. Muller, T. P. Burgin, and J. M. Tour, *Science* **278**, 252 (1997).
31. M. A. Reed, *Proc. IEEE* **87**, 652 (1999), and reference cited therein.
32. P. A. Smith, C. D. Nordquist, T. N. Jackson, T. S. Mayer, B. R. Martin, J. Mbindyo, and T. E. Mallouk, *Appl. Phys. Lett.* **77**, 1399 (2000).
33. Y. Cui and C. M. Lieber, *Science* **291**, 851 (2001).
34. For an overview, see special issue on carbon nanotubes, *Physics World* **13**, 6 (2000).
35. J. H. Schön, H. Meng, and Z. Bao, *Nature* **413**, 713 (2001).
36. N. D. Lang and Ph. Avouris, *Phys. Rev. B* **64**, 125323 (2001), and references cited therein.
37. L. A. Bumm, J. J. Arnold, M. T. Cygan, T. D. Dunbar, T. P. Burgin, L. J. Ones II, D. L. Allara, J. M. Tour, and P. S. Weiss, *Science* **271**, 1705 (1996).
38. J. Chen, W. Wang, M. A. Reed, A. M. Rawlett, D. W. Price, and J. M. Tour, *Appl. Phys. Lett.* **77**, 1224 (2000).
39. P. J. Dandliker, R. E. Holmlin, and J. K. Barton, *Science* **257**, 1465 (1997).
40. D. B. Hall, R. E. Holmin, and J. K. Barton, *Nature* **382**, 731 (1996).
41. M. R. Arkin, E. D. A. Stemp, R. E. Holmlin, J. K. Barton, A. Hormann, E. J. C. Olson, and P. F. Barbara, *Science* **273**, 475 (1996).
42. D. Porath, A. Bezryadin, S. de Vries, and C. Dekker, *Nature* **403**, 635 (2000).
43. H. W. Fink and C. Schonenberger, *Nature* **398**, 407 (1999).
44. Y. Okahata, T. Kobayashi, K. Tanaka, and M. Shimomura, *J. Am. Chem. Soc.* **120**, 6165 (1998).
45. A. Rakitin, P. Aich, C. Papadopoulos, Yu Kobzar, A. S. Vedeneev, J. S. Lee, and J. M. Xu, *Phys. Rev. Lett.* **86**, 3670 (2001).
46. J. S. Lee, L. J. P. Latimer, and R. S. Reid, *Biochem Cell Cio* **71**, 162 (1993).
47. J. Li, C. Papadopoulos, and J. M. Xu, *Nature* **402**, 253 (1999).
48. C. Papadopoulos, A. Rakitin, J. Li, A. S. Vedeneev, and J. M. Xu, *Phys. Rev. Lett.* **85**, 3476 (2000).
49. P. Solomon, D. Frank, and J. Appenzeller, private communication.
50. S. Tans, A. Verschueren, and C. Dekker, *Nature* **393**, 49 (1998).
51. R. Martel, T. Schmidt, H. R. Shea, T. Hertel, and Ph. Avouris, *Appl. Phys. Lett.* **73**, 2447 (1998).
52. R. Martel, H.-S. P. Wong, K. Chan, and Ph. Avouris, *Proc. Int. Electron Device Meeting*, IEEE, Piscataway, NJ, 2001, p. 159.
53. J. Appenzeller, J. Knoch, V. Derycke, R. Martel, S. Wind, and Ph. Avouris, *Phys. Rev. Lett.* **89**, 126801 (2002).
54. C. Zhou, J. Kong, E. Yenilmez, and H. Dai, *Science* **290**, 1552 (2000).
55. H. T. Soh, C. F. Quate, A. F. Morpurgo, C. M. Marcusa, J. Kong, and H. Dai, *Appl. Phys. Lett.* **75**, 627 (1999).
56. S. J. Wind, J. Appenzeller, R. Martel, V. Derycke, and Ph. Avouris, *Appl. Phys. Lett.* **80**, 3817 (2002).
57. A. Bachtold , P. Hadley, T. Nakanishi, and C. Dekker, *Science* **294**, 1317 (2001).
58. V. Derycke, R. Martel, J. Appenzeller, and Ph. Avouris, *Nano Letters* **1**, 453 (2001).
59. X. Liu, C. Lee, C. Zhou, and J. Han, *Appl. Phys. Lett.* **79**, 3329 (2001).
60. M. Bockrath, J. Hone, A. Zettl, P. L. McEuen, A. G. Rinzler, and R. E. Smalley, *Phys. Rev. B* **61** (2000) R10606.
61. R. Martel, V. Derycke, C. Lavoie, J. Appenzeller, K. K. Chan, J. Tersoff, and Ph. Avouris, *Phys. Rev. Lett.* **87**, 256805 (2001).
62. V. Derycke, R. Martel, J. Appenzeller, and Ph. Avouris, *Appl. Phys. Lett.* **80**, 2773 (2002).

International Journal of High Speed Electronics and Systems, Vol. 12, No. 2 (2002) 365–369

ELECTROTEXTILES

E. ETHRIDGE

Defense Advanced Research Projects Agency

D. URBAN

The Charles Stark Draper Laboratory

ElectroTexitles, the merging of electronics and textiles, is a new and novel technical area that leverages a fortuitous synergism between the electronics and textile industries. At some time in the future, we will look back upon what is happening now as turning point in the history of how electronic systems are fabricated. This session provides a summary of the technical challenges and some early technical results.

1. Introduction

The merging of electronics and textiles to provide unique functional products that are large-area, flexible, and conformable is a relatively new idea that leverages technology from two well-established industries. Electronic technology provides a wide variety of products that enable us to sense, filter, analyze, compute, decide, communicate and act. Chip scale integration has brought, and will continue to bring, increasing levels of functionality in a small form factor and will combine analog and digital circuitry that enables complete systems on a chip. Micromachined sensors and active pixel arrays with analog-to-digital converters and signal processing have already been achieved on a single chip. The typical manifestation of this functionality in products, however, is hard, rigid, rectangular artifacts that are mounted in or bolted to other hard, rigid, rectangular structures. Electronic systems, especially those that operate in or interact with the natural environment, should be able to conform to an environment that is characterized by irregular shapes and forms and is softer, more pliable and flexible. Both natural and artificial structures have significant unexploited area on which to place functional electronic systems if these systems are able to physically conform and can be manufactured cost effectively.

The production of textile products has evolved over several centuries and offers a wide variety of techniques to combine natural and artificial fibers and yarns into fabrics that have proliferated into every culture in the world. Woven, knitted, braided and non-woven textiles are found in clothing, rugs, curtains, wallpaper, roads, tunnels, space stations, and diapers and are even present on the Moon and Mars. They provide protection, structural integrity, and unique functionality in a form that can vary from flexible and compliant to hard and rigid. Textile technology has combined with fashion and art to make clothing and murals that provide style and beauty. Sophisticated looms, and knitting and braiding machines can produce large-area structures quickly and cost effectively.

The size of the textile industry, $500 billion annually, is approximately the same size as the electronic industry. Sophisticated manufacturing, automated design, and the existence of a strong equipment and materials infrastructure characterize both. The development of a new community of electronics and textile engineers, physicists, and material and computer scientists within industry, academia, and the government that are focused upon electro-textile science and technology can open new market opportunities and provide for functional systems that can be achieved in no other way. The Defense Advanced Research Projects Agency (DARPA), over the past two years, has conducted several focused studies by leading scientists and engineers to determine the feasibility of electro-textiles and to formulate a technical roadmap for subsequent development. The first public forum for ElectroTextiles was at the Advanced Workshop on Frontiers in Electronics (WOFE-2002) on January 6-11, 2002.

2. Technical Challenges

The basic technology and manufacturing techniques for the economical manufacture of large-area, flexible, conformable electronic systems needs to be developed in a systematic manner. Technology developments should not be focused on dense electronic systems, such as lap top computers, but instead focused on low power, large area applications where it is highly desirable to have many elementary sensors, actuators, logic, and power sources sparsely distributed over large conformal and compressible surfaces (*e.g.,* 1 to 100 meters).

The development of a multi-disciplinary infrastructure to establish the design tools and manufacturing processes to support this integration of capabilities into our everyday use of textiles and fabrics is also required. This capability will enable the integration of many formerly discrete physical devices/packages and functions into a compressible, homogeneous, intelligent textile system that will support rapid field deployment and adaptation plus provide a new set of revolutionary military as well as civilian uses. ElectroTextiles will have the revolutionary ability to sense, act, store, emit, and move while building from an existing low cost textile-manufacturing infrastructure. The underlying conceptual advance or enabling technology is the development and insertion of intelligent mixed signal yarns and textiles into everyday fabrics and fabric-based infrastructures.

A logical start is to make use of the physical properties and manufacturing capabilities of existing fibers, filaments, and yarn and then advance and develop new and novel functional fiber, filament, and yarn properties and capabilities. This should be done in conjunction with e-textile system concepts; embedding functions and applications; and the architectural implementation of electronics systems within everyday fabrics.

Currently electronics and electronic support systems and functions are encountered as separate and distinct rigid physical systems. We are virtually overwhelmed by the sheer number and breadth of electronic devices. In particular, the physical management and multiple packages that must be maintained to obtain the capabilities desired from this collection of systems becomes physically ponderous. Each system has a separate rigid physical structure that must be carried, supported, arranged, transported, and physically incorporated into stand-alone or as components of larger systems. These issues are very relevant due to the sheer number of separate physical devices, their individual bulk, and the limited ability to operate them collectively.

The ability to morph, control, and crumple the physical form factor of an intelligent electronic fabric system from 3D to 2D to 1D during transport, delivery, operation, and in-mission can provide truly revolutionary capabilities not possible before. The following examples illustrate this point: (1) an electronic fabric could be compressed and inserted into an artillery launcher, launched, and inflated into a tank decoy; (2) a self sustaining large antenna fabric (space sail) could be rolled up for easy delivery into space, then just unrolled to form a space sail with intelligent fabric performing physical self alignment; and (3) an intelligent fabric could be worn by a person that would perform health monitoring, mission communications, mission monitoring, and weight distribution.

3. A Technical Approach

An approach to the development of ElectroTextiles is to develop and demonstrate the basic technologies by integrating the necessary knowledge, design capabilities, and manufacturing processes from the materials, textile, microsystems, and information systems communities. There is a broad spectrum of functions that may be incorporated into textiles: (1) Information display; (2) Power sources/generation; (3) Sense; (4) Actuation; (5) Logic (passive, active); (6) Communication/connectivity; (7) Control/adaptation. An evaluation was performed through a series of studies and workshops of the current state of electro-textile research, development complexities, and potential for technology leveraging, plus identifying the key capabilities that must be developed to enable this new field to progress into realizable products.

Fundamental technology areas that need to be addressed are:

- fiber, filament, and yarn and micro-system materials
- yarn physical properties
- textile circuit design and assembly process
- textile manufacture process
- system level control and adaptation concepts
- robust algorithms and architectures
- network and logic functions
- operating system and resource management
- application, domains, and requirements

A strategy to systematically develop these technologies consists of: (1) Basic Yarn Component Development; (2) Textile Circuits; (3) Device Manufacturing CAD and Device Simulation; (4) System Simulation, Design, and Modeling; (5) System and Micro Architecture; and (6) Application Analysis and Concept Demonstrations.

3.1 *Basic Yarn Components Development*

Functional fibers, filaments, and yarns are the basic building blocks of electro-textiles. The textile industry has demonstrated a remarkable capability to incorporate both natural and manmade filaments into yarns and fabrics to satisfy a wide range of physical parameters (strength, absorbency, color, durability, *etc.*) that survive the manufacturing process and are tailored to specific application environments. Preliminary investigations indicate that electronic components can be fabricated within and/or on the surface of filaments and subsequently be woven into functional yarns and fabrics. Passive components such as resistors, capacitors, and inductors can be fabricated in several different manners. Diodes and transistors can be made on long, thin, flat strands of silicon or formed in a coaxial manner. Progress has been made in the development of fiber batteries and fiber-based solar cells. In addition, a variety of actuated materials (piezoelectric, *etc.*) can be made into multiple long strands (filaments) that can form yarns and subsequently be woven into fabric.

3.2 *Textile Circuits*

Basic yarn components discussed in 3.1, along with conventional filament/yarns, constitute the feedstock of the weaving process. Selectively fed into a loom and manipulated through advanced textile manufacturing processes, this feedstock can be woven into a complex variety of designs that result in a structurally sound, environmentally compatible fabric that provides electrical and mechanical functionality. Interconnection schemes, bussing structures, distributed power, conventional device attachment, and routing mechanisms need to be developed. We need to determine how current looms can be modified to address the particular needs for interconnecting, routing, and handling of the fiber components. Simple fabric electronic circuits need to be designed, fabricated, tested, and characterized. Tests should stress the survivability of the components during the manufacturing and interconnection process as well as environmental and operational stresses (flexibility, durability, *etc.*) of the composite.

3.3 *Device Manufacturing CAD and Device Simulations*

The electronics industry has demonstrated a remarkable ability to develop computer-aided design (CAD) tools for developing complex integrated circuits and printed wiring board products. The textile industry has demonstrated a similar computer-aided design capability for the design and development of advanced fabrics. Existing electronic CAD and textile CAD tools need to be tailored to the new e-textile manufacturing process. Device and circuit characterization (both electrical and mechanical parameters) developed in 3.1 and 3.2 should be used to develop macros and design libraries and be incorporated into the tailored e-textile design tools. This database can be used as a basis for developing device and circuit simulations that will feed into system simulation, design, and modeling.

3.4 *System Simulation, Design, and Modeling*

The key to performing early e-textiles technology trades is the ability to accurately predict expected behavior. The unique challenge is to model a range of relatively unpredictable behaviors associated with nano-cells and a connectivity infrastructure. An emphasis should be on the development, validation, and updating of early functional and micro-architecture models and simulation environments as the technology evolves. The models and simulation environments developed under this task can be used for e-textile system trades.

3.5 *Systems and Micro Architecture*

The basic micro architecture tenets that have evolved over time for conventional packaged silicon systems will no longer apply for e-textile systems. These basic design trades must be rethought in conjunction with the basic fiber and micro system enablers and constraints. System robustness is a major design question for e-textiles. The key to achieving a balanced design is to explore the cost at all levels: fibers, filaments, yarn, woven yarns, micro-architecture, connectivity, run time software, algorithms, and applications. There is a need to develop e-textile models, simulation tools and application analysis incorporating the real yarn developments and the use of these tools to explore the e-textile design space. These efforts can be combined to establish system and micro architecture approaches and demonstrable architectural implementations driven by the requirements of applications.

3.6 *Application Analysis and Concept Demonstrations*

ElectroTextiles will offer new approaches to build revolutionary electronic and information systems. Electronic textiles are not expected to compete with high-density, high-performance electronic systems typified by current computer or telecommunications products. Rather, electronic textiles are expected to have unique applications that are typified by the need for flexible and conformable systems or those that can take advantage of the availability of large surface areas (walls, parachutes, aircraft wings, ship hulls, etc.). The benefits of this technology will be shown in applications where there is no other way to achieve the desired results or in areas where e-textiles can demonstrate a significant cost advantage compared to other, more conventional, approaches.

4. Conclusion

ElectroTextiles is a new technical area that leverages and combines materials, design tools, and manufacturing processes from two large and growing industries, electronics and textiles. It requires the development of a new community of scientists and engineers from multidisciplines that can innovatively pull together novel and unique technologies and techniques to provide value added products. These products will be in the form of flexible, conformable, large-area electronic and information systems that either perform functions that cannot be performed today or perform these functions at a significantly reduced cost. The papers that follow illustrate some early and recent ideas and accomplishments in this exciting field.

International Journal of High Speed Electronics and Systems, Vol. 12, No. 2 (2002) 371–390

SEMICONDUCTOR THIN FILMS AND THIN FILM DEVICES FOR ELECTROTEXTILES

MICHAEL S. SHUR, SERGEY L. RUMYANTSEV, and REMIS GASKA

Electrical, Computer and Systems Engineering Department
Rensselaer Polytechnic Institute, Troy, NY 12180-3590, USA and
Biomedical and Information Technologies, Inc.
Latham, NY 12110, USA

We discuss the evolution from wearable electronics and conductive textiles to electrotextiles with embedded semiconducting films and semiconductor devices and review different semiconductor technologies competing for applications in electrotextiles. We also report on fabrication, characterization, and properties of nanocrystalline semiconductor and metal films and thin-film device structures chemically deposited on fibers, cloth, and large area flexible substrates at low temperatures (close to room temperature). Our approach is based on a new process of depositing polycrystalline CdSe (1.75 eV), CdS (2.4 eV), PbS (0.4 eV), PbSe (0.24 eV) and Cu_xS (semiconductor/metal) films on flexible substrates from the water solutions of complex-salt compounds. We have covered areas up to 8x10 inches but the process can be scaled up. The film properties are strongly affected by processing.

We fabricated a lateral solar cell with alternating $Cu_{2-x}S$ and nickel contact stripes deposited on top of a view foil. These sets of contacts represented "ohmic" and "non-ohmic" contacts, respectively. Then CdS films of approximately 0.5 μm thick were deposited on top. We also fabricated a "sandwich" type photovoltaic cell, where the CdS film was sandwiched between an In_2O_3 layer deposited on a view foil and a $Cu_{2-x}S$ layer deposited on top. Both structures exhibited transient response under light, with the characteristic response time decreasing with the illumination wavelength. This is consistent with having deeper localized states in the energy gap determining the transients for shorter wavelength radiation. (Slow transients related to trapping effects are typical for polycrystalline CdS materials.)

We also report on the photovoltaic effect in CdS/CuS films deposited on trylene threads and on a field effect in these films deposited on a flexible copper wire. CdS films deposited on viewfoils exhibit unique behavior under stress and UV radiation exposure with reproducible resistance changes of several orders of magnitude with bending up to 10 mm curvature.

Our results clearly demonstrate the feasibility of using this technology for photovoltaic and microelectronics applications for electrotextiles and wearable electronics applications.

1. Introduction

For many years, the progress in microelectronics has been associated with scaling of the minimum feature size of integrated circuits. This trend described by the famous Moore's law is doomed to end as this minimum feature size approaches the size of a single unit cell of silicon (see Fig. 1a). At the same time, the opposite trend of increasing the overall size of integrated circuits has emerged stimulated by the development of flat panel displays (see Fig. 1b).

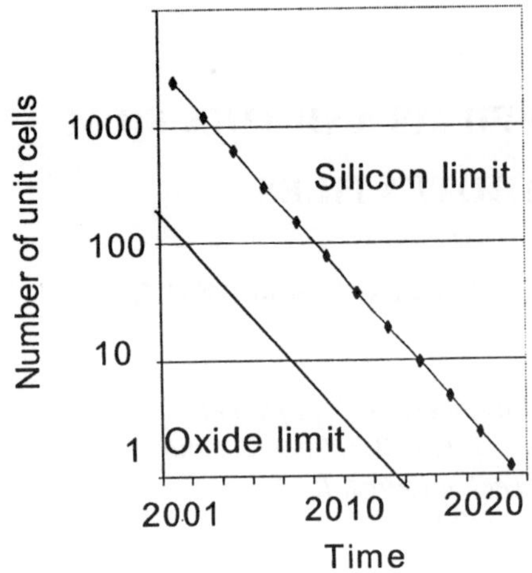

(a)

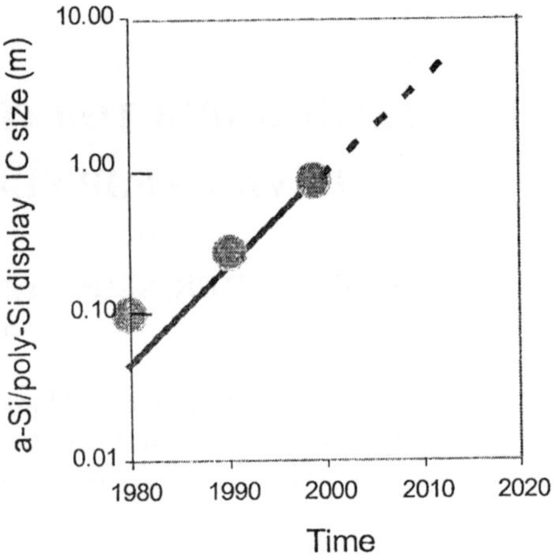

(b)

Fig. 1. Scaling trends.

Emerging electrotextile technology that aims to scale up the size of electronic circuits to the size of a human being and beyond exemplifies this scaling-up trend. We can distinguish three distinct directions in this field that might eventually merge in electrotextile systems: conducting textiles, wearable electronics, and semiconducting films and devices embedded into electrotextiles. In Section 2, we consider conducting textiles and wearable electronics. In Section 3, we discuss semiconducting technologies competing for applications in electrotextiles. Section 4 deals with A_2B_6 and A_4B_6 semiconducting films for electrotextile applications. Section 5 reviews emerging semiconductor devices on flexible substrates and cloth using A_2B_6 and A_4B_6 semiconducting films.

2. Conducting textiles and wearable electronics

Fig. 2 shows the SoftHeat™ warming blanket that uses a tape conducting heat evenly through the blanket. This blanket is sold by Comfort House[1], and this technology (also used in Soiree Automatic Mattress Pad) was developed THERMOSOFT International Corporation (http://www.thermosoftinternational.com).

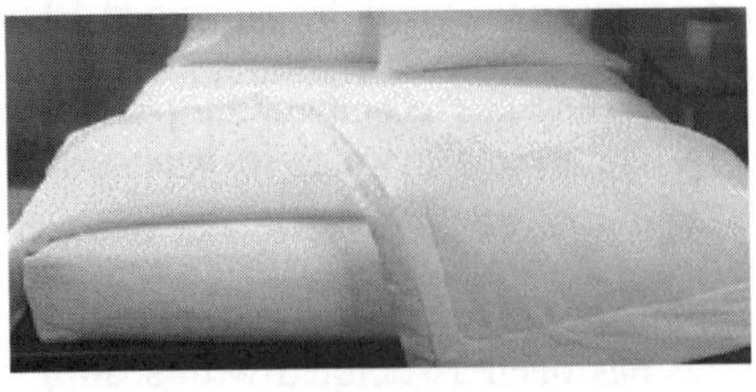

Fig. 2. SoftHeat™ warming blanket[1]

Fig. 3. Flexible cloth keyboard from ElekSen (http://www.eleksen.co.uk/flash/press_releases.shtml)[2]

ElekSen Company (formerly Electrotextiles) uses wires woven into cloth to produce pressure sensitive textiles. Pressing on the cloth deforms the wires changing their conductivity. Pressure response ranges can be range from 0.2 psi to 30 psi or higher with achievable output resolutions as high as 0.0001% of fabric size. Fig. 3 shows the flexible cloth keyboard for a personal digital assistant developed by Electrosen using this technology[2].

An example of defense applications of conducting textiles or flexible films is the tape used for decreasing the radar signature of a B2 bomber (see Figure 4).

Fig. 4. B2 bomber. All surfaces of the aircraft are covered with special paint and tape that create a sleek skin scattering radar signals.

Wearable electronics applications have emerged and have already found a variety of commercial and defense applications. Fig. 5 shows the cover of IEEE Spectrum magazine (October, 2000) devoted to wearable computers. MIT Media Lab has developed MIThril's body bus that supports USB interface and can support sensors and peripherals (Fig. 6).[3]

Fig. 5. IEEE Spectrum cover. October, 2000

Fig. 6. MIThril's body bus natively supports USB and sensors and peripherals (http://www.media.mit.edu/wearables/)

In August, 2000, Philips and Levi's reported on a wearable electronics jacket[4] that contained a body area network using wires integrated into the jacket design (see Fig. 7). This network allows the synchronous control of the new Philips Xenium GSM (Global System for Mobile communications) and Philips Rush MP3player through the use of a unified remote control. Reima developed a $563 waterproof snow jacket (Fig. 8) that incorporates loudspeakers, a microphone, and a pocket connector linked to a GSM [5].

Akseli Reho, research director at Clothing+ predicts that "after 10 years, every piece of clothing will include some electronics."

Defense applications of wearable electronics are exemplified by the Land Warrior system.[6] The Land Warrior infantryman (see Fig. 9) attaches computer/radio and GPS subsystems to his/her load-bearing frame. A hand grip attached to the chest acts as a computer mouse. The soldier has a radio and can send videos to his/her leader using a weapon-mounted camera. The leader version of this uniform has two radios and a flat panel display/keyboard. According to Maj. Gen. John C. Doesburg, Commander of the Army Soldier and Biological Chemical Command (SBCCOM), by 2010, the "dream soldier will have sensors built into a skintight uniform." This will signify a transition from wearable electronics to "sensitive skin" [7,8] and electrotextiles paradigm.

Fig. 7. Wearable electronics jacket from Philips and Levi[4]

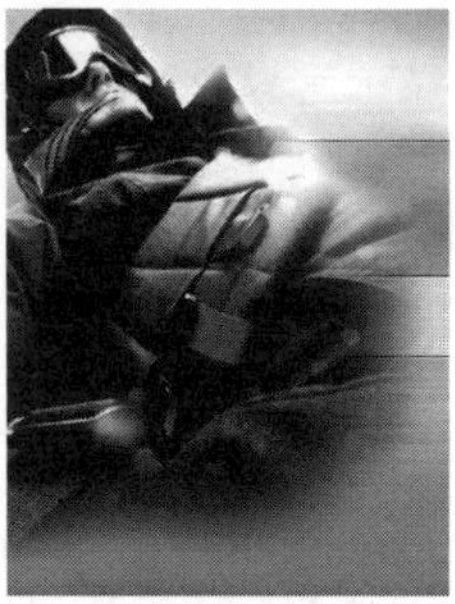

Fig. 8. Smart Clothing product, Reima Smart 3305, http://www.reimasmart.com/index.cfm?action=products&actionsub=top)

Fig. 9. Land Warrior

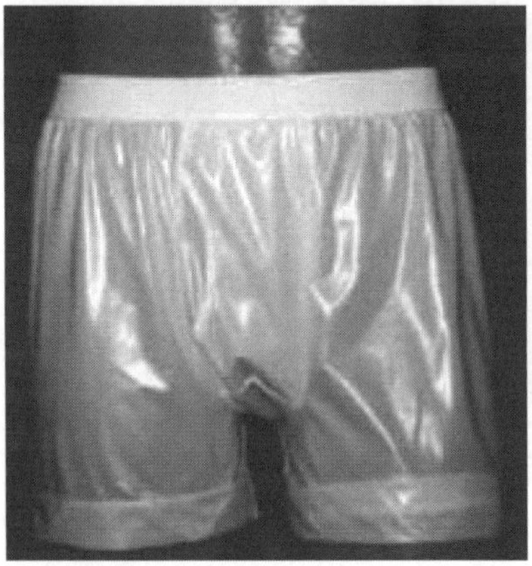

Fig. 10. Airtight underwear for flatulence with a replaceable charcoal filter[9](http://www.under-tec.com/store.htm).

Fig. 10 shows Under-Ease, airtight underwear with a replaceable charcoal filter that removes bad-smelling gases before they escape[9]. This device invented by Buck Weimer of Pueblo, Colorado was awarded IGNobel Prize in Biology (2001), which is supposed to be given to strange and useless inventions. In fact, this garment is a life saver for people with Krohn disease, since it allows them to function normally in a social environment.

3. Semiconducting technologies for applications in electrotextiles

Three key semiconductor technologies compete for the applications in electrotextiles: amorphous silicon technology, organic semiconductors, and A_2B_6 and A_4B_6 semiconductors that can be chemically deposited on a variety of different substrates at temperatures close to room temperature. One of the key challenges to the successful development of such "sensitive skin" is the development of large area sensor and detector arrays, which can be fabricated on flexible and, ideally, stretchable, substrates. At the present time, only amorphous silicon technology comes close to having such capabilities. Amorphous Si integrated circuits have been used to drive 25" flat panel displays.

As can be seen from Table 1, pentacene technology might successfully complete with a-Si technology, and it might be quite suitable for electrotextile applications. We should also mention the work on a-Si based flexible displays (see the WEB movie illustrating the operation of such a display)[10] and on flexible displays using organic LEDs (see Fig. 11[11]) These technologies also have promise for applications in electrotextiles.

Table 1. Comparison of a-Si and pentacene Thin Film Transistor Technologies (from [12])

Parameter	OTFT (single-crystal pentacene) [13]	Organic TFT (poly pentacene) [14]	α-Si TFT (on glass) [15]
Mobility [cm^2/Vs]	3.2	1.7	0.1 to 0.5
Drain current On/Off ratio	109	108	107
Subthreshold slope [V/dec]	0.06	0.4	1
Deposition T [°C]	Room temperature	60	110

Fig. 11. Flexible Organic LED Display. (Courtesy of Universal Display Corporation, Inc.[12])

4. A_2B_6 and A_4B_6 semiconducting films for electrotextile applications

Amorphous Si technology is one of the prime candidates for applications for the "sensitive skin" and electrotextiles, at least as one of the possible several layers forming the skin. However, one of the most important properties of such skin should be its tactile sensitivity. Polycrystalline films with grains oriented on average in the same direction might be used for both proximity and tactile sensors.

Our approach to fabricating semiconductor films for electrotextiles is based on the process of depositing polycrystalline (and/or nanocrystalline) CdSe (1.75 eV), CdS (2.4 eV), PbS (0.4 eV)[16], PbSe (0.24 eV) and Cu_xS (semiconductor/metal) films on flexible substrates or threads at temperatures close to room temperature. Large area surfaces can be covered. We have covered areas up to 8x11 inches (see Fig. 12) but the process can be scaled up. These semiconducting and metal films will serve as building blocks for thin-film technology, which will enable us to develop the sensitive skin arrays. Their properties are strongly affected by processing. For example, the dark resistance of the CdSe films can be reduced by more than five orders of magnitude using thermal annealing in the temperature range from 100 °C to 200 °C. The photosensitivity of PbS films can be increased by few orders of magnitude by annealing in the temperature range of 110-140 °C in optimized ambience. These new material systems are ideally suited for "sensitive skin" applications, since these films are suitable for development of optical, thermal, piezoelectric and pyroelectric [17] sensors.

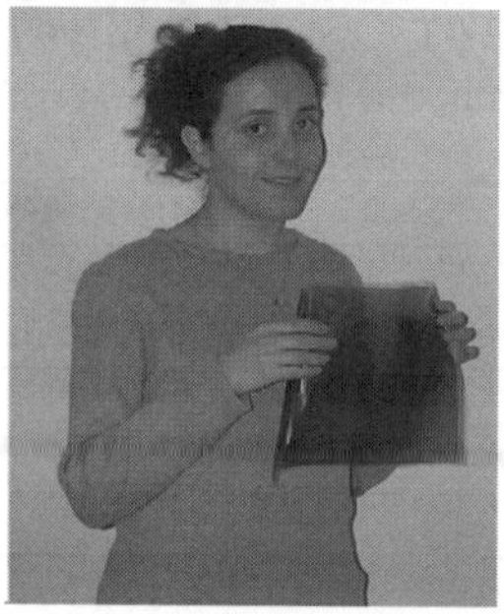

Fig. 12. CdS (a) and CuS (b) films on viewfoil [18]

The basic scheme of this technology is described below using CdSe films as an example[19]. Thin films (less than 1 μm) of CdSe are deposited using the water solutions of the corresponding complex-salt compounds containing necessary elements. The reducer of this chemical reaction is Se-sulfite, or Se-urea. The deposition of CdSe thin films can proceed only in the solutions with a certain pH. After the initial surface preparation, the substrates are immersed into the solution. The deposition is performed in the isothermal regime. The deposition time and temperature are adjusted to yield the desired thickness of the deposited film (the actual deposition temperature being close to room temperature).

The chemical deposition of the films is very simple, extremely reproducible and capable to cover very large surface areas. Low deposition temperature allows for the thin film deposition on thermally unstable substrates.

The chemical reaction used for the deposition of polycrystalline CdSe films is given by:

$$Cd^{2+}/citrat/_n + SeSO_3^{2-} + 2OH^- = CdSe + n/citrat/ + SO_4^{2-} + H_2O \quad (1)$$

The rate of the chemical reaction is determined by the dissociation constant and by the solubility product.

The deposition of CdSe is accompanied by other reactions. An intermediate compound, $Cd(OH)_2$, is formed in the solution and adsorbed on the substrate surface:

$$SeSO_3^{2-} + H_2O \leftrightarrow H_2Se + SO_4^{2-} \tag{2}$$

Se ions in the solution react with $Cd(OH)_2$ and yield CdSe, which is much less soluble. The reaction is autocatalytic and CdSe acts as a catalyst. The reaction described by Eq. (2) is reversible with the equilibrium constant, $K_c = 3.16\times10^{-32}$, given by:

$$K_c = \frac{[H_2Se]_e[SO_4^{2-}]_e}{[SeSO_3^{2-}]_e} = \frac{[H^+]_e^2[Se^{2-}]_e[SO_4^{2-}]_e}{[SeSO_3^{2-}]_e \cdot K_{H_2Se}}, \tag{3}$$

where index *e* stands for equilibrium concentrations of the components. The products of the reaction described by Eq. (2) are ions of SO_4^{2-} and H_2Se acid, which dissociates and yields ions of Se^{2-} :

$$H_2Se \leftrightarrow H^+ + HSe^- \tag{4}$$

with the dissociation coefficient $K_1 = 1.3x10^{-4}$, and

$$H_2Se^- \leftrightarrow H^+ + Se^{2-} \tag{5}$$

with the dissociation coefficient $K_2 = 10^{-11}$.

References [19, 20] describe properties of CdSe films chemically deposited from water solutions containing $CdCl_2$. For the characterization described below, the films were also deposited on ceramic substrates. Typical thickness of the films deposited in one cycle at room temperature was approximately 0.5 μm. The films exhibited polycrystalline structure with the average grain size around 0.12-0.15 μm. The results of X-ray studies showed that the inter-grain material consists of CdSe small-crystals with the size ranging from 0.007 μm to 0.01 μm. This inter-grain material plays a very important role for the improved adhesion and stability of the films deposited on flexible substrates. The films contain a mixture of amorphous, hexagonal and cubic crystal phases. The transition between the different phases was observed during thermal annealing (TA) of the films. These transitions along with the influence of the chemically active ambient during TA determine photoelectric characteristics of CdSe layers.

After the deposition, CdSe films exhibited a significant photosensitivity in a visible spectral range. Also, a large persistent photoconductivity was observed after a UV illumination. Fig. 13 shows the dark current of the 2x2 mm^2 samples. Contact pads for electric measurements were obtained by thermal deposition of SnO_2.

Spectral photoresponse of the deposited CdSe films was measured under CW illumination intensity of $3.8x10^{13}$ quanta/cm^2s. As-deposited CdSe films exhibited photosensitivity maximum at approximately 625 nm. The activation energy of the photocurrent, measured at 700 nm, varied from 1.6 eV to 1.7 eV.

Lux-ampere characteristics of the untreated films can be described by expression $I_{ph} \sim J^{\alpha}$, where J is illumination intensity and parameter α ranges from 0.5 to 1.0 from sample to sample.

As deposited, CdSe films exhibited persistent photoconductivity after UV illumination in the temperature range of 80-400 K. The ratio of $I_{ph}^{persist}/I_{dark}$ varied from 10^2 to 10^5 depending on the temperature. Fig. 14 shows time dependencies of I_{ph}/I_{dark} at different sample temperatures. The characteristic relaxation time was 1.5-1.8x10^2 sec. Our preliminary results showed that the conductance of the films after illuminating by UV light from a contact mask aligner increases and remains unchanged for more than a few hours.

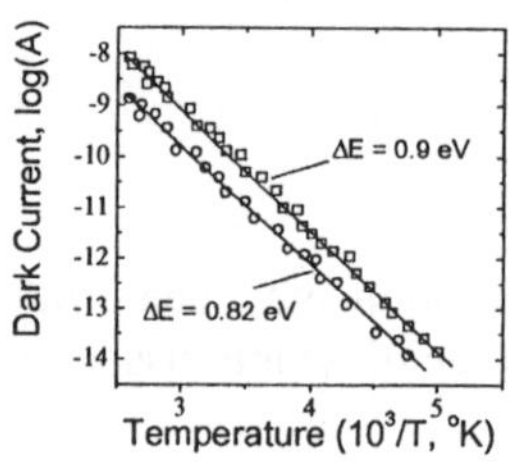

Fig. 13. Dark current of two chemically deposited CdSe films. Activation energy of the dark current varied from 0.8 eV to 0.9 eV. Measurements were performed before sensitizing treatment.[20]

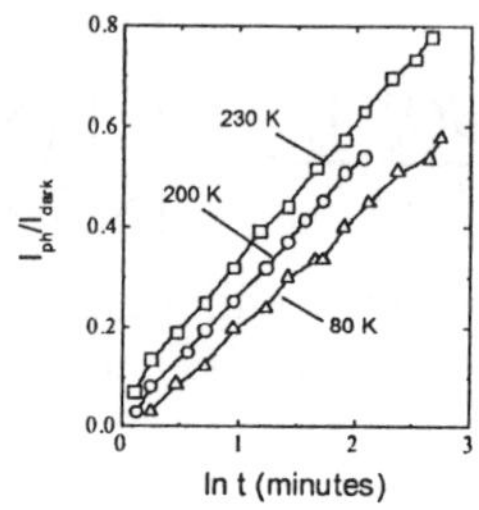

Fig. 14. Time dependence of photocurrent/dark current ratio for CdSe films. Samples were illuminated in spectral range of 350-400 nm.[20]

The spectral response of the films can be controlled using a special sensitizing treatment (see Fig. 15), which combines annealing and chemical processing. In order to increase photosensitivity, chemically deposited CdSe films were annealed at temperatures up to 200 °C in different chemically active ambiences. This allowed us to vary the dark resistance of the films by few orders of magnitude and to increase photosensitivity by 10^3-10^4 times (see Fig. 16).

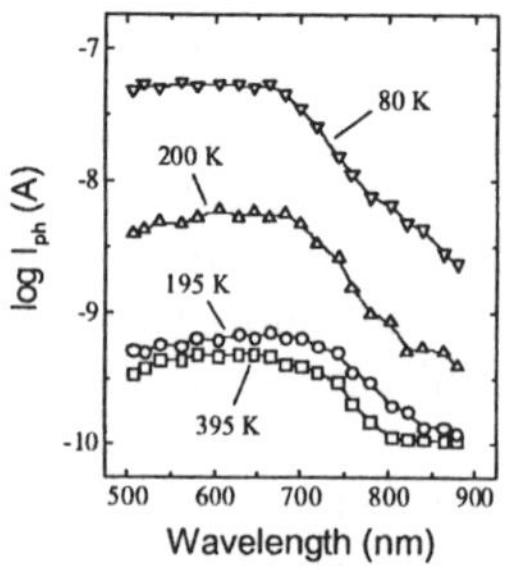

Fig. 15. Spectral dependencies of photocurrent of CdSe film at different temperatures after sensitizing treatment.[20]

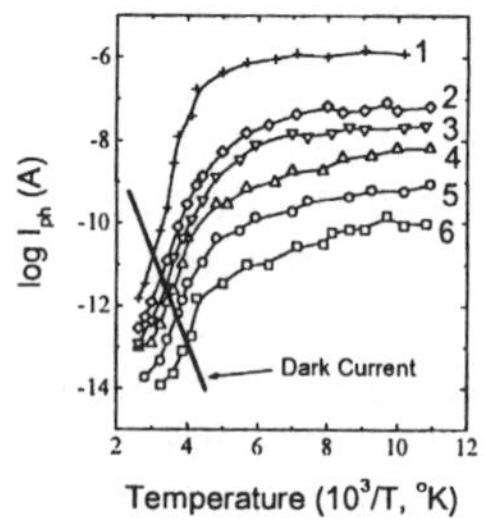

Fig. 16. Temperature dependencies of photocurrent of sensitized CdSe films at 700 nm wavelength for illumination intensities: 1- 200 lx; 2 - 50; 3 - 10; 4 - 1; 5 - 0.1; 6 - 0.01. Solid line shows temperature dependence of dark current.[20]

Fig. 12 a shows a CdS film deposited on a deposited on flexible view foil. As can be seen from the figure, we can cover large area flexible substrates.

A very interesting direction of research, which is now under way in our group, is the development of flexible heterostructures and homo and hetero *p-n* junctions (see Fig. 17).

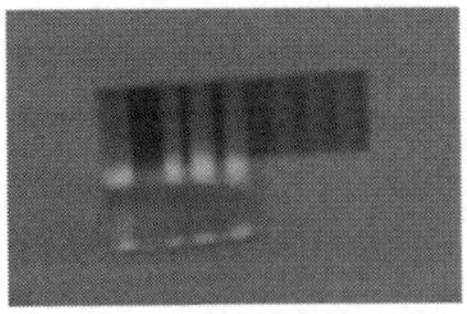

Fig. 17. CdS/PbS heterostructure with gold contacts on flexible substrate.

Using a similar process, we have also deposited Cu_xS on flexible substrates (see Fig. 12 b). Both metal and semiconductor films have been deposited (depending on the molar fraction of Cu). The characterization of these films is now in progress. By adjusting the concentrations of Cu^+ ions and Na_2S_n poly-sulfide and by using multiple deposition cycles we obtained metal CuS films with the resistivity as low as 100 ohm per square. The deposited thin films exhibited excellent adhesion to the polymeric substrate/film. The film thickness is controlled by the oxidation-reduction chemical reactions with different (changing) values of *x* and *y*:

$$Cu_{2-x}S + 2(x-y)Cu^+ = Cu_{2-y}S + (x-y)Cu^{2+} \quad (6)$$

$$Cu_{2-y}S + (x-y)/(2-x)S = (2-y)/(2-x)Cu_{2-x}S \quad (7)$$

where $x > y$.

We are also working on the chemical deposition, processing and characterization of $CdSe/Cu_xS$ multi-layered structures on flexible polymeric substrates/films. Metallic CuS films are expected to be suitable for fabrication of interconnects for sensor arrays.

We are now able to deposit a variety of the binary and ternary A_2B_6 and A_4B_6 compounds and heterostructures on a variety of flexible substrates (after a proprietary surface preparation process). Experiments are now under way to deposit CuS and CdSe on a nylon thread and on cloth. Fig. 18 illustrates the first key device building blocks for the development of this technology: the array of ohmic contacts on semiconductor thread. These contacts are used for testing the thread and for attaching interconnects to thread sensor arrays in a weaving process. Thread diameters vary from 0.2 mm to 1 mm.

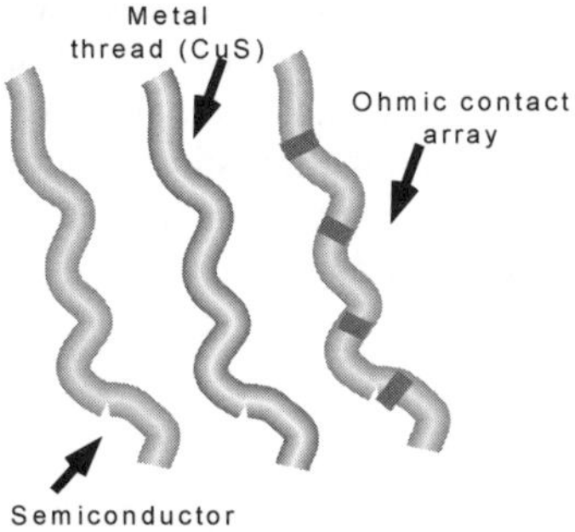

Fig. 18. Artist sketch of a piece of semiconductor thread and semiconductor thread with Ohmic contact array.

We have started to experiment with both nylon threads and metal (copper and steel) threads. This gives us a choice of having an insulating or conducting core.

The next step will involve the development of lateral and radial *p-n* junction on a thread. These *p-n* junctions will be characterized and assessed as future building blocks for a variety of semiconductor electronic and optoelectronic devices on a thread, including solar cells, photovoltaic detectors, and gate structures for Junction Thin Film Transistors.

Recently, we reported on the structural and electrical properties CdS films deposited on viewfoils and trylene threads. This study revealed that these films might serve as highly sensitive strain sensors.

The CdS films were deposited at temperatures 60-70 °C from solutions containing cadmium citrate complex-ions and thiourea. Because of low deposition temperatures, we were able to use a variety of flexible materials, including viewfoils, cloths, and threads. We covered the areas up to several tens of square inches with 0.2-0.5 μm thick CdS films. The layer thickness was measured using a profilometer on the selectively etched CdS layer[21]. Fig. 19 shows the result of such measurement for the ~0.2 μm thick CdS layer. More details of our film deposition process are described in Reference 7.

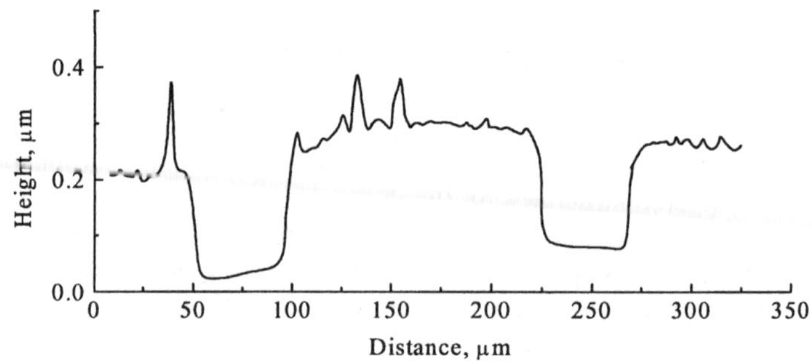

Fig. 19. The profilometer data used to find the CdS film thickness.

A Scanning Electron Microscope (SEM) study revealed that the CdS films deposited on the viewfoil are very homogeneous. As seen in Fig.20 a, no distinguishable structure of the layers has been found with the characteristic dimension of the fraction of microns. The structure of the layers deposited on the thread was more complicated. Fig.20 b shows the SEM picture of the CdS film deposited on thread. It is seen that the film is not uniform and contains some insertions and cracks.

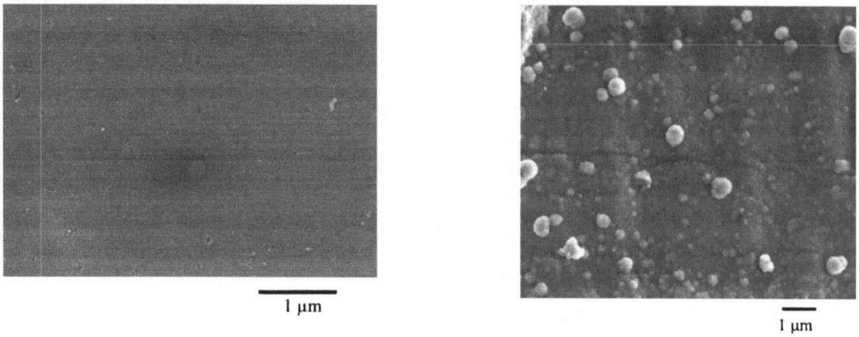

Fig. 20. SEM pictures of CdS film deposited on the viewfoil (a) and thread (b).[21]

Fig. 21 shows the X-ray diffraction curve of the CdS film deposited on viewfoil. Peaks on the diffraction curve at $2\theta = 26.5°$ and $2\theta = 28.2°$ correspond with good accuracy to the CdS hexagonal crystal structure with lattice constants $a = 4.14092 \pm 0.00032$ Å and $c = 6.7198 \pm 0.0012$ Å [22]. However some peaks on the diffraction curve at $2\theta > 30°$, which are also characteristic for the hexagonal form of CdS (for example, peaks corresponding to the orientations 110, 102, 112), are missing. One of the possible reasons for that is a systematic orientation of the crystallites. Therefore, the film might have pyroelectric properties.

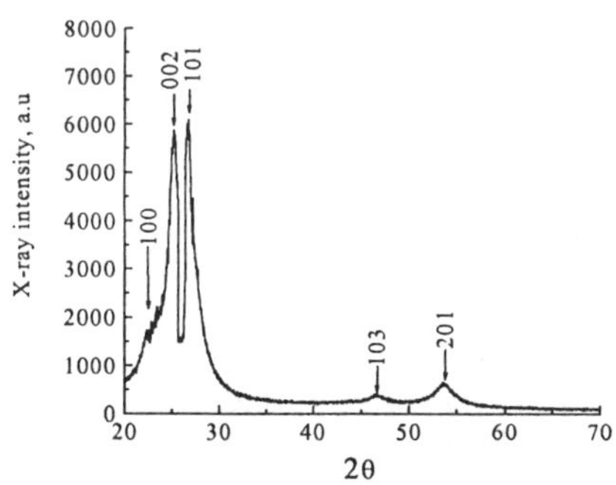

Fig. 21. X-ray diffraction curve of the CdS film deposited on viewfoil.[21]

In order to investigate the crystal structure further, the CdS films with reduced thickness were deposited on a nickel grid (Fig. 22). Since this grid itself was transparent for the electron beam of the High Resolution Transmission Electron Microscope (HRTEM) we were able to study the CdS structure in the corner as shown by arrow in Fig. 22. Fig. 23 a shows HRTEM picture with 400,000 magnification. As seen, the grain size is of the order of a few tens of nanometers. The surface of the grains is rough indicating that the grains are not perfect single crystals. Fig.23 b shows the HRTEM picture with a higher magnification, such that single crystallites and atomic layers in the grains can be distinguished. The size of the crystallites is about 3 to 5 nm. The crystallites are separated from each other by amorphous CdS.

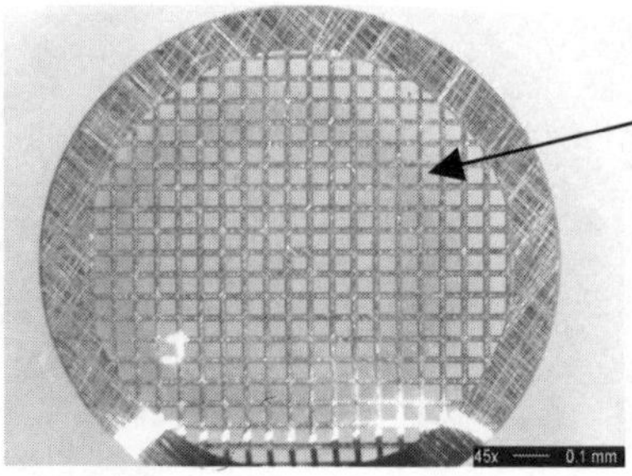

Fig. 22. Ni grid for HRTEM.[21]

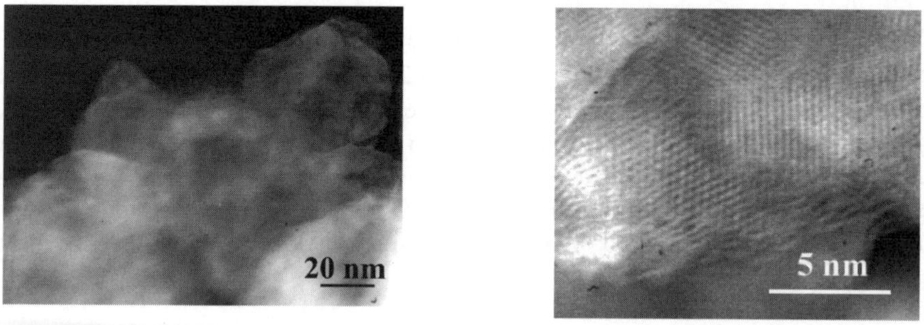

Fig. 23. HRTEM pictures (a and b) with different magnification.[21]

In order to study the electrical properties, gold interdigitized contacts were deposited on the top of the CdS film deposited on the viewfoil. Fig. 24 shows current voltage characteristics in the dark and under UV illumination with a xenon lamp. As seen, the resistance change under illumination is over seven orders of magnitude. The measurements of the spectral responsivity indicated that the spectral response is consistent with the bandgap of the hexagonal CdS of 2.45 eV. All films demonstrated a slow response to the illumination. Fig. 25 shows the time dependence of the current under illumination with a halogen lamp. As seen the current increase under illumination and the photoconductivity relaxation are characterized by different time constants. We also found that the characteristic time of the photoconductivity increase depends on the light wavelength: the smaller is the wavelength, the slower is the current increase under illumination.

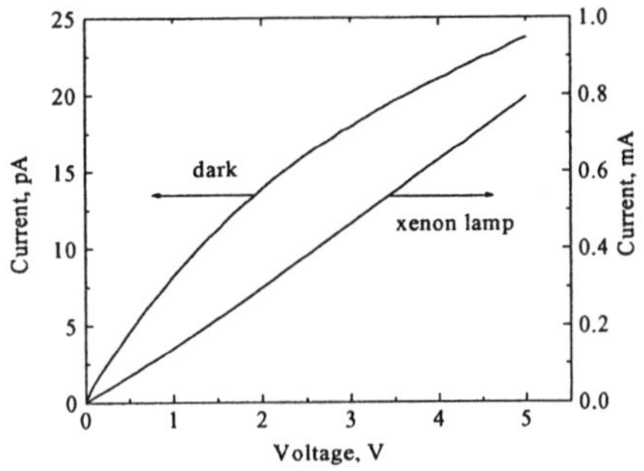

Fig.24. Current voltage characteristics in the dark and under UV illumination.[21]

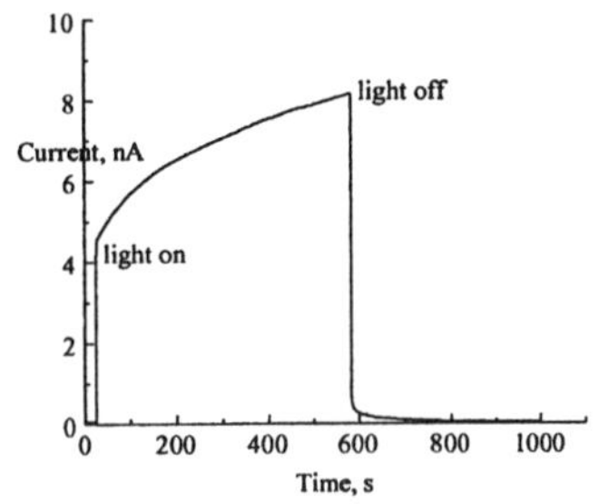

Fig. 25 Current response of the CdS film on viewfoil to the light of halogen lamp. DC voltage V=1V.[21]

These CdS films are very sensitive to the stress (tension) under UV illumination. In order to study the stress sensitivity we bent the viewfoil with the deposited CdS film as shown in the inset in Fig. 26.

The dependence of the structure resistance normalized to the unstressed resistance as a function of the reciprocal radius r of the film bending is shown in Fig. 26. Different symbols correspond to different measurement sessions on the same sample. The sample was illuminated with the diffused light of the xenon lamp. As seen, the resistance under tension increases by several orders of magnitude. Symbols at $1/r = 0$ (no strain) show the value of the initial resistance (R/R_0=1) and the value of the resistance after every measurement. These data show that the change in the resistance under stress is reproducible. This reproducibility might be explained by the film structure, which consists of nanocrystallites separated by elastic amorphous CdS (see Fig. 23 b).

Fig. 27 shows how the resistance under stress depends on time. The arrow in Fig. 27 shows the moment of time when the stress was applied. As seen the resistance under stress (tension) increases slowly and saturates at time $t \geq 150$ s. A similar behavior was observed when tension was removed.

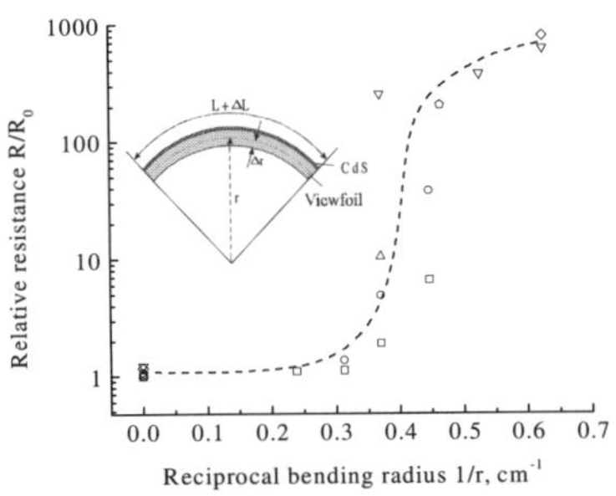

Fig. 26. Relative change in the film resistance under UV illumination versus reciprocal bending radius. The inset shows how the film was bent.[21]

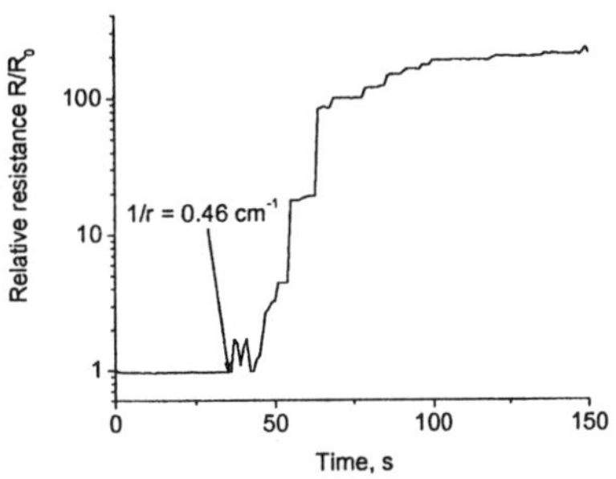

Fig. 27 Time dependence of the resistance under the tension.[21]

One of the most important characteristics of the piezoresistive sensors is the gauge factor defined as:

$$G = \frac{\Delta R / R_0}{\Delta L / L_0}, \tag{8}$$

where L is the sensor length. As follows from the inset in Fig. 26, $\Delta L/L \approx \Delta r/r$, where Δr=50μm is the half of the viewfoil thickness. We estimated the maximum value $G \approx 10^5$ for our CdS films, that is several orders of magnitude larger than that for metal ($G = 1 \div 2$) and regular semiconductor piezoresistive sensors ($G \approx 10^2$).

5. A_2B_6 and A_4B_6 semiconductor devices on flexible substrates and cloth

Our approach to fabricating solar cells is based on our new process of depositing polycrystalline CdS (with the bandgap of 2.4 eV) and Cu_xS (semiconductor/metal) films on flexible substrates at temperatures close to room temperature. The work is also under way to develop semiconductor threads and semiconductor cloth based on this technology. The devices built on these semiconducting and metal films will serve as building blocks for thin-film technology, which will enable us to develop the sensitive skin arrays. As mentioned above, these new material systems are ideally suited for "sensitive skin" and electro textile applications, since these films are suitable for development of optical, thermal, piezoelectric and pyroelectric [23] sensors.

The basic scheme of our film deposition process is described in Reference [24]. We fabricated two types of structures. Structure 1 represented a lateral solar cell with alternating $Cu_{2-x}S$ and nickel contact stripes deposited on top a view foil (see Fig. 28). These sets of contacts represented "ohmic" and "non-ohmic" contacts, respectively. Then CdS films of approximately 0.5 μm thick were deposited on top using the water solutions of complex-salt compounds.

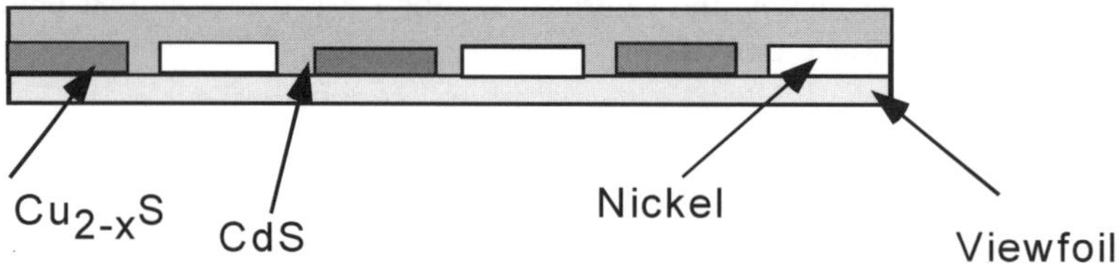

Fig. 28. Sample configuration. [25]

Structure 2 was a "sandwich" type photovoltaic cell, where the CdS film was sandwiched between an In_2O_3 layer deposited on a view foil and a $Cu_{2-x}S$ layer deposited on top.

Fig. 29 shows typical dark current-voltage characteristics measured for structure 1. As can be seen from the figure, the current-voltage characteristics are linear only at small current densities and have a small saturated dark current. Fig. 30 shows the spectral characteristic of the deposited CdS film. The maximum of the spectral response is consistent with the bandgap of CdS.

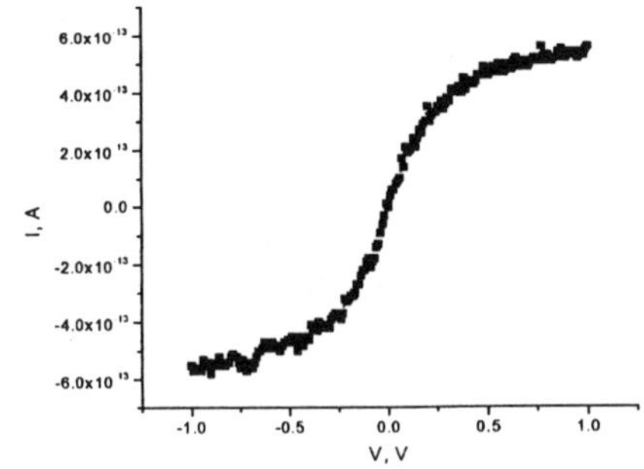

Fig. 29. Dark current-voltage characteristics.[25]

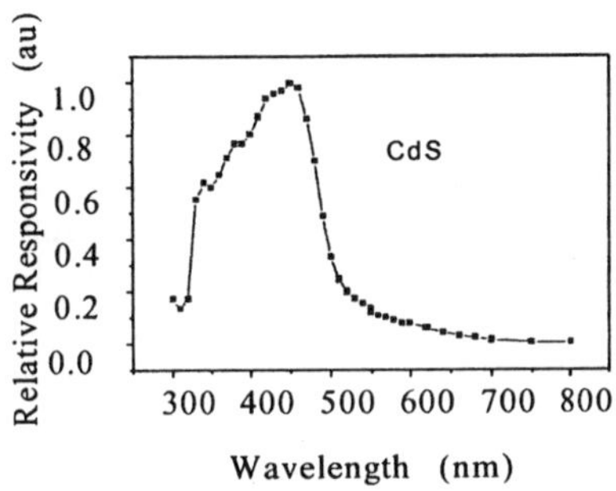

Fig. 30. Spectral responsivity.[25]

Fig. 31 shows the measured (a) and calculated (b) photovoltaic response. As seen from the figure, both the open circuit voltage and the fill factor are quite low.

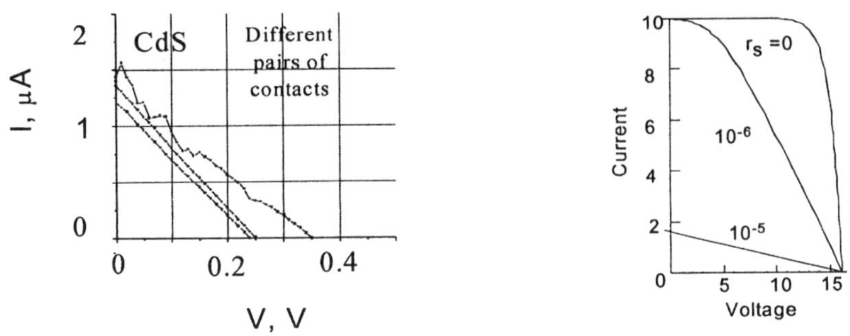

(a) (b)

Fig. 31. Measured current-voltage characteristics under light (a) and calculated dimensionless current-voltage characteristics under light for different values of r_s and zero shunt resistance, $r_{sh} = 0$ (b)[25]

The reason for the low fill factor is clear from the comparison with the calculated curves described by the following standard solar cell equation:

$$I = I_L - I_s\left[\exp\left(\frac{V + IR_s}{\eta V_{th}}\right) - 1\right] - \frac{V}{R_{sh}} \tag{9}$$

Here I is the solar cell current, I_L is the light-generated current, I_s is the dark saturation current, η is the ideality factor, $V_{th} = k_BT/q$ is the thermal voltage, k_B is the Boltzmann voltage, T is temperature, q is the electronic charge, R_s and R_{sh} are the series and shunt resistance, respectively. Eq. (9) can be re-written as follows

$$i = i_L - \exp(u + ir_s) + 1 - \frac{u}{r_{sh}} \tag{10}$$

where $i = I/I_s$, $i_L=I_L/I_s$, $u = V/(\eta V_{th})$, $r_s = R_sI_s/(\eta V_{th})$, and $r_{sh} = R_{sh}I_s/(\eta V_{th})$. As can be seen from Fig. 31 (b), the most probable reason for the low fill factor is a high series resistance.

Both structures exhibit transient response under light, with the characteristic response time decreasing with the wavelength of the illuminations. This is consistent with having deeper localized states in the energy gap determining the transients for shorter wavelength radiation. (Slow transients related to trapping effects are typical for polycrystalline CdS materials.)

These results clearly demonstrate a feasibility of fabricating CdS solar cells on flexible substrates that also might cover surfaces of complex geometry. These cells might find applications for autonomous power sources, especially for naval and marine applications, where such autonomous source placed, for example, a buoy might collect sunlight for different sun orientations throughout the day.

In Reference [26], we report on the photovoltaic effect in CdS/CuS films deposited on trylene threads. This technology should lead to the development of solar cells on textiles for harvesting solar energy using electro textiles and for providing renewable power source for wearable electronics. Using electron beam deposition, we deposited aluminum and nickel contacts on the trylene threads covered by the semiconductor layer. Fig. 32 shows the current voltage characteristics in dark and under illumination with a xenon lamp. The device configuration and dimensions are shown in the insert.

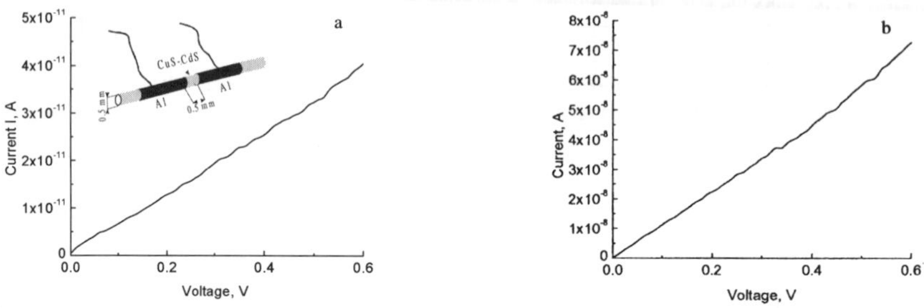

Fig. 32. Current-voltage characteristics in the dark (a) and under illumination (b) of the thread covered by CuS-CdS composition. The inset shows the sample configuration.[26]

The current under light is several orders of magnitude larger than in the dark showing a strong photovoltaic effect.

Fig. 33 compares the spectral characteristics of the photoresponse for the semiconducting thread with that for the CdS film deposited on the viewfoil. For comparison, we also show the energy gap for crystalline CdS[27]. As can be seen, our results are consistent with the expected absorption edge for this material.

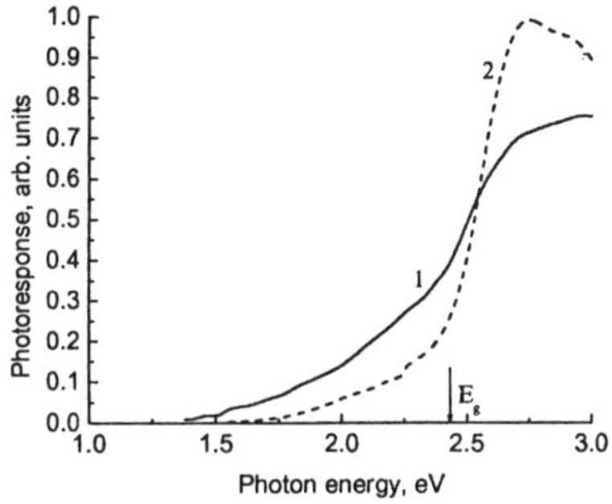

Fig. 33 Spectral characteristics of the photoresponse for the semiconducting thread (curve 1) and viewfoil (curve 2) with deposited CdS film.[26]

In Fig. 34, line 1 shows the current-voltage characteristic of the semiconducting thread with two different contacts (aluminum and nickel) under illumination with the xenon lamp. For comparison, we also show the current-voltage characteristics of a conventional CdS solar cells [28] (lines 2 and 3) and of the CdS photovoltaic element fabricated on the viewfoil[25] (line 4). In this figure, the light generated current is normalized to its maximum value (the actual measured values at V=0 were 5 nA, 1 μA and 50 μA for thread, viewfoil and conventional photovoltaic element, respectively). As can be seen, the photovoltaic voltage for the semiconducting thread is close to that for the CdS solar cell. A relatively low fill factor can be explained by a large series resistance.

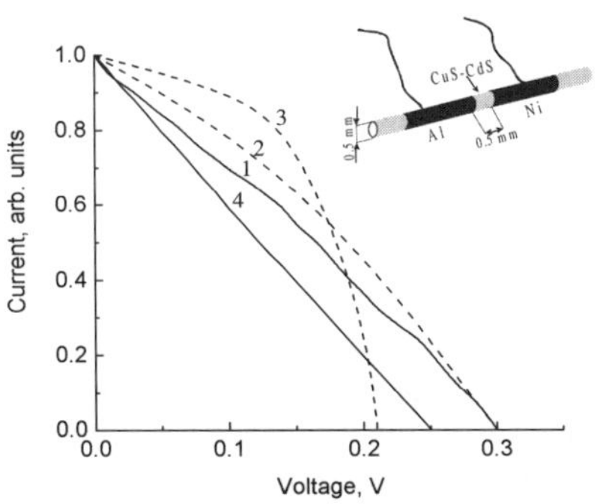

Fig. 34. Current voltage characteristics under illumination for semiconducting thread (curve 1), conventional CdS-based solar cells (curves 2 and 3), and for semiconducting viewfoil (curve 4). The currents are normalized to their maximum values at V=0.[26]

The structure on the thread did not exhibit slow transient response as we found before for the CdS structure on the viewfoil.

These results clearly demonstrate a feasibility of fabricating CdS solar cells on cloth. We have fabricated semiconducting cloth samples with identical contacts that show photovoltaic effect. Fig. 35 shows current voltage characteristics of the nylon cloth with deposited CdS in the dark and under illumination with the xenon lamp.[26]

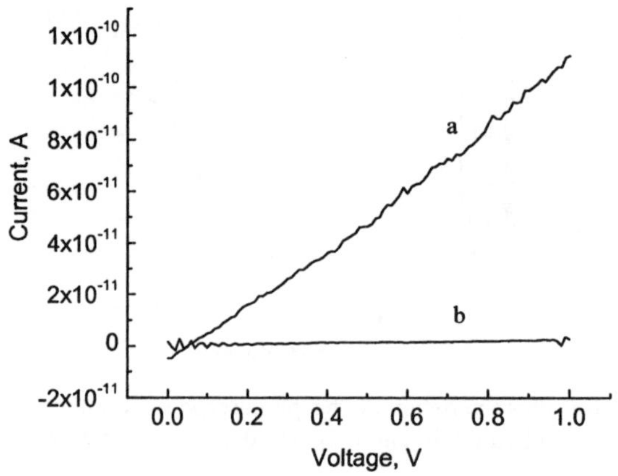

Fig. 35. Current-voltage characteristics under light (a) and in the dark (b) for semiconducting cloth.[27]

6. Conclusions

The conducting textiles and wearable electronics have already found commercial and defense applications. Semiconductor materials and devices embedded into electrotextiles are now emerging and preliminary results are very promising. However, many challenges have to be addressed to develop practical electrotextile technology. These challenges include the development of reliable low temperature deposition technology for depositing semiconducting films on threads and cloth, the development of flexible/stretchable metal and transparent metal films, the demonstration of semiconductor heterostructures on threads and cloth, the development of ultra large area (length) continuous material deposition and processing. As the first step, basic device building blocks have to be demonstrated and optimized. These device building blocks include flexible, stretchable ohmic and Schottky contacts, p-n junctions, Metal-Insulator-Semiconductor structures, and heterostructure diodes. Our group recently demonstrated field effect in PbS deposited on a flexible metal wire. The development of thin film transistors is now under way. Once the materials and device problems are solved, a variety of important issues related to systems, subsystems, energy harvesting, and communication on cloth and on "sensitive skin" will have to be addressed. Some of these problems have been discussed in References 7,8. However, the work has just begun. Quoting Samuel Butler, "Our minds want clothes as well as our bodies."

Acknowledgment

The work at Biomedical and Information Technologies, Inc. was supported by DARPA and administered by the U.S. Army Soldier and Biological and Chemical Command (SBCCOM) – Natick under Contract No. DAAD16-00-C-9290 (Project Monitor Dr. Elana Ethridge).

References

1 http://www.comforthouse.com/comfort/blankets1.html
2 http://www.electrotextiles.com/www/site/press/pr_06112000b_popup.html (http://www.eleksen.co.uk/flash/press_releases.shtml)
3 http://www.media.mit.edu/wearables/
4 http://www.research.philips.com/pressmedia/releases/000801.html
5 http://www.reimasmart.com/index.cfm?action=products&actionsub=top
6 http://www.defenselink.mil/news/Sep1998/n09111998_9809117.html
7 V. Lumelsky, M. S. Shur, and S. Wagner, Editors, Sensitive Skin, World Scientific, ISBN 981-02-4369-3, 2000
8 V. Lumelsky, M. S. Shur, S. Wagner, Sensitive Skin, IEEE Sensors Journal, Volume 1, No. 1, pp. 41-51 (2001)
9 http://www.under-tec.com/store.htm
10 http://nina.ecse.rpi.edu/shur/OTFT Movie (062601).mpg
11 http://www.universaldisplay.com/foled.php
12 P. Necliudov, Modeling and Characterization of Organic Pentacene Thin Film transistors, Ph. Dissertation, Rensselaer Polytechnic Institute (2001)
13 J.H. Schon, et al., "On the intrinsic limits of pentacene field-effect transistors", Organic Electronics, vol.1, no.1 p. 57-64
14 D. J. Gundlach et al., "High-mobility, low-voltage organic thin film transistors", 1999 International Electron Devices Meeting Technical Digest, pages 111-114, December 1999
15 C.S. Yang, et al., "Stability of low-temperature amorphous silicon thin film transistors formed on glass and transparent plastic substrates", Journal of Vacuum Science & Technology B, vol.18, p. 683-689
16 J. Sinius et al., "Photovoltaic thin layer detector for radiation in near IR region", Radio electronics, Vilnius, Technika, 30 (3), 170 (1994)
17 J. Sinius et al., "IR-detectors on PVF2 films", Radio electronics, Vilnius, Technika, 29 (3), 88 (1993)
18 J. Sinius, R. Gaska, and M. S. Shur, Flexible Semiconductor Films for Sensitive Skin, in Proceedings of the 2000 IEEE/RSJ International Conference on Intelligent Robotics and Systems, IEEE Catalog Number :00CH37113C, ISBN: 0-781-6351-5, IEEE (2000) pp. 1511-1515
19 J. Soltys, J. Sinius and A. Sinius, "Kinetics of structural transitions in CdSe thin film", European Conference "Surface Science", April 1984, UK.
20 R. Baltramiejunas, R. Gaska, E. Kuokstis, J. Sinius, "The influence of laser annealing on optical properties and photoluminescence in CdSe layers", Fifth Conference on Physics and Applications of A2B6 Compounds", Vilnius, Lithuania, June 1983.
21 M. S. Shur, S. Rumyantsev, R. Gaska, B. Q. Wei, R. Vajtai, P. M. Ajayan, and J. Sinius, "Structural and transport properties of CdS films deposited on flexible substrates", Accepted to Solid State Electronics, 2002
22 JCPDS International Center for Diffraction Data, Powder Diffraction File, 6-314
23 J. Sinius et al., "IR-detectors on PVF2 films", Radio electronics, Vilnius, Technika, 29 (3), 88 (1993)

24 J. Sinius, R. Gaska, and M. S. Shur, Flexible Semiconductor Films for Sensitive Skin, published in Proceedings of IROS2000, October 30 - November 5, 2000, Kagawa University, Takamatsu, Japan (see http://iros2000.iis.u-tokyo.ac.jp/)

25 M. S. Shur, J. Sinius, R. Gaska, and S. Rumyantsev, Photovoltaic Effect in CdS on Flexible Substrate, Electronics Letters, vol. 37, No. 8, pp. 518-519 (2001)

26 M. S. Shur, G. Gaskiene, S. Rumyantsev, R. Rimeika, R. Gaska, and J. Sinius, Photovoltaic Effect in Threads Covered with CdS, Electronics Letters, Vol. 37, No. 16, pp. 1036-1038 (2001)

27 B. Batlogg, A. Jayaraman, J. E. Van Cleve, and R. G. Maines, "Optical absorption, resistivity, and phase transformation in CdS at high pressure", *Phys. Rev. B*, vol 27, pp.3920-3923 (1983)

28 R. K. Pandey, S. Mishra, S. Tiwari, P. Sahu, and B. P. Chandra, "Comparative study of performance of CdTe, CdSe and CdS thin films-based photoelectrochemical solar cells", *Solar Energy Mater. & Solar Cells*, 2000, **60**, pp.59-72

International Journal of High Speed Electronics and Systems, Vol. 12, No. 2 (2002) 391–399

ELECTROTEXTILES: CONCEPTS AND CHALLENGES

SIGURD WAGNER, EITAN BONDEROVER,
WILLIAM B. JORDAN, and JAMES C. STURM

Department of Electrical Engineering and POEM, Princeton University,
Princeton, New Jersey 08544, USA

Electrotextiles (e-textiles) are fabrics made from yarns that carry electronic components. Circuits are made from such yarns by weaving. We describe the technical development of the concept of e-textiles, and briefly discuss three challenges: connectivity, materials and fabrication, and wear.

Keywords: Electrotextile; wearable electronics; woven circuits.

1. The Concept of Electrotextiles

The concept of electrotextiles, or e-textiles, has evolved from the concept of wearable computers. Wearable electronics have been demonstrated in the US[1,2] and in Europe.[3,4] Clothing companies have started to market products that carry hybrid mounted electronics,[5] and the U.S. Armed Forces have conducted programs to develop uniforms that incorporate electronics.[6]

Wearable electronics and electronic textiles lie at the two ends of a spectrum that ranges from added-on electronics to components integrated with textile yarn to true integration as shown in Fig. 1.

A first-generation wearable computer enables an engineer to compare blueprints with a workpiece in real time, as illustrated by Fig. 2 from the Carnegie Mellon program.[1] The application is clearly defined, and while the system is wearable, it is not integrated into the clothing of the user.

In the second generation of wearable computing, the system is surface mounted on clothing in a hybrid package. This approach is illustrated by Fig. 3, taken from

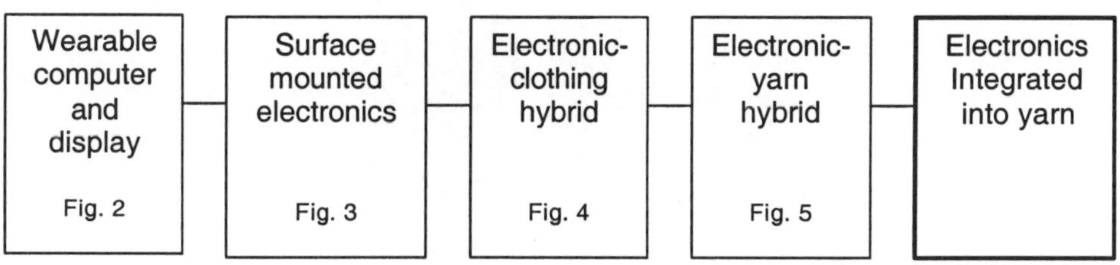

Fig. 1. A continuous spectrum of integration of electronics into textiles exists between wearable computers on the left and electronic components integrated into yarn on the right.

Fig. 2. Wearable electronics are body mounted but not integrated. This kind of wearable computer represents the first generation of practical, body-worn electronics.[1]

the MIT MediaLab's Mithril project,[2] and this is the concept adopted today by leisure fashion companies.[5] The system components are distributed over a vest that has a secondary function as clothing, in the same way that a tool belt functions as a clothing belt. This approach is prominent at the IEEE Symposia on Wearable Computers.[7]

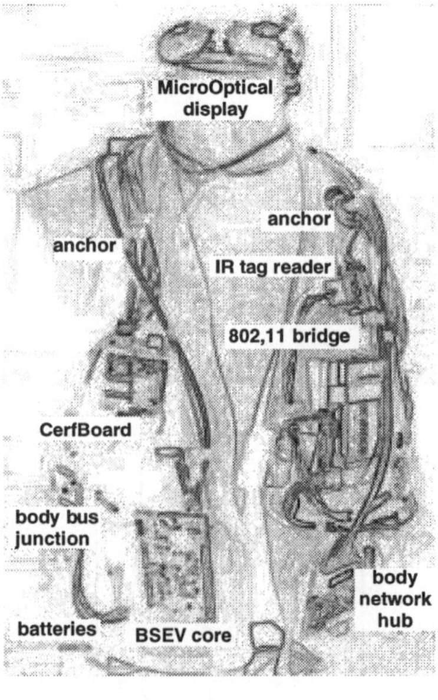

Fig. 3. Surface mounted electronics are the next step toward integrated e-textiles.[2] Surface mounting is entering commercial leisure products.

Fig. 4. A measure of integration is achieved with lighting attached to dresses and in jewelry. Here electronics serves to enhance the appeal of clothing, instead of just using it as a carrier.[8]

In the third generation of wearable electronics the electronic and the clothing functions are equally important, as illustrated by the two demonstrations from the MIT MediaLab and International Fashion Machines shown in Fig. 4.[8] Both the dress and the necklace meet clothing and esthetic functions and they carry lighting and sensing functions. This third generation is an important step toward true electronic textiles.

From this point on the concept of wearable electronics blends into that of electronic textiles. A case in point are early efforts to add an electronic function to e-textile yarn, as illustrated by Fig. 5, which shows a fabric made of plain silk thread warp and a weft of silk thread wrapped with copper foil.[9] This is an important conceptual advance in that the integration is brought to below-device level, but it still is a hybrid. It combines the separate free-standing functions of textile yarn and copper conductor.

Nearly all of these initiatives approach electrotextiles from the systems level. They attempt to answer the question how known, or desirable, electronic functions

Fig. 5. The weft yarn of this fabric is silk wound with copper foil. This is an example of hybrid electronics at the yarn level.[9]

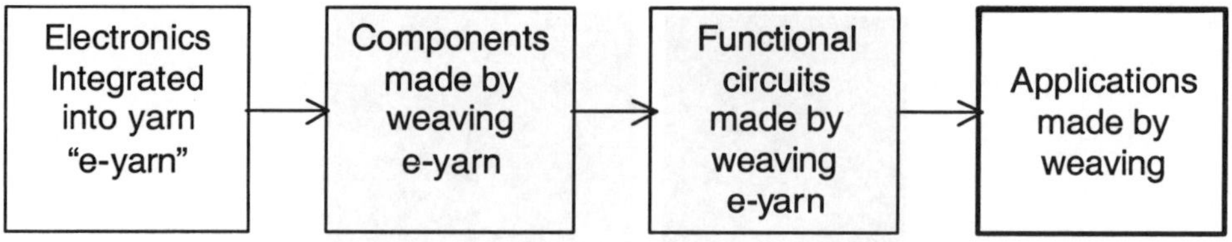

Fig. 6. Integration from the yarn level (rightmost frame of Fig. 1 and leftmost frame above) to components and functional circuits made by weaving the yarns, and eventually to applications such as smart actuating textiles.

can be worn on the body in comfort. This applications oriented approach indeed is needed to target early product opportunities, and the current pressing need for pervasive security may give birth to the first generation of clothing integrated sensor electronics. However, the hybrid approach responds to the task of marrying electronics and textile technologies at a superficial level, quite literally. Electronic patches are applied on fabric or are inserted into pockets. Stand-alone components that would function by themselves — clothes, DVD players — are combined.

E-textiles represent integration at the yarn level. E-yarns are used to make functional circuits, and are made functional by weaving. This approach is illustrated in Fig. 6, which in a sense is the inverse of Fig. 1. E-textiles take the approach taken by the I.C. industry, which first developed component technology and then proceeded to ever higher integration.

Wearable electronics follow the microprocessor model. Their components are miniaturized and are applied in high functional density. E-textiles follow the flat panel display model. They are large-area electronics with low functional density. Because early e-textiles are expected to be built of low performance components, it is likely that e-textiles will be hybridized with silicon integrated circuits.

2. Technical Challenges

We select three challenges for discussion: connectivity, materials and fabrication, and wear.

2.1. *Connectivity*

A large range of techniques exist for making fabric from yarn.[10] These techniques represent a variety of possible ways for making electric interconnects. The circuit engineer seeks to make permanent interconnects to ensure high manufacturing yield and operational reliability.

A simple example for taking advantage of the connectivity of woven fabric is illustrated by Fig. 7, which shows schematically how solar cells made on different yarns may be connected in series by weaving. The colored and black segments are solar cells and the gray areas are insulating portions of the fibers. In Group A, fiber segments are electrically connected on both sides, except for an insulator between

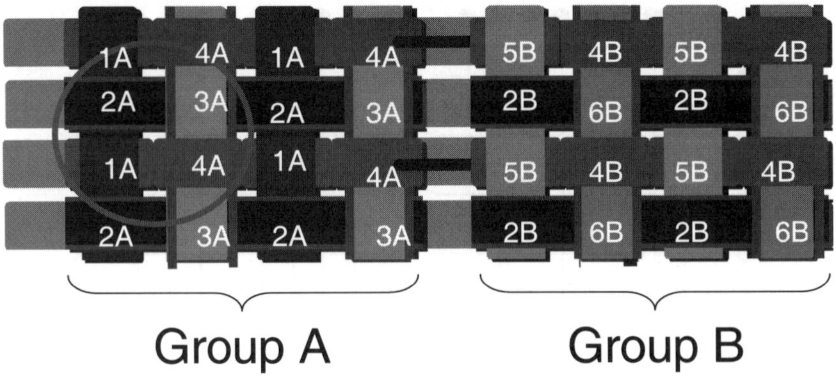

Fig. 7. Series connection of solar cells by weaving. Groups A and B are separate circuits.

1A and 4A. Thus fiber segments 1A through 4A are in series in numerical order. Therefore, the voltages add, and the fibers/geometries must be engineered such that the photocurrent in each fiber segment is the same. To connect Groups A and B in series, the top contact of 4A is connected to the bottom contact of 4B, as shown. Note that all fibers may require patterning since the series combinations must be horizontal-vertical-horizontal-vertical-etc. to adjust to the orthogonal nature of the weave. To connect the groups in parallel, the bottom of fiber segment 1A must be connected to the bottom of segment 5B (which would add much complexity to the weave), while the top of 4A would be conveniently connected to the top of 4B. Several groups could be connected in series to achieve a desired voltage, and then several of these series groups could be connected in parallel to achieve a desired current. To reduce possible shading effects, care should be taken to combine as few groups (or as small an area) as possible in series, and then to connect the parallel groups with diodes.

When fabric is draped the yarns experience shear. As the number of fixed interconnects is increased, the fabric stiffens and its drapability is diminished. This is illustrated by Fig. 8. To preserve drapability one can design the fabric such that it is connected only at its edges, as illustrated by the left panel of Fig. 8. Doing that requires the design and fabrication of yarn that allows addressing a large number of cells placed on the yearn with a small number of wires. One example of a cell design for fiber is shown in Fig. 9. This cell contains a (pressure or light) sensor. It is addressed by sequential sampling. Each cell contains only 17 transistors, and the fiber needs to carry only 6 wires.

This circuit can be used to sequentially sample each cell in the line. The circuit of every cell is identical. An arbitrary number of cells can be in each string, but the driver circuit will have to be configured to accommodate the exact number of cells. The D flip flops pass the input at D to the output at Q when the clock input C receives a high signal from the clock. Therefore, with this Master (D_M)–Slave (D_S) configuration, the address pulse will propagate from D_M to Q_M/D_S when C_M is

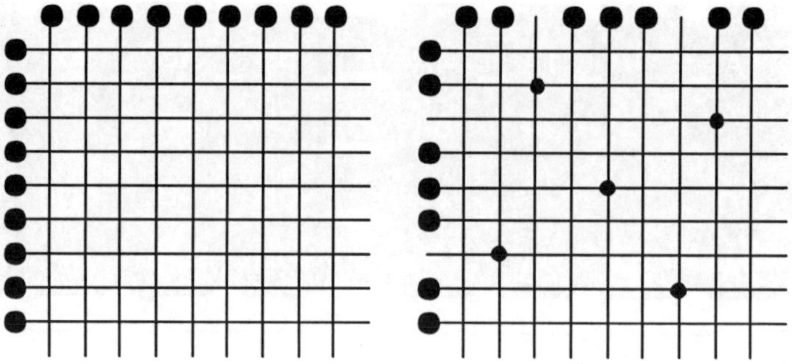

Fig. 8. As the connectivity of a fabric is increased, it becomes stiff and its drapability is reduced.

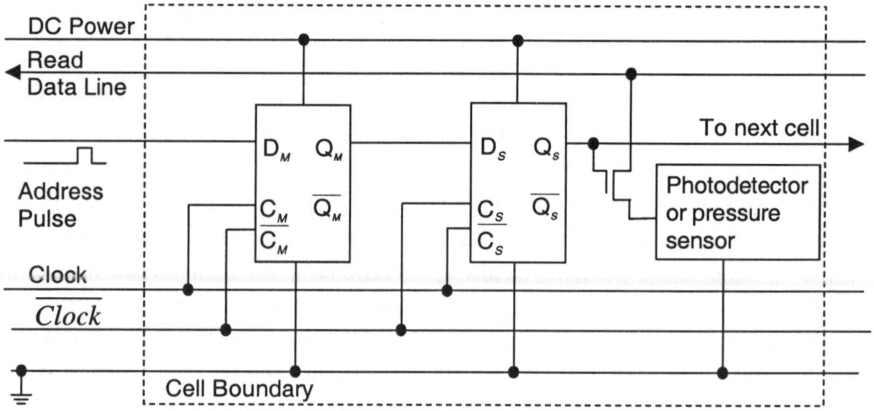

Fig. 9. Cell of an on-yarn circuit used for sequential sampling of sensors mounted on the yarn. Each D flip flop contains 8, plus 1 pass transistor = 17 per cell.

high, but cannot pass to Q_S since C_S is low. During the second half of each clock cycle, D_M cannot pass the signal since C_M is low, but D_S will pass the signal to Q_S (and D_M of next cell) since C_S is high, and so on. When the address pulse appears at Q_S, the transistor is turned on, passing the data to the data line to be read by the read circuit on the driver chip. The address pulse will propagate on down the line, moving from one cell to the next with each clock cycle, causing data from each successive cell to be sent to the read circuit for every low portion the clock period.

We expect that designing small wire and component count cells will be an important task in the development of e-textiles. This need is similar to the need for low-transistor count intelligent pixels, which also is on the research agenda for ASICs, such as on-pixel processing for CMOS based camera sensors.[11]

2.2. *Materials and fabrication*

Silicon based thin film transistors can now be fabricated at temperatures as low as 150°C, and can be made on foils of glass, steel, and of certain organic polymers.[12] While all of these materials can be made into yarn, and might be employed in

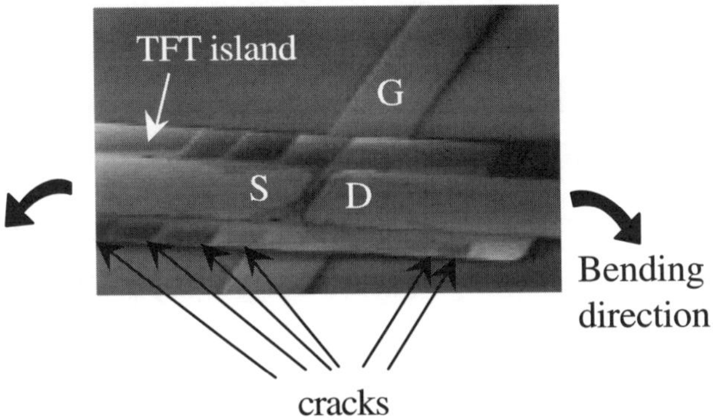

Fig. 10. Scanning electron micrograph of an a-Si:H thin film transistor that failed electrically by mechanical fracture of a SiN_x layer.[13]

specialty fabric, the most attractive yarn materials are organic polymers. The temperature of 150°C separates high-temperature polymers from the large volume polymers used for consumer applications. Therefore one target of e-yarn fabrication is to reduce the fabrication temperature. Several groups of semiconductors are available, including silicon processed at ultralow temperatures, II–VI compound semiconductors processed by wet chemical and electrochemical techniques, and organic semiconductors. None of these semiconductor materials has reached maturity in low temperature processing. While present attention is focused on obtaining these various semiconductors with good device properties, future concerns common to each will be stable dielectrics and ohmic contacts.

2.3. *Wear*

2.3.1. *Bending*

When used in e-textile, an e-yarn will be subjected to bending, stretching and shear. The consequences of bending on the performance of a-Si:H based thin film transistors have been studied in some detail. A-Si:H TFTs become less susceptible to bending to a given radius when the substrate is made thinner. While bending affects the electrical performance, the TFTs fail by mechanical fracture of one of their component layers, as is evident from Fig. 10.[13] It also appears that long-time fatigue may cause failure at lower strains than needed to fracture a TFT initially.

2.3.2. *Stretching*

Yarn may be stretched during wear. The transistor material will tolerate tensile strain up to relatively small values. The technique developed to protect the semiconductor circuits from mechanical failure is illustrated in Fig. 11. The circuits are placed on hard islands, here a-Si:H on SiN_x, such that the strain developed in the stretched fiber is concentrated between the islands.[14] Design rules for island size and thickness, and for interisland spacing, have been developed.

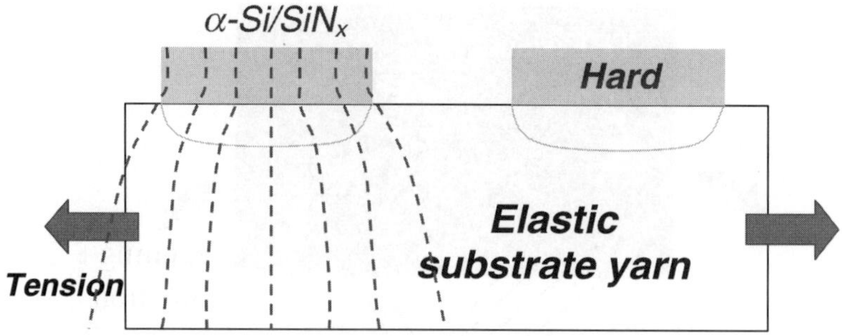

Fig. 11. Concept of hard device islands on elastic substrate. Design rules have been developed for sizing the islands such that their strain remains below a critical value while the yarn is stretched.

No standard approach to making stretchable interconnects exists at present.

2.3.3. *Shear*

Shearing yarns against each other during draping has two consequences. One is the making and unmaking of nonfixed electrical interconnects. The ensuing need for reconfiguring the local circuit from newly interconnected components will depend on some neural net capability of the e-textile. The second consequence is abrasive wear associated with shear friction. This is another aspect of e-textile that needs to be understood and prevented.

Acknowledgments

This work is supported by DARPA, NSF, and the New Jersey Commission on Science and Technology.

References

1. http://www-2.cs.cmu.edu/afs/cs.cmu.edu/project/vuman/www/home.html.
2. http://www.media.mit.edu/wearables/diagram.html.
3. http://www.design.philips.com/smartconnections/index.html.
4. http://www.iswc.ethz.ch/.
5. I-wear consortium: http://www.starlab.org/bits/intell_clothing/.
6. http://ginsburg.ipm.virginia.edu/Summer/. See the slides on the Land Warrior 2020/2025 program.
7. http://www.computer.org/proceedings/iswc/0795/0795toc.htm.
8. E. R. Post, M. Orth, P. R. Russo, and N. Gershenfeld, "E-broidery: Design and fabrication of textile-based computing", *IBM Syst. J.* **39**, 3 & 4 (2000) 840, http://www.research.ibm.com/journal/sj/393/part3/post.html; M. Orth, E. Cooper, and D. Lockwood, "International Fashion Machines", http://www.ifmachines.com/design_firefly-dress.html.
9. E. R. Post and M. Orth, *MIT MediaLab publication*, http://computer.org/proceedings/8192/pdf/81920167.pdf.
10. S. Adanur, "Wellington Sears handbook of industrial textiles", Technomic Publ. Co., Lancaster, PA, 1995. Fig. 4.1 on p. 88.

11. D. X. D. Yang, B. Fowler, and A. El Gamal, *IEEE J. Solid-State Circuits* **34** (1999) 348.
12. http://www.ee.princeton.edu/~asg/.
13. H. Gleskova, S. Wagner, and Z. Suo, *Appl. Phys. Lett.* **75** (1999) 3011.
14. P.-H. I. Hsu, M. Huang, S. Wagner, Z. Suo, and J. C. Sturm, *Mat. Res. Soc. Symp. Proc.* **621** (2000).

International Journal of High Speed Electronics and Systems, Vol. 12, No. 2 (2002) 401–419

STRAIN ENERGY BAND ENGINEERING APPROACH TO AlN/GaN/InN HETEROJUNCTION DEVICES

ASIF KHAN, J. W. YANG, G. SIMIN

Department of Electrical Engineering, University of South Carolina, Columbia, South Carolina 29208

R. GASKA

Sensor Electronic Technology, Inc., Latham, NY 12110

M. S. SHUR

CII 9015, Rensselaer Polytechnic Institute, 110 8th Street, Troy, NY 12180

Our new Strain and Energy Band Engineering approach allows us to control, modulate or eliminate the strain thereby significantly enhancing the performance of electronic and optoelectronic devices on SiC and sapphire substrates using a new-pulsed atomic layer epitaxy procedure. New types of high power high temperature transistors with improved performance were fabricated. XRD, PL and AFM spectra are used to establish a high structural and optical quality of the quaternary multiple quantum wells. We have also demonstrated the use of PALE deposited quaternary AlInGaN multiple quantum wells in the active region of a UV LED with emission ranging from 305 nm to 304 nm.

Introduction

Unique materials properties of GaN-based semiconductors stimulated a great deal of interest in research and development in materials growth and optoelectronic and electronic devices using this semiconductor system. An AlGaN/GaN Heterostructure Field Effect Transistor is one of the most promising electronic devices, which has been a topic of intensive investigations since the first report in 1991. Solar blind photodetectors, blue and UV light emitting diodes, piezo- and chemical sensors are other examples of unique devices that can be fabricated from nitride semiconductor materials. The major problems with growth and fabrication of III-N materials and devices are the lack of bulk GaN substrates and a large lattice mismatch in the heterostructures with large variations of Al or In fractions, which, in turn, create very high built-in electric fields and strain in the layers, resulting in high density of defects and cracks. Our new Strain and Energy Band Engineering approach allows us to control, modulate or eliminate the strain thereby significantly enhancing the performance of electronic and optoelectronic devices.

Strain energy band engineering approach

Recent progress in the development of AlGaN/GaN materials and devices for high-power, high-temperature applications has been remarkable. The record microwave power of over 100 W for a single device reported by CREE Inc. at WOCSEMADD 2002 as well as high-temperature and low-noise performance [1] of AlGaN/GaN Heterostructure Field Effect Transistors (HFETs) hold promise of far surpassing GaAs-based devices. A superior electron transport in GaN [2] an excellent thermal properties of SiC (used as a substrate in the best GaN-based devices) allow us to enhance current-carrying capability [3] and to reduce self-heating effects in GaN-based devices [4]. However, the microwave power levels achieved for conventional GaN-based HFETs are still far below what is expected from DC performance of GaN-based HFETs.[5] Device stability, reliability, transient effects (such as current slump) also present a problem. Partially, this problem might be related to the quality of the heterointerface [6] that deteriorates with an increased molar fraction of Al because of strain.

Large lattice mismatches between GaN, AlN and InN and the strong piezoelectric and pyroelectric effects in these materials significantly impact the electrical and optical properties of III-N heterojunction devices. Piezoelectric constants of GaN are 4-5 times larger than GaAs piezoelectric constants when they are compared using the wurtzite - zinc blende transformation procedure. [7] Nearly all HEMT devices reported to date use strained GaN-AlGaN junctions with alloy compositions below 35 % and 15-20 nm thick barriers to avoid exceeding the critical thickness. Strain and differences in spontaneous polarization produce piezoelectric and polarization doping with about $1\text{x}10^{13}$ cm^{-2} sheet carriers. On the other hand, strain might facilitate the creation of traps leading to long-term instabilities observed in conventional AlGaN/GaN HEMTs.

An importance of lattice mismatching and strain-relating issues for UV light-emission and photodetector devices is illustrated in the Fig. 1. In order to achieve a "Solar-Blind" operation at the wavelength of 300 nm the material with the Al fraction of around 28% is needed (Fig. 1(a)). However, due to the lattice mismatching, the critical thickness of the AlGaN layer with 28% Al composition grown over GaN buffer is less than 150 Å. This thickness is neither sufficient for providing a full light absorption in photodetectors nor for fabricating a multiple quantum well light-emitting device.

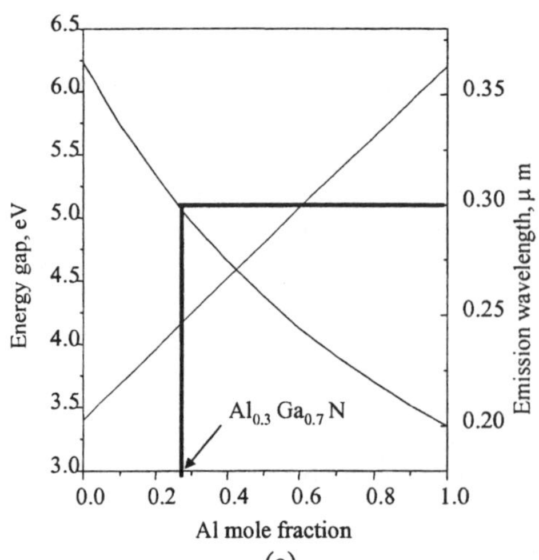

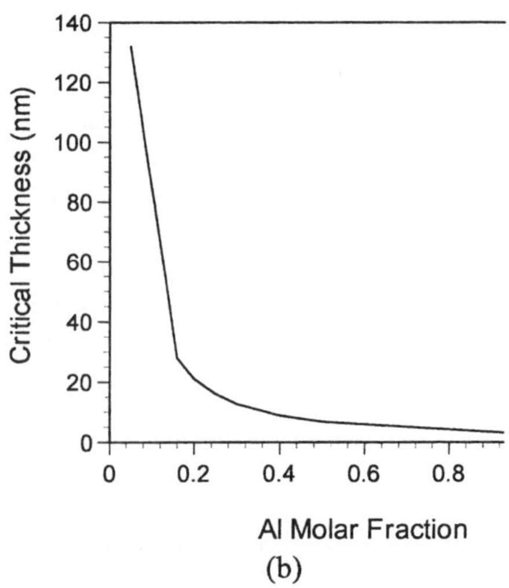

Fig. 1 The emission wavelength (a) the critical thickness (b) dependencies on the Al fraction in AlGaN alloy.

Our objective is to develop a new class of nitrides -- quaternary AlInGaN/InGaN heterostructures that should allow us to independently control strain and conduction band discontinuities [8]. This approach should find many applications in GaN-based electronic and optoelectronic devices. Fig. 2 shows lattice mismatch and band offset in AlGaN and InGaN grown on GaN. Based on a linear extrapolation of lattice constants as functions of molar fractions, we estimate that quaternary $Al_xIn_yGa_{1-x-y}N$ layers with Al/In mole fraction ratio of 5 should be nearly lattice matched to GaN. On the other hand, the band discontinuity is affected much less. For example, strain free growth of 50% AlGaN on GaN requires 10% of In. For the same 10% In the band offset will remain as high as 1.2 eV.

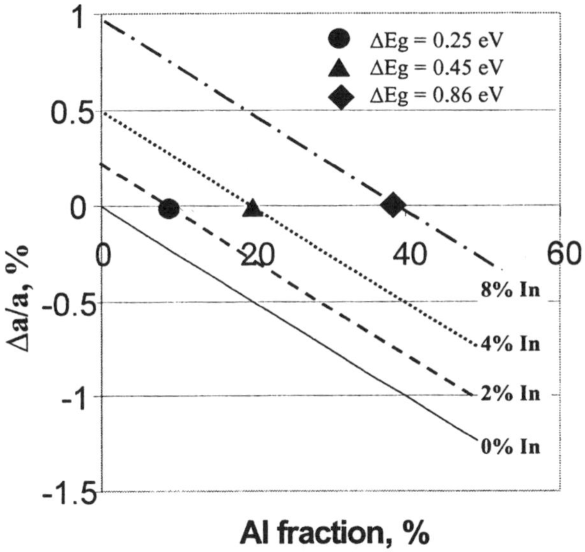

Fig. 2. Relative lattice mismatch as a function of Al molar fraction for different molar fractions of In. Bullets show the band gap offsets for lattice-matched Al-In compositions (after [8]).

$Al_xIn_yGa_{1-x-y}N$-GaN heterojunctions over sapphire and 6H/4H SiC substrates were grown using a low pressure MOCVD. Several series of $Al_xIn_yGa_{1-x-y}N$-GaN heterojunctions were deposited with different Al-precursor flows. For each series, the Ga-precursor flow was kept constant while varying the In-flow. The thickness of the GaN and the quaternary $Al_xIn_yGa_{1-x-y}N$ layers were respectively kept at 100 and 200 nm. SIMS and EDAX measurements were then used to determine the In-Ga-Al profiles and their atomic percentages; see Fig. 3. These data clearly establish the incorporation of indium fractions as large as 6% into AlInGaN films. Electron Microprobe data agree with SIMS data.

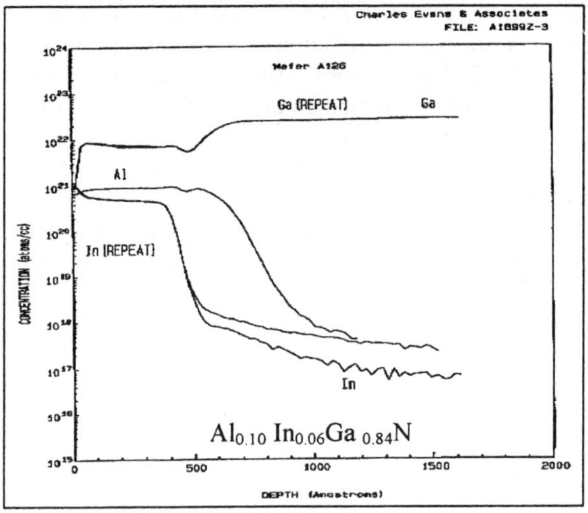

Fig. 3. SIMS data that show up to 6% incorporation of indium (after [8])

Using RHEED measurements we also confirmed the single crystal nature of all the $Al_xIn_yGa_{1-x-y}N$ layers for our growth conditions. For the entire $Al_xIn_yGa_{1-x-y}N$-GaN heterojunctions lattice mismatch and bandgaps were measured using X-ray (θ-2θ) diffraction and optical transmission versus wavelength. The bandgap values from transmission were in good agreement with the measured room temperature PL peak positions using the 4th harmonic of a Q-switched YAG:Nd^{3+} laser radiation (incident photon energy 4.66 eV, pulse duration 10 ns) for optical excitation.

The X-ray, SIMS, EDAX and the PL data were then used to estimate the alloy compositions for the quaternary $Al_xIn_yGa_{1-x-y}N$ layers. Our data clearly indicate that the addition of In changes the lattice constant difference (between GaN and the quaternary) about 3-4 times faster than the bandgap difference. However, for the nearly lattice matched $Al_{0.09}In_{0.02}Ga_{0.89}N$-GaN junction, the PL measurements yielded the bandgap difference of only 120 meV, which significantly smaller from what was expected using the linear interpolation of the band offsets for a fixed molar fraction of Al.

Our Atomic Force Microscopy data show that surface morphology improves with incorporation of indium. The RMS roughness changes from 0.7 to 0.2 nm with the increase of In incorporation from 0 to 2 %.

Several $Al_xIn_yGa_{1-x-y}N$-GaN junctions with barrier thickness less than 50 nm and varying alloy compositions (x from 0.1 to 0.2 and y from 0.02 to 0.04) were also grown. For these heterojunctions, room temperature sheet-carrier density and mobility values ranging from 1-3 x 10^{13} cm^{-2} and 1000-1500 $cm^2V^{-1}sec^{-1}$ were measured. The mobility values increase by about a factor of 5 upon cooling to 77 K. This establishes the presence of the two-dimensional electron gas at the $Al_xIn_yGa_{1-x-y}N$-GaN heterojunction interface and confirms a high quality of heterointerfaces.

Pulsed Atomic Layer Epitaxy (PALE) approach

Recently, using AlGaN/AlGaN MQWs in the active region several groups [9,10,11] have demonstrated UV LEDs with peak emission wavelength around 340-350 nm. In contrast with transistors, the AlGaN/GaN based optoelectronic devices suffer from poor material quality and high built-in strain due to large piezo- and spontaneous polarization charges. The strain energy band engineering (SEBE) approach reviewed above allows tailoring the built-in strain [12] and significantly improving the quantum well emission properties by using the quaternary AlInGaN material system [13, 14]. These improvements result not only from the strain management but also from an overall material quality improvement by the In- incorporation in the ternary AlGaN layers [15]. For our LEDs, we employed a unique pulsed atomic layer epitaxy (PALE) approach for depositing the AlInGaN layers for the quaternary active region. PALE allows for an accurate control of the active layer composition and thickness [16] and reduces the single crystal epitaxy temperatures thereby increasing the In-incorporation in the quaternary AlInGaN layers with high Al-mole fractions. This is an outcome of the increased precursor mobility in the PALE process. Using PALE deposited high Al-mole fraction AlInGaN layers we now report on a *p-n* junction LEDs with peak emission wavelength tuned from 305 nm to 340 nm by adjusting the AlInGaN alloy composition (different number of Al-, In- and Ga- pulses in the unit growth cell [16]).

For all the $Al_xIn_yGa_{1-x-y}N$ depositions, we used trimethyl aluminum (TMA), trimethylgallium (TMG), trimethyl indium (TMI) and NH_3 as the precursors and the basal plane sapphire substrates. Prior to the quaternary layer, a 250 Å thick AlN buffer layer and a 1.5 μm thick intrinsic *i*-GaN layer were grown using conventional low-pressure MOCVD. The growth pressures for the two layers were kept at 40 Torr whereas their growth temperatures were 450 °C and 1000 °C respectively. Quaternary $Al_xIn_yGa_{1-x-y}N$ layers were then grown at 760 °C by 150 repeats of a unit cell. The unit cell had a growth sequence as outlined in [16].

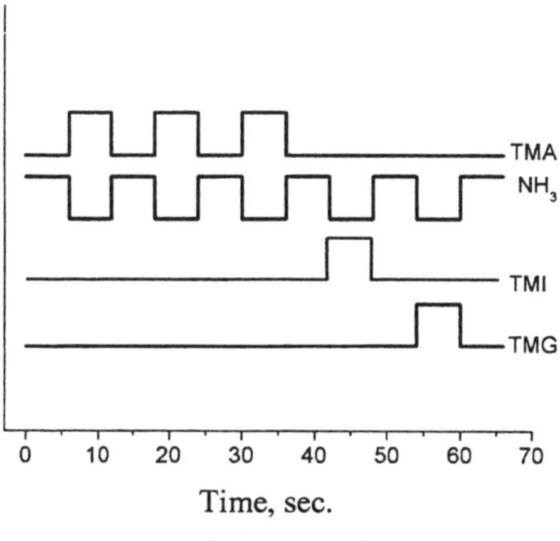

Fig. 4. A representative growth unit cell of PALE. We denote this cell as $(3_{Al}, 1_{In}, 1_{Ga})$, or simply as (3,1,1) (after [16])

As seen, 6 sec. long pulses of the precursors TMA, TMI, TMG and NH_3 were alternately introduced in the low pressure MOCVD reactor. An ammonia pulse always followed the metal organic pulses. As an example, the unit cell in Fig. 4 consisted of 3 repeats of Al and N pulses, followed by one In and N and one Ga and N pulse. We refer to the resulting $Al_xIn_yGa_{1-x-y}N$ layer as the $(3_{Al},1_{In},1_{Ga})_{150}$ or simply as a $(3,1,1)_{150}$ layer. The subscript denotes the number of unit cell repeats. The precursor fluxes in individual pulses were adjusted in such a way that the deposited thickness in each unit cell as determined from the total layer thickness divided by the number of unit cell repeats, was around 6 Å.

In order to ascertain that the PALE approach yields high quality quaternary AlInGaN layers and also allows for a simple composition control, a $(3,3,1)_{150}$ layer was grown first. We then used Rutherford Backscattering (RBS) spectra to determine the composition of this $(3,3,1)_{150}$ layer to be about 2% of In, 36% of Al and 62% of Ga. Using atomic force microscopy (AFM) we measured the root mean square (RMS) roughness for this layer to be only 9 Å. This compares favorably with the RMS roughness value for typical MOCVD grown AlInGaN layer with similar Al mole fraction. Quaternary layers with unit cell configurations of (3,0,1), (3,1,1), (3,3,1) and $(3,3^+,1)$, (3, 6,1) were then deposited. (Here the superscript + denotes pulses with the In- metalorganic flux doubled) In Fig. 5, we plot the relative indium composition in these layers as determined from energy dispersion analysis of x-ray (EDAX) measurements. As seen, the indium signal for the quaternary layers scales linearly with the number of In-pulses in the unit-cell and the indium flux (within the pulse). The data included in Fig. 5 clearly establish the viability of the PALE process in accurately controlling the quaternary AlInGaN material composition in a wide range. Using the RBS profiles for the $(3,3,1)_{150}$ sample, we also confirmed the thickness of this quaternary layer to be approximately 1000 Å.. The RBS data clearly showed a simultaneous presence of Al, In, and Ga in the quaternary layer. Similar observations were also made from the secondary ion mass spectroscopy profiling data.

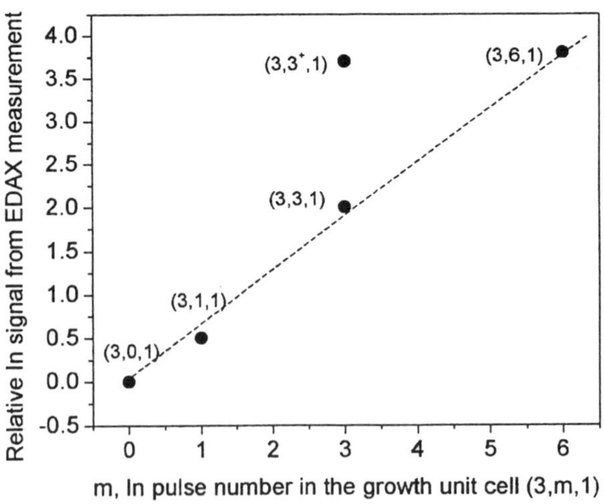

Fig. 5. Relative EDAX signals versus the number of In-pulses in unit cell (after [16]).

Fig. 4 shows the room temperature PL spectra for the quaternary $Al_xIn_yGa_{1-x-y}N$ films with unit cell configurations of (3,0,1), (3,1,1), (3,3,1), $(3,3^+,1)$, (2,2,1), (1,3,1) and $(1,3^+,1)$. The growth temperature for all samples was 760 °C. We have also included the data for a 1000 Å thick quaternary layer deposited at the same temperature (760 °C) but using a conventional low pressure MOCVD process and precursor flows similar to that of the (1,3,1) sample of Fig. 6. In addition, the data show the PL signal for PALE quaternary layers to be much stronger than that of the conventional MOCVD grown sample. Also, the PALE samples have no deep level emissions. Fig. 6 data further establish that addition of Al- in the unit cell blue-shifts the PL peak positions. The addition of In-, on the contrary, results in a red shift. This is to be expected from the known band gaps of AlN, InN and GaN. Comparing the PL signal strength from the (3,0,1) and the (3,1,1) samples we conclude that the addition of indium also improves the optical quality of the films. This observation confirms the results reported earlier for quaternary AlInGaN layers grown by a conventional MOCVD process. Furthermore, in contrast to conventional $Al_xGa_{1-x}N$ films, the PL peak intensity for the quaternary layers does not degrade with increasing the Al- mole fraction. This is true even for the material with a cutoff wavelength of 280 nm.

The PL peak positions for the PALE samples also coincide with the measured band-edge cut-off wavelengths. This confirms that the PL of the PALE deposited quaternary AlInGaN layers at 760 °C to be from band-to-band emissions.

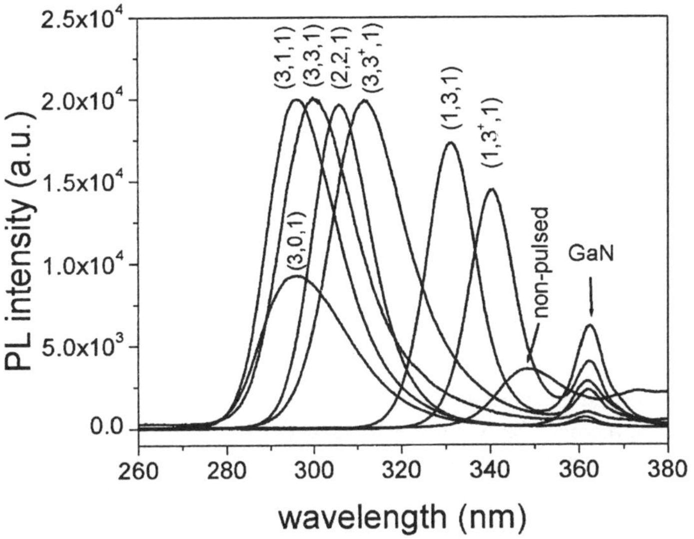

Fig. 6. Room temperature excimer laser excited PL spectra of the quaternary samples (after [16]).

Quaternary AlInGaN based deep UV LEDs

The PALE grown MQWs can be successfully employed as an active layer for LEDs [17, 18, 19, 20]. The LED peak emission wavelength can be tuned by adjusting the AlInGaN alloy composition (changing the number and flux of precursor pulses). For this three sets of LED structures were deposited over sapphire having a 0.8 μm thick n^+-AlGaN layer followed by a 30 period n^+-AlGaN/AlGaN superlattice with a periodicity of 30 Å. The n^+-AlGaN layer had an Al-mole fraction of 22 %, 26 % and 36 % for the 340 nm, 320 nm and the 305 nm LED structures. The n-AlGaN was capped with the n^+-AlGaN/AlGaN superlattice with Al-compositions of 16 % - 20 %, 20 % - 25 % and 25 % - 30 %, respectively. These values were selected to have a transparency for the LED peak emission wavelengths. On top of n-superlattice the MQW active layer was grown using

PALE. For these three sets of LEDs the following MQW unit growth cell configurations were deposited: (1) (Al_2, In_2, Ga_1)/(Al_3, In_3, Ga_1), (2) (Al_2, In_{2+}, Ga_1)/(Al_3, In_{3+}, Ga_1), (3) (Al_1, In_3, Ga_1)/(Al_{1+}, In_{3-}, Ga_1). The subscripts represent the number of repeats of the precursor pulses in the unit cells. The + and – signs denote an increase or decrease of precursor flux in the pulse by a factor of two. These unit cell configurations correspond to the well/barrier compositions of $Al_{0.25}In_{0.02}Ga_{0.73}N$/ $Al_{0.31}In_{0.01}Ga_{0.68}N$, $Al_{0.20}In_{0.04}Ga_{0.76}N$/ $Al_{0.26}In_{0.06}Ga_{0.68}N$, $Al_{0.1}In_{0.03}Ga_{0.87}N$/ $Al_{0.13}In_{0.02}Ga_{0.85}N$ to give a band-to-band photoluminescence (PL) emission at 305 nm, 320 nm, and 340 nm, respectively. Finally, the active layers were then capped with a 10 period p-superlattice and p-GaN. The n-ohmic contact was made to the n-AlGaN layer after accessing it via a reactive ion etched mesa of different geometries. The Pd/Au and Ti/Al/Ti/Au were used for p- and for n-contacts, respectively.

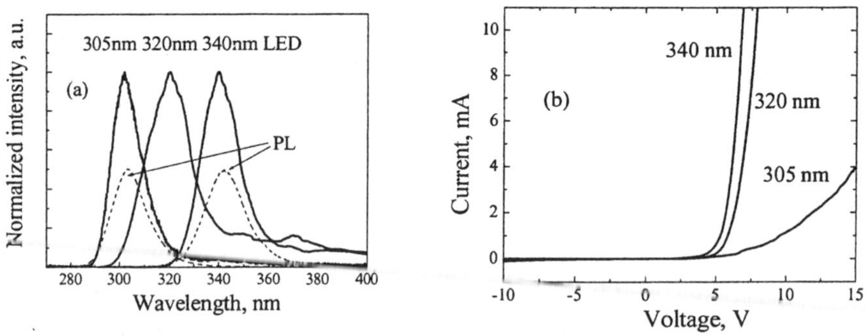

Fig. 7. (a) Room temperature dc EL spectra of LEDs with three different MQW configurations. Dotted lines show RT PL spectra of MQWs optimized for 305 nm and 340 nm, (b) The I-V curves of LEDs (after [17])

In Fig. 7(a) we include the room temperature (RT) emission spectra for the LEDs with the three different MQW configurations. Also included are the room temperature PL spectra. As seen the electroluminescence (EL) emission peaks are at 305 nm, 320 nm and 340 nm and there is a good agreement between the PL and the EL spectra. These data clearly show the viability of our PALE approach for tuning the LED emission wavelengths by simply adjusting the number of precursor pulses in the well and the barrier layers of the unit growth cell. In the Fig. 7 (b) we show the current-voltage characteristics for the 20x2000 μm stripes of the 305, 320 nm and the 340 nm LEDs. A turn-on at 5.5 V and a series resistance of about 50 Ω was measured for the 320 and the 340 nm devices. However, the series resistance for 305 nm LED was as high as 600 Ω. Room temperature transmission line model (TLM) measurements indicate this increase to be from the lateral resistance from the low doping in the bottom n-AlGaN layer and the n-AlGaN/AlGaN SL.

In Fig. 8, we plot the emitted power for the three LED types having identical stripe geometry. From these data it is easy to see that the emitted powers rapidly decrease with peak emission wavelengths. We believe this to result from the poor quality of the buffer n^+-AlGaN layer and not enough doping of top and bottom AlGaN layers. In addition for the 305 nm LED the heating from the high-series resistance also decreases the emission efficiency. Finally, PALE approach followed by a flip-chip device packaging to avoid strong device overheating allowed us to obtain a power values as high as 15 μW at and

230 μW for 320nm emitting LEDs under 50 mA dc and 800 mA pulse pumping, respectively.

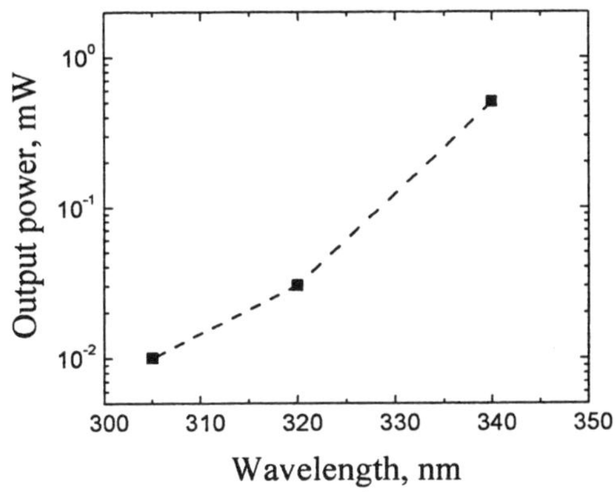

Fig. 8. The LED output power vs. peak emission wavelength (after [17])

Recently using AlN/AlGaN superlattice (SL) defect filtering [21] we grew thick (> 2 μm) n^+-$Al_{0.2}Ga_{0.8}N$ layers with low dislocation densities. Using them as buffer layers we have also reported on deep UV LEDs on sapphire with emission at 324 nm and powers as high as 4 mW. Combining a new epilayer design with the low defect density n^+-$Al_{0.4}Ga_{0.6}N$ buffer layers, we fabricated sub-milliwatt power AlGaN *p-n* junction deep UV LEDs with emission at 285 nm [22].

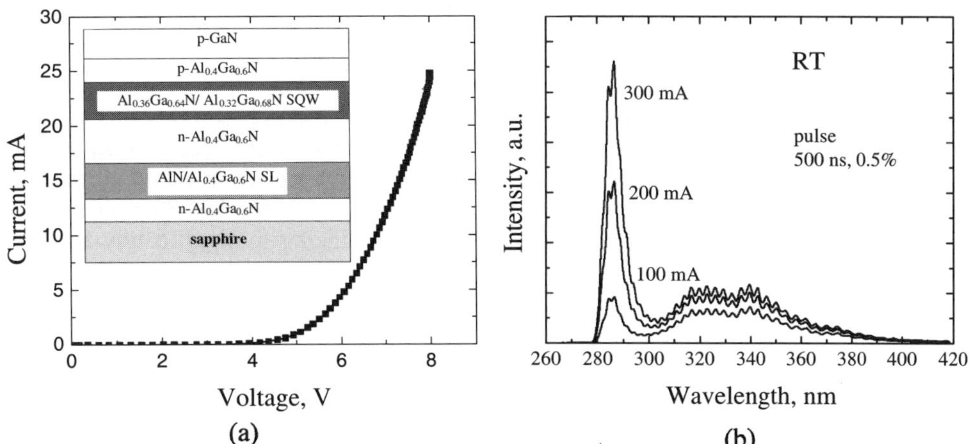

Fig. 9. The I-V (a) and spectral (b) characteristics of 285 nm deep UV LED (after [22]).

The epilayer structure of our 285 nm LED is shown in inset to Fig. 9. As seen, it consists of a 0.2 μm thick $Al_{0.4}Ga_{0.6}N$ layer that is deposited over base plane sapphire using conventional low-pressure metalorganic chemical vapor deposition (LPMOCVD). This is followed by a 10 period AlN (20Å) /$Al_{0.4}Ga_{0.6}N$ (300Å) SL for strain relief and dislocation filtering and a 1.8 μm thick Si-doped n^+-$Al_{0.4}Ga_{0.6}N$ buffer layer. As reported earlier the use of the strain relief AlN/AlGaN superlattices results in a reduction of threading dislocations by a factor of 5 and thus enables the deposition of 1.8 μm thick n^+-$A_{0.4}Ga_{0.6}N$ layers without cracking. It also significantly improves the emission characteristics of the active layers by reducing the nonradiative recombination. The device active region consisted of an $Al_xGa_{1-x}N$ (x = 0.36, 100 Å) / $Al_xGa_{1-x}N$ (x = 0.32, 30 Å) / $Al_xGa_{1-x}N$ (x = 0.36, 100 Å) single quantum well (SQW) which was capped with a Mg-doped p-$Al_{0.4}Ga_{0.8}N$ (200 Å) and a p^+-GaN (500 Å) layer. All layers of the structure were deposited at 1050 °C and 76 Torr.

Square geometry *p-n* junction devices were then fabricated using a reactive ion etched mesa to access the bottom n^+-$Al_{0.4}Ga_{0.8}N$ layer. A Ti(20 Å)/Al(100 Å)/ Ti(200 Å)/Au(2000 Å) and Ni(20 Å)/Au(200 Å) were used for the *n*- and *p*- contact metals. The contact anneal procedures were identical to our earlier reports' [21].
In Fig. 9(a) we include the current-voltage (I-V) characteristics of a 200 μm×200 μm device. As seen the device turns on at 6 V with a series resistance of about 74 Ω. A major contribution to this value comes from the lateral spread resistance of the n-$Al_{0.4}Ga_{0.6}N$ layer. The device emission spectra as a function of pulsed pump current are shown in Fig. 9(b). For these measurements, the emitted light was collected through the sapphire substrate. As seen, at low pump currents, the emission spectra consist of two peaks, one at 285 nm and the other at 340 nm. As the pump currents are increased, the peak at 340 nm rapidly saturates while that at 285 nm shows a near linear increase. We believe the peak at 340 nm to arise from deep levels in the bottom n-$Al_{0.4}Ga_{0.6}N$ layer. This is validated by photoluminescence (PL) and cathodoluminescence (CL) measurements that will be reported separately. The peak at 285 nm is the band edge emission from the quantum well. This conclusion is based on observing identical peaks in excimer laser (193nm) excited PL from test structures without the p-$Al_{0.4}Ga_{0.6}N/p^+$-GaN layers. For these PL measurements the pump is incident from the top (SQW side).

The 285 nm LEDs were then bonded to pads on ceramic carriers and mounted on TO-39 headers. The emitted powers were then measured under CW and pulsed-pumping. A calibrated integrating sphere and an UV-enhanced Si detector with a 285 nm bandpass filter were used for these measurements. From the data in Fig. 10 we see a peak CW-power of 10 μW was measured which rapidly saturated at pump currents in excess of 60 mA. We believe current crowding to be primarily responsible for this saturation. Under pulse pumping no power saturation was observed even up to pump currents as high as 400 mA. At 400 mA we measured a power of 0.15 mW for the 285 nm emission. We expect a better device geometry and proper heat sinking to significantly improve these power levels.

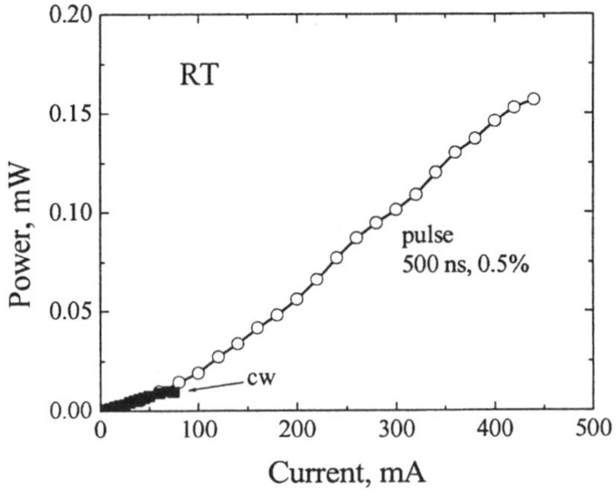

Fig. 10. CW and Pulsed output optical power of the 285 nm deep UV LED (after [22]).

Double heterostructure AlGaN/InGaN/GaN field effect transistors

To date all the reported high microwave power heterojunction field-effect transistors [23, 24] (HFETs), metal- oxide HFETs (MOSHFETs [25, 26] and metal-insulator HFETs (MISHFETs) [27] have employed AlGaN/GaN heterojunctions. Typical epilayer structures for these devices comprise of a few micron thick insulating GaN buffer layer and approximately 200-400 A thick AlGaN barrier layer with Al- molar fractions ranging from 10 to 40%. The 2D electron gas channel forms at the GaN-AlGaN heterojunction interface. In spite of the impressive results, the AlGaN/GaN heterojunction epilayer design has several limitations.

First, the carrier confinement in this design is only due to the self-consistent triangular potential quantum well at the interface. Both at high positive or negative bias, carriers can easily spill over into the buffer GaN or the barrier AlGaN layers. This carrier spill over increases low frequency noise [28] and decreases transconductance. In addition, the spilled over carriers get trapped at defect levels that commonly occur in GaN-AlGaN layers and thus give rise to slow transient processes and an RF-current collapse. Finally, for the GaN-AlGaN design a large gate voltage swing leads to a significant strain modulation both in the buffer and in the barrier layers. As we recently reported [29], the change in polarization charges resulting from this strain modulation also contributes to the current collapse.

We have proposed a new AlGaN/InGaN/GaN double heterojunction structure for high power microwave devices [30]. We demonstrate 1.5 μm gate length DHFET devices with the saturation currents of about 0.7 -0.9 A/mm, pulsed RF output power of 6.3 W/mm with only a 4 dB gain compression at saturation. For these DHFET devices, no current-collapse effect was observed in either the pulsed I-V or the RF output power characteristics.

An AlGaN/GaN(InGaN)/AlGaN double heterostructure design has previously been discussed in [31, 32]. In the reference [31], one-dimensional Poisson and Schrödinger equations were solved to determine the effect of the InGaN layer on the bandgap profile. However, the polarization charges, known to be of critical importance for nitride-based heterostructures, were not accounted for. In the reference [32], an improvement was observed in the low temperature mobility (200 K) but not in the room temperature mobility. Also, no DHFET devices were fabricated or reported on. The use of the bottom AlGaN layer in the double heterostructure designs of references [31, 32] deteriorated the surface morphology. It also created an additional parasitic second 2D channel (at the lower AlGaN/GaN interface), which in turn deteriorates the device transconductance.

In contrast, our proposed design (250 Å $Al_{0.25}Ga_{0.75}N$ / 50 Å $In_{0.1}Ga_{0.9}N$ / 0.4 μm i-GaN / i-SiC) eliminates the formation of the parasitic second channel, improving the structure morphology and simplifying the growth procedure. The band diagrams and the electron concentration distributions in our DHFET and a regular $Al_{0.25}Ga_{0.75}N$/GaN HFET structure (same as DHFET but without the InGaN layer) at different gate voltages are shown in Fig. 11(a, b). These were simulated using the 1D Poisson solver [33] and the material constants from reference [34]. As seen from the Fig. 11, the combination of a lower bandgap and polarization charges results in a much better confinement of the 2D electrons in the DHFET structure. As a result of this confinement, in a wide range of gate voltages, the electrons are localized within the 50 Å InGaN quantum well. In contrast, for the base-line HFET, especially at high negative gate voltages, there is a significant electron concentration tail in the GaN buffer layer. A better confinement should yield superior microwave and transient performance and a better linearity for the DHFET device. Also compared to HFET, the potential barrier between the DHFET InGaN channel and AlGaN barrier layer increases by about 1.4 eV at zero gate-bias, significantly reducing the electron spillover. In the DHFET, there is also an additional potential barrier (see Fig. 11) between the InGaN channel and the underlying i-GaN buffer layer. This potential barrier is depleted and this prevents recharging of the traps adjacent to the 2D channel.

We also performed 2D simulations using G-PISCES simulator from Gateway Inc. In Fig. 11 (c, d) we compare the electron distributions in a DHFET and HFET with 5 μm source - drain openings and 1 μm gate length at V_{GS} = -2 V and V_{DS} = +30 V. As seen the electron confinement in the DHFET device is nearly ideal while the regular HFET demonstrates a significant electron spill over into the underlying GaN buffer layer.

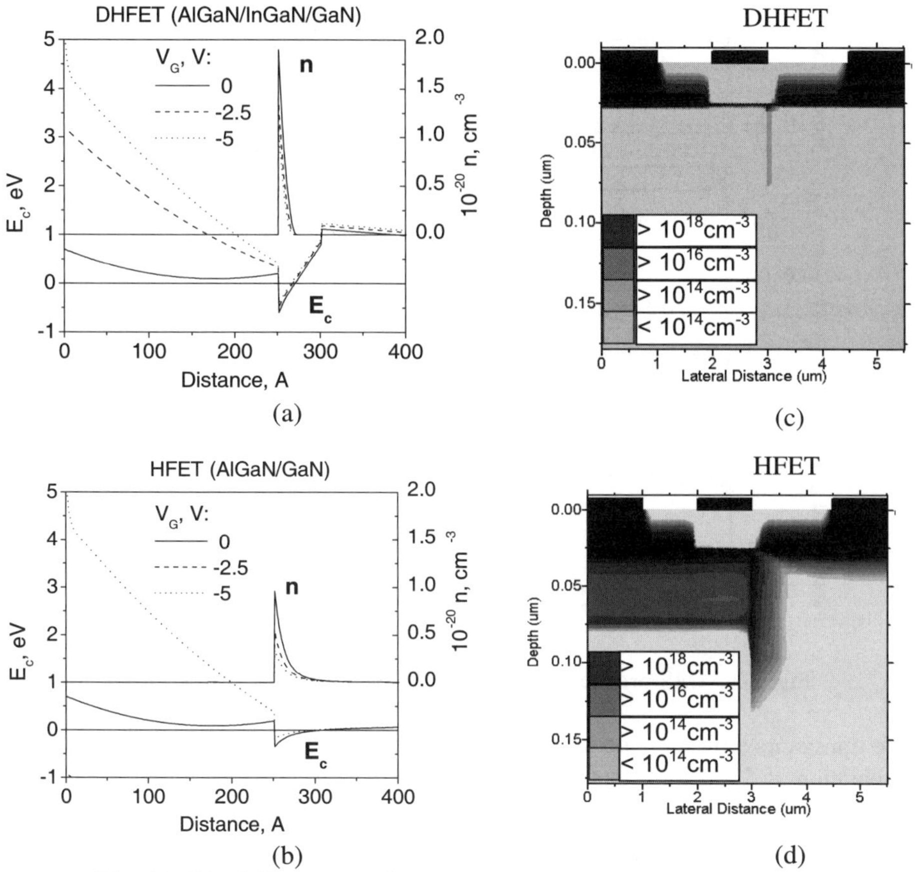

Fig. 11. Band diagrams and electron density distribution in DHFET (a, c) and conventional HFET (b, d) devices (Figures (a, b – after [30])

The epilayers for the DHFET structure were deposited using conventional low-pressure metal organic chemical vapor deposition (LP-MOCVD). The growth pressure was kept fixed at 76 Torr. The growth temperatures for the GaN buffer, InGaN channel and the AlGaN barrier layers were 1000 °C, 760 °C and 1100 °C respectively. Both the InGaN and the AlGaN layers were n-doped with the bulk electron concentration approximately 2×10^{17} cm^{-3} and 1×10^{18} cm^{-3}, respectively. The room temperature Hall mobility and sheet carrier concentration for the deposited DHFET structure were measured to be 730 cm^2/V-s and 1.1×10^{13} cm^{-3}, respectively. In Fig. 12, we show the SIMS profiles of Ga, N, Al, and In concentration in the DHFET epilayer structure. These data clearly confirm the existence of the 50 Å thick InGaN layer. Note that, due to different ionization efficiencies, these data do not provide an absolute value of the Al or the In fractions in the individual layers of the structure. The In- fraction of the InGaN layer of the DHFET ($In_{0.1}Ga_{0.9}N$) was determined by photoluminescence (PL) measurements. For these measurements, we used a GaN/InGaN multiple quantum well structure that was deposited over a 0.4 μm thick GaN layer on an i-SiC substrate. The growth conditions of the InGaN well for the MQW and the DHFET were kept identical.

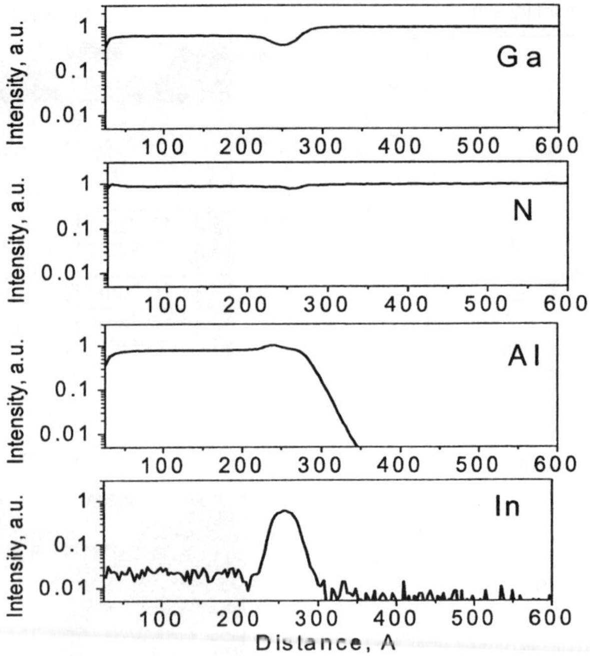

Fig. 12. SIMS profiles for the DHFET epilayer structure (after [30])

The transistors with the gate length and width of 1.5 μm and 200 μm and a source-drain separation of 5 μm were then fabricated. Ti(200Å)/Al(500Å)/Ti(200Å)/Au(1500Å) was used for source and drain contacts which were annealed at 850 °C for 1 min in nitrogen ambient. Pt/Au Schottky gates were then metallized. A reactive ion etched mesa was used for the device isolation. This device fabrication procedure is similar to one that we reported on earlier for HFETs and MOSHFETs [26]. From the Transmission Line Model (TLM) measurements, the contact resistance in the DHFET was found to be 2×10^{-5} Ohm–cm^2, which is close to that for our conventional AlGaN/GaN HFET devices. The saturation drain current (at zero gate bias) and the threshold voltage values varied from 0.7 A/mm to 0.9 A/mm and -5V to -7 V across a 2" wafer.

Fig. 13 shows the DHFET pulsed transfer characteristics (solid lines) and the transient drain current measured when the gate voltage pulse has returned to zero (dashed lines). The pulsed drain current between the gate voltage pulses when the gate voltage has returned to zero is referred to as the pulsed "return" current in Fig. 13.

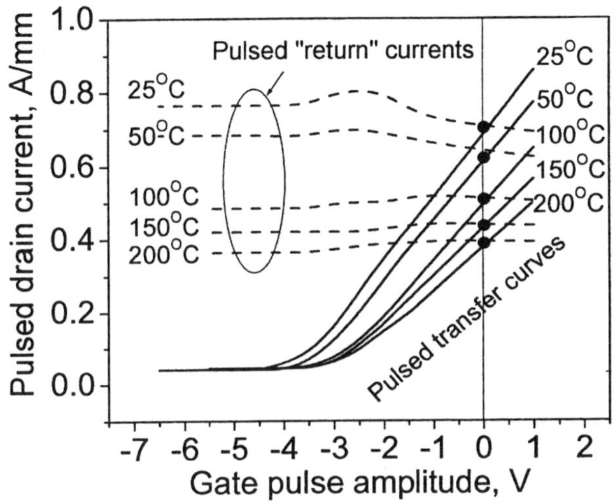

Fig. 13. Pulsed transfer characteristics of DHFET at different temperatures (after [30])

These pulsed measurements were made at different temperatures as a function of the gate voltage amplitude. For the pulse measurements, rectangular shaped 2 μs long gate-voltage pulses with a 0 V offset were used. The drain voltage was kept fixed at + 15 V DC. As we have shown before [27, 29], the difference between the pulsed "return" current and the DC drain current (at V_g=0) is an excellent measure of the current collapse. We found this to agree well with the degree of current collapse as measured from the rf-power characteristics. As seen from the data of Fig. 13, for the DHFET, the pulsed "return" current (at zero pulsed gate voltage) is nearly equal to the DC current at zero gate bias for any gate voltage pulse amplitude. This is true for temperatures from 25 °C up to 200 °C. In contrast, for devices with current-collapse, not only is the return current less then the zero gate-bias DC current, but it also shows a strong temperature dependence [35]. The data of Fig. 13 therefore clearly establish the absence of any current collapse in the DHFET devices. This we believe results from three key factors. First, due to the excellent carrier confinement in the InGaN quantum well (QW) there is no electron spill over into the GaN buffer. Therefore no electrons are trapped that could give rise to slow transients and the accompanying current collapse. Second, the current collapse is linked to the piezoelectric charge variations due to the changes of the electric field in the channel. As can be seen from the band diagrams of Fig. 11, the electric field in the conventional AlGaN/GaN HFET channel changes dramatically, from a high value at a positive or zero gate bias to a very small value near or below threshold. In contrast, the electric field in the InGaN quantum well in the DHFET remains nearly constant, since it is mainly controlled by the bandgap discontinuities and polarization charges at the InGaN/GaN and AlGaN/InGaN heterointerfaces. This reduction of field variation thereby reduces the current collapse. Finally, the strain in the DHFET AlGaN barrier layer is almost completely compensated by the InGaN layer. This is similar to our earlier reports on quaternary AlInGaN-GaN heterostructures [8]. In the InGaN based DHFET design, the cumulative strain in the AlGaN/InGaN bilayer structure on GaN buffer is close to zero at any gate-voltage due to the lattice matching resulting from the use of In. Thus, there are no significant changes in strain or polarization charges resulting from gate voltage modulations.

We also measured the microwave performance of InGaN channel DHFETs. The cutoff frequency, f_T, was measured to be 14 GHz for DHFET. This corresponds to an $fT \times L_G$ product of about 21 GHz-μm, which is very close to the highest values reported to date. The RF output power was measured both in CW and pulsed mode in the class A regime at 2 GHz. Maury Microwave automated tuners were used for the load- and source pull tuning. In the pulsed mode, both the drain bias and the output of RF generator were modulated by 2 μs pulses in order to reduce the effects of device overheating. The dependencies of the output RF power and the gain on the input RF power for both the CW and pulse modes of operation are shown in Fig. 14. For these measurements the new modification of DHFET devices with an additional 100 Å thick SiO_2 insulating layer between the gate electrode and AlGaN barrier were used. These Metal-Oxide-Semiconductor DHFET (MOSDHFET) [36] devices exhibit the gate leakage current 4 – 6 orders of magnitude lower than that of Schottky gate transistors, therefore allowing for larger gate voltage swing and higher output powers.

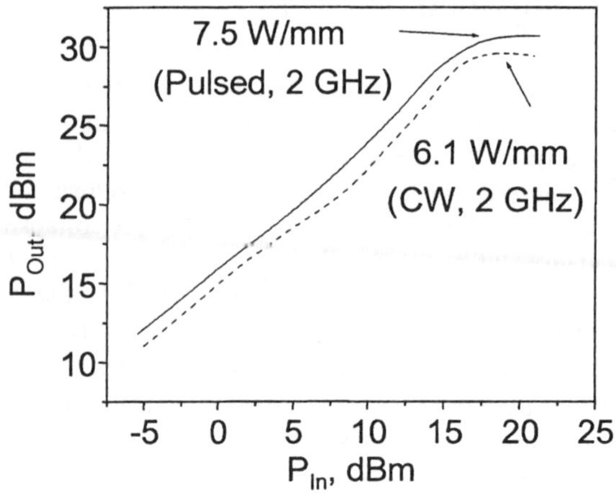

Fig. 14. RF Output power of MOSDHFET devices in CW and pulsed modes measured at 2 GHz (after [36], ©IEEE 2002)

As seen, the MOSDHFET demonstrates a high output RF power in both the CW and the pulsed operation modes. For CW operation with a drain-bias of V_d = 25 V, the maximum RF power was 7.5 W/mm, while in the pulsed mode it was 6.1 W/mm. These values of output power are very close to those expected from the device DC I-V characteristics. For example, under CW operation, the saturation current of $I_{DS} \approx 1.2$ A/mm was measured for the MOSDHFETs at the gate bias $V_G = +5V$. The knee voltage was $V_{Kn} \approx 6$ V. For the drain bias $V_D = 27$ V (used for CW RF power measurements), the estimated maximum RF power is $P_{out} = (V_{DS} - V_{Kn}) \times I_{DS}/4 = 6.3$ W/mm. This estimation is very close to the actual saturation RF output power of 6.1 W/mm confirming the absence of the current collapse. In addition, the MOSDHFETs demonstrated a greatly improved linearity. The gain compression at saturation was as low as 4 dB. The typical gain compression values recently published for conventional AlGaN/GaN HFET[37,38], are considerably higher, about 8 dB. We attribute this improved device linearity to a much better carrier confinement in AlGaN/InGaN/GaN DH structure and the absence of current collapse.

In conclusion, we report on the use of Energy Strain Band Engineering approach for deposition high quality quaternary AlInGaN MQW structures over sapphire and SiC on using a new-pulsed atomic layer epitaxy procedure. New types of high power high temperature transistors with improved performance were fabricated. XRD, PL and AFM spectra are used to establish a high structural and optical quality of the quaternary multiple quantum wells. We have also demonstrated the use of PALE deposited quaternary AlInGaN multiple quantum wells in the active region of a UV LED with emission ranging from 305 nm to 304 nm.

This work at USC was supported by the Missile Defense Agency (MDA) under Army SMDC contract DASG60-98-1-0004, monitored by Mr. Terry Bauer, Dr. Brian Strickland and Dr. Kepi Wu. The work at RPI was supported by the DoD Multidisciplinary University Research Initiative (MURI) program administered by the Office of Naval Research under Grant N00014-98-1-0654 (Project Monitors Drs. Harry Dietrich and John Zolper). The work at SET, Inc. was supported by MDA under SBIR contract F33615-01-C-1911 (monitored by G. Gillispie of WPAFB), and Army SMDC contract DASG60-00-C-0088 (monitored by F. Clark), and by ONR under contract N00014-01-C-0195 (monitored by Y.S. Park)

References

1 M. E. Levinshtein, S. L. Rumyantsev, R. Gaska, J. W. Yang, and M. S. Shur, Appl. Phys. Lett., **73,** 1089 (1998).

2 Q. Chen, J. W. Yang, M. A. Khan, A. T. Ping, and I. Adesida, Electronics Letters **33**, 1413 (1997).

3 R. Gaska, J. Yang, A. Osinsky and M. S. Shur, IEEE Electron Device Letters, **19**, 89 (1998). ©IEEE 1998

4 F. G. McIntosh, K. S. Boutros, J. C. Roberts, S. M. Bedair, E. L. Piner and N. A. El-Masry, Appl. Phys. Lett., **68,** 40 (1996).

5 M. S. Shur, Wide Band Gap Semiconductors. Good Results and Great Expectations, pp. 279-290, in Proceedings of NATO Advanced Research Workshop, Ile de Bendor, France, July (1995), Future Trends in Microelectronics. Reflection on the Road to nanotechnology, edited by Serge Luryi, Jimmy Xu, and Alex Zaslavsky, NATO ASI series, Series E: Applied Sciences - vol. 323, Kluwer Academic Publishers, Dordrecht/ Boston/ Lisbon (1996)

6 M.E. Levinshtein, F. Pascal, S. Contreras and W. Knap, S. L. Rumyantsev, R. Gaska, J.W. Yang, and M. S. Shur, Low frequency Noise in GaN/GaAlN heterojunctions, Appl. Phys. Lett. 73, No. 23, pp. 3053-3055, June 8 (1998)

7 M. S. Shur, A. D. Bykhovski, R. Gaska, and A. Khan, GaN-based Pyroelectronics and Piezoelectronics, in Handbook of Thin Film Devices, Volume 1: Hetero-structures for High Performance Devices, Edited by Colin E.C. Wood, Handbook edited by Maurice H. Francombe, pp. 299-339, Academic Press, San Diego, 2000

8 M. Asif Khan, J. W. Yang, G. Simin, Hans zur Loye, R. Bicknell-Tassius, R. Gaska, M. S. Shur, G. Tamulaitis, and A. Zukauskas, "Energy Band/Lattice Mismatch Engineering in Quaternary AlInGaN/GaN Heterostructures", Physica Status Solidi, v. 176 (1), pp. 227-230 1999

9 T. Nishida, H. Saito, and N. Kobayashi, Appl. Phys. Lett. **78**, (2001) 399.

10 A. Kinoshita, H. Hirayama, M. Ainoya, Y. Aoyagi, A. Hirata, Appl. Phys. Lett. **77**, (2000) 175.

11 N. Otsuka, A. Tsujimura, Y. Hasegawa, G. Sugahara, M. Kume, Y. Ban, Jpn. J. Appl. Phys. **39**, (2000) L445.

12 M. Asif Khan, J. W. Yang, G. Simin, R. Gaska, M. S. Shur, Hans-Conrad zur Loye, G. Tamulaitis, A. Zukauskas, D. J. Smith, D. Chandrasekhar, and R. Bicknell-Tassius, Appl. Phys. Lett. **76**, (2000) 1161.

13 J. P. Zhang, J. W. Yang, G. Simin, M. Shatalov, M. Asif Khan, M. S. Shur, R. Gaska, Appl. Phys. Lett. **77**, (2000) 2668.

14 M. Shatalov, A. Chitnis, V. Adivarahan, A. Lunev, J. Zhang, J. W. Yang, Q. Fareed, G. Simin, A. Zakheim, M. Asif Khan, R. Gaska, M. S. Shur, Appl. Phys. Lett. 78, (2001) 817.

15 G. Tamulaitis, K. Kazlauskas, S. Jursenas, A. Zukauskas, M. A. Khan, J. W. Yang, J. Zhang, G. Simin, M. S. Shur, and R. Gaska, Appl. Phys. Lett. **77**, (2000) 2136.

16 J. P. Zhang, E. Kuokstis, Q. Fareed, H. Wang, J. W. Yang, G. Simin, M. Asif Khan, R. Gaska and M. S. Shur, Appl. Phys. Lett. **79**, (2001) 925.

17 M. Asif Khan, Vinod Adivarahan, Jian Ping Zhang, Changqing Chen, Edmundas Kuokstis, Ashay Chitnis, Maxim Shatalov, Jin Wei Yang, Grigory Simin. Stripe Geometry Ultraviolet Light Emitting Diodes at 305 nanometers using quaternary AlInGaN multiple quantum wells, Jpn. J. Appl. Phys. Vol. 40 (2001) pp. L 1308–L 1310

18 Ashay Chitnis, Vinod Adivarahan, Maxim Shatalov, Jianping Zhang, Michael Gaevski, Wu Shuai, Radhika Pachipulusu, Jason Sun, Kirill Simin, Grigory Simin, Jinwei Yang and Muhammad Asif Khan. Submilliwatt Operation of AlInGaN based Multifinger-Design 315nm LED Over Sapphire Substrate "Jpn. J. Appl. Phys. Vol. 41 (2002) L320-L322 Part 2, No. 3B, 15 March 2002 "

19 M. Shatalov, J. P. Zhang, A. Chitnis, V. Adivarahan, J. W. Yang, G. Simin, M. Asif Khan, Deep Ultraviolet Light Emitting Diodes using Quaternary AlInGaN Multiple Quantum Wells, to be published in IEEE Selected topics in QE, April 15th, 2002

20 Ashay Chitnis, Jian Ping Zhang, Vinod Adivarahan, Wu Shuai, Jie Sun, Maxim Shatalov, Jin Wei Yang, Grigory Simin, Muhammad Asif Khan 324 nm Light Emitting Diodes with Milliwatt Powers. JJAP (Letters) to be published April 15, 2002

21. J. P. Zhang, H. M. Wang, M. E. Gaevski, C. Q. Chen, Q. Fareed, J. W. Yang, G. Simin, M. Asif Khan, Crack-free thick AlGaN grown on sapphire using high temperature AlN/AlGaN superlattice for strain management. To be published in APL, May 13, 2002

22 V. Adivarahan, J. Zhang, A. Chitnis, Wu Shuai, Jie Sun, R. Pachipulusu, M. Shatalov, M. Asif Khan, Sub-milliwatt Power III-N Light Emitting Diodes at 285 nm. To be published in JJAP (Express Lett.), 15 April, 2002

23 S. T. Sheppard, K. Doverspike, W. L. Pribble, S. T. Allen, J. W. Palmour, L. T. Kehias, and T. J. Jenkins, "High-Power Microwave GaN/AlGaN HEMT's on Semi-Insulating Silicon Carbide Substrates," IEEE Electron Dev. Lett, **20**, 161, (1999).

24 N.X. Nguyen, M. Micovic, W.-S. Wong, P. Hashimoto, L.-M. McCray, P. Janke and C. Nguyen, High performance microwave power GaN/AlGaN MODFETs grown by RF-assisted MBE, Electron. Lett., **36** (2000), 468

25 M. Asif Khan, X. Hu, G. Simin, A. Lunev, and J. Yang, R. Gaska and M. S. Shur, AlGaN/GaN Metal-Oxide-Semiconductor Heterostructure Field Effect Transistor, IEEE Electron Device Letters, **21** (2000), 63. ©IEEE 2000

26 M. Asif Khan, X. Hu, G. Simin, and J. Yang, R. Gaska and M.S. Shur, AlGaN/GaN Metal-Oxide-Semiconductor Heterostructure Field Effect Transistors on SiC Substrates" Appl. Phys. Lett. **77**, (2000), 1339

27 X. Hu, A. Koudymov, G. Simin, J. Yang, M. Asif Khan, A. Tarakji, M. S. Shur and R. Gaska, Si_3N_4/AlGaN/GaN-Metal-Insulator-Semiconductor Heterostructure Field Effect Transistors, Appl. Phys. Lett, v.79, p.2832-2834 (2001)

28 S. L. Rumyantsev, N. Pala, M. S. Shur, and R. Gaska, M. E. Levinshtein, M. Asif Khan, G. Simin, X. Hu, and J. Yang , Low frequency noise in GaN metal semiconductor and metal oxide semiconductor field effect transistors", Journ. Appl. Phys. **90**, (2001), 310

29 G. Simin, A. Koudymov, A. Tarakji, X. Hu, J. Yang and M. Asif Khan, M. S. Shur and R. Gaska, Induced Strain Mechanism of Current Collapse in AlGaN/GaN Heterostructure Field-Effect Transistors. Appl. Phys. Lett. 79, 2651-2653 (2001)

30 G. Simin, X. Hu, A. Tarakji, J. Zhang, A. Koudymov, S. Saygi, J. Yang, M .A. Khan, M. S. Shur, and R. Gaska, AlGaN/InGaN/GaN Double Heterostructure Field-Effect Transistor, Jpn. J. Appl. Phys. Vol.40 No.11A pp. L1142 - L1144 (2001)

31 S. Imanaga and H. Kawai, Novel AlN/GaN insulated gate heterostructure field effect transistor with modulation doping and one-dimensional simulation. J. Appl. Phys., **82** (1997), 5843

32 N. Maeda, T. Saitoh, K. Tsubaki, T. Nishida and N. Kobayashi, Jpn. J. Appl. Phys., **38** (1999), L799

33 1D Poisson/Schrödinger solver program developed by Dr. Gregory Snider, University of Notre Dame, http://www.nd.edu/~gsnider/.

34 O. Ambacher, J. Smart, J. R. Shealy, N. G. Weimann, K. Chu, M. Murphy, W. J. Schaff, L. F. Eastman, R. Dimitrov, L. Wittmer, and M. Stutzmann, W. Rieger and J. Hilsenbeck Two-dimensional electron gases induced by spontaneous and piezoelectric polarization charges in N- and Ga-face AlGaN/GaN heterostructures, Journ. Appl. Phys., **85** (1999), 3222

35 R. Dietrich, A. Wieszt, A. Vescan, H. Leier, Joan M. Redwing, K. S. Boutros, K. Kornitzer, R. Freitag, T. Ebner and K. Thonke, Current limitation after pinch-off in AlGaN/GaN FETs. MRS Internet J. Nitride Semicond. Res. **5**, 2 (2000)

36 G. Simin, A. Koudymov, H. Fatima, J. Zhang, J. Yang, M. Asif Khan, X. Hu, A. Tarakji, R. Gaska and M. S. Shur, SiO2/AlGaN/InGaN/GaN Metal-Oxide-Semiconductor Double Heterostructure Field - Effect Transistors, Submitted to IEEE EDL (2002)

37 S. Binary, K. Ikossi, J. A. Roussos, W. Kruppa, D. Park, H. Dietrich, D. Koleske, A.E. Wickenden, and R. L. Henry, Trapping Effects and Microwave Performance in AlGaN/GaN HEMTs, IEEE Trans. Electron. Dev. **48** (2001), 465.

38 E. M. Chambes, J. A. Smart, T. Prunty, and J.R. Shealy, Microwave Performance of AlGaN/GaN Metal Insulator Semiconductor Field Effect transistors on Sapphire substrates, IEEE Trans. on Electron. Dev., **48** (2001), 416

25 M. Asif Khan, X. Hu, G. Simin, A. Lunev, J. Yang, R. Gaska and M. S. Shur, AlGaN/GaN Metal Oxide Semiconductor Heterostructure Field Effect Transistor, IEEE Electron Device Letters 21 (2000) [illegible]
26 M. Asif Khan, X. Hu, G. Simin, J. Yang, R. Gaska and M. S. Shur, AlGaN/GaN metal-oxide-semiconductor heterostructure field-effect transistors on SiC substrates, Appl. Phys. Lett. 77 (2000) 1339.
[illegible]

International Journal of High Speed Electronics and Systems, Vol. 12, No. 2 (2002) 421–428

SEMICONDUCTOR UV SOURCES AND DETECTORS: SOME NON-CONSUMER APPLICATIONS

ELIAS MUÑOZ
ISOM-IEL, ETSI Telecomunicación, Univ. Politécnica de Madrid
Ciudad Universitaria, 28040-Madrid, Spain

UV emitters and photodetectors based on wide band-gap semiconductors are being investigated and may soon become commercially available. Solid state lighting and information storage are two main applications in the consumer area for these new semiconductor devices. Presently, III-nitrides seem to be the most promising materials for such near UV semiconductor devices. In this work some non-consumer applications are indicated. Biophotonics appears to be a very promising area for such devices.

1. Introduction

The commercial success of blue light emitting diodes (LED) and the subsequent development of blue laser diodes (LD) based on (In,Ga)N alloys, has launched the research on III-nitrides to be one of the main efforts of the solid state community. Similar devices are also being pursued as based on wide band-gap II-VI semiconductors, although developments are being made at a much slower pace. As a natural extension, UV emitters and detectors are being intensively investigated. White light, solid–state lighting (SSL), using the light from a UV-emitting device to excite an appropriate combination of phosphors may produce a good color rendering. Higher-density information storage can also be obtained by using UV LDs.

AlGaN UV photodetectors with Al mole fractions exceeding 0.4, showing detectivities surpassing Si devices, and with high contrast ratios (above- to below- band-gap responsivity ratio), have been reported by several laboratories. As compared to the bulky, expensive and slow UV lamps, Nichia has just announced GaN/AlGaN UV LDs at 365 nm, and UV LEDs emitting from 400 nm to the 300 nm frontier have been already reported.[1,2]

In this work we discuss briefly the impact of these UV sources and detectors on various applications, exploring with some emphasis the near UV biophotonics. It is suggested that life and many physical-chemical processes have adapted to the solar emissions reaching us, making the near UV spectrum, down to 280 nm, a very important optical range.

2. UV emitters and detectors based on nitrides

The successful InGaN/GaN/AlGaN quantum well (QW) structures used for blue emission generation, have been evolving to single and multiple QW emitters with emission in the UV (LDs and LEDs). Monochromatic, or rather monochromatic, these UV sources are becoming available at the laboratory and industrial development levels. White SSL and information storage are the main forces driving such efforts. For the first time, LEDs emitting from 400 nm to the 300 nm frontier have been achieved (Figure 1).[2]

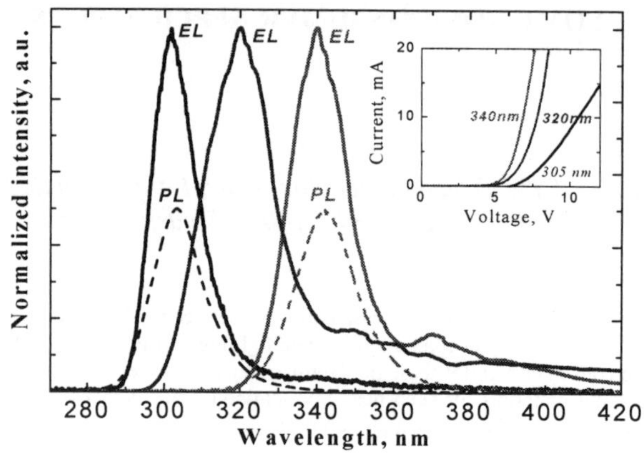

Figure 1.-Electroluminescence of AlInGaN-based light emitting diodes grown over sapphire substrates. [2]

These UV sources that can exploit all the fast modulation techniques developed for visible and IR semiconductor LDs and LEDs, allowing very powerful noise rejection techniques in spectroscopy, fluorescence and imaging applications.

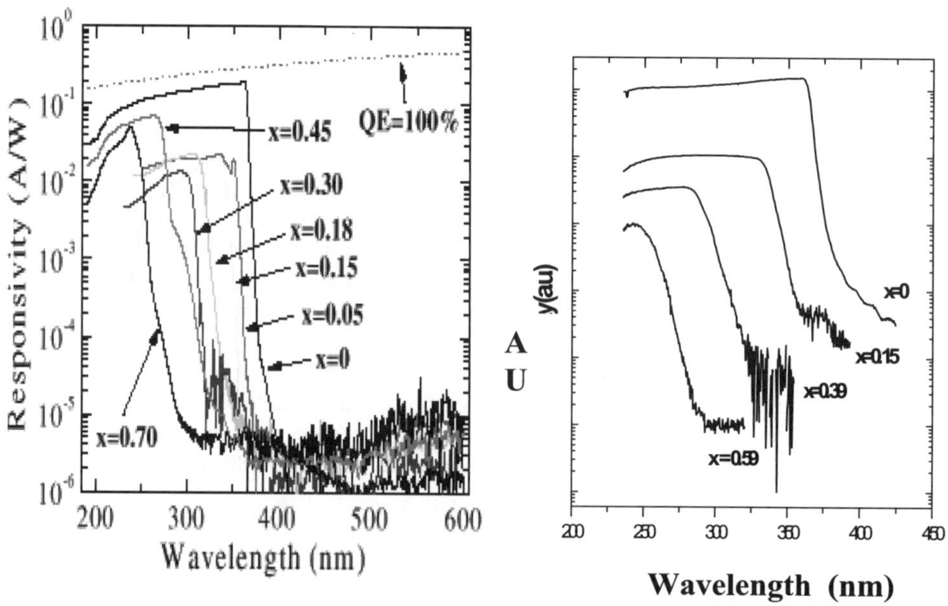

Figure 2.- Responsivity of AlGaN photodetectors: (a) p-i-n devices grown by MOVPE on sapphire; (b) m-s-m devices grown on Si (111) by MBE.[3,4]

Concerning UV detectors based on III-nitrides, all basic structures have been fabricated. They are intrinsically visible-blind, with contrast ratios above three orders of magnitude. The so-called solar-blind detectors (cutoff at, or below, 280 nm) have been reported by various laboratories. One important point is that properly designed p-i-n structures may have a narrow window response in the UV, avoiding the need of any external filter. Recently, resonant cavity detectors, avalanche photodiodes, and UV imagers have started to be developed. UV photon counting has been already reported. Subnanosecond responses can also be achieved. As an illustration, Figure 2.a shows the responsivity of AlGaN p-i-n detectors for various Al compositions, grown by metal organic vapour phase epitaxy (MOVPE) on sapphire substrates, while Figure 2.b shows the spectral response (broad band detectors) obtained in metal-semiconductor-metal (MSM) structures grown by molecular beam epitaxy (MBE) on Si (111) substrates.[3,4]

The point we want to address is: these new sources and detectors, may them have a significant impact on other applications besides white SSL and information storage?. Of special interest is the field of biosciences and biotechnology. These disciplines are today an interesting frontier for Physics, Electrical Engineering and Photonics. Besides, we will briefly explore the area of water quality monitoring and combustion control.

3. Near UV biophotonics

Present life on Earth is a result of evolution and adaptation since the *big-bang* days. It seems that during the initial formation times very energetic UV photons reached our surface planet (*UV springs from the Sun and other starts*). Afterwards, nature has been adapting through natural selection, genetic variants, UV-damaging DNA process, repairing mechanisms, etc., to the solar UV photons reaching us.

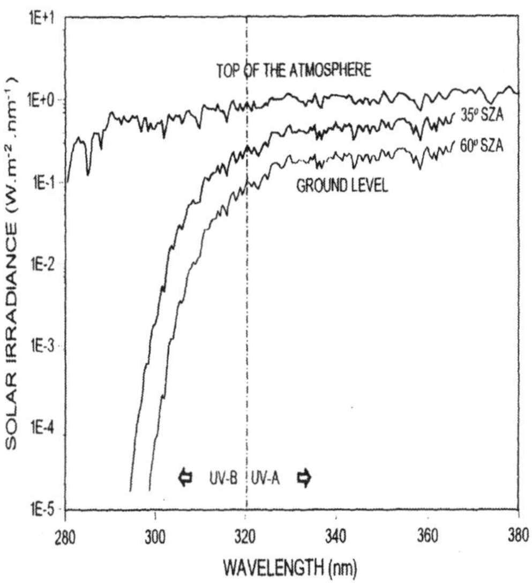

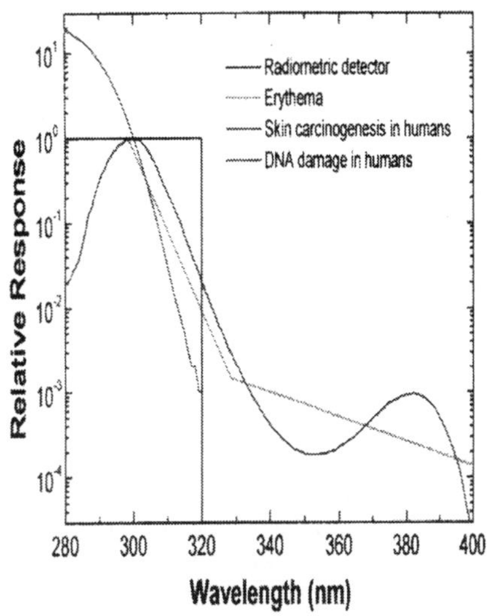

Figure 3.-(a) UV region of the solar irradiance; (b) its biological action spectra.

Meanwhile, the Earth atmosphere was being formed. Living organisms developed protection pigments and layers, furs, shells, etc., against the UV radiation not absorbed in the upper layers of the atmosphere. Specific flavonoids and aminoacids absorb, protect, plants, algae, etc., against UV photons. *So, the surviving Life has adapted to the final solar UV spectrum reaching us!.* [5]. Figure 3.a reminds us that atomic N and O, diatomic nitrogen and oxygen and later the O_3 layer, absorb the most energetic solar radiation in a number of overlapping bands, and only a band of medium energy photons from 280 nm to 320 nm (UV-B), plus the UV-A band (320 nm-400 nm) irradiate the Earth surface. In other words (Figure 3.b), surviving living organisms have specific photon action spectra, but in all cases photons with wavelengths shorter than 280 nm produce very severe damage and DNA break (250 nm). As the photon energy lowers, damage decreases exponentially with wavelength, roughly, with details specific for each specie. For example, as compared to the UV-B band, very long exposures to the UV-A are needed before human skin develops erythema or even skin cancer (Figure 4.b). Specific AlGaN UV detectors for erythema action have been developed.[6]

In other words, *life today and many physical-chemical processes on Earth have adapted to the near UV, the 290nm-400nm range. Thus, this is the main range of interest in UV biophotonics, and, fortunatelly, this is the range where III-nitrides offer today both emitters and filter-like detectors.* More energetic photons tend to have a rather severe damaging action.

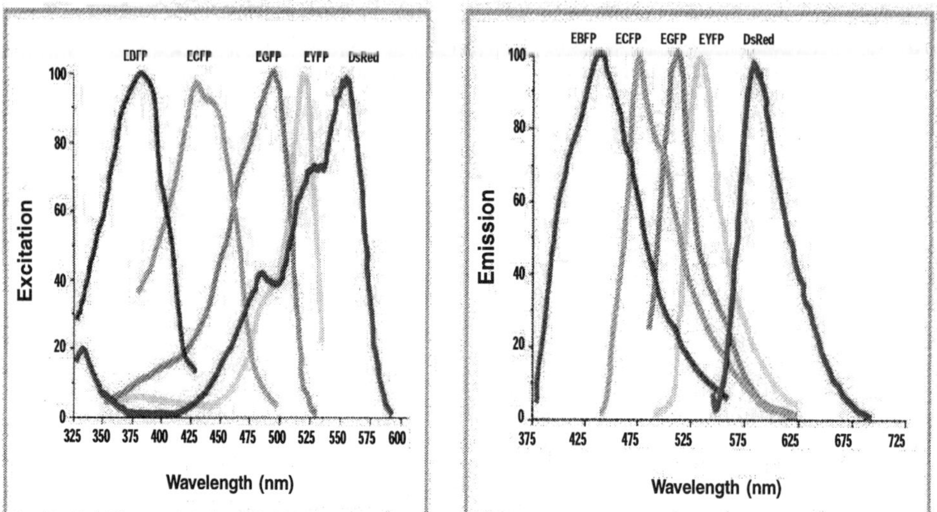

Figure 4.-The excitation and emission spectra of enhanced blue, cyan, red, yellow and DsRed fluorescence proteins.[8]

Because this matching, nitride-based UV emitters and detectors can play a significant role in biophotonics. As a first example, let us consider marking (tagging) cells, molecules, organisms, etc., by fluorescent proteins (FP) and dyes. Fluorescence is a well established technique in analytical chemistry, but it was in 1970 when the discovery of a green fluorescent protein (GFP) in *Aequorea Victoria (*a Pacific jelly fish) launched fluorescent techniques to be probably the most important advance in cell biology. This non-invasive technique is becoming important in bioscience, biotechnology and in the medical laboratory practice. When 390 nm light excites GFP it responds by emitting light in the 510-700 nm range. This molecule may be coupled (attached) to other proteins, or antibodies, for example, without altering their biochemical behaviour or the GFP fluorescence response. By introducing these tailored hybrid molecules into a cell, their presence can later be detected by fluorescence.[7]

Excitation, GFP fluorescence detection, and discrimination from other fluorescent species, become today a problem of careful selection of optical filtering. A full set of fluorescent proteins in the visible is now available (Figure 4).[8] Nitride devices may help significantly to improve selectivity (Figure 4), sensitivity, and to extend to shorter wavelengths the practical range of FP .

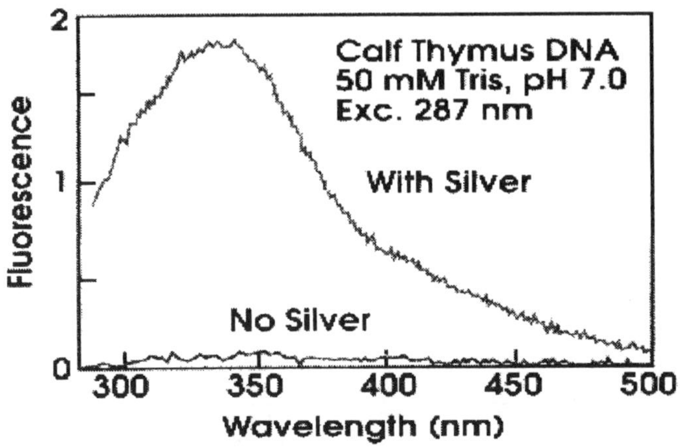

Figure 5.-The presence of metallic particles can boost the fluorescence of DNA dramatically (several thousandfold). No need to attach fluorescent tags.[10]

A similar extension to the UV has been made in fluorescent marking dyes to stain the nuclei of live cells.[9] Excitation at 350 nm is used for nuclear stain. DNA has a weak intrinsic autofluorescence (330 nm emission) , but it can be enhanced in the presence of metal particles or periodic metal surfaces (fluophore-metal interaction), as shown in Figure 5.[10] Fluorescence of the skin (295-450 nm) seems to help to distinguish a cancerous tissue, and fluorescence also helps to show how cancer drugs move into cancer cells and to develop gene sequencing bio-chips based on using the near UV region.[11]

4. Absorption of near UV light in water and contamination detection

Figure 6 depicts the absorption of light by seawater. Although it is obviously dependent on the water origin, it indicates that the near UV photons have a good transmission in clean waters.[12] It also reflects, for example, the importance of the solar UV band for the development of algae and, in general, of aquatic life.

On the other hand, aromatic organics, pesticides, detergents, etc., have absorption bands or fluorescence emission in the 250 –450 nm. By using UV LDs and LEDs, electrical modulation and pulse excitation techniques can be used. High sensitivity, portable, low consumption, near UV absorption systems can be built. Although non-specific in general, water contaminant detection systems can be fabricated based on nitride devices.[13]

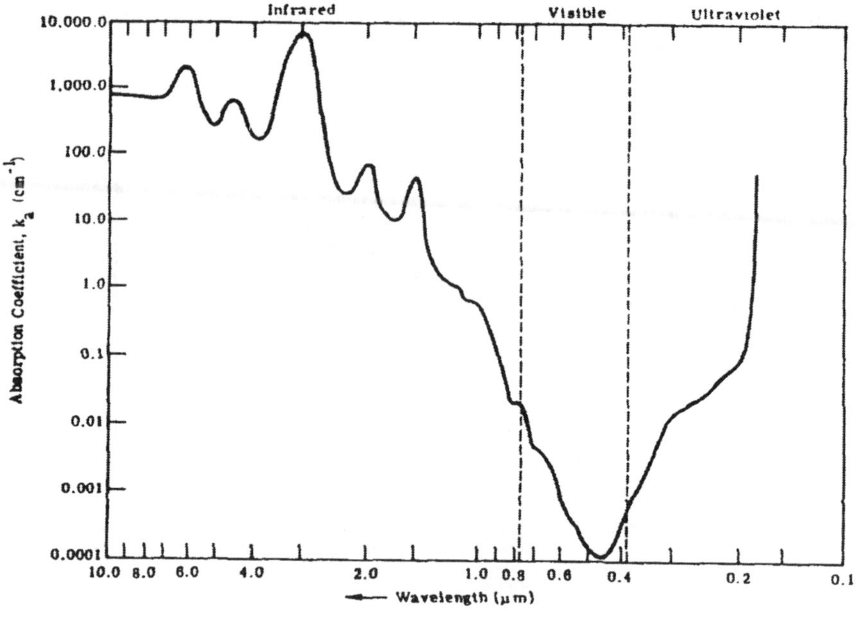

Figure 6.-Optical absorption coefficient of seawater.

5. Combustion control

Figure 7 shows typical flame emissions from natural gas and oil combustion. The fluorescence emission from two radicals (OH^-, CH^-) at 306 nm and 430nm, respectively, are the two main, resolved peaks in the short wavelength region. Flame spectra are similar for the combustion of other hydrocarbon fuels.[6] The relative heights of the OH^- and CH^- emissions can be studied by present AlGaN and InGaN detectors, respectively. Proper sources can also allow to study the combustion processes. Again, the near UV region becomes of interest.

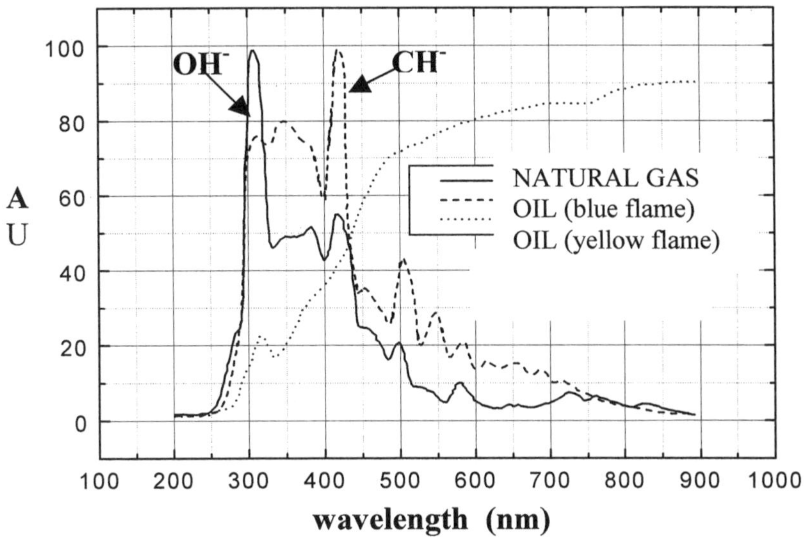

Figure 7.-Emission spectra of some standard hydrocarbon flames.

6. Conclusions

Present developments in III-nitrides UV devices indicate that LEDs emitting at 300 nm, LDs at 365 nm, and detectors with cutoff wavelengths as short as 250 nm, are being obtained. Although lighting, printing and information storage are foreseen as the most important application areas, these UV frontier devices may have an important impact in biophotonics and in other areas related with physico-chemical processes. Some of these non-consumer applications have been briefly indicated.

Life, living organisms, have adapted to survive under the solar UV radiation that reaches us presently (280-400 nm). In this band, for the most energetic photons, their biological action increases exponentially as wavelength decreases, roughly. One of the areas where near UV devices may have an important contribution, is excitation and fluorescence detection from marking or tagging fluorescent proteins and dyes. The evolution, movements, attachements, etc., of proteins and molecules tagged with FP can be detected by fluorescence. Monochromatic, or rather-monochromatic, sources, easy to modulate, and band-pass detectors will have a significant impact in near UV biophotonics.

Seawater, clean waters, have a good transmission properties in the near UV region. Again, sea life near the water surface is markedly dependent on the arriving solar UV photons. Important organic contaminants, toxic, have absorption bands in the 250-450 nm region. Nitride UV devices will allow improved, non-specific, contaminant detection systems.

This same UV region is also important in flame emission from hydrocarbon fuels, and nitride devices may help to develop sensors for combustion research and control.

Concerning the 250-280 nm region, that requires devices with rather high Al mole fractions, it is related to very high damage in cells and tissues (sterilization, bacteria killing), thus having a more specific, narrower area of application.

Acknowledgements

The support of project MCyT-PTR-1995-0466-OP and of MCC (Dr. S. Alvarez), are acknowledged. I also thank LEDEX (L. Liu) for kindly supplying samples.

References

1. T. Mukai, S. Nagahama, T. Yamamoto, M. Sano, "Expanding emission wavelength on nitrides light-emitting devices", ISBLLED (2002).
2. A. Khan, J. Yang, G. Simin, R. Gaska, M. Shur, "Strain Energy Energy Band Engineering Approach to AlN/GaN/InN Heterojunction Devices", invited paper, WOFE (2002).
3. R. McClintock, P. Sandvik, K. Mi, F. Shahedipour, A. Yasssan, C. Jelen, P. Kung and M. Razeghi, SPIE (2001), Photodetectors Materials and Devices VI, at press.
4. J. L. Pau, E. Muñoz, E. Monroy, M. A. Sánchez-García, and E. Calleja, "Solar-blind AlGaN-based UV photodetectors grown on Si (111) substrates", SPIE (2002), Photodetectors Materials and Devices VII.
5. J. Withgott, Natural History (2001) 38.
6. E. Muñoz, E. Monroy, J. L. Pau, F. Calle, F. Omnès, and P. Gibart, "III nitrides and UV detection", J. Phys. Condens. Matter **13** (2001) 7115-7137.
7. M. Chalfie and S. Kain, eds., *GFPs: Proteins, Properties, Applications and Protocols*. First Edition, Wiley &Sons (1998).
8. Clontech Laboratories Inc., Palo Alto, California, USA.
9. Molecular Probes Inc., Eugene, Oregon, USA.
10. J. R. Lakowicz, I. Gryczynski, Y. Shen, J. Malicka, and Z. Gryczynski, "Intensified Fluorescence", Photonics Spectra (2001) 96-104.
11. 9.-R. A. Lewis, "Mouse models melanoma", Biophotonics Int. (2001) 38-39.
12. R. A. J. Litjens, T. I. Quikenden, and C. G. Freeman, "Visible and near-ultraviolet absorption spectrum of liquid water", Appl. Optics **38** (1999) 1216-1223.
13. F. Calle, J. C. Martínez, J. Crespo, E. Monroy and E. Muñoz, "Viability of GaN-based optoelectronic systems for water quality monitoring", OPEL (2001), Barcelona.

International Journal of High Speed Electronics and Systems, Vol. 12, No. 2 (2002) 429–437

OPTIMIZATION OF WHITE ALL-SEMICONDUCTOR LAMP FOR SOLID-STATE LIGHTING APPLICATIONS

ARTŪRAS ŽUKAUSKAS
Institute of Materials Science and Applied Research, Vilnius University, Saulėtekio 9-III, LT-2040 Vilnius, Lithuania

RIMANTAS VAICEKAUSKAS and FELIKSAS IVANAUSKAS
Faculty of Mathematics and Information Science, Vilnius University, Naugarduko 24, LT-2006 Vilnius, Lithuania

MICHAEL S. SHUR and REMIS GASKA
Department of Electrical, Computer, and Systems Engineering, Rensselaer Polytechnic Institute, Troy, New York 12180, USA

A stochastic method is used to optimize white all-semiconductor lamps composed of tree, four, and five primary light-emitting diodes (LEDs) with Gaussian spectra and realistic line widths typical of AlGaInP and AlGaInP LEDs. Within an optimal set of central wavelengths and relative powers, a trade-off between luminous efficacy and the general color rendering index can be performed. The quadrichromatic all-semiconductor lamp is shown to meet most needs of *deluxe* general lighting. However, practical implementation of this lamp requires substantial improvement of LEDs in the yellow-green region (~570 nm).

1. Introduction

Present lighting technology relies on tungsten incandescent and compact fluorescence lamps for residential use, fluorescence lamps for work environment, and sodium lamps for street lighting. Today, 21% of energy use is in lighting, and, perhaps, half of this energy or more can be saved by switching to efficient and cold solid-state lighting sources. Solid-state lighting will use visible and UV LEDs that are projected to reach lifetimes exceeding 100,000 hours. From traffic lights to road signs, from automobile taillights to outdoor displays, from landscape to accent lights, solid-state light sources have already arrived as harbingers of the next lighting revolution.[1]

The creation of appropriate sources of white light is the ultimate goal of the solid-state lighting technology. The efficiency of white phosphor-conversion LEDs (now approximately up to 20 lm/W, already twice of that for incandescent lamps) is expected to reach 50 lm/W by year 2010. However, all-semiconductor white lamps based on additive color mixing can exhibit even higher efficiency by using present state-of-the-art red-to-yellow AlGaInP LEDs[2] and blue-green AlInGaN LEDs.[3] Optimization of such polychromatic LED modules is one of the most important problems of the emerging solid-state lighting technology. For general lighting applications, the optimization of the lamp spectrum involves a trade-off between the ability to excite visual sense (luminous efficacy) and the ability to properly reproduce colors of the illuminated objects (color rendering).[4–7]

In this paper, we use a stochastic method of optimization of a white-light source that relies on additive color mixing of the emissions from colored light emitting diodes (primary LEDs).[5,6] We present results on optimized spectral power distribution (SPD) for all-semiconductor lamps composed of four primary LEDs with the line widths typical of present AlGaInP and AlInGaN technologies. We discuss the feasibility of the quadrichromatic lamp, which meets most general lighting needs.

2. Obtaining White Light from Colored LEDs

Consider a primary LED of type j with a normalized Gaussian SPD

$$S_j(\lambda) = c_j \exp\left[-4\ln(2)\left(\frac{\lambda-\lambda_j}{\Delta\lambda_j}\right)^2\right], \tag{1}$$

where c_j is the normalization constant, λ_j is the central wavelength, and $\Delta\lambda_j$ is the line width (full width at half magnitude, FWHM). The color of the LED is described by chromatic coordinates

$$x_j = \int_{380\,\mathrm{nm}}^{780\,\mathrm{nm}} \bar{x}(\lambda) S_j(\lambda)\, d\lambda, \tag{2a}$$

$$y_j = \int_{380\,\mathrm{nm}}^{780\,\mathrm{nm}} \bar{y}(\lambda) S_j(\lambda)\, d\lambda, \tag{2b}$$

and

$$z_j = \int_{380\,\mathrm{nm}}^{780\,\mathrm{nm}} \bar{z}(\lambda) S_j(\lambda)\, d\lambda, \tag{2c}$$

where $\bar{x}(\lambda)$, $\bar{y}(\lambda)$, and $\bar{z}(\lambda)$ are the color matching functions defined by the International Commission on Illumination (CIE; Commission Internationale de l'Éclairage). The normalization constant c_j in Eq. (1) is to be set to hold the condition $x_j + y_j + z_j = 1$.

Our previous calculations[5,6] employed an average line width of 30 nm for all central wavelengths of primary LEDs. However, in practical high-brightness LEDs, the line width depends on the materials system used and varies with central wavelength. In AlGaInP LEDs, the emission is due to band-to-band transitions with the almost constant line width of $1.8k_BT$ on the photon-energy scale. On the wavelength scale, this translates to $\Delta\lambda(\text{nm}) = 1.25 \times 10^{-7}\ \lambda^2 T$. In AlInGaN LEDs, the emission is due to carriers localized on the band potential fluctuations caused by nonuniformity of the spatial distribution of the alloy constituents and indium segregation. Typically, the width of the emission line increases with increasing In molar fraction and only weakly depends on temperature. Data from Ref. 8 (points in Fig. 1) suggests that the line width increases almost linearly with the central wavelength as $\Delta\lambda(\text{nm}) = 0.202\lambda - 66$.

It is to be noted that in AlGaInP LEDs, the luminous efficiency rapidly decreases for wavelengths below 590 nm because of the proximity of the direct-to-indirect band cross-over. Again, in AlInGaN LEDs, the luminous efficiency drops above 500 nm because of degradation of the InGaN alloy with increased In molar fraction.[8] As the result, AlGaInP LEDs are more efficient above 570 nm, whereas AlInGaN LEDs are preferred below 570 nm. Therefore, here we use an empirical line width vs central wavelength dependence with an offset at 570 nm (bold lines in Fig. 3). Above 570 nm, the line width typical of AlGaInP LEDs with the junction temperature of 373 K was used. Below 570 nm, the line width typical of AlInGaN LEDs was employed.

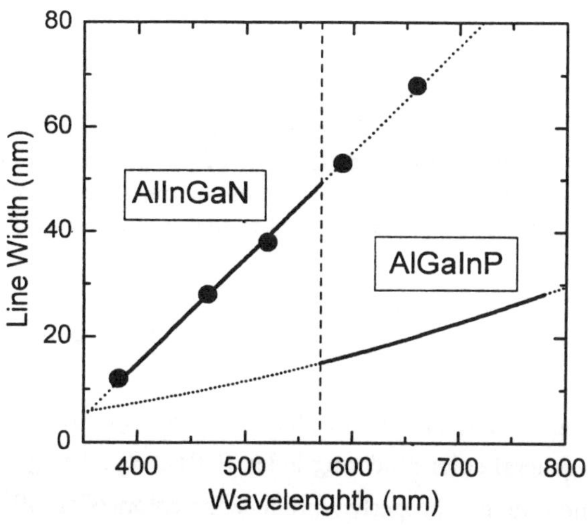

Fig. 1. Line width dependence on central wavelength in practical high-brightness LEDs. AlInGaN LEDs: points, experimental data from Ref. 8; dotted line, approximation. AlGaInP LEDs: dotted line, $1.8k_BT$ line width for $T = 373$ K. Bold lines, approximation used in the present work.

A white lamp with the chromaticity (x_w, y_w) can be composed of n primary LEDs ($n \geq 2$). The resulting "white" SPD is

$$S_w(\lambda) = \sum_{j=1}^{n} I_j S_j(\lambda), \tag{3}$$

where the relative contributions I_j must satisfy the color mixing equations

$$\sum_{j=1}^{n} x_j I_j = x_w, \tag{4a}$$

$$\sum_{j=1}^{n} y_j I_j = y_w, \tag{4b}$$

and

$$\sum_{j=1}^{n} I_j = 1. \tag{4c}$$

Given that the central wavelengths λ_j and relative contributions I_j are known for all j, the relative power of the jth LED is

$$p_j = c_j I_j \Delta\lambda_j \Big/ \sum_{j=1}^{n} c_j I_j \Delta\lambda_j . \tag{5}$$

To find the central wavelengths and relative contributions, the $2n$-dimensional parameter space $\{\lambda_1, \ldots, \lambda_n, I_1, \ldots, I_n\}$ is subjected to three color-mixing constraints [Eqs. (4a,b,c)]. This results in solutions that are on the feasible-vector hypersurface with dimensionality $2n - 3$. In other words, even for two primary LEDs, there exist an infinite number of solutions that yield white light of a predetermined chromaticity. To properly select the wavelengths and relative contributions of the primary LEDs, an optimization problem should be solved. The optimization aims at maximizing two figures of merit. One is luminous efficacy, K, which represents the amount of visual stimuli per unit power of the radiant flux:

$$K = 683\ \text{lm/W} \times \int_{380\ \text{nm}}^{780\ \text{nm}} V(\lambda) S_w(\lambda) d\lambda \Big/ \int_0^{\infty} S_w(\lambda) d\lambda , \tag{6}$$

where $V(\lambda)$ is the spectral sensitivity for photopic vision specified by CIE. Another figure of merit is the general color rendering index (CRI), R_a. The general CRI describes the ability of a white source to properly reproduce the colors of the illuminated objects. It is derived from the SPD by using a procedure CIE 1995 (Ref. 9), which estimates the colorimetric shift with respect to the reference source for eight test samples with the specified reflectivity spectra.

Typically, efficacy and the general color rendering index can be increased only at the expense of each other and a trade-off between these two figures of merit should be performed. For this purpose we introduce an objective function

$$F_{\sigma}(\lambda_1,\ldots,\lambda_n,I_1,\ldots,I_n)=\sigma K+(1-\sigma)R_a, \tag{7}$$

where σ is the weight that controls the trade-off between the efficacy and color rendering. In these terms, the optimization problem reduces to finding the global maxima of the objective function for different values of σ.

3. Method of Optimization

For two or three primary LEDs, the optimization can be performed by exhaustive search inside the feasible-vector hypersurface. For instance, for two primary LEDs, one can run over all possible values of λ_1 and find the maxima of the objective function. For three primary LEDs, exhaustive search is still possible, although the feasible-vector hypersurface is three-dimensional. However, for four and more primary LEDs, running over all feasible vectors becomes too time consuming.

To facilitate the optimization for $n>3$, we introduced a routine that, instead of searching within the entire space of feasible parameters, finds only the sets of the parameters that relate to global maxima of the objective function. The routine relies on stochastic moving the vector $[\lambda_1,\ldots,\lambda_n,I_1,\ldots,I_{n-3}]$ within the component intervals $\lambda\in[380,780]$ and $I\in[0,1]$. At each point, the rest three intensities are calculated from the color-mixing equations and the objective function is evaluated. By stochastic variation of the components by small increments, a local maximum of the objective function is found. Once a local maximum is found, the components are perturbed by random increments and the variation starts at a new point. After *N* cycles of variation and perturbation, a final vector with the highest value of the objective function is determined. To find the global maximum of the objective function, the number of cycles is repeatedly doubled until the objective function stops increasing for all weights σ and the resulting points in the (K,R_a) phase distribution form a smooth dependence (an optimal boundary).

4. Results and Discussion

Figure 2 presents the optimal boundaries for the (K,R_a) phase distribution obtained for quadrichromatic lamps. The results are calculated for two white chromaticities with different color temperatures (CTs), which correspond to CIE standard sources A (tungsten at 2856 K) and B (direct sunlight, 4870 K). Each point represents a set of four central wavelengths and four relative powers.

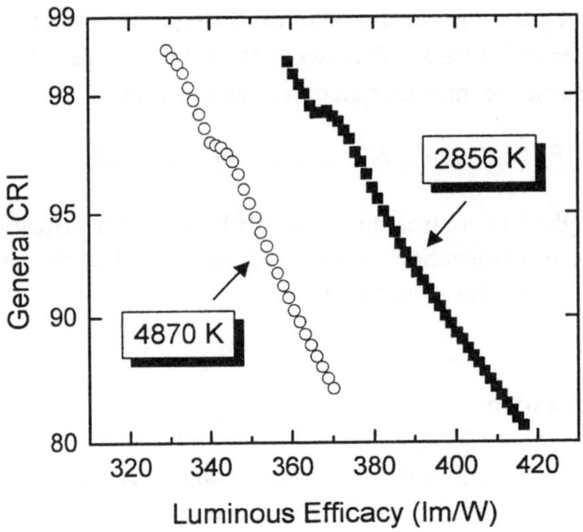

Fig. 2. Optimal boundaries of the (K, R_a) phase distribution for quadrichromatic lamps at two color temperatures.

Each optimal boundary shown in Fig. 2 links the point of the highest efficacy with the point of the highest general CRI. The boundaries indicate that for a particular color temperature, luminous efficacy can be increased only at the expense of the general color rendering index and vice versa. (Kinks are due to the crossover from narrow lines in AlGaInP LEDs to wide lines in AlInGaN LEDs around 570 nm, see Fig. 3.) Moving along an optimal boundary, a trade-off between efficacy and the general CRI can be performed. Roughly, to increase the value of the general CRI from 80 to 98 points, from 10 to 15% of efficacy is to be bargained away. It is also seen that a lower color temperature results in a higher luminous efficacy at a constant value of the general CRI. However, regardless of the color temperature, the quadrichromatic lamps sustain the general CRI in excess of 80 points. The obtained results indicate that this kind of white all-semiconductor lamp is able to meet most needs of general lighting. (For comparison, typical values of R_a for *deluxe* fluorescent lamps are around 80 points.)

To provide guidelines for assembling all-semiconductor white lamps, materials systems required for primary LEDs are to be discussed. Figure 3 displays variation of central wavelengths with the general CRI for trichromatic and quadrichromatic lamps for $R_a > 50$. Trichromatic lamps with $R_a \approx 80$ can be assembled of AlGaInP red (610 nm) LEDs and AlInGaN green (540 nm) and blue (460 nm) LEDs. At these wavelengths, state-of-the-art AlGaInP and AlInGaN LEDs with external quantum efficiencies of 50% (Ref. 2) and 20%,[3] respectively, are available owing to mature technology based on metalorganic chemical vapor deposition (MOCVD) technology. White lamp assembled from such LEDs can exhibit wall-plug efficiencies in excess of 90 lm/W.

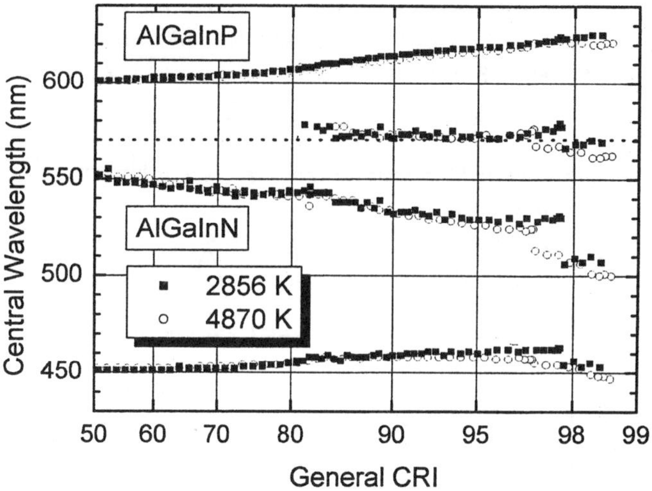

Fig. 3. Variation of peak wavelength with the general CRI value for all-semiconductor white lamps composed of three, four and five primary LEDs. Filled points, 2856 K color temperature; open points, 4870 K color temperature.

Quadrichromatic lamps are seen to require red AlGaInP (610–625 nm) LEDs and green (500–540 nm) and blue (450–460 nm) AlInGaN LEDs. However, the fourth primary LED exactly matches the crossover region of 570 nm (yellow-green). In this region, the quantum efficiencies of both AlGaInP and AlInGaN LEDs do not exceed 2% (Ref.10) because of the materials structure and technological constraints discussed in Section 2. This shortcoming is the principle technological obstacle for practical implementation of efficient quadrichromatic lamps with high color rendering properties. Introduction of the fifth primary LED results in a marginal increase of the general CRI up to 99 points.[5,6] However the quintichromatic lamp still requires the yellow-green primary LED.

It is to be noted that the results presented refer to luminous efficacy that is the property rather of SPD than of practical lamp. Practical all-semiconductor lamps, which are assembled of primary LEDs with different radiant efficiencies, η_{ej}, should be optimized for luminous efficiency. The luminous efficiency of a polychromatic lamp, η_υ, relates to luminous efficacy, K, as

$$\eta_\upsilon = K \Big/ \sum_{j=1}^{n} p_j / \eta_{ej} , \tag{8}$$

and the objective function translates into

$$F'_\sigma(\lambda_1, \ldots, \lambda_n, I_1, \ldots, I_n) = \sigma\, \eta_\upsilon + (1-\sigma) R_a . \tag{9}$$

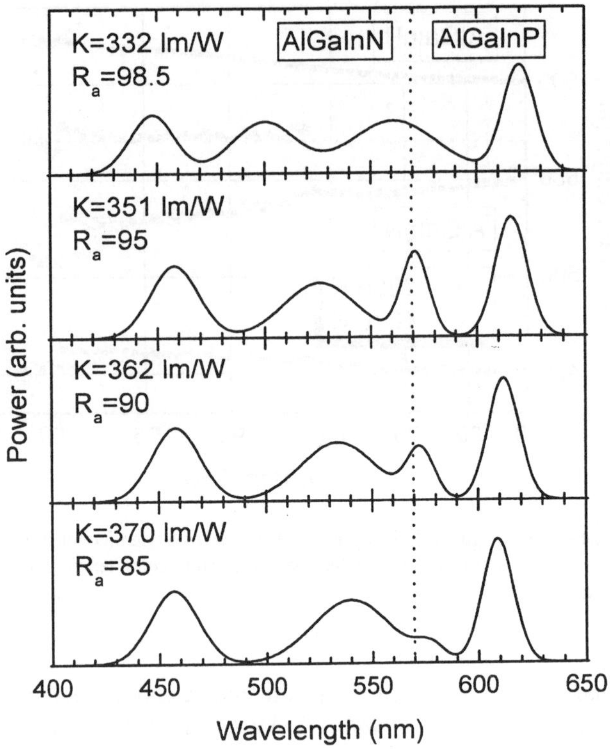

Fig. 4. Model spectra of white emission from quadrichromatic lamp (CT = 4870 K) for different points on the optimal boundary of the (K,R_a) phase distribution (see Fig. 2).

For primary LEDs with the radiant efficiencies that differ negligibly ($\eta_{ej} \approx \overline{\eta}_e$ for all j), the new objective function yields almost the same results of optimization as discussed above with the luminous efficiency $\eta_v = \overline{\eta}_e K$. This means that, roughly, the average radiant efficiency of colored LEDs around 30% is capable in sustaining the luminous efficiency of quadrichromatic lamps in excess of 100 lm/W. However, because of the yellow-green abyss at the 570-nm wavelength, this value might be noticeably lower.[7]

5. Conclusions

In conclusion, we introduced an optimization routine for a white all-semiconductor lamp assembled of an arbitrary number of primary LEDs. Based on the solution of the optimization problem, we provided guidelines for developing of the quadrichromatic lamp, which meets most needs of general lighting. We point out the need of improvement of the performance of yellow-green emitters that are required for implementation of high-quality multichip lamps with four or more primary LEDs

Acknowledgement

One of the authors (A. Ž.) acknowledges support from the NRC Twinning Program.

References

1. A. Žukauskas, M. S. Shur, and R. Gaska, Introduction to Solid-State Lighting, Wiley, New York, 2002, 220 p.
2. M. R. Krames, M. Ochiai-Holcomb, G. E. Höfler, C. Carter-Coman, E. I. Chen, I.-H. Tan, P. Grillot, N. F. Gardner, H. C. Chui, J.-W. Huang, S. A. Stockman, F. A. Kish, M. G. Craford, T. S. Tan, C. P. Kocot, M. Hueschen, J. Posselt, B. Loh, G. Sasser, and D. Collins, "High-power truncated-inverted-pyramid $(Al_xGa_{1-x})_{0.5}In_{0.5}P/GaP$ light-emitting diodes exhibiting >50% external quantum efficiency," Appl. Phys. Lett. **75** (1999) 2365--2367.
3. J. J. Wierer, D. A. Steigerwald, M. R. Krames, J. J. O'Shea, M. J. Ludowise, G. Christenson, Y.-C. Shen, C. Lowery, P. S. Martin, S. Subramanya, W. Götz, N. F. Gardner, R. S. Kern, and S. A. Stockman, "High-power AlGaInN flip-chip light-emitting diodes," Appl. Phys. Lett. **78** (2001) 3379--3381.
4. R. Mueller-Mach and G. O. Mueller, "White light emitting diodes for illumination," Proc. SPIE **3938** (2000) 30--41.
5. A. Žukauskas, F. Ivanauskas, R. Vaicekauskas, M. S. Shur, and R. Gaska, "Optimization of multichip white solid-state lightning source with four or more LEDs," Proc. SPIE **4445** (2001) 148--155.
6. A. Žukauskas, R. Vaicekauskas, F. Ivanauskas, R. Gaska, and M. S. Shur, "Optimization of white polychromatic semiconductor lamps," Appl. Phys. Lett. **80** (2002) 234--236.
7. N. Narendran, N. Maliyagoda, L. Deng, and R. M. Pysar, "Characterizing LEDs for general illumination applications: Mixed-color and phosphor-based white sources," Proc. SPIE **4445** (2001) 137--147.
8. T. Mukai, M. Yamada, and S. Nakamura, "Characteristics of InGaN-based UV/blue/green/amber/red light-emitting diodes," Jpn. J. Appl. Phys. **38** (1999) 3976--3981.
9. CIE, "Method of measuring and specifying colour rendering properties of light sources," Publication No. 13.3 (1995).
10. M. R. Krames, G. Christenson, D. Collins, L. W. Cook, M. G. Craford, A. Edwards, R. M. Fletcher, N. Gardner, W. Goetz, W. Imler, E. Johnson, R. S. Kern, R. Khare, F. A. Kish, C. Lowery, M. J. Ludowise, R. Mann, M. Maranowski, S. Maranowski, P. S. Martin, J. O'Shea, S. Rudaz, D. Steigerwald, J. Thompson, J. J. Wierer, J. Yu D. Basile, Y.-L. Chang, G. Hasnain, M. Hueschen, K. Killeen, C. Kocot, S. Lester, J. Miller, G. Mueller, R. Mueller-Mach, J. Rosner, R. Schneider, T. Takeuchi, and T. S. Tan, "High-brightness AlGaInN light-emitting diodes," Proc. SPIE **3938** (2000) 2--12.

International Journal of High Speed Electronics and Systems, Vol. 12, No. 2 (2002) 439–448

SYSTEM IMPACT OF SILICON CARBIDE POWER DEVICES

BURAK OZPINECI[1,3], LEON M. TOLBERT[1,2], SYED K. ISLAM[1], Md. HASANUZZAMAN[1]

[1]*Department of Electrical and Computer Engineering*
The University of Tennessee, Knoxville, TN 37996-2100

[2]*Oak Ridge National Laboratory*
P.O. Box 2009, Oak Ridge, TN 37831-6472

[3]*Oak Ridge Institute for Science and Education*
Oak Ridge, TN 37831-0117

The emergence of silicon carbide- (SiC-) based power semiconductor switches, with their superior features compared with silicon- (Si-) based switches, has resulted in substantial improvements in the performance of power electronics converter systems. These systems with SiC power devices have the qualities of being more compact, lighter, and more efficient; thus, they are ideal for high voltage power electronics applications such as a hybrid electric vehicle (HEV) traction drive. More research is required to show the impact of SiC devices in power conversion systems. In this study, findings of SiC research at Oak Ridge National Laboratory (ORNL) including SiC device design and system modeling studies, will be given.

1. Introduction

Presently, almost all the power electronics converter systems use silicon- (Si-) based power semiconductor switches. The performance of these switches is approaching the theoretical limits of the Si material. Another material, silicon carbide (SiC), with superior properties compared with Si, is a good candidate to be used in the next generation of power devices.

SiC power devices, with their close-to-ideal characteristics, bring great performance improvements to power converter applications. Some of these advantages compared with Si-based power devices are as follows:

- SiC unipolar devices are thinner, and they have lower on-resistances. At low breakdown voltages (~50V), these devices have specific on-resistances of 1.12μΩ, around 100 times less than their Si counterparts. At higher breakdown voltages (~5000V), this goes up to 29.5mΩ, 300 times less than comparable Si devices.[1] With lower R_{on}, SiC power devices have lower conduction losses; therefore, the converters have higher overall efficiency.

- SiC-based power devices have higher breakdown voltages because of their higher electric breakdown field; e.g., Si Schottky diodes are commercially available typically at voltages lower than 300 V, but the first commercial SiC Schottky diodes are already rated at 600 V.
- SiC has a higher thermal conductivity (4.9 W/cm-K for SiC and 1.5 W/cm-K for Si), and SiC power devices have a lower junction-to-case thermal resistance, $R_{th\text{-}jc}$ (0.02 K/W for SiC and 0.06 K/W for Si); thus device temperature increase is slower.
- SiC can operate at high temperatures. SiC device operation at up to 600°C is mentioned in the literature.[2] Si devices, on the other hand, can operate at a maximum junction temperature of only 150°C.
- SiC is extremely radiation hard; i.e. radiation does not degrade the electronic properties of SiC.
- Forward and reverse characteristics of SiC power devices vary only slightly with temperature and time; therefore, they are more reliable.
- SiC-based bipolar devices have excellent reverse recovery characteristics.[3] With less reverse recovery current, the switching losses and EMI are reduced, and there is less or no need for snubbers.
- Because of low switching losses, SiC-based devices can operate at higher frequencies (>20 kHz) not possible with Si-based devices in power levels of more than a few tens of kilowatts.

Although SiC has these advantages compared with Si, the present disadvantages limit its widespread use. Some of these disadvantages are

- Low processing yield because of micropipes. The best wafers available have <1/cm^2, but they are more expensive than the typical wafer with <10/cm^2.
- High cost–$50 for a 600 V, 4 A Schottky diode (similar Si pn diode <<$1).
- Limited availability (only Schottky diodes at relatively low power are commercially available).
- Need for high temperature packaging techniques that have not yet been developed.

At Oak Ridge National Laboratory (ORNL), a project team consisting of materials, device, and systems researchers in cooperation with The University of Tennessee-Knoxville, Auburn University, and Vanderbilt University are building SiC power MOSFETs. The reason for having a diverse team of people from different disciplines is to develop better devices by encouraging interaction between the researchers who build the devices and the ones who use them. The final power MOSFETs will be used in power electronics converter systems for automotive applications to demonstrate the benefits of SiC-based power devices. One of the selected systems for this project is a traction drive.

2. SiC Devices

2.1 Schottky and pn junction diodes:

The Schottky diode is the least-complex SiC power device. Schottky diodes have been fabricated by forming a metal contact with doped SiC samples. Four different SiC samples were prepared with characteristics given in Table I. All p-type contacts are Al-Ni, and n-type contacts are Au-As-Ni alloy.

Table I: Characteristics of the SiC samples

Sample	A	B	C	D
Schottky diode (p-type epi)	×	×		
pn diode (p-type epi on n-type substrate)			×	×
Implanted with boron and silicon	×		×	
Implanted with boron only		×		×

The *I-V* plots for samples A and D in Figures 1 and 2 do not show any extra resistance due to the new contact material. This means that Au-As-Ni alloy makes a good contact for n-SiC. Note that based on an extensive literature search, Au-As-Ni contacts are a novel approach for SiC device applications.

Doping densities have been calculated from the slope of the *C-V* characteristic plots. The *C-V* plots for samples A and B are shown in Figures 3 and 4, respectively. The measured values of the activated doping densities are sample A, 5.6×10^{15} cm^{-3}, sample B, 1.02×10^{16} cm^{-3}, sample C, 8.6×10^{17} cm^{-3}, and sample D, 1.3×10^{18} cm^{-3}. These values are similar to the applied doping, another sign of good contact.

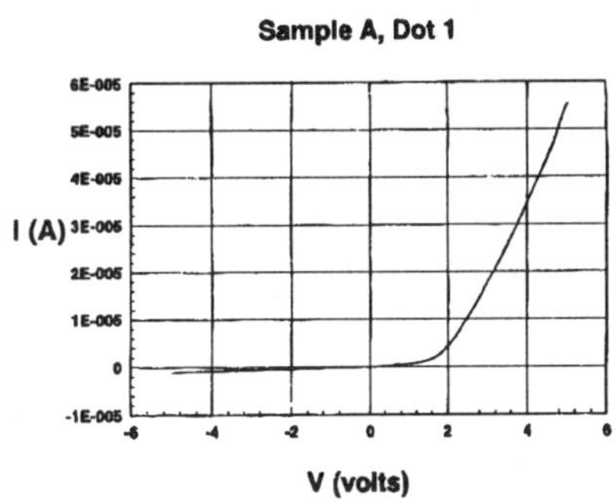

Figure 1. *I-V* Characteristic of sample A.

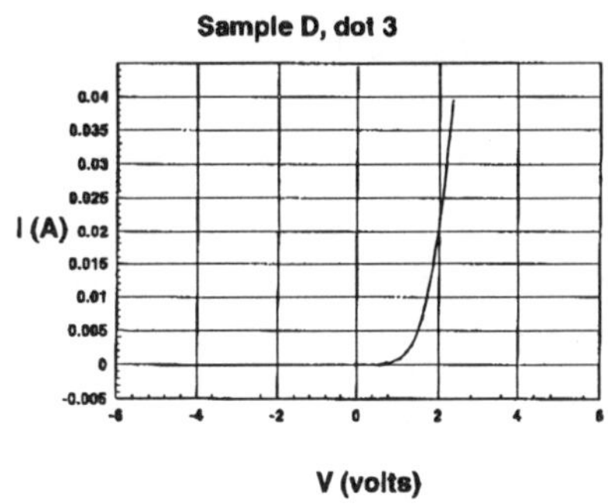

Figure 2. *I-V* Characteristic of sample D.

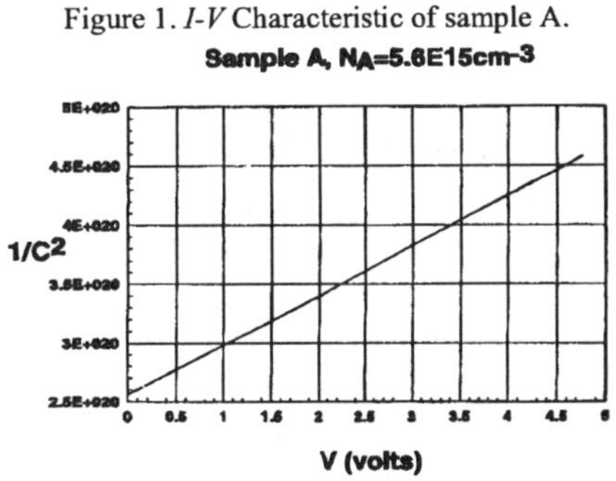

Figure 3. *C-V* plot of sample A.

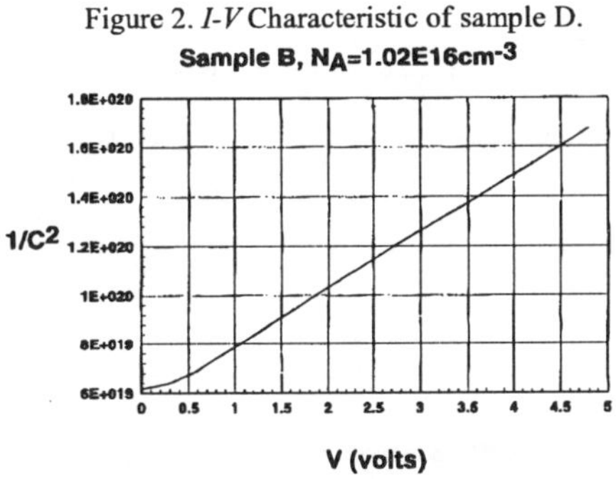

Figure 4. *C-V* plot for sample B.

2.2 MOSFETs:

The double implanted metal-oxide semiconductor (DIMOS) field effect transistor has been frequently used in high-voltage power electronics applications.[4,5] The performance of a DIMOS device is limited by the quasi-saturation behavior in its characteristics. It is shown that such an effect is due to carrier velocity saturation because of the high electric field, low impurity concentration in the drift layer, and narrow p-body spacing.[6] A detailed study has been made of the vertical DIMOS, and an analytical model has been developed. Figure 5 shows the DIMOS device, identifying different regions of operation. An analytical model is developed from regional analysis of carrier transport in the channel and drift regions. The channel is defined by the compensation between the p-body and n-source lateral diffusion. The current/voltage characteristic in the triode region is given by

$$I_{ch} = \frac{W\mu_{neff}}{2L[1+(\mu_{neff}/2v_{sat}L)V_{ch}]} V_{ch}\left[2C_{ox}(V_{GS}-V_T)-(C_{ox}+C_{do})V_{ch}\right] \tag{1}$$

where W is the width of the transistor, μ_{neff} is the effective electron mobility, v_{sat} is the saturation drift velocity, V_{ch} is the channel voltage, V_{GS} is the gate-source voltage, V_T is the threshold voltage, C_{ox} is the oxide capacitance, and C_{do} is the body depletion capacitance.

The drift region is divided into three parts: an accumulation region A, a drift region B with a varying cross section area, and a drift region C with constant cross section. Voltage across each of these regions is given by

$$V_A = \int_0^{W_j+W_d} E_y dy = \frac{I_D(W_j+W_d)}{W(L_{diff}qN_d\mu_{neff})-I_D/E_C} \tag{2}$$

$$V_B = \frac{I_D}{WqN_d\mu_{neff}\cot\alpha}\log\left[\frac{WqN_d\mu_{neff}(L_{diff}+L_P)-I_D/E_C}{WqN_dL_{diff}\mu_{neff}-I_D/E_C}\right] \tag{3}$$

$$V_C = \frac{I_D(W_t-W_j-W_d-L_P\tan\alpha)}{W(L_{diff}+L_P)qN_d\mu_{neff}-I_D/E_C} \tag{4}$$

where I_D is the drain current and E_c is the critical electric field.

The total drift region voltage $V_{drift} = V_A + V_B + V_C$, and the voltage across the drain to source $V_{DS} = V_{drift} + V_{ch}$. The current/voltage characteristics are shown for the analytical model in Figure 6.

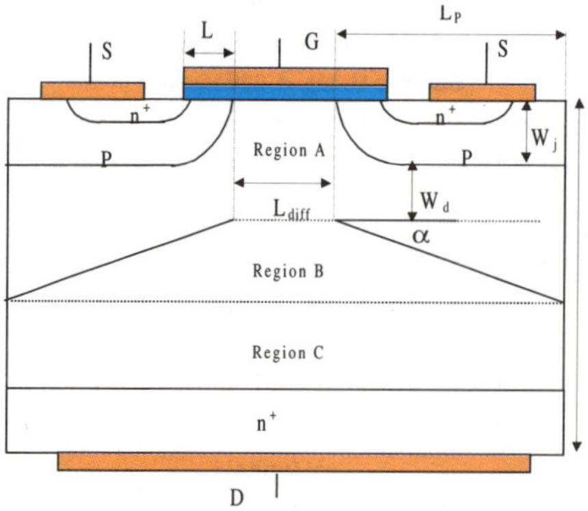

Figure 5. Vertical DIMOS (Regions A, B, and C form the drift region).

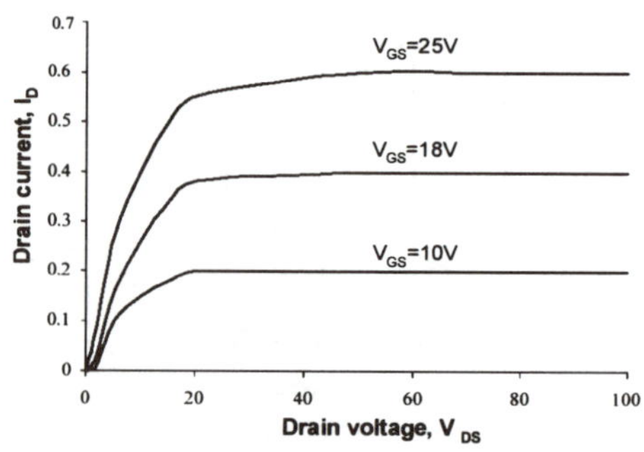

Figure 6. Current-voltage characteristics for DIMOS structure.

3. System Modeling

The main objective of system modeling is to show some of the system-level benefits of using SiC devices in power converter applications, such as the large reduction in the size, weight, and cost of the power conditioning and/or thermal management systems. The selected system for this study is a hybrid electric vehicle (HEV) traction drive. A typical electric traction drive schematic for an HEV is shown in Figure 7. The schematic consists of a battery, a three-phase inverter (dc/ac converter), and an induction motor.

The inverter consists of six MOSFETs and six diodes, two of each on a phase leg. The upper and lower switches on a phase leg turn on and off alternately at frequencies at or below 20kHz with varying duty cycles. The switching instants of the MOSFETs are determined using the popular sinusoidal pulse-width modulation (SPWM) technique. The resulting voltage waveform, v_{ao}, for phase leg A is a chopped waveform with only two values, either $+V_{dc}/2$ or $-V_{dc}/2$, and with as low a harmonic content as possible. The other two phase legs supply similar voltages to the machine, but they are $2\pi/3$ and $4\pi/3$ radians apart from v_{ao}. The induction motor, with its high magnetizing inductance, acts like a filter; thus, the motor currents, unlike the voltages, are sinusoidal.

To evaluate any transportation system, its operation while the vehicle is accelerating, decelerating, stopped, etc., should be considered. For this reason, it is required to simulate the HEV and observe the response of the traction drive over a driving cycle. Using ADVISOR (ADvanced VehIcle SimulatOR), an HEV model is simulated in ADVISOR over the federal urban driving schedule (FUDS), which is a standard 1369 second velocity profile of an average person's vehicle on the way to work.[7]

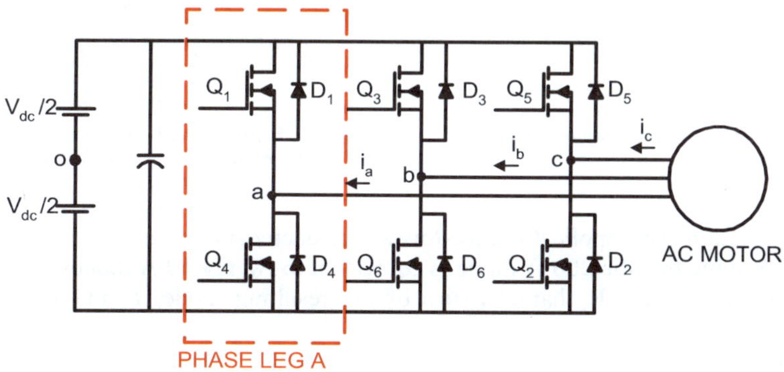

Figure 7. Electric traction drive.

The induction motor in this study is rated at 31 kW, 230 V, 4-pole, 3000 rpm. The peak current passing through the switches is 136 A, and the peak voltage across them is the battery voltage V_{dc}, which is 300 V in this case. Therefore, the MOSFETs and diodes have to be rated at least 400 V and 200 A. The Si MOSFETs are not available in this power range, and SiC MOSFETs are not yet available. For comparison purposes in this study, we will assume that Si MOSFETs exist at high power, too.

The following section will explain the device loss modeling approach.[8]

4. Diode Modeling

4.1 Conduction losses:

The circuit in Figure 8 is built to find the *I-V* characteristics of the diodes. The dc voltage supply is varied, and the diode forward voltage and current are measured. This test is carried out at several temperatures up to 250°C.

The results for both Si pn and SiC Schottky diodes are given in Figure 9, in which it can be seen that the forward voltage of the SiC diode is higher than that of the Si diode. This is expected because of SiC's wider bandgap. Another difference between these two diodes is their high-temperature behavior. As the temperature increases, the forward characteristics of the Si diode change severely while those of the SiC diode stay confined to a narrow region. Note that the pn diode (negative) and the Schottky diode (positive) have different polarity temperature coefficients; that is why the slope of the curve at higher currents is increasing in the Si diode case and decreasing in the SiC diode case with the temperature increase.

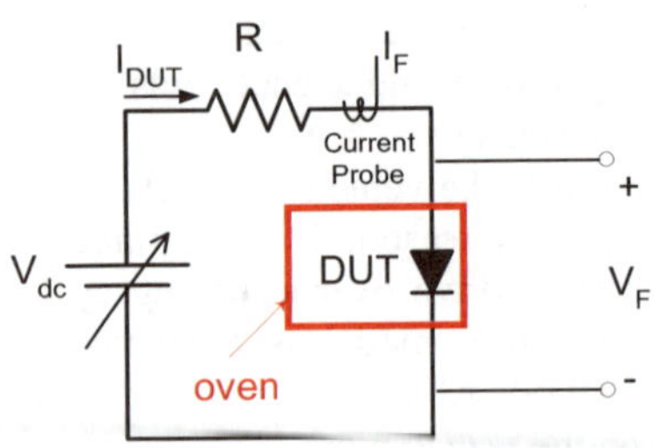

Figure 8. *I-V* characterization circuit.

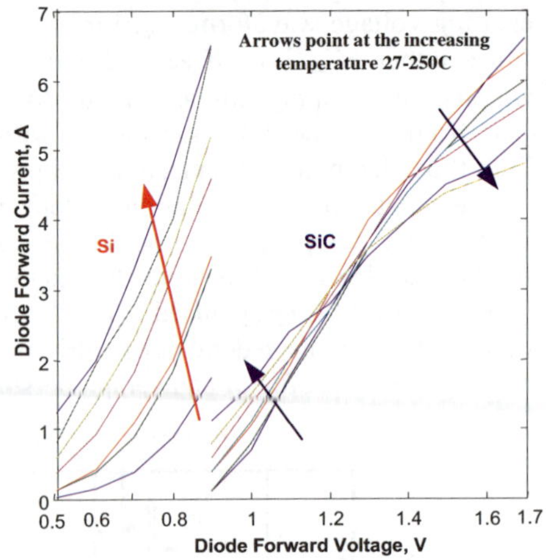

Figure 9. Experimental *I-V* characteristics of the Si and SiC diodes in an operating temperature range of 27 C to 250°C.

For the traction drive application, the forward characteristics of 200 A rated diodes are required. Assume that the 200 A diode is equivalent to twenty 10 A diodes connected in parallel. Then, in the *I-V* characteristics of the resulting diode, I_s and R_s should be changed accordingly.

The diode equation is

$$I = I_S\left(e^{q\left(V - IR_S\right)/nkT} - 1\right). \tag{5}$$

I_s is directly and R_s is inversely proportional to the area of the device. For the 200 A diode, the area is increased 20 times; therefore, I_s increases 20 times and R_s decreases 20 times. The resulting diode equation is

$$I = 20 \times I_S\left(e^{q\left(V - I\frac{R_S}{20}\right)/nkT} - 1\right). \tag{6}$$

If a line is drawn along the linear high-current portion of the *I-V* curves of (6) extending to the x-axis, the intercept on the x-axis is V_D and the slope of this line is $1/R_D$.

After these parameters are obtained for each diode at different operation temperatures, the conduction losses can be calculated using (7).[8]

$$P_{cond,D} = I^2 \cdot R_D \cdot \left(\frac{1}{8} - \frac{1}{3\pi} M\cos\phi\right) + I \cdot V_D \cdot \left(\frac{1}{2\pi} - \frac{1}{8} M\cos\phi\right). \quad (7)$$

where M is the modulation index and ϕ is the power factor.

The conduction losses for an SiC diode and for an Si diode are plotted in Figures 10 and 11, respectively. These plots show that for low temperatures (T < 55°C), the conduction loss of the SiC diode is less than that of the Si diode, and vice versa for higher temperatures (T < 55°C). This is because the series resistance of the Schottky diode is increasing while that of the pn diode is decreasing. This increase seems to be a disadvantage in the SiC Schottky diode case; however, note that the Si diode cannot withstand temperatures over 150°C.

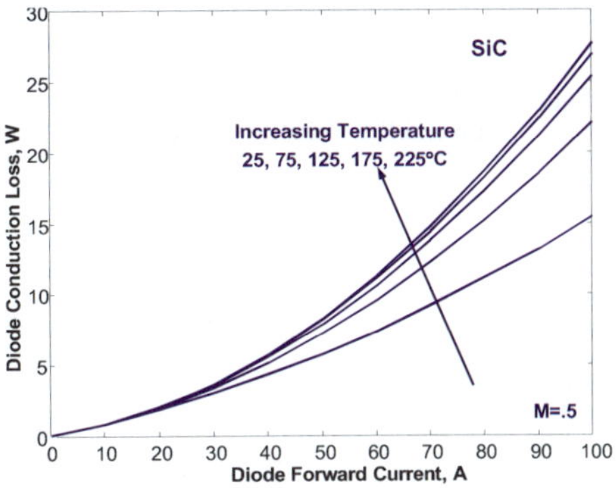

Figure 10. Conduction losses for a 200A SiC diode.

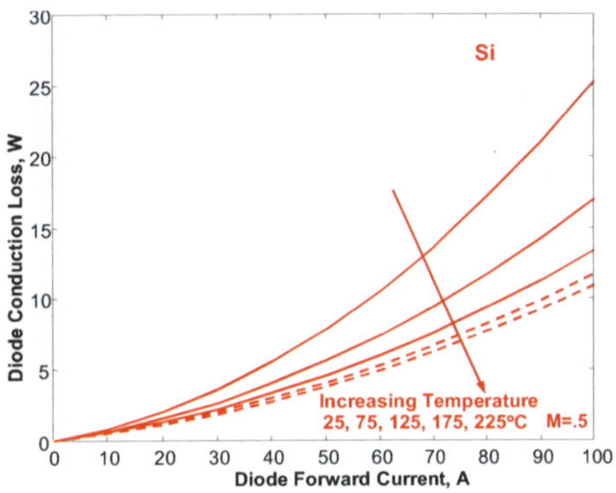

Figure 11. Conduction losses for a 200A Si diode.

4.2. Switching losses:

The most important part of the diode switching loss is the reverse recovery loss. The rest of the losses are negligible. Reverse recovery losses, in this paper, will be calculated experimentally.

For this purpose, the chopper circuit in Figure 12 was built. Here, the main switch Q is turned on and off at 1 kHz with a duty ratio of 75%. When Q is on, the diode D is off, and the current is forced by the dc supply to build up through the load and Q. When Q turns off, the load current starts flowing through D and the load. In this mode, the dc supply is out of the loop, and therefore the current starts decreasing. After a while, Q is turned on again and then D turns off. The typical diode turn-off waveforms are given in Figure 13. These experimental waveforms show that the Si diode switching losses are almost three times more than those of the SiC diode.

The peak reverse recovery current, I_R, and the switching loss, P_{sw}, of the diodes are measured at different operating temperatures with varying load currents. The results are plotted in Figures 14 and 15. In Figure 14, I_R of the Si diode is higher than that of the SiC diode at any operating temperature. As the temperature increases, the difference increases because the I_R of the Si diode increases with temperature, but that of the SiC diode stays constant. Note that the Si diode failed when operating at 150°C and 4.5 A while the SiC diode survived that temperature and failed at a higher 250°C and 4 A.

To obtain a 200 A diode model for the traction drive, assume as before that 20 of the

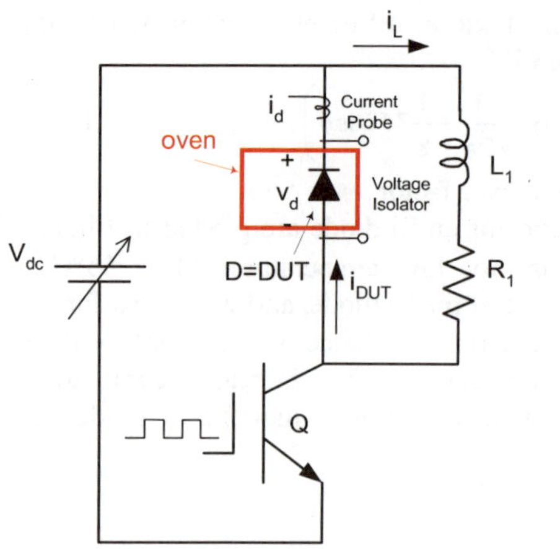

Figure 12. Reverse recovery loss measurement circuit.

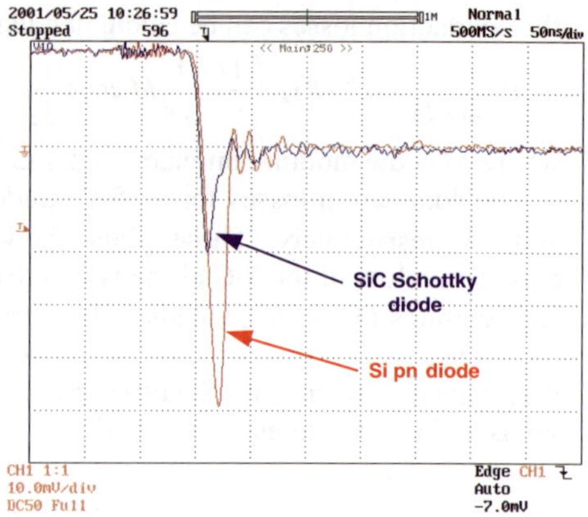

Figure 13. Typical reverse recovery waveforms of the Si pn and SiC Schottky diode (2 A/div.).

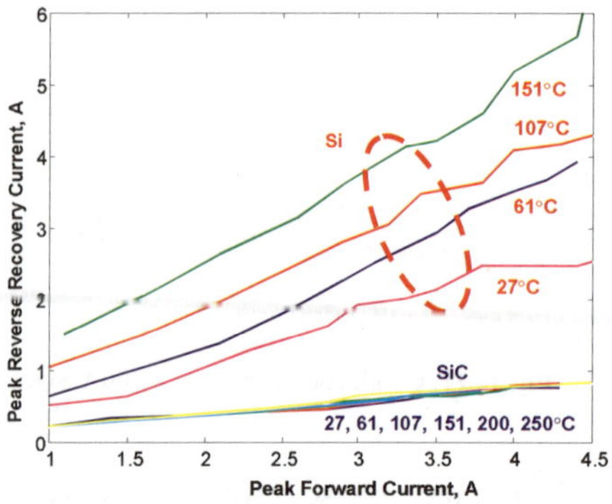

Figure 14. Peak reverse recovery values with respect to the forward current at different operating temperatures.

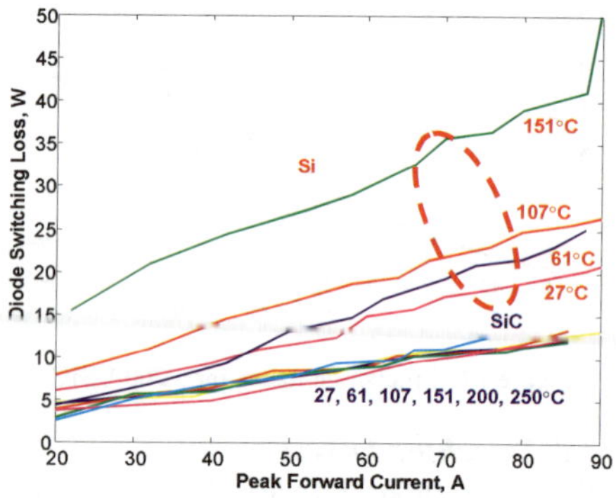

Figure 15. Diode switching loss of 200A diode at different operating temperatures.

diodes tested here are connected in parallel to form a 200 A diode. The switching losses multiplied by 20 gives the switching losses of the 200 A Si and SiC diodes (Figure 15).

5. Results

The HEV traction drive is simulated over the FUDS cycle using the model developed in the previous sections. The loss profiles of a diode and a MOSFET in the drive are shown in Figure 16. SiC diode losses are lower than Si diode losses mostly because the SiC diode has lower reverse recovery losses. On the other hand, SiC MOSFET losses are lower because the switching losses are similar but SiC MOSFET conduction losses are lower. The reason for lower conduction losses is the lower specific on-resistance [$R_{on,sp}$ (Si) = 180×10^{-3} Ω-cm^2, $R_{on,sp}$ (SiC) = 0.3×10^{-3} Ω-cm^2].[1]

Total energy loss (six diodes and six MOSFETs) is 925 W·s for the Si inverter and 338 W·s for the SiC inverter over the FUDS cycle. The corresponding efficiency of the Si inverter is 80–85%, while that of the SiC inverter is 90–95%. This is a 10% increase in the average efficiency. As a result, the battery in the HEV with the SiC inverter will need less charging than the one with the Si inverter.

The loss profiles in Figure 16 are fed to the thermal models of the devices. The resulting junction temperature profiles are shown in Figure 17. Natural air-cooled heatsinks are used to limit the junction temperature to 150°C for Si and 175°C for SiC. The latter temperature limit is found on the datasheet of the Infineon SiC Schottky diode used in this study.[9] The resulting heatsink volumes and masses for each device and each inverter are given in Table II. Using SiC devices instead of their Si counterparts in an HEV traction drive reduces the size and weight of the heatsink to one-third. Note that a heatsink usually occupies one-third the volume of the converter and weighs more than the electronics.

Theoretically, SiC devices can work at higher temperatures. If new packaging techniques are developed such that these higher temperatures could be used as the junction temperature limits, then the amount of cooling required would be less, and more weight and volume savings would be possible.

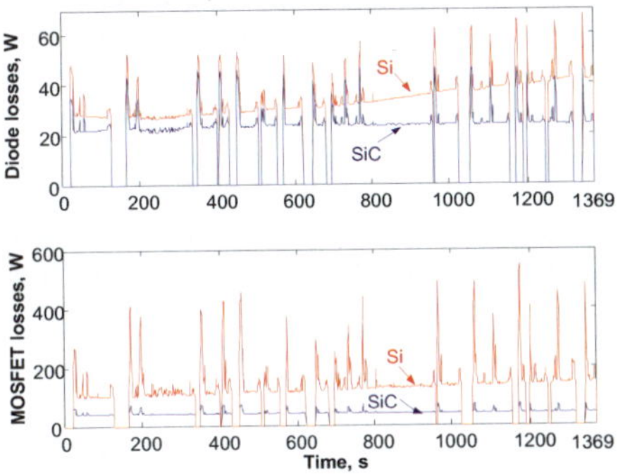

Figure 16. Total loss profile for a diode (top) and a MOSFET (bottom).

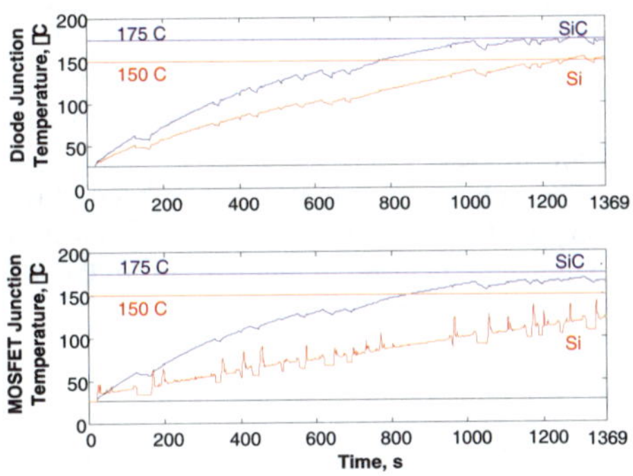

Figure 17. Junction temperature profiles of the diodes and MOSFETs in the three-phase inverter.

Table II: Heatsink mass and volume for each device and inverter

	Volume (cm^3)	**Mass (g)**
Si diodes	444	1200
SiC diodes	162	450
Si MOSFETs	1554	4200
SiC MOSFETs	444	1200
Si inverter	***1998***	***5400***
SiC inverter	***606***	***1650***

6. Conclusions

Even the first SiC devices show the superiority of SiC compared with Si. One of the challenges for building SiC power devices is the contact material. In this paper a new contact material, Au-As-Ni alloy, is introduced. The *I-V* characteristics of the SiC Schottky diodes built with this contact material show that it makes solid and reliable contact with SiC.

System studies show that power electronics systems using SiC power devices are on average 10% more efficient because of the low losses of the SiC power devices. Moreover, with their high-temperature operation capability, they need less stringent cooling requirements. In the HEV traction drive studied here, using SiC power devices saves 1392cm^3 of space and 3.75kg of weight.

The weight reduction and efficiency increase result in an increase in the fuel economy of the vehicle and longer lifetime for the battery.

Note that SiC technology is still in its infancy. More studies like the one in this paper are required to forecast the impact of SiC power devices. When this technology matures, for power devices in the medium-to-high power range, the future will be SiC.

Acknowledgements

This project is supported by ORNL and the U.S. Department of Energy's Office of Transportation Technology Freedom Car Research Program.

The authors would like to thank Michael Krach and Michael Brandt from Infineon Technologies AG for supplying the SiC Schottky diode samples, and Mustafa Saglam from the Technical University of Darmstadt for contacting them about the samples. The authors also would like to thank Dr. O. W. Holland of ORNL for providing SiC samples and Dr. F.C. Jain of the University of Connecticut for providing facilities and assistance for material processing, fabrication, and testing of SiC diodes.

References

1. M. Bhatnagar and B. J. Baliga, "Comparison of 6H-SiC, 3C-SiC, and Si for power devices," *IEEE Trans. on Electron Devices*, 40 (3), pp. 645–655, March 1993.
2. K. Shenai, R. S. Scott, and B. J. Baliga, "Optimum semiconductors for high-power electronics," *IEEE Transactions on Electron Devices*, 43 (9), pp. 1811–1823, Sept. 1989.
3. A. Elasser, M. Kheraluwala, M. Ghezzo, R. Steigerwald, N. Krishnamurthy, J. Kretchmer, and T. P. Chow, "A comparative evaluation of new silicon carbide diodes and state-of-the-art silicon diodes for power electronic applications," *IEEE IAS Annual Meeting Conference Proceedings*, pp. 341–345, 1999.
4. B. J. Baliga, *Modern Power Devices*, Wiley, New York, 1987.
5. D. A. Grant and J. Gower, *Power MOSFETs: Theory and Application*, Wiley, New York, 1989.
6. M. Darwish, "Study of the quasi-saturation effect in VDMOS transistors", *IEEE Transactions on Electron Devices*, vol. ED-33, no. 11, pp. 1710–1716, 1986.
7. http://www.ctts.nrel.gov/analysis/advisor.html
8. B. Ozpineci, L. M. Tolbert, S. K. Islam, and Md. Hasanuzzaman, "Effects of silicon carbide (SiC) power devices on PWM inverter losses," *The Annual Conference of the IEEE Industrial Electronics Society (IECON'01)*, pp. 1187–1192, 2001.
9. http://www.infineon.com/cgi/ecrm.dll/ecrm/scripts/prod_cat.jsp?oid=-8681

International Journal of High Speed Electronics and Systems, Vol. 12, No. 2 (2002) 449–458

LOW FREQUENCY NOISE IN GALLIUM NITRIDE FIELD EFFECT TRANSISTORS

SERGEY L. RUMYANTSEV[a] and MICHAEL S. SHUR
Rensselaer Polytechnic Institute, Troy NY 12180-3590

REMIS GASKA
Sensor Electronic Technology, Inc., 21 Cavalier Way, Latham, NY 12110

MICHAEL. E. LEVINSHTEIN
The Ioffe Physical-Technical Institute of Russian Academy of Sciences, 194021, St. Petersburg, Russia

M. ASIF KHAN, GRIGORY SIMIN, and J. W. YANG
University of South Carolina, Columbia, SC, 29208

We report on experimental study of the low frequency noise in GaN-based Field Effect Transistors. In both GaN Metal Semiconductor Field Effect Transistors (MESFETs) and AlGaN/GaN Heterostructure Field Effect Transistors (HFETs), the main noise sources are located in the channel. Gate voltage dependence of noise in MESFETs complies with the Hooge formula and indicates the bulk origin of noise. The dependencies of the Hooge parameter, α, on sheet electron concentration n_s in HFETs are extracted from measured drain current fluctuations taking into account the contact resistance, and the resistance of the ungated regions of the transistors. At low channel concentrations α is inversely proportional to n_s ($\alpha \sim 1/n_s$). This dependence as well as the temperature dependence of noise might be explained by electron tunneling from the 2D gas into the traps in the bulk GaN or AlGaN.

1. Introduction

Wide band-gap semiconductors such as GaN, InN, AlN and their ternary and quaternary alloys have been extensively investigated in recent years because of their potential microwave and optoelectronic applications. Devices based on this materials system can be used in high-temperature, high-power density, high frequency, and radiation-hard electronic circuits[1, 2].

Research efforts are currently focused on the development of GaN/GaAlN field-effect transistors, ultraviolet visible-blind photodetectors, blue, violet and UV light emitting diodes, blue lasers, pyroelectric and piezoelectric sensors.

The level of the low-frequency noise is a good indicator of material quality and an important figure-of-merit for many electronic and optoelectronic devices. It has been shown that the value of the low-frequency noise is highly sensitive to the presence of crystal imperfections and, under certain conditions, can be used as a diagnostic tool for device quality and reliability.

---

[a] On leave from the Ioffe Institute, 194021 St-Petersburg, Russia

On the other hand, when GaN-based transistors are used as oscillators or mixers in microwave and optical communication systems, their low frequency noise is one of the major factors determining the phase noise characteristics. In order to reduce the phase noise, the current noise has to be low in the entire operation range of the gate biases.

2. Experimental details

We studied the low frequency noise in four different types of GaN-based transistors: Metal Semiconductor Field Effect Transistors (MESFETs), Metal Oxide Semiconductor Field Effect Transistors (MOSFETs), Heterostructure Field Effect Transistors (HFETs), and Metal Oxide Semiconductor Heterostructure Field Effect Transistors (MOSHFETs)

The structures were grown by low-pressure Metal Organic Chemical Vapor Deposition (MOCVD) on (0001) sapphire or semi-insulating 4H-SiC substrates.

The AlGaN/GaN heterostructures consisted of a 50-nm-thick AlN buffer layer, 0.4-μm-thick GaN layer, followed by $Al_{0.2}Ga_{0.8}N$ barrier layer, which was doped with silicon to approximately 2×10^{18} cm^{-3}.

In GaN MESFETs the deposition of approximately 2 μm of nominally undoped GaN was followed by the growth of a Si-doped GaN channel. The thickness and doping level of the channel (extracted from capacitance-voltage characteristics) were ~60 nm and 10^{18} cm^{-3}, respectively.

Prior to the MOSFET and MOSHFET fabrication, a 7 nm SiO_2 layer was deposited on a part of the heterostructures using Plasma Enhanced Chemical Vapor Deposition (PECVD).

The transistors had the source-drain spacing of 4-5 μm, the gate length, L, of 1-1.5 μm, and a gate width, W, in the range of 50 -150 μm. The Transmission Line Model (TLM) structures and HFETs were fabricated on the same wafers.

A low-frequency noise was measured in the frequency range from 1 Hz to 100 kHz with the sources grounded.

3. Results and Discussion

The noise spectra of drain current fluctuations S_{Id} at room temperature had the form of $1/f^{\Gamma}$ noise with Γ close to unity (Γ= 1.0-1.15) for all types of devices. At low drain biases, V_d< 1V, the spectral noise density S_{Id} was proportional to the square of the drain voltage $S_{Id} \sim V_d^2$, as expected.

At elevated temperatures, the generation-recombination (GR) noise was observed in low-frequency noise measurements of GaN/GaAlN HFETs and MOS-HFETs. The extracted activation energy was $E_a \sim 0.8$ - 1.0 eV. The analysis of the noise amplitude showed that noise might come from thin (30nm) AlGaN barrier layer[3]. The estimates of the level parameters based on this assumption resulted in reasonable values of the capture cross section $\sigma_n \approx (10^{-12} - 10^{-13})$ cm^2 and the trap concentration $N_t \approx 5\times10^{16}$cm^{-3}.

Many different noise sources in FETs might be important including the contribution of the gate leakage current, contact noise, bulk noise, surface noise, and the fluctuations of the Schottky barrier space charge region (SCR) width[4]. Depending on the device structure, different noise mechanisms are responsible for the main contribution to the overall noise[5-7].

3.1. Contribution to noise from gate leakage current

The contribution of the gate leakage current fluctuations to the output drain current noise of AlGaN/GaN HFETs was studied by three different methods[8].

First, we compared the low frequency noise properties of the AlGaN/GaN HFETs and MESFETs with noise properties of MOSHFETs and MOSFETs. The MOS-structures had an extremely low gate leakage current and the difference in gate current between oxide and non-oxide structures was several orders of magnitude.

Second, the gate current fluctuations were measured directly in AlGaN/GaN HFETs. The appropriate analysis of the results also allowed us to calculate the contribution of the gate current fluctuations to the output noise.

Third, the correlation between the gate and drain current fluctuations were measured and analyzed.

Our results showed that in the GaN-based HFETs with a high gate leakage current (I_g/I_d)>10^{-3}-10^{-4}, where I_g is the gate current and I_d is the drain current), the gate current fluctuations can significantly contribute to the output noise. For all further measurements the transistors with the low gate leakage current have been selected. In such devices the contribution of the gate leakage to noise was negligible.

3.2. Contact contribution to noise

To determine the contribution of the contact noise to the measured noise spectra, the noise measurements were performed on the Transmission Line Model (TLM) structures[9]. The measurements using TLM structures showed that the contact resistance R_c in MESFET and MOSFET structures was negligible compared with the channel resistance. Assuming that the contributions of the contact noise and of the noise from the GaN layer are not correlated and taking into account that the contact resistance is much smaller than the resistance of the GaN layer, the spectral noise density of the current fluctuations S_I/I^2 can be expressed as:

$$\frac{S_I}{I^2} = \frac{S_{Rc} + S_{GaN}}{R_{GaN}^2}, \qquad (1)$$

where S_{Rc} and S_{GaN} are the spectral noise densities of the contact resistance and of the GaN layer resistance fluctuations, respectively, R_{GaN} is the resistance of the GaN layer between the pads of the TLM structure. In the limiting case when the contact noise is dominant ($S_{Rc}>>S_{GaN}$), the spectral noise density, S_I/I^2, should be proportional to L_1^{-2}, where L_1 is the distance between the TLM contact pads:

$$\frac{S_I}{I^2} = \frac{S_{Rc}}{R_{GaN}^2} \propto \frac{1}{L_1^2}. \qquad (2)$$

In the opposite limiting case, when $S_{Rc}<<S_{GaN}$, the spectral noise density of the GaN layer resistance fluctuations is proportional to the reciprocal volume of the GaN layer

(the bulk origin of the noise) or to the reciprocal GaN area (surface origin of the noise) between the contact pads. In both these cases:

$$\frac{S_I}{I^2} = \frac{S_{GaN}}{R_{GaN}^2} \propto \frac{1}{L_1}. \qquad (3)$$

Figure 1 shows the dependence of the relative spectral noise density of the current fluctuations on the distance L_1 between the pads of the TLM structure. Since this dependence is close to the $1/L_1$ law, we conclude that contacts do not contribute much to the overall noise. Using the same method we found that in HFETs and MOSHFETs transistors contacts do not contribute much to noise as well.

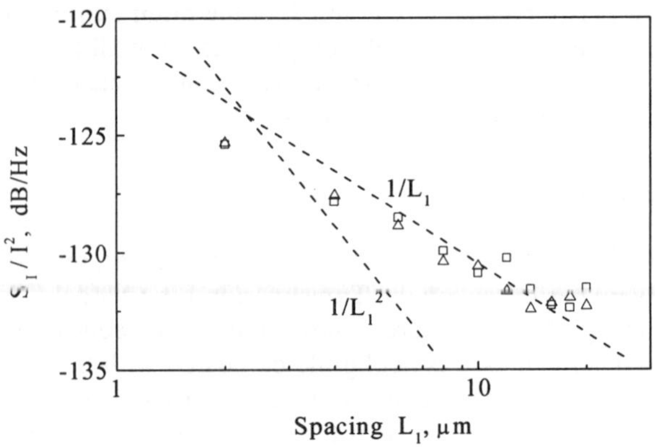

Fig.1. The dependence of the relative spectral noise density S_I/I^2 on the distance L_I between the pads of TLM structure. Frequency of analysis $f = 200$ Hz.

3.3. Gate voltage dependence of noise and location of the noise sources (MESFETs and MOSFETs)[9]

The data points in Fig.2 correspond to the experimental results for the dependence of noise on the drain current for MESFETs and MOSFETs at a constant drain bias. Within an experimental error, the noise behavior for both types of transistors was identical. This indicates that SiO_2 film deposited in order to fabricate MOSFETs does not affect much the noise properties. As seen from the figure, the relative spectral noise density S_{Id}/I_d^2 decreases with the drain current increase, i.e. with the gate voltage increase.

Let us first consider the location of the noise sources at the surface of GaN in the source -gate and drain - gate regions.

For the noise sources located at the surface, the relative spectral noise density of the short circuit drain current fluctuations can be presented in the following form:

$$\frac{S_{Id}}{I_d^2} = \frac{S_{Rds}}{R_{ds}^2} \frac{R_{ds}^2}{(R_{ds} + R_{Ch})^2}, \qquad (4)$$

where R_{Ch} is the channel resistance which depends on the gate voltage, V_g, $R_{ds}=R_d+R_s$ is the resistance of source-gate and drain-gate regions, S_{Rds} is the spectral noise densities of the R_{ds} fluctuations. In this case, the noise gate voltage dependence is determined by the dependence of R_{Ch} on V_g.

According to Eq. (4), noise S_{Id}/I_d^2 should decrease with the R_{Ch} increase, i.e. with the drain current decrease. Since the experimental dependence exhibits the opposite trend, the measured noise can not be explained by the surface noise.

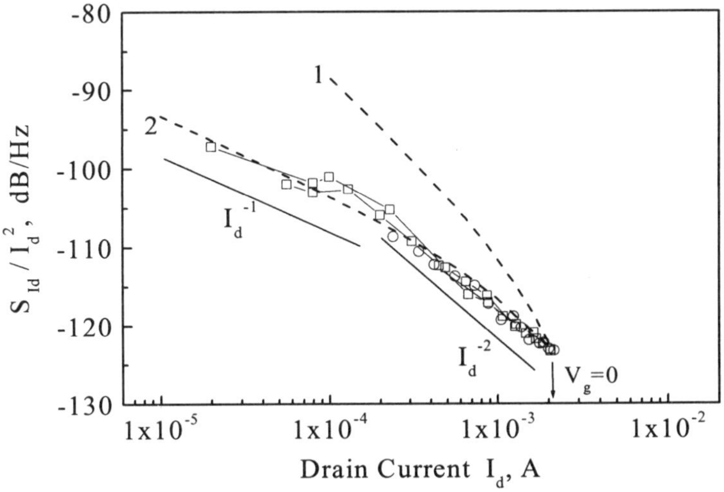

Fig.2. The dependence of the of the relative spectral noise density of the drain current fluctuations on drain current. Drain voltage $V_d = 0.5$V. Frequency of analysis $f = 200$ Hz. Different symbols show data for MESFETs and MOSFETs. Line 1 was calculated according the Lauritzen model[4]

A specific mechanism of a low frequency noise in MESFETs was analyzed by Lauritzen[4]. He assumed that the fluctuations of the charge state of the levels inside the depletion region of a *p-n* junction or of a Schottky barrier result in the fluctuations of the depletion region width, *W*, and, consequently, in the fluctuations of the channel width and the channel resistance.

In Fig.2 line 1 was calculated in accordance with the Lauritzen model[9]. As seen in Fig.2, this noise mechanism predicts a much faster decrease of noise with current compared with that observed experimentally.

Van der Ziel[10] assumed that fluctuations of the carrier density in the bulk of the channel are responsible for the noise in FETs. Curve 2 was calculated using this approach. As can be seen from the figure the experimental data can be fitted quite well using the bulk noise model.

Since it appears that the noise sources in the transistors under investigation are located in the bulk of GaN layer, the Hooge parameter $\alpha = (S_I/I^2)Nf$ (*N* is the total number of carriers in the sample) can be used to characterize the noise level[11]. We estimated $\alpha = (2\text{-}3)\times10^{-3}$ for the entire range of the gate voltages. These values of α are at least one order of magnitude smaller than those reported previously for bulk GaN[12].

3.4. Gate voltage dependence of noise and location of the noise sources (HFETs and MOSHFETs)

Using the techniques described above, we made sure that neither the contact noise nor the noise from gate leakage current contributed much to the total output HFET noise.

Fig. 3 shows the dependencies of the relative spectral noise density S_{Id}/I_d^2 on drain current for several HFETs and MOSHFETs at constant drain voltage V_d=0.5V. Dashed line shows the attempt to fit experimental assuming channel location of the noise sources and Hooge parameter $\alpha = 10^{-3}$. As seen, the experimental dependence decreases much steeper than calculated one. This trend is well known[13-16] and it can be phenomenologically described via the dependence of the Hooge parameter, α, on the gate voltage.

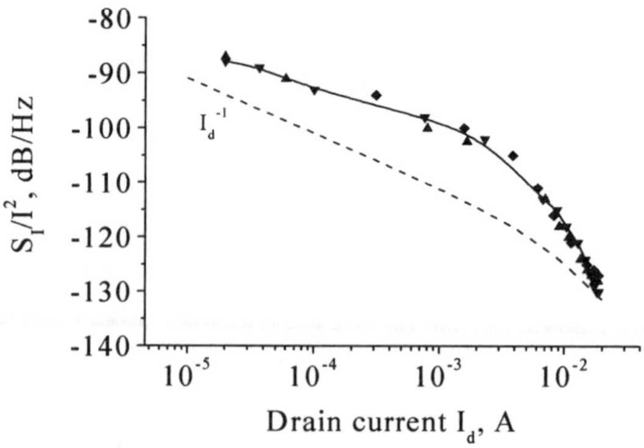

Fig.3 The dependence of the relative spectral noise density of the drain current fluctuations on drain current for HFETs and MOSHFETs. Drain voltage V_d = 0.5V. Frequency of analysis f = 200 Hz. Dashed line was calculated assuming channel location of the noise sources and Hooge parameter $\alpha = 10^{-3}$.

Fig.4 shows the dependence of the Hooge parameter α on the sheet channel concentration n_s, extracted for three different transistors taking into account contact resistance, and resistance of the ungated regions of the transistors[17].

In spite of difference in the absolute values of the parameter, α, the dependencies of α on channel concentration, n_s, are identical for all structures. At low channel concentrations, α decreases with the increase of n_s as $\alpha \sim 1/n_s$, reaches a minimum and then increases with a further increase of n_s.

The dependence $\alpha \sim 1/n_s$ is often observed in n-channel Si MOSFETs[18]. It is generally accepted that the most probable mechanism for this noise is the electron tunneling from the channel into the oxide. This noise mechanism was first proposed by McWhorter[19] in 1957. Since we found the same behavior in AlGaN/GaN HFETs, the noise observed might be linked to the similar noise mechanism: electron tunneling from the 2D gas into the traps in GaN or AlGaN.

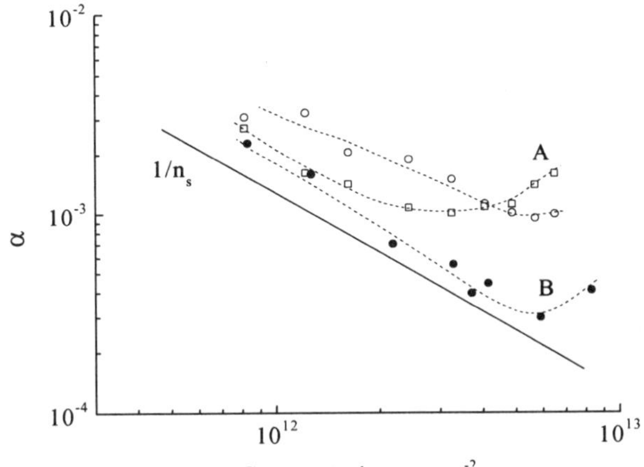

Fig. 4 The dependencies of the Hooge parameter α on 2D sheet channel concentration n_s for *A* and *B* GaN/GaAlN HFETs. Different symbols for *A* transistor represent the data for different samples.

3.5. Low frequency noise in GaN/AlGaN heterostructure field effect transistors at cryogenic temperatures

Fig.5 shows the temperature dependences of noise at different frequencies of analysis. Within the temperature range from 50K to 200K a broad maxima have been observed on temperature dependences of noise. Although this temperature dependence of noise looks somewhat similar to that typical for GR noise, the noise observed cannot be attributed to the simple GR noise for two main reasons: (I) the position of the maxima practically does not depend on frequency of analysis, (II) the noise spectra at all temperatures $T > 50K$ are close to the 1/f noise.

An extremely broadened GR-like noise was found in AlGaAs/GaAs heterostructures at low temperatures in ref. [20], and was attributed to the real-space transfer from the two-dimensional electron gas to the AlGaAs layer. Another possible mechanism which could account for the observed temperature dependence of noise is tunneling of electrons from the 2D gas to the levels in GaN or AlGaN.

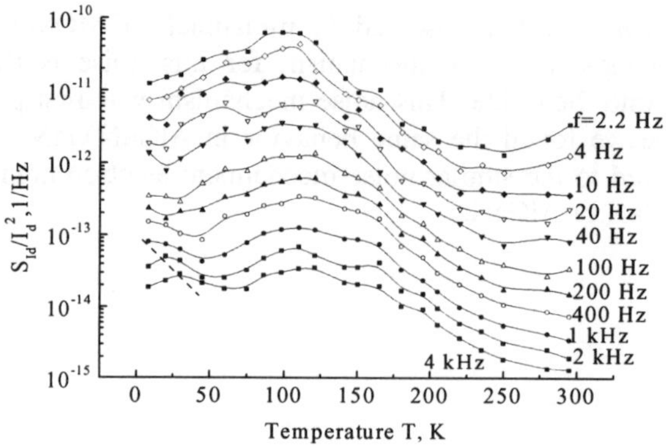

Fig.5. Temperature dependences of noise at different frequencies for the doped channel HFET.

4.Conclusions

GaN-based transistors (both MESFETs and HFETs) are characterized by rather low level of the 1/f noise with α values of ~ $10^{-3}\div10^{-5}$. These magnitudes of α are comparable with those for GaAs/GaAlAs HFETs.

The measurements of the low frequency noise on GaN MESFETs, MOSFETs, HFETs and MOSHFETs showed that the noise properties of all these devices are similar and that the drain and source contacts do not contribute much to the low frequency noise. Analysis of the gate current fluctuations showed that gate current gives only small contribution to the low frequency noise, if any. The dependence of the noise on gate voltage in GaN FETs indicates that the noise originates from the bulk of GaN in the channel and in the source to gate and drain to gate regions.

The, concentration dependence of Hooge parameter α and temperature dependence of noise in HFETs can be explained by the electron tunneling from 2D channel to GaN or AlGaN.

Acknowledgements

The work at RPI was supported by the Office of Naval Research. Project monitor Dr. Harry Dietrich. The work at USC was supported by the Ballistic Missile Defense Organization (BMDO) under Army SSDC contract DASG60-98-1-0004, monitored by Dr. Brian Strickland and Dr. Kepi Wu. The work at SET, Inc. was supported by BMDO under SBIR Phase II contract F33615-01-C-1911 and monitored by Dr. J. Gillespie of WPAFB.

References

1. S. J. Pearton, F. Ren, "GaN electronics", Advanced Materials **12**, (2000) 1571 -1580
2. G. Simin, X. Hu, N. Ilinskaya, J. Zhang, A. Tarakji, A. Kumar, J. Yang, Asif M. Khan, R. Gaska, M. S. Shur, "Large periphery high-power AlGaN/GaN metal-oxide-semiconductor heterostructure field effect transistors on SiC with oxide-bridging", IEEE Electron Device Letters, **22**, (2001) 53-55

3. S. L. Rumyantsev, N. Pala, M. S. Shur, and E. Borovitskaya, A. P. Dmitriev and M. E. Levinshtein, R. Gaska, M. Asif Khan, J. Yang, X. Hu, and G. Simin, "Generation-Recombination Noise in GaN/ AlGaN Heterostructure Field Effect Transistors", IEEE Trans. on ED, **48**, (2001) pp.530-534
4. P. O. Lauritzen, "Low-frequency generation noise in junction field effect transistors", Solid State Electr., **8**, (1965) 41-58.
5. J. Graffeuil, K. Tantrarongroj, and J. F. Sautereau, "Low frequency noise physical analysis for the improvement of the spectral purity of GaAs FETs oscillators", Solid State Electr., **25**, (1982) 367-374,
6. M. Chertouk and A. Chovet, "Origins and characterization of low-frequency noise in GaAs MESFETs grown on InP substrates", IEEE Trans. Electr. Dev., **43**, (1996) 123-129
7. L. Dobrzanski and Z. Wolosiak, "On the origin of low frequency noise in GaAs metal-semiconductor field-effect transistors", J. Appl. Phys., **87**, (2000) 517-521
8. S. L. Rumyantsev, N. Pala, M. S. Shur and R. Gaska, M. E. Levinshtein, M. Asif Khan, G. Simin, X. Hu, and J. Yang, "Effect of gate leakage current on noise properties of AlGaN/GaN field effect transistors", J. of Appl. Phys., **88**, (2000) 6726-6730
9. S. L. Rumyantsev, N. Pala, M. S. Shur, R. Gaska, M. E. Levinshtein, M. A. Khan, G. Simin, X. Hu, J. Yang, "Low frequency noise in GaN metal semiconductor and metal oxide semiconductor field effect transistors", J. Appl. Phys., **90**, (2001) 310-314
10. A. van der Ziel, " Carrier density fluctuation noise in field-effect transistors", Proc. IEEE, (1963) 1670
11. F. N. Hooge, "1/f noise sources", IEEE trans. Electron Dev., **41**, (1994) 1926-1935
12. M. E. Levinshtein, S. L. Rumyantsev, D.C. Look R. J. Molnar, M. Asif Khan, G. Simin, V. Adivarahan , and M. S. Shur, "Low frequency noise in n-GaN with high electron mobility", J. Appl. Phys., **86**, (1999) 5075-5078.
13. J. A. Garrido, B. E. Foutz, J. A. Smart, J. R. Shealy, M. J. Murphy, W. J. Schaff, and L. F. Eastman, "Low-frequency noise and mobility fluctuations in AlGaN/GaN heterostructure field-effect transistors," Appl. Phys. Lett., **76,** (2000) pp.3442-3444
14. S. L. Rumyantsev, N. Pala, M. S. Shur, R. Gaska, M. E. Levinshtein, M. A. Khan, G. Simin, X. Hu, and J. Yang, "Effect of gate leakage current on noise properties of AlGaN/GaN field effect transistors", J. Appl. Phys., **88**, (2000) 6726-6730
15. J. Berntgen, K. Heime, W. Daumann, U. Auer, F.-J. Tegude, and A. Matulionis, "The 1/f noise of InP based 2DEG devices and its dependence on mobility", IEEE Trans Electron Devices, **46**, (1999) 194-203
16. M. A. Py, and H.-J. Buehlmann, "Evidence for screening effects on the 1/f current noise in GaAs/AlGaAs modulation doped field effect transistors", J. Appl. Phys. **80**, (1996) 1583-1593
17. S. L. Rumyantsev, N. Pala, M. S. Shur, and R. Gaska, M. E. Levinshtein, P. A. Ivanov, M. Asif Khan, G. Simin, X. Hu, and J. Yang, "Concentration dependence of the 1/f noise in AlGaN/GaN Heterostructure Field Effect Transistors", Semicond. Science and Techn, 2002, to be published
18. L. K. J. Vandamme, X. Li, and D. Rigaud, "1/f noise in MOS devices, mobility or number fluctuations?", IEEE Transaction Electron Device*s* **41**, (1994) 1936 – 1945

19. A. L. McWhorter, "1/f noise and germanium surface properties", in Semiconductor Surface Physics, R.H. Kingston, ed., Philadelphia PA, Univ. of Pennsylvania Press, 1957, pp.207-228
20. L. Ren and M. R. Leys, "1/f noise in an $Al_XGa_{1-X}As/GaAs$ heterostructure between 77 and 300 K", Physica B, **192**, (1993) 303-310

International Journal of High Speed Electronics and Systems, Vol. 12, No. 2 (2002) 459–468

HOT-PHONON LIMITED ELECTRON ENERGY RELAXATION IN AlN/GaN

A. MATULIONIS*, J. LIBERIS, L. ARDARAVIČIUS
Semiconductor Physics Institute, A. Goštauto 11
Vilnius 2600, Lithuania

J. SMART
RF Micro Devices, 10420 Harris Oaks Blvd.
Charlotte, North Carolina 28269, USA

D. PAVLIDIS, S. HUBBARD
Department of Electrical Engineering and Computer Science
The University of Michigan, 1301 Beal Avenue
Ann Arbor, Michigan 48109–2122, USA

L. F. EASTMAN
School of Electrical and Computer Engineering, Cornell University,
425 Phillips Hall, Ithaca, New York 14853–5401, USA

Microwave noise technique is applied to study energy dissipation in an AlN/GaN heterostructure containing a two-dimensional electron gas channel. Measurements of the dissipated power and the noise temperature are performed at 80 K lattice temperature in the electric field range up to 40 kV/cm. The energy relaxation time is found to decrease from 40 ps to 0.55 ps when the bias is increased. The experimental data are discussed in the electron temperature approximation assuming electron energy dissipation on optical phonons and hot-phonon effects. Dependencies of the hot-phonon number and the hot-phonon temperature on the hot-electron temperature are deduced. The frequency cutoff imposed by the limited energy dissipation through optical phonons is estimated.

Keywords: Microwave noise, AlN/GaN, high electric field, hot electrons, hot phonons, energy relaxation time, cutoff frequency

1. Introduction

The best high-power performance at microwave frequencies has been demonstrated on high electron mobility transistors prepared from nitride heterostructures [1]. At a high power, hot-electron effects become important, and investigation of hot-electron energy dissipation is among major issues. The energy dissipation can be studing using different experiments, including those of femtosecond optical pump and probe [2], of luminescence [3], of Shubnikov-de Haas magnetoresistance [4], of thermal pulse propagation [5], of microwave signal mixing [6], and of microwave noise [7, 8,

*Corresponding author E-mail: matulionis@uj.pfi.Lt

9]. Ultra-short relaxation time of 5–10 fs for optical-phonon emission by a high-energy electron in GaN is deduced from heat pulse phonon propagation experiments [5]. Essentially longer hot-carrier relaxation time constants are obtained using the pump and probe technique: 0.5 ps for GaN [2], 0.2 ps for GaN [10], and 0.5 ps for InGaN [11]. The difference may be caused by hot-phonon effects appearing in the laser experiments at a high level of excitation.

Shubnikov-de Haas [12] and noise [13] experiments have been succesfully performed on AlGaN/GaN heterostructures containing two-dimensional electron gas (2-DEG) channels subjected to electric field applied in the plane of electron motion. The magnetoresistance experiments at low temperatures [12] are well suited for investigation of acoustic-phonon emission by electrons. This energy dissipation mechanism dominates when the electron temperature is low. Unlike this, when the electron temperature is high, the main energy dissipation mechanism is optical-phonon emission. Since the SdH technique fails when the electron and the lattice temperatures are high, the noise technique is preferable for high-power studies: there is no temperature limitation for microwave noise experiments. Indeed, hot-electron noise and energy dissipation have been studied in AlGaN/GaN heterostructures at room temperature in the field range up to 10 kV/cm [13]. The field range is limited by an onset of real-space transfer noise. The real-space transfer can be eliminated by substituting AlN for AlGaN.

The main aim of the work is to investigate energy dissipation on optical phonons in AlN/GaN heterostructures at strong electric fields corresponding to high electron temperatures.

2. Definitions

The applied voltage and the current supply energy to the electron system. The input power gained by an average electron can be estimated as:

$$P_{\mathrm{g}} = IU/N_{\mathrm{e}}, \tag{1}$$

where I is the current, U is the voltage applied to the channel, and N_{e} is the number of electrons in the channel. The number can be estimated from the sheet electron density $n_{2\mathrm{D}}$, and the surface area A of the 2-DEG channel: $N_{\mathrm{e}} = n_{2\mathrm{D}}A$.

The gained power is dissipated through inelastic collisions of the electrons with phonons or in some other way, for example, through exit of hot electrons into contacts. Under a steady state, the gained power equals the dissipated power, $P_{\mathrm{g}} = P_{\mathrm{d}}$. Because of a high density of electrons in the 2-DEG channel, it is reasonable to assume that the electron distribution is governed by the electron temperature T_{e}. In this case, for an ideal 2-DEG in the energy relaxation time approximation, the dissipated power (per electron) is:

$$P_{\mathrm{d}} = k_{\mathrm{B}}(T_{\mathrm{e}} - T_0)/\tau_{\varepsilon}, \tag{2}$$

where e is the elementary charge, τ_{ε} is the energy relaxation time.

The electron temperature is the measure of the chaotic motion of hot electrons, and it can be estimated from hot-electron noise measurements. The equivalent noise temperature T_{n} is related to the measured spectral intensity of the available noise power dP_{a}/df according to:

$$dP_{\mathrm{a}}/df = k_{\mathrm{B}}T_{\mathrm{n}}, \tag{3}$$

where f is the frequency and k_{B} is the Boltzmann constant.

In the electron temperature approximation, one has [9]:

$$\begin{aligned} T_{\mathrm{n}\perp} &= T_{\mathrm{e}}, \\ T_{\mathrm{n}\parallel} &= T_{\mathrm{e}}\{1 + C(T_{\mathrm{e}})\}, \end{aligned} \tag{4}$$

where $T_{\mathrm{n}\perp}$ is the transverse noise temperature measured in the 2-DEG plane in the direction perpendicular to the applied electric field, $T_{\mathrm{n}\parallel}$ is the longitudinal noise temperature measured in the direction of the field, and $C(T_{\mathrm{e}})$ is the correction term that takes into account the noise from the sources appearing in the bias direction only. Real-space-transfer noise, intervalley transfer noise, impact ionization noise, and some others can be important at high electric fields [14]. In the range of sub-threshold fields for these sources, the main correction goes for electron temperature fluctuations. For a linear dependence of current on voltage, the correction term $C(T_{\mathrm{e}})$ due to the electron temperature fluctuations is zero. The correction term remains small until the deviations from the Ohm law become essential. As a result, at not too high densities of 2-DEG [15], one can assume that $T_{\mathrm{n}\parallel} \cong T_{\mathrm{e}}$ in the range of fields where the other sources of noise and the associated non-linearity of the current-voltage characteristic are not important.

3. Energy dissipation on optical phonons

For emission of an optical phonon the electron energy should exceed the phonon energy, and the energy loss on optical phonons dominates at electron temperatures exceeding certain critical value. For GaN at 1.5 K lattice temperature the critical electron temperature is 70 K [5]. This value can be somehow higher in a quantum-well channel where electron scattering on acoustic phonons is not quasielastic [16].

At a low lattice temperature ($T_{\mathrm{L}} \ll \hbar\omega_0/k_{\mathrm{B}}$) in the electron temperature approximation, the mean power dissipated by an electron through spontaneous emission of optical phonons can be estimated as follows:

$$P_{\mathrm{d}} = \frac{\hbar\omega_0}{\tau}\exp\left(-\frac{\hbar\omega_0}{k_{\mathrm{B}}T_{\mathrm{e}}}\right), \tag{5}$$

where the second factor accounts for the number of electrons able to emit optical phonons, $\hbar\omega_0$ is the optical-phonon energy (the energy lost in an event of emission) and $1/\tau$ is the number of emission events per second (spontaneous emission rate). Equation (5) neglects hot phonons and stimulated emission.

Supposing that phonons are present in the phonon states involved in the optical-phonon emission, the stimulated emission takes place in addition to the spontaneous one. In this case, the electron can absorb an optical phonon as well. As a result, the net dissipated power is:

$$P_{\rm d} = (1 + N_{\rm ph})\frac{\hbar\omega_0}{\tau}\exp\left(-\frac{\hbar\omega_0}{k_{\rm B}T_{\rm e}}\right) - N_{\rm ph}\frac{\hbar\omega_0}{\tau}, \quad (6)$$

where the first term takes into account the spontaneous and stimulated emission, and the second term stands for the absorption.

4. Experimental

The investigated AlN/GaN structures consisted of a strained 10 nm AlN layer deposited onto a $\sim 1\mu$m GaN buffer layer grown onto an Al_2O_3 substrate. For details on growth using low-pressure organometalic vapor phase epitaxy and the heterostructure characterization see [17-19]. Evaporated Ti/Al/Ti/Au metalization was used for ohmic contacts. A polarization and strain induced 2-DEG channel formed in GaN. The 2-DEG contained $n_{2D} \sim 5 \cdot 10^{12}$cm^{-2} sheet electron density. Samples for noise measurements were cut from transmission line model (TLM) patterns.

The 2-DEG channels were subjected to a pulsed electric field applied parallel to the interface. The microwave noise power and the current were measured in the direction of the applied field at 80 K temperature. The 10 GHz frequency was high enough to register noise due to hot-electron velocity fluctuations avoiding contributions from generation–recombination and flicker noise. The available noise power of the hot-electron source was determined from the data on the emitted noise power taking into account sample mismatch, contribution of noise sources outside the 2-DEG channel (due to the contact resistance), and contributions of microwave circuit elements (for details see [13]).

5. Dissipated power, noise temperature, and energy relaxation time

Figure 1 presents the experimental dependence of the power gained by the 2-DEG from the applied electric field. The gained power increases by approximately 10^6 times when the electric field is increased by 10^3 times. The exact proportionality to the E^2 (solid line in Fig. 1) would mean that Ohm's law held. In fact, the deviations from Ohm's law are present. However, the deviations are not essential in the investigated range of fields. This gives the experimental evidence that the assumption $T_{\rm n} \cong T_{\rm e}$ is applicable in our case.

The hot-electron noise measurements provide us with the data on the equivalent noise temperature. It is given as a function of the gained power in Fig. 2. When the electron temperature is low, a small increase in the gained power changes the electron temperature considerably. This steep increase turns into a slower one when the chaotic energy of the electrons $k_{\rm B}T_{\rm e}$ approaches the optical-phonon energy $\hbar\omega_0$. In AlN/GaN heterostructure, the energies of the interfacial and confined modes of the optical phonons emitted by the 2-DEG electrons are not well known, but it is reasonable to assume that they are close to that of the optical phonon in GaN. (The

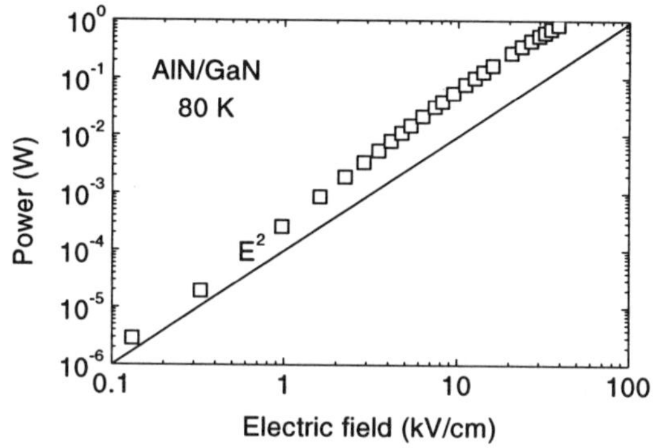

Fig. 1. Dependence of gained power on electric field in a 2-DEG channel of AlN/GaN heterostructure at 80 K lattice temperature: squares are experimental data, line corresponds to the squared dependence. The sample electrode area is $50 \cdot 50\ \mu m^2$, the interelectrodal distance is 3 μm.

optical-phonon energy $\hbar\omega_0 = 92$ meV in GaN [5]).

Sometimes it is convenient to consider the dissipated power in terms of the energy relaxation time. As mentioned, there is the steep increase of the noise temperature with the gained power at a low bias (Fig. 2). This means that the power dissipation is not efficient, and, according to Eq. (2), this corresponds to a long relaxation time

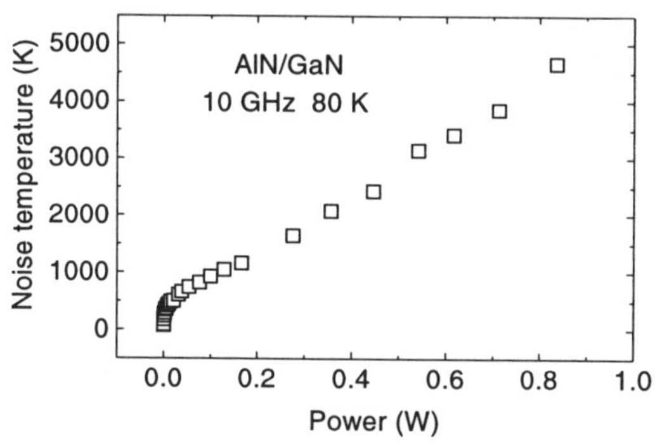

Fig. 2. Dependence of longitudinal noise temperature on gained power for the AlN/GaN 2-DEG channel at 80 K lattice temperature.

τ_ε at a low bias. Figure 3 presents the experimental values of the energy relaxation time obtained, after Eqs. (1) and (2), according to the expression

$$\tau_\varepsilon = \frac{k_B(T_n - T_0)}{P_d}, \tag{7}$$

where $P_d = P_g = IU/N_e$. The energy relaxation time decreases down from the values close to 40 ps at the lowest bias and reaches 0.55 ps at the highest bias.

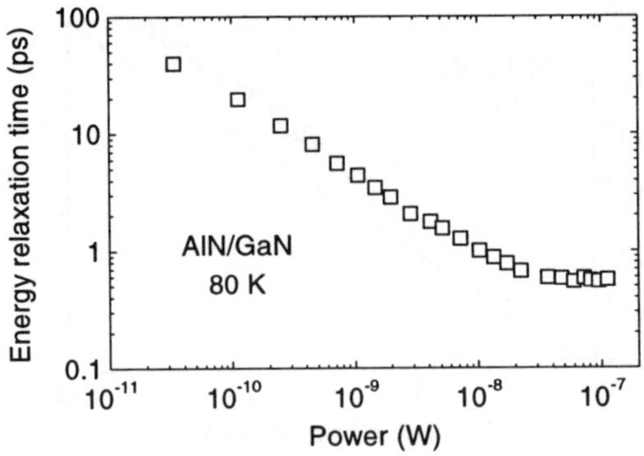

Fig. 3. Dependence of energy relaxation time on power gained by an average electron for the AlN/GaN 2-DEG channel at 80 K lattice temperature. The electron number is $N_e = 7.5 \cdot 10^6$.

6. Discussion

6.1. *Hot-phonon number and temperature*

Under steady state, the gained power equals the dissipated one. According to Eq. (5), the dissipated power is an exponential function of the inverse electron temperature. Therefore, it is convenient to plot the logarithm of the dissipated power as a function of the inverse temperature. In Fig. 4, symbols stand for the experimental data, and solid line is plotted after Eq. (5) where τ = 10 fs and $\hbar\omega_0$ = 0.1 eV (cf. available data for GaN [5]).

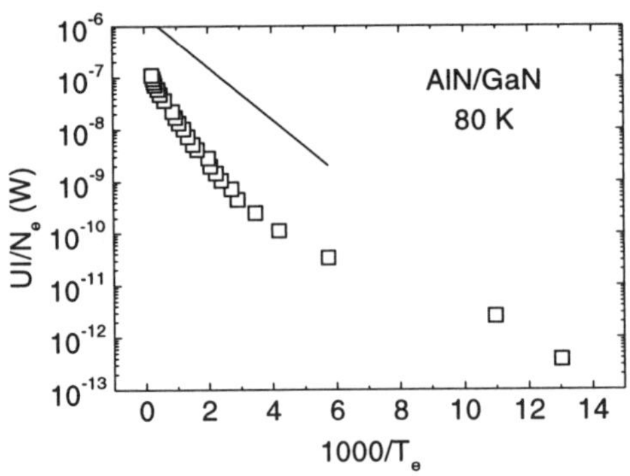

Fig. 4. Dependence of the power gained by an average electron on the inverse electron temperature for the AlN/GaN 2-DEG channel at 80 K lattice temperature: squares are experimental data, line is Eq. (5) with τ = 10 fs and $\hbar\omega_0$ = 0.1 eV. $N_e = 7.5 \cdot 10^6$.

The experimental energy dissipation rate is essentially lower than the rate predicted by Eq. (5). This misfit would be even larger if one took into account more energy loss mechanisms. Some correction could be introduced assuming that the

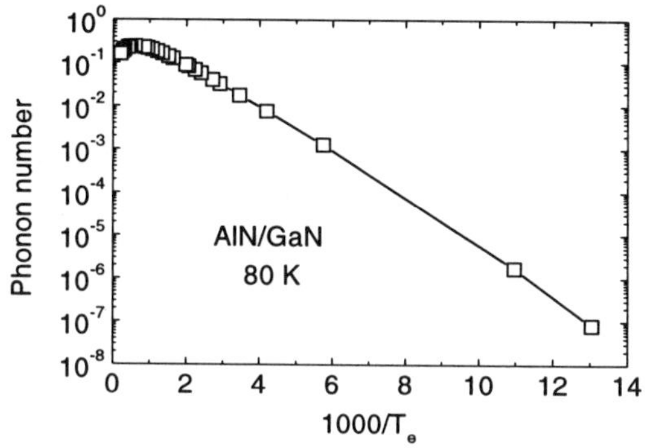

Fig. 5. Dependence on the inverse electron temperature of the equivalent phonon number (per state involved in electron–phonon interaction) for the AlN/GaN 2-DEG channel at 80 K lattice temperature.

interfacial phonons and the phonons located in the quantum well have different energies and different spontaneous emission rates than the phonons located in bulk GaN. However, this leads to corrections that do not eliminate the misfit.

We think that the assumption $N_{\mathrm{ph}} \ll 1$ is not acceptable for the interpretation of the experimental data. Thus, one should use Eq. (6) rather than Eq. (5). Supposing that the optical-phonon energy and the spontaneous emission rate are known, Eq. (6) can be solved in respect to N_{ph}, that is, in respect to some average number of the optical phonons in the phonon states that are involved in the electron–phonon interaction. For $\tau = 10$ fs and $\hbar\omega_0 = 0.1$ eV, the experimental

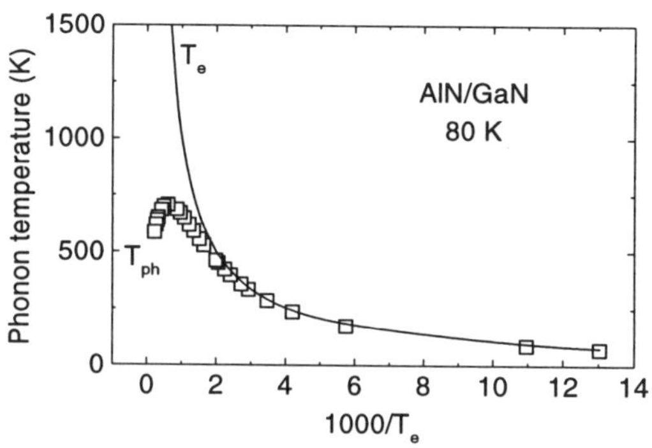

Fig. 6. Dependence on the inverse electron temperature of the equivalent optical-phonon temperature in the phonon states involved in electron–phonon interaction (squares), compared with the electron temperature (line), for the AlN/GaN 2-DEG channel at 80 K lattice temperature.

results of Fig. 4 and Eq. (6) lead to the equivalent phonon number plotted in Fig. 5. The phonon number is low at low electron temperatures. However, values over 0.2 are obtained at the electron temperatures exceeding 1000 K.

The results on the phonon number can be interpreted in terms of the optical-phonon temperature. Figure 6 compares the optical-phonon temperature (symbols) with the electron temperature (line). It is noteworthy, that the temperatures of the electrons and the emitted optical phonons are very close each other when the electron temperature is not high. On the other hand, the curves spread apart at the electron temperature exceeding 500 K. Moreover, the optical-phonon temperature starts decreasing as the electron temperature exceeds 1000 K. To our knowledge, a similar hot-phonon behavior has not been reported.

There is a simple explanation for the decrease of the equivalent optical-phonon temperature at very high electric fields. At a not-too-high electric field optical-

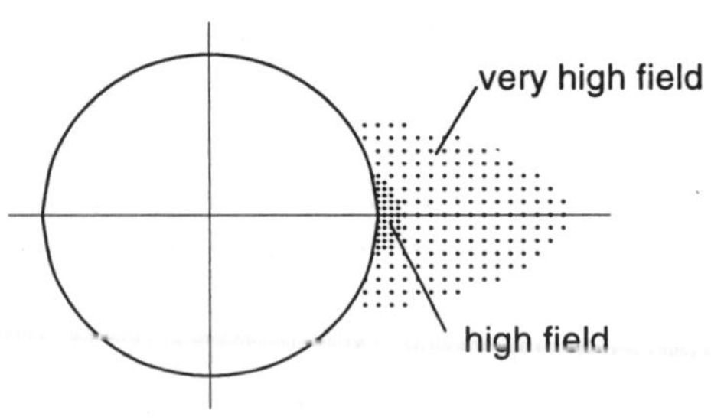

Fig. 7. The plane of electron momenta showing the isoenergetic circle $\varepsilon = \hbar\omega_0$ (solid line) and the electron states before optical-phonon emission (dots).

phonon emission take place when the electron energy $\varepsilon \cong \hbar\omega_0$ (densely distributed dots in Fig. 7). As a result, all emitted optical phonons look similar, they crowd in a small area of their momenta space, and their number in these phonon states is high. On the other hand, at a *very high electric field*, the electrons reach energies that exceed the optical-phonon energy considerably. In this case, the electron states before the emission spread over a wide area of their momenta plane (rarely distributed dots in Fig. 7). As a result, the momenta space occupied by the emitted phonons is large. Since the space increases with the electric field faster than the total number of the emission events per unit time, the optical-phonon overpopulation in the involved phonon states tends to decrease when the electric field is increased beyond some critical value.

6.2. *Cutoff frequency*

We have demonstrated that the hot phonons have an effect on the energy dissipation rate and the energy relaxation time. Data of Fig. 4 illustrate that the energy dissipation rate is much lower than that due the spontaneous optical-phonon emission rate estimated using Eq. (5). Accordingly, Fig. 3 shows that the minimum value of 0.55 ps obtained for the energy relaxation time exceeds considerably the inverse rate of the spontaneous optical-phonon emission (the inverse rate is near 10 fs in GaN when estimated from the heat-pulse propagation and other experiments [5]). Let us recall that 0.5 ps value for the relaxation time has been obtained for GaN [2] and InGaN [11] in the laser experiments.

Hot phonons seem to have an effect on the limit of the cutoff frequency imposed by the electron energy dissipation. The cutoff frequency f_c can be estimated using the energy relaxation time data according to $f_c = 1/(2\pi\tau_\varepsilon)$. Figure 8 presents the

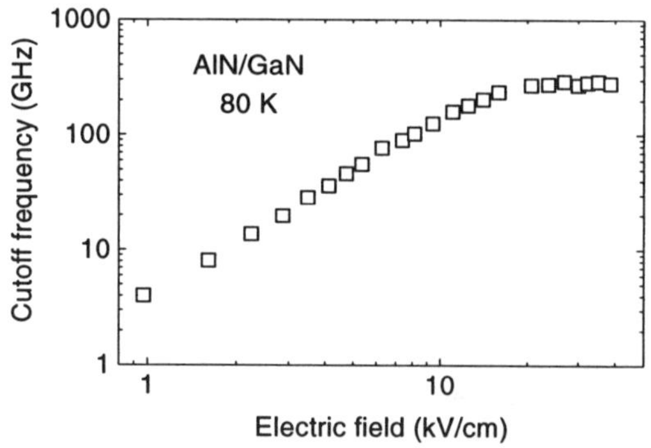

Fig. 8. Dependence of the cutoff frequency on electric field for the AlN/GaN 2-DEG channel at 80 K lattice temperature.

results for the investigated AlN/GaN 2-DEG channels. According to data of Fig. 8, the channels support a high-power operation with cutoff frequency near 270 GHz. However, because the energy relaxation is quite slow at a low bias, an essentially lower cutoff frequency is expected for a low-power operation.

7. Summary

Electron energy dissipation on optical phonons is studied in AlN/GaN channels subjected to a high electric field applied in the 2-DEG plane. When the bias is increased, the electron energy relaxation time decreases until saturates at fields exceeding 20 kV/cm. The saturation is associated with hot-phonon effects. A non-monotonous dependence of the hot-phonon temperature on the bias is observed.

Acknowledgements

The authors from Vilnius acknowledge support from the Office of Naval Research (Award No. N00014-01-1-0828) and from the Lithuanian National Foundation for Science and Education (Contract No. V-019-2001). One of us (A.M) acknowledges support from Army Research Laboratory-European Research Office, United States Army (PO No. N68171-02-M-5070). Work is also supported by ONR (Contract No. N00014-92-J-1552, N00014-00-1-0879) and DARPA/ONR (Contract No. N00014-99-1-0513).

References

1. L. F. Eastman, "Performance limits of AlGaN/GaN HEMT's". In this volume
2. C. -K. Sun, Y. -L. Huang, S. Keller, U. K. Mishra, and S. P. DenBaars, "Ultrafast electron dynamics study of GaN", *Phys. Rev. B* **59** (1999) 13535–13538.
3. J. Shah, "Ultrafast luminescence studies of carrier relaxation and tunneling in semiconductor nanostructures" in: *Hot Carriers in Semiconductor Nanostructures: Physics & Applications*, ed. Jagdeep Shah, Academic Press, Boston, 1992, pp. 279–312.
4. K. Hirakawa and H. Sakaki, "Energy relaxation of two-dimensional electrons and the deformation potential constant in selectively doped AlGaAs/GaAs heterojunctions", *Appl. Phys. Lett.* **49** (1986) 889–891.
5. N. M. Stanton, A. J. Kent, A. V. Akimov, P. Hawker, T. S. Cheng, and C. T. Foxon, "Energy relaxation by hot electrons in n-GaN epilayers", *J. Appl. Phys.* **89** (2001) 973–979.
6. A. Verevkin, E. M. Gershenzon, G. N. Gol'tsman, N. G. Ptitsina, G. M. Chulkova, K. S. Smirnov, and R. Sobolewski, "Direct measurements of energy relaxation times in two-dimensional structures under quasi-equilibrium conditions", *Material Science Forum,* **384-385**, (2002) 107–116.
7. H. Hartnagel, R. Katilius R., and A. Matulionis, *Microwave Noise in Semiconductor Devices*, Wiley, New York, 2001, 312 p.
8. A. Matulionis, "Noise, hot carrier effects", in: *Wiley Encyclopedia of Electrical and Electronics Engineering*, ed. John G. Webster, Wiley, New York, 1999, vol. 14, pp. 410–428.
9. A. Matulionis, V. Aninkevičius, J. Liberis, I. Matulionienė, J. Berntgen, K. Heime, H. L. Hartnagel, "Hot-electron energy relaxation, noise, and lattice strain in InGaAs quantum well channels", *Appl. Phys. Lett.* **74** (1999) 1895–1897.
10. H. Ye, G. W. Wicks, and P. M. Fauchet, "Hot electron relaxation time in GaN", *Appl. Phys. Lett.* **74** (1999) 711-713.
11. C. -K. Sun, F. Vallée, S. Keller, J. E. Bowers, and S. P. DenBaars, "Femtosecond studies of carrier dynamics in InGaN", *Appl. Phys. Lett.* **70** (1997) 2004-2006.
12. K. J. Lee, J. J. Harris, A. J. Kent, T. Wang, S. Sakai, D. K. Maude, and J.-C. Portal, "Investigation of phonon emission processes in an AlGaN/GaN heterostructure at low temperatures", *Appl. Phys. Lett.* **78** (2001) 2893–2895.
13. A. Matulionis, J. Liberis, L. Ardaravičius, M. Ramonas, I. Matulionienė, J. Smart, "Hot-electron energy relaxation time in AlGaN/GaN" *Semicond. Sci. Technol.* **17** (2002) (in press).
14. A. Matulionis, V. Aninkevičius and J. Liberis, "Hot-electron velocity fluctuations in 2-DEG channels", *Microelectronics Reliability* **40** (2000) 1803–1814.
15. J. Liberis, L. Ardaravičius, and A. Matulionis, "Longitudinal and transverse noise of hot electrons in 2DEG channels", in: *Noise in Physical Systems and 1/f Fluctuations ICNF 2001*, ed. Gijs Bosman, World Scientific, Singapore, 2001, pp. 513–516.
16. P. J. Price, "Hot electrons in a GaAs heterolayer at low temperature", *J. Appl. Phys.* **53** (1982) 6863–6866.
17. A. Eisenbach, E. Alekseev, and D. Pavlidis, "Growth and Characterization of AlN/GaN HFETs" *Ninth Biennial Workshop on Organometallic Vapor Phase Epitaxy (OMVPE '99)*, Ponte Vedra Beach, Florida, May 23-27, 1999, Session X, Devices II.
18. E. Alekseev, A. Eisenbach, and D. Pavlidis, "Low Interface State Density AlN/GaN MISFETs" *Electronics Letters* **35** (1999) 2145–2146.
19. S. Hubbard, D. Pavlidis, V. Valiaev, and A. Eisenbach, "Growth and Characterization of AlN/GaN Heterostructures Using Low-Pressure Organometallic Vapor Phase Epitaxy" *Electronic Materials Conference*, University of Notre Dame, Indiana, June 27-29, 2001.

International Journal of High Speed Electronics and Systems, Vol. 12, No. 2 (2002) 469–478

POLAR-OPTICAL PHONON ENHANCEMENT OF HARMONIC GENERATION IN SCHOTTKY DIODES

B. Gelmont[1], D. Woolard[2], and S. Chen[1]
[1] University of Virginia, Charlottesville, VA
[2] U.S. Army Research Laboratory, ARO, RTP, NC

Polar-optical phonons have an important influence on polar semiconductor device behavior, in polar crystals at very high frequency, because they directly perturb the effective material permittivity in the vicinity of the transverse polar-optical vibration frequency. This phenomenon leads to dramatic nonlinear effects on the resistive and the reactive physics at the Schottky barrier interface. Hence this physical effect provides a mechanism with the potential for producing significant levels of power at higher harmonics. We have established a simulator that combines the modified harmonic-balance circuit analysis technique with a polar-optical phonon hydrodynamic transport model to determine the maximum power generation and/or power efficiency in the second harmonic. An abrupt junction model that includes the polar-optical phonon dynamics in a depletion layer was developed and utilized in this study. Simulation results have revealed a dramatic influence on the second harmonic power of the Schottky diode at the transverse polar-optical frequency and in the vicinity of one-half of the transverse polar-optical frequency. The transverse polar-optical frequency in GaAs is 7.78 THz and this is a rather high for practical applications. Bismuth trisulfide (Bi_2S_3) has been identified as a polar material that has the transverse polar-optical frequency on the order of 300 GHz, which is in a range more suited to THz-frequency multiplier applications. The dependence of the second harmonic power of the Schottky diode on the frequency demonstrates two enhanced peaks in the vicinity of the transverse polar-optical frequency. These preliminary results suggest that Bi_2S_3 Schottky diode based multipliers have potential as enhanced sources within the terahertz regime.

1. Introduction

Polar-optical phonons play a significant role at high frequencies in introducing a frequency dependent permittivity that dramatically affects the nonlinearity in capacitance of a material near its transverse polar-optical vibration frequency. Specific to multiplier devices, these enhanced nonlinearities can be harnessed to produce higher harmonic generation of very high frequency signals yielding better output power and output efficiency in multiplier circuits. As will be demonstrated, this effect might be useful in more efficiently extending multiplier-based devices into the terahertz regime. The Schottky diode continues to be an important device in high frequency harmonic generation due to the fundamental advantages in its basic structure. This paper demonstrates the influences introduced by polar-optical vibrations and how they can be applied in obtaining higher efficiency in Schottky diodes. The incorporation of polar-optical phonon physics into the Schottky diode abrupt junction model was developed and applied in this work. Simulation of a GaAs-based Schottky diode multipliers was performed to show the dramatic enhancement of output efficiency in the vicinity of the polar-phonon frequency. However, it was desirable to find materials with approximately the same forbidden gap as GaAs, where we expect the Schottky diode barrier height to be of the same order as in GaAs, but with a lower transverse polar-phonon frequency. Hence, to practically utilize these novel nonlinear effects related to polar-optical vibration within terahertz region, new materials were required. One of the candidates is Bi_2S_3, which has the transverse polar-optical frequency on the order of 300 GHz. Results from a study on a Bi_2S_3-based Schottky diode multiplier will also be presented to highlight the phonon effect at a lower polar-phonon frequency. Hence, this paper reports on simulations for GaAs and Bi_2S_3 materials systems applied towards Schottky diode based harmonic multipliers.

2. Dielectric Permittivity dependence on frequency

It is known that the perturbations of polar-optical phonons dramatically change the effective material permittivity of polar semiconductors in the vicinity of its transverse polar-optical vibration frequency[1]. The effective semiconductor permittivity with the inclusion of lattice vibrations is a function of angular frequency given by

$$\varepsilon = \varepsilon_{\rm real} + i\varepsilon_{\rm imag}, \tag{1}$$

where

$$\varepsilon_{\rm real} = \varepsilon_\infty + \frac{(\varepsilon_0 - \varepsilon_\infty)(1-\omega^2/\omega_{\rm to}^2)}{(1-\omega^2/\omega_{\rm to}^2)^2 + (2\gamma\omega/\omega_{\rm to}^2)^2}, \tag{2a}$$

$$\varepsilon_{\rm imag} = \frac{(\varepsilon_0 - \varepsilon_\infty)(2\gamma\omega/\omega_{\rm to}^2)^2}{(1-\omega^2/\omega_{\rm to}^2)^2 + (2\gamma\omega/\omega_{\rm to}^2)^2}, \tag{2b}$$

with ε_0 as the zero-frequency electric permittivity, ε_∞ as the high-frequency permittivity, ω_{to} is the transverse polar-optical frequency, and 2γ is the phonon damping factor. In the case of GaAs at room temperature, the physical parameters are known to be $\omega_{to}/2\pi$=7.78THz, $\omega_{lo}/2\pi$=8.47THz, $2\gamma = 7.0 \times 10^{-3}\omega_{to}$, $\varepsilon_0 = 12.9\varepsilon_{vac}$, and $\varepsilon_\infty = 10.89\varepsilon_{vac}$ where ε_{vac} is the dielectric permittivity in a vacuum[2]. The dependence of permittivity on frequency for GaAs is shown in Fig. 1.

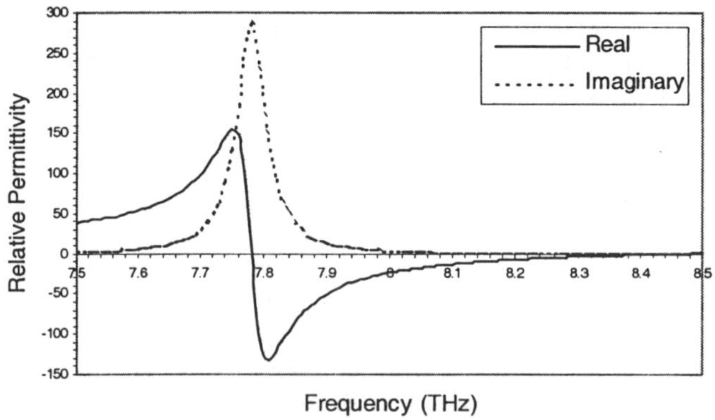

Fig. 1. Relative Permittivity dependence on frequency of GaAs at room temperature

It is apparent that the frequency of operation in transport-based device applications is significantly lower than the transverse polar-phonon frequency ω_{to} found in GaAs. For example, very high frequency spectroscopic studies of interstellar molecular clouds need to be measured between 200GHz to 1600GHz frequency range, such as the case for C and CH^+ transition lines. Hence, it is desirable to find materials with approximately the same forbidden gap as GaAs, where we expect the Schottky diode barrier height to be of the same order as in GaAs but with a lower transverse polar-phonon frequency. Thus far, preliminary polar crystal material candidates have been screened for reasonable transport properties (i.e., adequate values of mobility) and lower transverse polar-phonon frequencies. Bismuth trisulfide (Bi_2S_3) has been identified as a polar material that meets such criteria. The physical parameters used to determine this dependence were obtained from experimental data of transverse optical mode parameters for Bi_2S_3 experimental crystals[3]. They are known to be $\omega_{to}/2\pi$=300GHz, $\omega_{lo}/2\pi$=514GHz, $\varepsilon_0=100\varepsilon_{vac}$, and $\varepsilon_\infty=34.05\varepsilon_{vac}$. Following from equations (1) and (2), the frequency dependence of permittivity in Bi_2S_3 is shown in Fig. 2 and Fig. 3 for 2γ factors equal to $0.1\omega_{to}$ and $0.01\omega_{to}$ respectively. In both cases the peak permittivity is shifted and lowered significantly to 300GHz , which is well below that of GaAs.

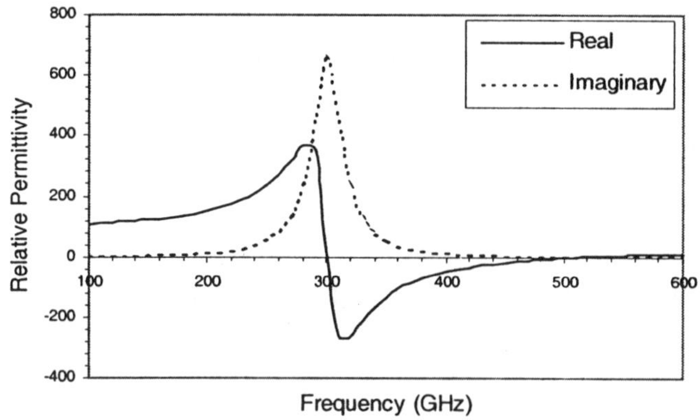

Fig. 2. Relative permittivity dependence on frequency of Bi_2S_3 at room temperature, $2\gamma=0.1\omega_{to}$

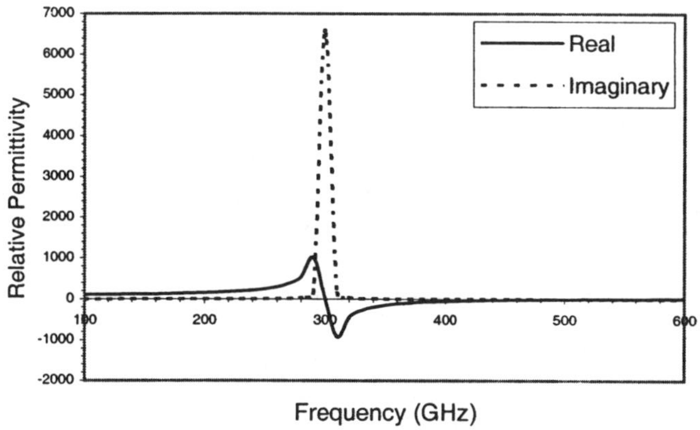

Fig. 3. Relative permittivity dependence on frequency of Bi_2S_3 at room temperature, $2\gamma=0.01\omega_{to}$

3. Increase of nonlinear effects and their utilization in multiplier

Frequency multipliers utilize nonlinear two-terminal devices to realize higher harmonic generation. Schottky diodes are feasible devices to be used as multipliers because of their voltage-dependent depletion-layer space charge. The Schottky diode can be utilized as a varactor, taking advantage of the nonlinear capacitance that can be used to convert a harmonic input signal into a periodic output signal containing harmonic components at multiples of the input frequency. Using the abrupt junction model given in Fig. 4,

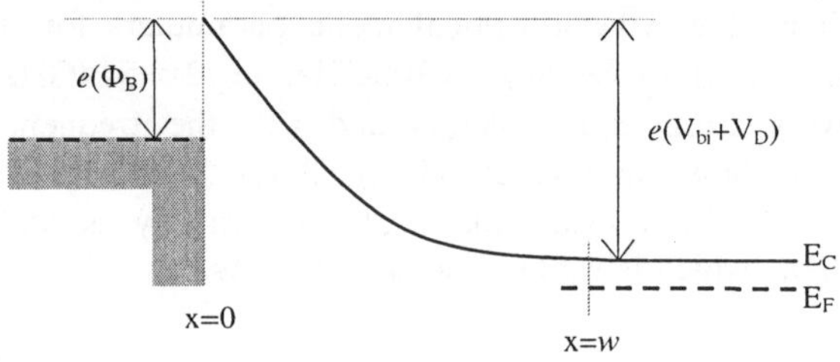

Fig. 4. The abrupt junction model and where the potential barrier is given by the dark solid curve.

the width (w) of the depletion region is given by

$$w = \left[\frac{2\varepsilon}{eN_d} (V_{bi} + V_D) \right]^{\frac{1}{2}}, \tag{3}$$

where N_d is the uniform donor doping density inside the epitaxial layer, ε is the permittivity of the semiconductor, V_{bi} is the built in potential barrier, e is the electron charge ,and V_D is the applied reverse-bias voltage (which is a function of time). Since the junction capacitance is only attributed to the charge in the depletion layer, the junction capacitance is

$$C_B = \frac{\varepsilon}{w}, \tag{4}$$

Thus, the abrupt Schottky diode approximation may be interpreted as a varactor, essentially a nonlinear capacitor with variable width depending on the applied voltage. At high frequencies without polar-optical phonons, the capacitance is a function of the nonlinearities due the changes in the depletion width as a function of time. Polar semiconductors possess intrinsic dipole vibrations that are responsible for the singularity of the dielectric permittivity that occurs at ω_{to}. From a physical standpoint, any device with a polar semiconductor between its contacts will result in a singularity as a function of the external input frequency. It is apparent then that the enhancement of the nonlinear capacitance is due to polar-optical phonons. The introduction of polar-optical phonons magnifies the nonlinearities in the vicinity of ω_{to} by the changes in the permittivity and therefore adds another nonlinear factor to the junction capacitance. These enhancements in the nonlinearity can be expected when any input harmonics are in the vicinity of the transverse polar-phonon frequency. These facts suggest a potential for enhancing the second harmonic generation in multiplier devices.

The simple equation, Eq. (3)-(4) can be used only when neglecting the dependence of the permittivity on frequency. However, it is understood that the spatial variation in the electric field will be influenced by polar-optical phonons in the neighborhood of the resonant vibrational frequency. A more detailed theory of the Schottky diode operation, taking into account polar-phonon effects, has been previously developed[4]. In addition to the Gauss's Law in the depletion region

$$\nabla \cdot \vec{D} = \nabla \cdot \left(\varepsilon_\infty \vec{F} + P \right) = eN_d , \tag{5}$$

where D is the displacement field, an equation for the vibrating polarization field can be written as

$$\ddot{\vec{P}} + 2\gamma \dot{\vec{P}} + \rho_m \omega_{lo}^2 \vec{P} = (\omega_{lo}^2 - \omega_{to}^2) \vec{D} , \tag{6}$$

Eq. (5) requires that the displacement field be a linear function of the one-dimensional z-coordinate space inside the depleted region. Furthermore, since Gauss's law requires the one-dimensional displacement field D to be identically zero everywhere outside the depletion region, defined here to be between z=0 and z=w, it is such that

$$D(z) = e N_d \left(z - w(t) \right), \quad 0 \le z \le w(t) , \tag{7}$$

where w is the width of the completely depleted layer. It can be found that the abrupt junction approximation barrier height is given by

$$V_b(t) = \frac{eN_d}{2\varepsilon_o} \left\{ w^2(t) + 2\frac{\varepsilon_o - \varepsilon_\infty}{\varepsilon_\infty} w(t)\left(w(t) - B(t)\right) \right\} , \tag{8}$$

where V_b, is the voltage across the barrier and B(t) is the stationary-state time-dependent component of lattice displacement obtained from

$$\ddot{B}(t) + 2\gamma \dot{B}(t) + \omega_{lo}^2 B(t) = \omega_{lo}^2 w(t) , \tag{9}$$

When including the emission current, the total current across the diode is given by

$$I_{total} = -eN_d A \frac{dw}{dt} - I_{sat} \left[1 - \exp\left(\frac{e(V_{bi} - V_b)}{kT} \right) \right] , \tag{10}$$

where A is the anode area. The saturation current, denoted by I_{sat} is given by

$$I_{sat} = AeN_c \left(\frac{kT}{\pi m_e} \right)^{\frac{1}{2}} \exp\left(-\frac{e\Phi_B}{kT} \right) , \tag{11}$$

where N_c is the effective density of states in the conduction band and m_e is the effective mass of the material, T is the temperature, k is the Boltzmann constant, and Φ_B is the barrier height .

The current voltage characteristics can be found from diode circuit equation

$$R_s I = V_b + V_D - V_{bi} , \tag{12}$$

where

$$R_s = L/(e\mu N_d A) , \tag{13}$$

μ is the mobility and L is the epilayer thickness.

4. Modified Harmonic Balance simulation with inclusion of polar-optical phonons

Simulation of the dramatic nonlinearities due to polar-optical phonons at the transverse polar-phonon frequency involved the use of a modified harmonic balance (HB) technique for computational efficiency. The following provides insight and justification into the use of the modified harmonic balance technique.

In general, two major criteria are important in designing varactor multipliers at very high frequencies. The first is an accurate physical device model and the second is the ability to couple the device model into the proper matching impedance circuit. Indeed, the majority of the analytical and computational effort involved in designing efficient, high frequency multipliers entail the matching of the nonlinear diode to the embedding circuit[5]. Popular harmonic-balance algorithms such as the accelerated fixed point[6] and multiple reflection[7] algorithm have been derived to improve the matching of the various harmonics between nonlinear device and linear circuit. One significant drawback in incorporating both criteria in a single algorithm is the large number of iterations that are required to optimize power-efficiency of the nonlinear element over both embedding impedance and local oscillator (LO) voltage. Many multiplier circuit designs make use of multiple nonlinear devices for high frequency generation, and thus the computational effort is made even more intense.

The simulation used in this research utilizes part of a two-step modified HB algorithm to enable computational efficiency. This modified HB algorithm introduces a new optimization constraint that separates the device from the embedding circuit. Specifically, this procedure allows the current/voltage harmonics of the nonlinear device to be optimized independent of the external circuit. Polar-optical phonon physics can be easily integrated into the device model. Once the device has been optimized, the optimum impedance of the embedding circuit can then be derived from simple analysis.

The basic technique is outlined in the text that follows. Invoking the normal assumptions that the nonlinear diode-element of a doubler experiences first and second harmonic voltage excitation, it is possible to express the total instantaneous voltage across the diode as

$$V_D(t) = -|V_0| + V_1\cos(\omega t) + V_2\cos(2\omega t + \theta_2), \tag{14}$$

where, ω is the fundamental angular frequency and V_0, V_1, V_2, and θ_2 is the dc, first harmonic voltage, second harmonic voltage, and second harmonic phase respectively. The assumption that these harmonics are known and presented to the diode, only require that we assume that the first and second harmonic embedding impedances are matched and that the higher-harmonics are subject to short-circuit. While a traditional HB algorithm is able to treat arbitrary embedding impedances, these imposed conditions are ones that are optimal for the final design. Hence, they are completely consistent with an optimization of the varactor-doubler. The main goal of the technique now is to maximize either the second harmonic power available from the diode, or the overall efficiency of the diode-doubler. The first-harmonic power delivered to the diode is

$$P_1 = \frac{V_1 I_{C1}}{2}, \tag{15}$$

and the available second-harmonic power from the diode is

$$P_2 = -\frac{1}{2}V_2\left(I_{C2}\cos(\theta_2) - I_{S2}\sin(\theta_2)\right) \tag{16}$$

The relations given in Eqs. 15 and 16 can now be used to optimize multiplier operation. The first step in the problem is to maximize either P_2 or P_2/P_1 over the four-dimensional space (V_0, V_1, V_2, θ_2). Here the current harmonics I_{C1}, I_{C2}, and I_{S2} are supplied by Fourier transforms of time-domain simulations from the physical diode model according to

$$I_D(t) = I_0 + \sum_{k=1}\{I_{Ck}\cos(k\omega t) + I_{Sk}\sin(k\omega t)\} \tag{17}$$

At this point, the optimization problem complexity is simplified tremendously. The conventional HB method will maximize the nonlinear diode-circuit problem over LO voltage (dc and ac) and over the real and imaginary parts of the first and second harmonic embedding impedance which is a six-dimensional space problem.

Once the optimization over the diode-voltages has been completed, the matching of the embedding circuit impedance can be easily determined[8]. The second step of the procedure is to derive the optimum LO voltage, which can also be easily determined. It is sufficient for the purposes of this research for just the first step to be made since the effect of the polar-optical phonon physics at ω_{to} is desired. Hence, maximum P_2 is sought within the vicinity of ω_{to} at the device level.

5. Simulation results

The modified harmonic balance simulation was first utilized to determine the input/output power efficiency of GaAs at its transverse polar-phonon frequency. This simulation was performed using the physical parameters from the UVA 6P4 GaAs Schottky Barrier Varactor. These parameters are listed in Table 1. The dispersion constants used were mentioned previously in Section 2. The results are shown in Figure 5, where efficiency is plotted with dependence on frequency, frequency being normalized to f_{to}. It can be observed the significant boost in efficiency beginning at f_{to} (indicated by the number one on the independent axis in Figure 5) peaking at approximately 1.04 and again at 1.12 normalized frequency. This suggests strong coupling of the displacement field to the polar-optical phonons within the barrier depletion region. The results also show a significant drop in efficiency at roughly half the phonon frequency.

Table 1. Parameters used in simulation of GaAs Schottky Diode

	Value	**Unit**
N_d	3.50E16	cm^{-3}
N_c	4.70E17	cm^{-3}
Mobility	8380	$cm^2/V\text{-}s$
Length of epitaxial layer	1000	nm
Φ_B	1.03	eV
Operating Temperature	300	K

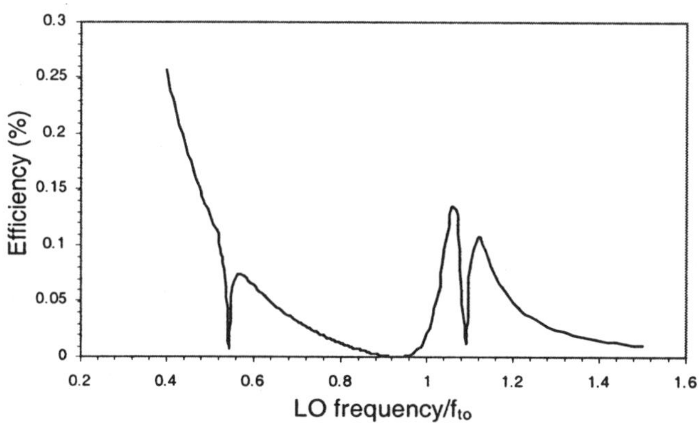

Fig. 5. Efficiency dependence on normalized frequency for GaAs Schottky diode

Simulation of the Bi_2S_3 material was also performed to determine any enhancements in efficiency. Table 2 lists the parameters used in the simulation of the Bi_2S_3 material based on experimental data discussed previously. Figure 6 and 7 show the results of simulations using a phonon damping factor of $2\gamma=0.1\omega_{to}$ and $2\gamma=0.01\omega_{to}$ respectively. It is clear that the damping factor plays a critical role in the lattice vibrations at the polar-phonon frequency. A larger damping factor appears to suppress the nonlinearities at $\omega_{to}/2\pi$=300GHz as shown in Fig. 6. A smaller damping factor yields two enhanced peaks in the vicinity of ω_{to}, similar to the results for GaAs.

Table 2. Parameters used in simulation of Bi_2S_3 Schottky Diode

	Value	Unit
N_d	4.00E18	cm^{-3}
N_c	2.86E19	cm^{-3}
Mobility	1000	cm^2/V-s
Φ_B	1.00	eV
Operating Temperature	30	K

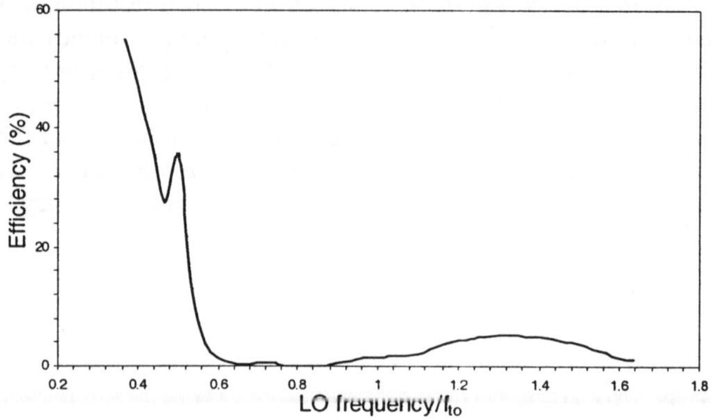

Fig. 6. Efficiency dependence on normalized frequency for Bi_2S_3 Schottky diode, $2\gamma=0.1\omega_{to}$

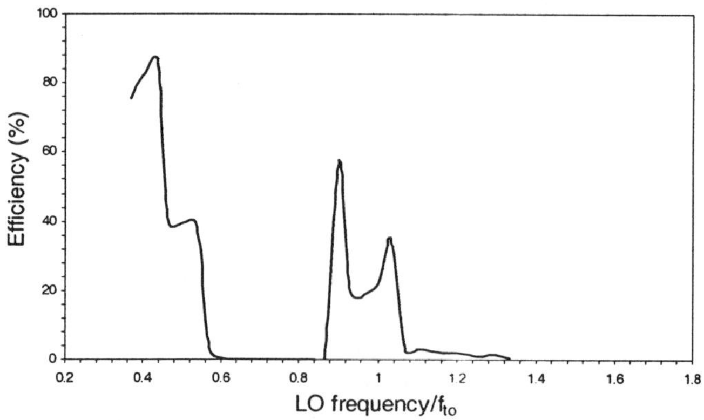

Fig. 7. Efficiency dependence on normalized frequency for Bi_2S_3 Schottky diode, $2\gamma=0.01\omega_{to}$

Presence of the polar-optical phonon effect in the vicinity of ω_{to} is more clearly visible when looking at the second harmonic power versus frequency as shown in Fig. 8 and 9 for $2\gamma=0.1\omega_{to}$ and $2\gamma=0.01\omega_{to}$ respectively. Here, an anode area of 6.45 μm^2 was used from experimental data[9]. Although the larger damping factor decreases the power over the simulated frequency range, the relative increase in power at ω_{to} is still quite noticeable.

Recently 200GHz, 400GHz, and 800GHz multipliers have been designed and tested at JPL where GaAs Schottky diodes were utilized[9,10]. In particular, the 400 GHz multiplier diode demonstrated a maximum calculated efficiency of approximately 38%. In addition, the 800 GHz multiplier diode demonstrated a maximum calculated efficiency of approximately 13%. Note that material quality is the primary factor that determines the γ parameter used in our simulations. Also, it is very reasonable to expect that γ factors in Bi_2S_3 should be achievable on the order of those observed for GaAs if high quality material growth is employed. Hence, the results obtained for Bi_2S_3 diodes in Fig. 7 and 9, where a γ factor in excess of that known for high-quality GaAs was utilized, indicate that enhanced multiplier performance can be expected in this new material system.

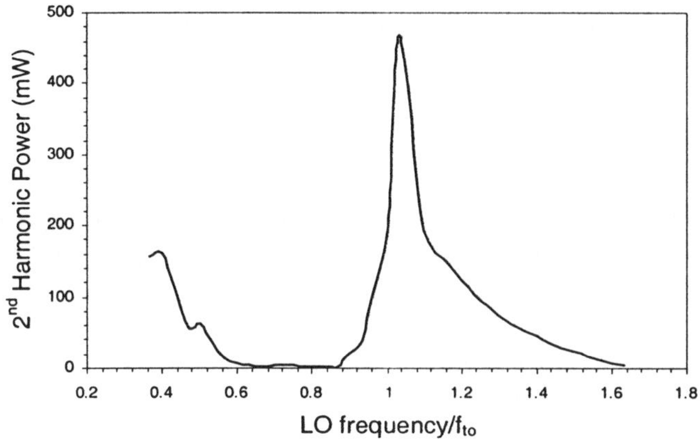

Fig. 8. Second harmonic power dependence on normalized frequency for Bi_2S_3 Schottky diode, $2\gamma=0.1\omega_{to}$

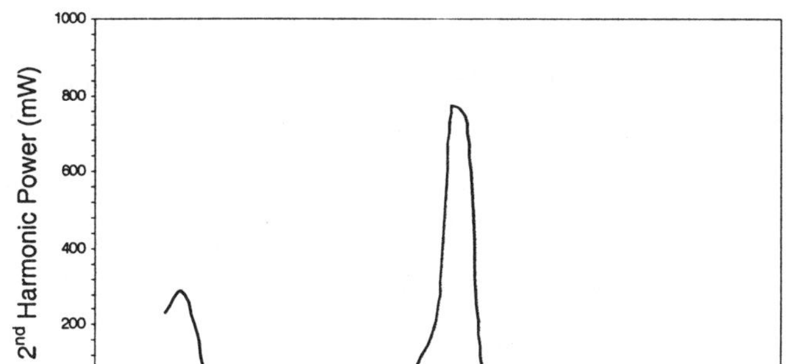

Fig. 9. Second harmonic power dependence on normalized frequency for Bi_2S_3 Schottky diode, $2\gamma=0.01\omega_{to}$

6. Conclusions

This paper demonstrates that dramatic improvements in harmonic generation can be expected from Schottky diode based multipliers near the polar-optical phonon resonance frequency. Here, modified harmonic balance simulations and new physics-based models were used to demonstrate significant improvements in efficiency (and power) of the second harmonic generation capability of GaAs and Bi_2S_3 polar materials. Furthermore, these studies suggest that Bi_2S_3-based Schottky diode offer important advantages for the enhancement of multiplier source performance within the terahertz frequency regime. Finally, these studies motivate further investigation of use of phonon effects and phonon engineering towards the improvement of very high frequency devices and components.

Acknowledgements

The authors wish to acknowledge the US Army Research Office as well as the National Science Foundation for their support in this endeavor.

References

1. M. Born and K.H Huang, Dynamic Theory of Crystal Lattices (Clarendon, Oxford, 1954)
2. S. Iwasa, L. Balslev, and E. Burstein, "Physics of Semiconductors", 7th International Conference on Physics of Semiconductors, p. 1077, 1964
3. A. Kajokas, J. Grigas, "Origin of anomalies of physical properties of Bi_2S_3 crystals", Lithuanian Physics Journal , Vol. 39, No. 1, pp. 38-45, 1999
4. B. L. Gelmont, D. L. Woolard, T.W. Crowe and R. J. Mattauch , "Resonant Polar-Optical Phonon Modulation of the Schottky Barrier Charging Dynamics" , Phys. Rev. B., 61, 23, p. 15939-51 (2000)
5. S. A. Maas, Nonlinear Microwave Circuits, (Artech House, MA, 1988)
6. G. B. Tait, "Efficient Solution Method for Unified Nonlinear Microwave Circuit and Numerical Solid-State Device Simulation," IEEE Microwave Guided Wave Letters, 4, pp. 420-422, 1994
7. P. H. Siegel, A. R. Kerr, and W. Hwang, "Topics in the Optimization of Millimeter-Wave Mixers, " NASA Technical Papers, no. 2287, 1984
8. C. C. Lee, B.L. Gelmont, D.L. Woolard, "A Modified Harmonic-Balance Analysis of Schottky Diode Multipliers based upon a hydrodynamic transport model", Tenth International Symposium on Space Terahertz Technology, Charlottesville, Va., March 1999
9. E. Schlecht, G. Chattopadhyay, A. Maestrini, A. Fung, S. Martin, D. Pukala, J. Bruston, I. Mehdi,"200,400, and 800 GHz Schottky Diode Substrateless Multiplier: Design and Results", IEEE MTT-S Digest, 2001, pp. 1649-1652
10. G. Chattopadhyay, E. Schlecht, J. Gill, S. Martin, A. Maestrini, D. Pukala, F. Maiwald, and I. Mehdi, "An 800 GHz Broadband Planar Schottky Balanced Doubler", 9th International Conference on Terahertz Electronics, Charlottesville, Va., October 2001

International Journal of High Speed Electronics and Systems, Vol. 12, No. 2 (2002) 479–489

ENVIRONMENTAL SENSING OF CHEMICAL AND BIOLOGICAL WARFARE AGENTS IN THE THz REGION

ALAN C. SAMUELS*, DWIGHT L. WOOLARD†, TATIANA GLOBUS‡, BORIS GELMONT‡, ELLIOTT R. BROWN§, JAMES O. JENSEN*, RICHARD SUENRAM¶, and WILLIAM R. LOEROP*

**Edgewood Chemical and Biological Center, Aberdeen Proving Ground, MD*

†Army Research Laboratory, Research Triangle Park, NC

‡University of Virginia, Charlottesville, VA

§University of California at Los Angeles, Los Angeles, CA

¶National Institutes of Standards and Technology, Gaithersburg, MD

Discrimination between hazardous materials in the environment and ambient constituents is a fundamental problem in environmental sensing. The ubiquity of naturally occurring bacteria, plant pollen, fungi, and other airborne materials makes the task of sensing for biological warfare (BW) agents particularly challenging. The spectroscopic properties of the chemical warfare (CW) agents in the long wavelength infrared (LWIR) region are important physical properties that have been successfully exploited for environmental sensing. However, in the case of BW agents, the LWIR region affords less distinction between hazardous and ambient materials. Recent studies of the THz spectroscopic properties of biological agent simulants, particularly bacterial spores, have yielded interesting and potentially useful spectral signatures of these materials. It is anticipated that with the advent of new THz sources and detectors, a novel environmental sensor could be designed that exploits the peculiar spectral properties of the biological materials. We will present data on the molecular spectroscopy of several CW agents and simulants as well as some THz spectroscopy of the BW agent simulants that we have studied to date, and discuss the prospectus with regard to detection probabilities through the application of sensor system modeling.

Keywords: Millimeter waves; microwave spectroscopy; remote sensing; spectral prediction; chemical warfare agent detection; biological warfare agent detection.

1. Introduction

1.1. *Chemical agents*

Microwave spectroscopic techniques offer some very unique advantages in the arena of chemical warfare (CW) agent detection.[1–3] The technique relies on the unique and specific rotational spectral lines of chemical structures in order to identify and quantify chemical materials. An effort is underway to adapt existing instrumentation for operation in real time and sample directly from the atmosphere without the need for filtering or pre-separation, as the rotational spectral bands are generally quite peculiar to each molecular structure, and linewidths are extremely narrow

($\leq$ 10 KHz). It should be noted that the technique is an extractive gas technique, which, by itself, does not apply directly to standoff measurement. However, the application of microwave spectroscopy to the cataloging of CW agent microwave spectra and their associated spectral assignments provides a unique tool for the prediction of rotational signatures of the measured agents in high frequency regions of the microwave band, where remote sensing may be feasible. While pressure broadening of microwave signatures reduces their intensity and instills the requirement for deconvolution of the observed spectra, the fact that microwave remote sensing is indifferent to the presence of water, smoke, or dust in the atmosphere preserves the attractiveness of this technique in the battlefield CW detection scenario.

1.2. *Biological agents*

Environmental monitoring for hazardous materials of biological origin presents a significant technical and technological challenge. Currently, technologies that are fielded or in development for field biological warfare agent monitoring rely on harvesting a sample of the atmosphere to be monitored and processing the sample through sophisticated assays to effect identification. Less sophisticated optical techniques have been developed that rely on back-scattering principles and/or on ultraviolet fluorescence to identify potential threat aerosols; however these systems are bulky, power hungry, and limited in their capability to discriminate between harmful versus innocuous biological materials.

We have been investigating the feasibility for monitoring biological materials using infrared spectroscopy. Naumann has demonstrated that the use of careful data analysis of the mid- and long wave infrared spectra of biological microorganisms affords a significant capability to discriminate between species.[4] Additionally, the theoretical predictions of van Zandt *et al.*[5–7] as well as the experimental measurements of Woolard *et al.*[8,9] have supported the notion that biological materials exhibit spectral properties in the far infrared and millimeter wave region of the electromagnetic spectrum. In mid- and long wave infrared measurement, the phenomenology is of course related to the chemical composition of the material under study. Specifically, absorption bands observed in the mid- and long wave infrared region are associated with the fundamental molecular vibration modes of the various functional groups that constitute the proteins and other biomolecules in the microorganism. While these basic building blocks are common to all materials of biological origin, the multicomponent manifold of spectral signatures that result from the spectroscopic investigation of intact microorganisms, spores, and biomaterials may demonstrate a unique quality that can be correlated to the type of organism.

In the far infrared and millimeter wave spectral region, the data are even more sparse. This is primarily due to experimental difficulties with measuring the spectroscopy of materials at these wavelengths. The underlying phenomenology is purported to be the low frequency vibrations associated with long-range structure in large molecules and molecular lattices, such as the helical structure of deoxyribonucleic acid and the secondary and tertiary structure of proteins. The natural frequency of vibrations of this nature has been predicted to be below 30 cm^{-1} which is more commonly known as the gigahertz and terahertz region of the time domain.

2. Experimental Methods and Results

Experimental rotational spectra were acquired using the Fourier transform microwave technique described elsewhere.[10–12] Experimental far infrared spectra of the biological materials were acquired using the Fourier transform infrared technique described elsewhere.[13] The molecular spectroscopy of two nerve agents, Sarin and Soman, are presented in Figs. 1 and 2. We note the extremely rich structure and uniqueness of the rotational spectral structure attributed to the specific molecular geometry of the species under study. The far infrared spectra of two closely related bacteria species, *Bacillus subtillus* and *Bacillus thuringiensis*, are presented in Fig. 3. Note that despite the close similarity of the two species, the spectra are clearly distinctive.

3. Broadening and Predicted Spectra Under Ambient Conditions

Pressure broadening of the rotational transitions critically diminishes both their intensity and their uniqueness in serving as a signature for the chemical agents in air at atmospheric pressure. However, recent investigations at Argonne National

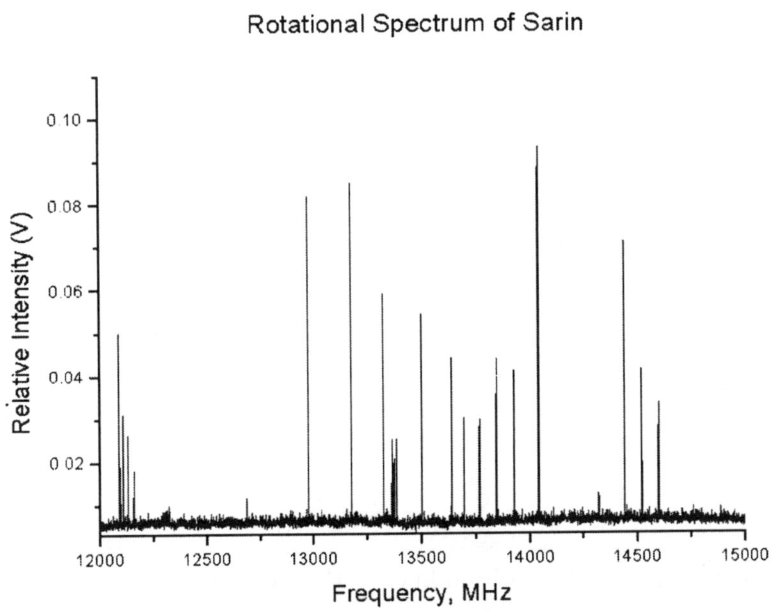

Fig. 1. Fourier transform microwave spectrum of Sarin.

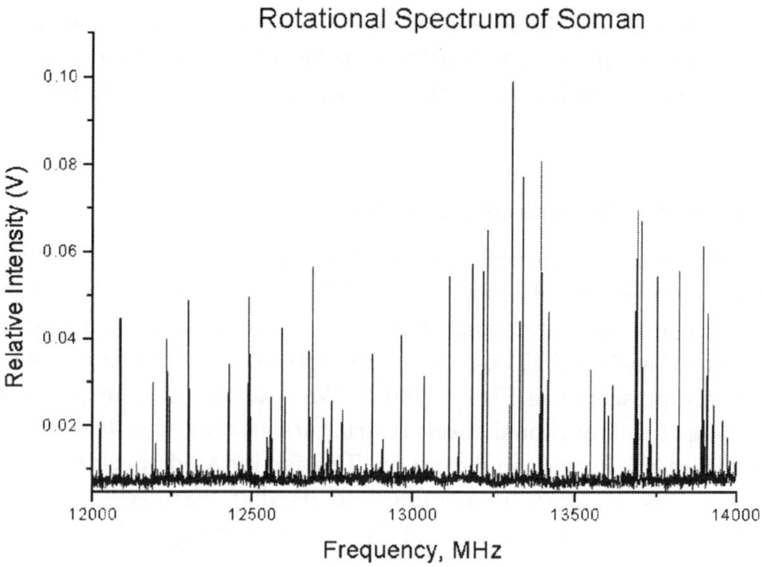

Fig. 2. Fourier transform microwave spectrum of Soman.

Laboratory have demonstrated that it is possible to partially mitigate pressure broadening at atmospheric pressures by measuring at higher frequencies and wider sweep-widths.[14] The improvements in sensitivity at higher frequencies help to offset the reduced intensities of the transitions, and the wide sweep widths (in excess of 100 GHz) make it possible to collect pressure-broadened spectral lines. Several atmospheric "window regions", in which the propagation of millimeter wave radiation is not severely damped by the presence of water vapor and other atmospheric components, exist at high frequency.[15] The density of the spectral lines in the complex molecules studied herein are such that observation of the individual lines under atmospheric broadened conditions is ruled out. Nevertheless, the characteristic nature of the overall sequence of individual lines may contribute to a distinguishable resultant spectrum under pressure broadened conditions.

To get a first estimate of the expected appearance of the Sarin and Soman millimeter wave spectra under these conditions, we predicted the rotational spectra of these two nerve agents in the atmospheric window region that spans the neighborhood of ~ 200–320 GHz. We then applied the following Lorentzian line-shape function to account for atmospheric pressure broadening.[16] The overall pressure-broadened absorption coefficient may be defined as $\alpha(v)$:

$$\alpha(v) = \sum_{i=1}^{m} \frac{\alpha_{\text{int},i}\Delta v}{(v - v_i)^2 + (\Delta v)^2}\,, \tag{1}$$

where m is the number of transition lines in the frequency range of interest, $\alpha_{\text{int},i}$ is the integrated absorption peak of an observed line in the low pressure limit as

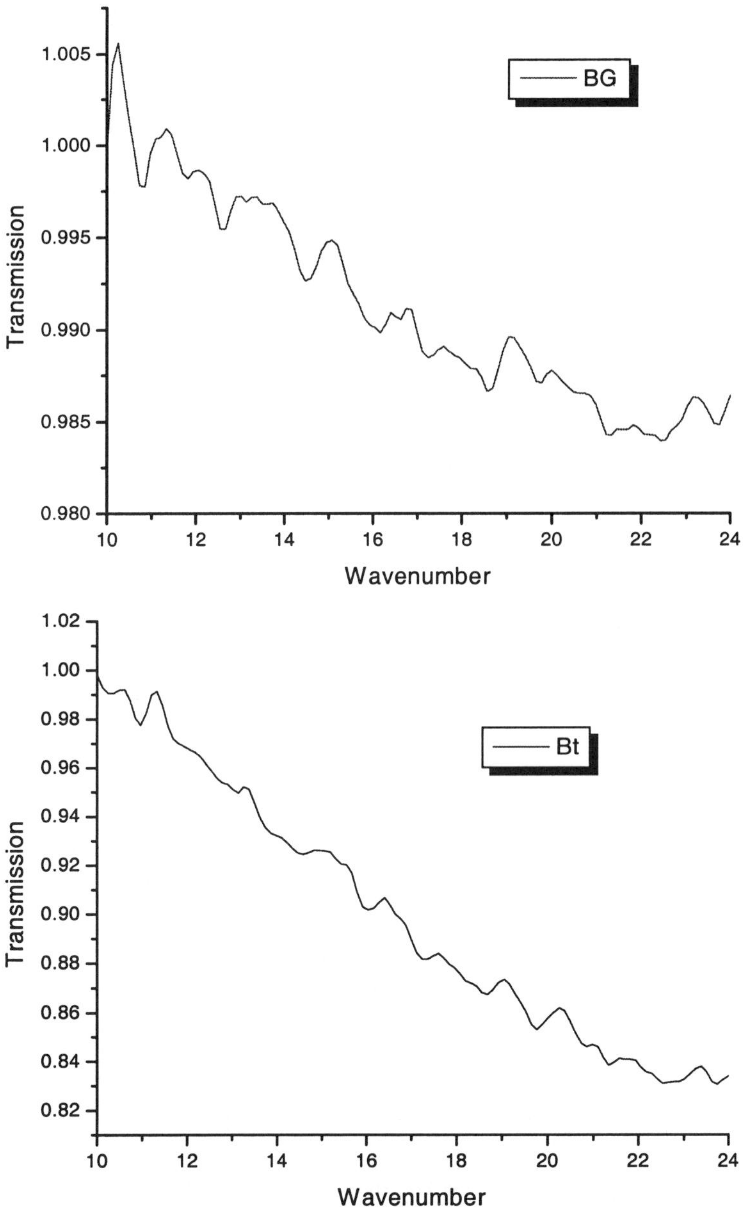

Fig. 3. Fourier transform infrared spectra of *B. subtillis* (BG) and *B. thuringiensis* (Bt).

predicted from the assignments derived from the FTMW experiment, and Δv is the linewidth at atmospheric pressure (typically in the range of 3–5 GHz for medium-sized organic molecules).[17] The predicted spectra for Sarin and Soman are presented in Fig. 4. Despite the broadening of the spectral lines, the two absorption traces retain certain overall characteristic features.

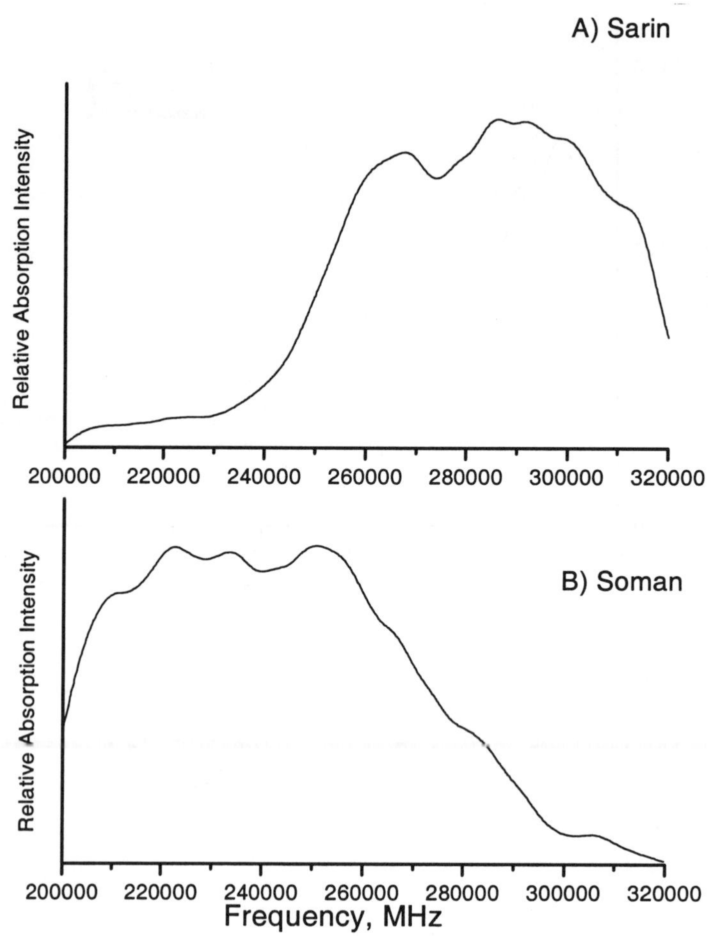

Fig. 4. Predicted broadened spectra at high frequency using experimental Sarin and Soman parameters. Linewidth set at 5 GHz for the purpose of the simulation (see text for discussion).

4. Predictions of the Probability of Detection of Bacteria

The results given in Fig. 3 do reveal spectral characteristics that are clearly resolvable with approximately 5–15% variation between maximum and minimum. These spectral data may be used to assess the overall sensitivity of a sensor system based upon the phonon-signature detection approach. Indeed, a collaborative research project is already underway to determine sensitivity limits of various sensor architectures (e.g., differential-absorption spectrometer) that utilize existing electronic components.

The data presented allows for sensitivity and ranging assessments of actual remote detection systems operating under realistic battlefield conditions (i.e., atmospheric propagation subject to water vapor and trace pollutants) and this effort is actively part of a collaborative research project. In fact, these investigations are already underway and preliminary estimates for three remote detection schemes (i.e., active bistatic (differential-absorption) spectrometer, passive incoherent radiometer, and passive coherent radiometer) have been performed. These studies utilized commercial-off-the-shelf (COTS) sensor components and realistic architectures. While the project is actively developing models for the inclusion of atmospheric propagation effects (e.g., absorption by water vapor and trace pollutants), the initial work focused on establishing sensitivity, range, and minimum detectable agent concentrations under ideal conditions. The study was conducted using the *Bacillus subtillus* spectra data given in Fig. 3. The results of the preliminary study suggest that the differential-absorption spectrometer (DAS) was the only approach that yielded very high sensitivity, and as will be shown, such performance was attainable even with a room temperature detector.

The DAS study utilized a photomixer-based system (with an assumed output power of 1 mW) where the agent under test was detected via a long-range resonant cavity formed by a distant reflecting mirror — see Fig. 5. The system architecture in Fig. 5 was assessed for sensing over a cavity length of 1000 meters. The active sensing for this system was estimated by modeling the differential-absorption between the phonon resonant peak at 20.65 cm^{-1} (619.5 GHz) and the absorption minimum at 20.0 cm^{-1} (600 GHz). The DAS was assessed for three types of direct detecting bolometers, namely, a Golay cell (@ 300 K), a superconducting edge (HTC) detector (@ 77 K) and a Silicon composite (@ 4.2 K). Signal-to-Noise ratios (S/N) were generated for three particle densities as a function of the agent cloud depth. The system was first assessed for a perimeter-defense type scenario where we assume some latitude in the deployment in the system (i.e., entire THz spectrometer system is in a fixed and stationary position). Figure 6 gives the system performance assuming a very large receiver aperture (i.e., 1000 cm^2) for the three

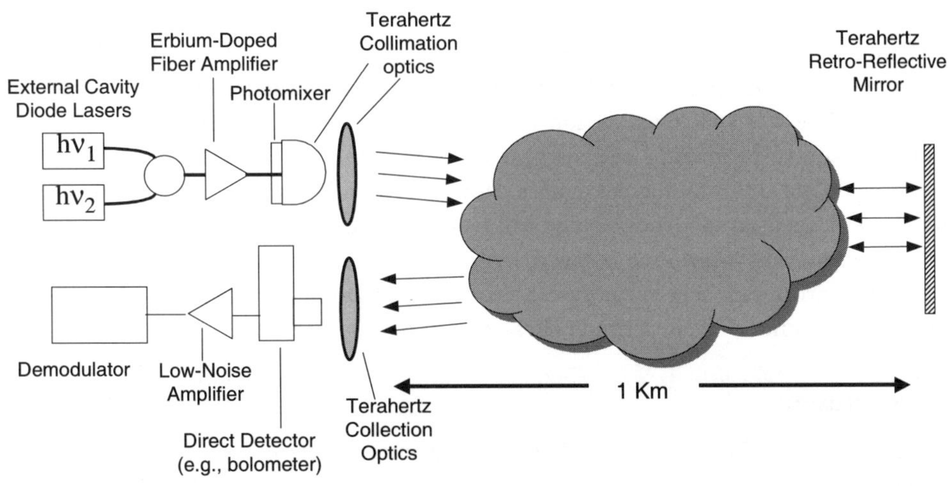

Fig. 5. The architecture used for the THz frequency Active Differential Absorption Spectrometer.

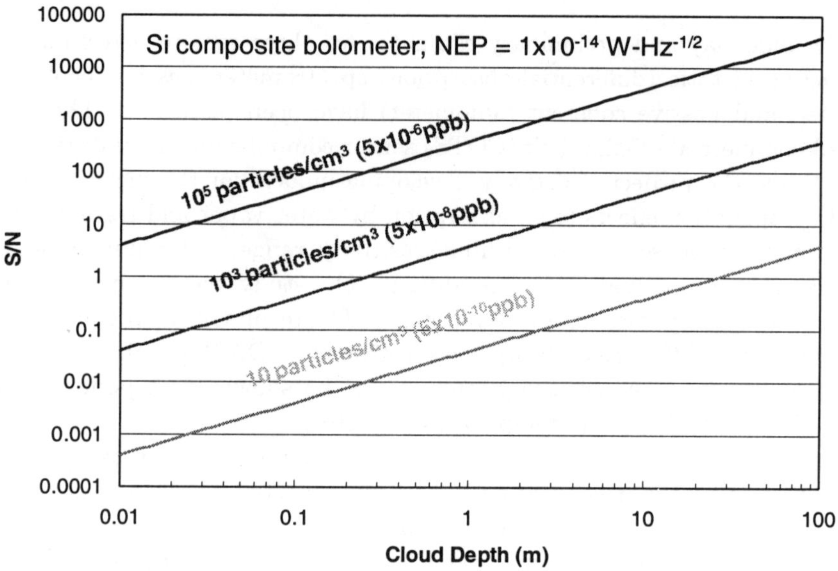

Fig. 6. Signal-to-Noise ratios (S/N) for a perimeter-defense type (stationary) system with a receiver aperture 1000 cm^2. Results are given as a function of *Bacillus subtillus* cloud-depth and particle-density.

types of direct detection. As shown, the results indicate an expectation of good detection performance (i.e., $S/N > 1$) for clouds of larger than 10 meters in depth and this is true for room temperature detection if the agent density is greater than 10^3 cm^{-3} (or 10 000 particles per liter). Hence, these results suggest that a THz-frequency spectrometer offers promise for a perimeter-defense system where issues of detector beam pointing is not a limiting issue (i.e., and allows for a large receiver aperture).

The utility of the system for application to more general remote detection scenarios was also considered. The overall detection performance is reduced, however, ~ 25 meter clouds with agent densities greater than 10^3 cm^{-3} are still detectable with nitrogen-cooled detectors. Furthermore, room temperature detection of 30 meter thick clouds is still feasible for 10^5 cm^{-3} agent densities. Hence, these initial results are quite promising and suggest there is potential for THz-frequency spectroscopic detection of biological agents. However, an adequate determination of sensitivity will need to incorporate the influence of atmospheric attenuation and also consider scenarios where one is faced with discriminating a target species within a realistic environment where many other interferent agents are expected be present. Indeed, these effects may degrade these estimates significantly.

5. Conclusion

The rich information and unambiguous rotational spectral response observed for the nerve agents Sarin and Soman demonstrate the analytical utility of microwave spectroscopy as an extractive gas technique. No other analytical technique affords such a high level of certainty while obviating the need for a pre-separation process. The implications of these results for standoff detection, however, are less obvious. Pressure broadening of the rotational transitions critically diminishes both their intensity and their uniqueness in serving as a signature for the chemical agents in air at atmospheric pressure. The improvements in sensitivity at higher frequencies help to offset the reduced intensities of the transitions, and the wide sweep widths (in excess of 100 GHz) make it possible to collect pressure-broadened spectral lines. Furthermore, it is possible to predict the expected spectra of a molecule at high frequencies by starting with the rotational constants determined by experimental measurements such as the FTMW technique described herein. Using the molecular rotational constants reported herein we have predicted the expected wide band millimeter wave response of the CW agent isomers at atmospheric pressure.

We have collected infrared, far infrared, and millimeter wave data on a variety of bacterial spores and some common environmental constituents. The data demonstrate a significant level of promise for infrared spectroscopy in particular to be developed into a unique and reasonably specific detector for biological materials in the ambient environment. A quantitative analysis of the data and signal to noise levels achievable in currently available instrumentation is underway to establish the ultimate detection limits possible and the obstacles that must be addressed in order to achieve a real-time environmental monitor for threat aerosol concentrations in a field environment.

Although the far infrared results presented here demonstrate very low extinction, the approach may yet have merit in a mode in which the sensor employs a pre-concentrator to accumulate solid particles from the atmosphere, perhaps even directly onto a polycarbonate substrate. When an adequate amount of material has been collected, distinct and potentially highly unique spectral signatures of the materials will emerge, as demonstrated with the higher concentration BG film data in Fig. 5(c). These signatures are likely to be even more specific than the mid-infrared spectra by virtue of the fact that the secondary structure of the constituent molecules in the material is responsible for the observed spectral lines. Additionally, it is possible to employ coherent sources and diode heterodyne detectors with resonator cavities to key in on low levels of signals, potentially bringing out the weak spectral lines with highly enhanced sensitivity. These measurements should be taken as an early demonstration of the fact that there is some exploitable phenomenology in the far infrared spectral region.

Acknowledgments

This work was funded by the Defense Advanced Research Projects Agency Special Projects Office, and the U.S. Army Edgewood Chemical and Biological Center technical base and in-house laboratory independent research program.

References

1. A. R. Hight-Walker, R. D. Suenram, A. C. Samuels, J. O. Jensen, M. W. Ellzy, J. M. Lochner, and D. Zeroka, "Rotational spectrum of Sarin", *J. Molec. Spectrosc.* **207** (2001) 77–82.
2. D. F. Plusquellic, R. D. Suenram, B. Mate, J. O. Jensen, and A. C. Samuels, "The conformational structures and dipole moments of ethyl sulfide in the gas phase", *J. Chem. Phys.* **115** (2001) 1–11.
3. R. D. Suenram, F. J. Lovas, D. F. Plusquaellic, A. Lesarri, Y. Kawashima, J. O. Jensen, and A. C. Samuels, "Fourier transform microwave spectrum and *ab initio* study of dimethylmethylphosphonate", *J. Molec. Spectrosc.*, submitted July 2001.
4. D. Naumann, C. P. Schultz, and D. Helm, "What can infrared spectroscopy tell us about the structure and composition of intact bacterial cells?", *Infrared Spectroscopy of Biomolecules*, eds. H. H. Mantsch and D. Chapman, Wiley, New York, 1996, pp. 279–310.
5. V. K. Saxena, L. L. van Zandt, and W. K. Schroll, "Atomic motions and high frequency cutoff in biological macromolecules", *Chem. Phys. Lett.* **164**, 1 (1989) 82.
6. V. K. Saxena and L. L. van Zandt, "Local modes in a DNA polymer with hydrogen bond defect", *Biophys. J.* **67** (1994) 2448.
7. W. N. Mei, M. Kohli, E. W. Prohovsky, and L. L. van Zandt, "Acoustic modes and nonbonded interactions of the double helix", *Biopolymers* **20** (1981) 833.
8. D. L. Woolard, T. Koscica, D. L. Rhodes, H. L. Cui, R. A. Pastore, J. O. Jensen, J. L. Jensen, W. R. Loerop, R. H. Mittleman, and M. C. Nuss, "Millimeter wave-induced vibrational modes in DNA as a possible alternative to animal tests to probe for carcinogenic mutations", *J. Applied Toxicology* **17**, 4 (1997) 243–246.
9. D. L. Woolard, E. J. Crowne, T. R. Globus, M. Bykhovskaia, J. L. Hesler, T. W. Crowe, N. Paravastu, A. C. Samuels, J. O. Jensen, J. L. Jensen, and W. R. Loerop, "Feasibility of submillimeter-wave technology for the identification of biological warfare agents", *4th Joint Workshop on Standoff Detection for Chemical and Biological Agents*, Williamsburg, VA, October 1998.
10. R. D. Suenram and A. M. Andrews, "Microwave spectroscopy", Chapter 14, *Experimental Methods in the Physical Sciences*, Vol. 29B, eds. F. B. Dunning and R. G. Hulet, Academic Press, 1996, pp. 273–296.
11. R. D. Suenram, J.-U. Grabow, A. Zuban, and I. Leonov, "A portable, pulsed-molecular beam, Fourier-transform microwave spectrometer designed for chemical analysis", submitted to *Rev. Sci. Instrum*, 1999.
12. R. D. Suenram, A. R. H. Walker, D. Woolard, W. Wiebach, A. C. Samuels, and J. O. Jensen, "Using Fourier transform microwave spectroscopy to identify chemical warfare agent signatures", *Proc. 1998, On-Site Analysis Conf.*, Orlando, Florida, January 1998.
13. A. C. Samuels, J. O. Jensen, W. R. Loerop, M. W. Ellzy, J. M. Lochner, D. Paterno, K. K. Ong, D. L. Woolard, and T. Globus, "Infrared, far infrared, and millimeter wave spectroscopic measurements on biological materials: Bacterial spores, pollens, and mold", *Proc. 5th Joint Conf. Standoff Detection for Chemical and Biological Defense*, Williamsburg, VA, September 2001, Science and Technology Corporation.
14. N. Gopalsami, S. Bakhtiari, and A. C. Raptis, "Development of a millimeter-wave sensor for environmental monitoring", *SPIE* **2842** (1996) 274.
15. E. Serabyn, E. W. Weisstein, D. C. Lis, and J. R. Pardo, "Submillimeter Fourier-transform spectrometer measurements of atmospheric opacity above Mauna Kea", *Appl. Opt.* **37**, 12 (1998) 2185.

16. G. Birnbaum, "Microwave pressure broadening", *Intermolecular Forces*, ed. J. O. Hirschfelder, Advances in Chemical Physics, Vol. 12, John Wiley and Sons, 1967.
17. N. Gopalsami, Argonne National Laboratory, Private Communication.

International Journal of High Speed Electronics and Systems, Vol. 12, No. 2 (2002) 491–500

THZ DETECTION BY RESONANT 2-D PLASMONS IN FIELD EFFECT DEVICES

XOMALIN PERALTA et al.

WOJTEK KNAP et al.

We demonstrate the resonant detection by the two-dimensional electron plasma waves in gated two-dimensional electron gas. Such detection was demonstrated for two devices: for a field effect transistor (FET) THz detector, where the standing 2-D plasmon is tuned to the frequency of the THz radiation by varying the gate bias and for double quantum well field effect transistors with a periodic grating gate, which exhibit a rich spectrum corresponding to standing 2-D plasmons under the metal part of the periodic gate.

The terahertz part of the electromagnetic spectrum lies in the gap in solid state technology between semiconductor electronics and photonics. Tunable solid state sources and detectors are key elements for the development of this part of the spectrum. There is a large variety of terahertz incoherent and coherent detectors - hot-electron bolometers[1,2,3], composite bolometers, superconducting-insulator-superconducting (S-I-S) junctions[4,5], pyroelectric detectors[6], Schottky diodes[7,8] and photoconductive detectors[9]. Tunability is demonstrated in magnetic field tuned photoconductivity from hydrogenic donors in lightly doped GaAs[10] and InSb.[11] Here we explore the potential of developing detectors tuned by plasmon resonances in field effect structures.[12,13] Tuning the detector response by gate voltage is more attractive for applications than tuning with a magnetic field.

In this paper, we describe two different approaches. First, we report on a field effect transistor (FET) THz detector, where the standing 2-D plasmon is tuned to the frequency of the THz radiation by varying the gate bias. Second, we fabricate double quantum well field effect transistors with a periodic grating gate, which exhibit a rich spectrum corresponding to standing 2-D plasmons under the metal part of the periodic gate.

Both sets of experiments demonstrate the feasibility of the resonance tunable detection of THz radiation by plasmons in a gated two dimensional electron gas.

Plasma wave detectors in a gated FET

There exists a great interest in new solid state devices that could detect and emit terahertz radiation in a selective and tunable way. The difficulty in detecting and generating terahertz radiation using solid state devices is related to two factors: i) the transit times in the devices are typically much longer than the period of the terahertz oscillations and ii) the quanta of terahertz radiation are much smaller than room or even liquid nitrogen thermal energy. However, plasma waves in a gated two dimensional electron gas can propagate with much larger velocity than the electron drift velocity and their excitation is not linked to any type of electronic transition. Hence, they can be excited even at elevated temperatures[14]. Allen et al. observed infrared absorption[15], and Tsui et al. reported weak infrared emission related to plasma waves in silicon inversion layers[16]. Burke et al. showed that the impedance of a high mobility transistor exhibits maxima at the fundamental plasma frequency and its harmonics[17].

The theory in Ref. [5] predicted that a constant drain-to-source voltage might be induced by electromagnetic radiation coupled into the channel of a field-effect transistor (FET). In a high mobility channel, this constant drain-to-source voltage should have a resonant dependence on the radiation frequency, with the resonant frequency being proportional to the square root of the surface carrier density and inversely proportional to the gate length. With carrier densities on the order of $10^{12}cm^{-2}$ and the gate length dimensions below 1 μm the FET plasma frequency can reach the THz range[1]. The carrier density can be tuned by the gate-to-source voltage. This implies that a FET may be used as a resonant tunable detector of THz radiation.

Plasma waves that may propagate propagate in a FET channel have a linear dispersion law, $\omega(k) = sk$. Here s is the plasma wave velocity that depends on carrier density, and k is the wave vector. The velocity of the plasma waves, s, is typically on the order of 10^8 cm/s, which is much larger than the drift velocity of the two-dimensional (2D) electrons in the FET channel. This is why the propagation of plasma waves can be used for new regimes of FET operation, with a much higher frequency than for conventional, transit-time limited devices.

In a FET with a given length L the values of plasma frequencies are discrete and given by[1] $\omega_N = \omega_0 (1+2N)$, where $\omega_0 = \pi s/2L$, and N= 0, 1,2,....

A FET, biased by the gate-to-source voltage and subjected to electromagnetic radiation, can develop a constant drain-to-source voltage, which has a resonant dependence on the radiation frequency $f=\omega/2\pi$ with the maxima at the plasma oscillation frequencies $f_N=\omega_N/2\pi$ .The plasma wave velocity depends on the carrier density in the channel , n ,and the gate to channel capacitance per unit area C, $s=(e^2n/mC)^{1/2}$, where e is the electron charge, m is the electron effective mass. In the gradual channel approximation, the carrier density in the channel is related to the gate voltage as $n=CU_0/e$. U_0 is the gate to channel voltage swing that is defined as $U_0=U_g - U_{th.}$, where U_g is the gate voltage and U_{th} is the threshold voltage. In this case, the fundamental plasma frequency can be expressed by an approximate relation $f_o=\omega_0/2\pi=(eU_0/m)^{1/2}/4L$. This relation leads to two important consequences: i) a sufficiently short (sub-micron) FET can operate as a THz detector and ii) the frequency of this detector can be tuned by the gate voltage.

The dimensionless parameter determining the detector behavior is $\omega\tau$, where $\omega=2\pi f$ and f is the radiation frequency. Two opposite cases ($\omega\tau >> 1$ and $\omega\tau << 1$) has to be considered. When $\omega\tau >> 1$, the FET can operate as a resonant detector. When $\omega\tau << 1$, the plasma oscillations are overdamped and the detector response is a smooth function of ω as well as of the gate voltage (broadband detector)[18]. There have been a few experimental observations of the non-resonant detection[19]. However, the final experimental confirmation of the possibility of resonant detection of THz radiation and the theoretical description of the nonresonant signal close to the threshold voltage were open questions for a long time.

The FETs used in the experiments were commercial GaAs/AlGaAs field effect transistors (Fujitsu FHR20X) with the gate length L = 0.15 μm. They were removed from packages and mounted on a quartz plate in order to avoid any parasitic interference and reflections -like in.

The photo-response measurements were performed with 200 GHz and 600 GHz radiation systems based on a 100 GHz Gunn diode with a frequency doubler (for the 200 GHz system) or doubler and tripler (for the 600 GHz system). The maximum output power was about 3 mW for the 200 GHz system and 0.3 mW for the 600 GHz system, respectively. Attenuators were used in order to limit the power at the focused spot to about 0.1 mW in order to avoid heating effects. A waveguide finished with a cylindrical cone was connected to the output of the Gunn diode to out-couple the radiation. The radiation beam was then focused by a parabolic mirror and led through the cryostat window to the sample, which was placed on a quartz substrate glued to the cold finger of the closed cycle cryostat. The radiation beam diameter at the sample holder was ~1mm – a dimension that was much larger than the device size. No special coupling antennas were used and the terahertz radiation was coupled to the device through metallization pads. In each measurement, we tested a few parabolic mirror orientations and a few incident angles in order to verify possible interference effects. We found that different beam geometry/incidence angles changed only the amplitude of the signal and not its spectral shape.

The radiation intensity was modulated with a mechanical chopper (30–300 Hz range) and the open-circuit source drain voltage was measured by a voltage preamplifier (1 MΩ input resistance) followed by a lock-in amplifier. A regulated temperature system allowed for the temperature stabilization to within 1K over the range from 8-300 K.

In Fig.1, we show a few examples of typical experimental data. The drain–source voltage photoinduced by the 600 GHz radiation in GaAs/AlGaAs FET, is shown as a function of the gate voltage, U_g. At the lowest temperature (8K), one can also see a resonant feature in the response (marked by an arrow in Fig.1) that is superimposed on the broad background. It is due to a resonant detection of 600GHz radiation. The resonant response appears at lowest temperatures, since, at these temperatures, the electron scattering time increases and we reach $\omega_0\tau=1$ - the resonant detection condition. The resonant detection was reported recently in (see Ref. 20) and will described in more details later.

Let us first concentrate on the nonresonant part of the signal. For relatively large positive gate voltage swings, U_0, the detector response decays as $1/U_0$. For negative values of U_0 (U_g smaller than the threshold voltage), the experimentally observed response decreases, so that the response always reaches a maximum value close to the FET threshold . This maximum changes its position and amplitude with the temperature but not with the frequency of the radiation.

In order to understand the origin of the maximum one has taken into account the leakage current from the metal gate to the 2D electron gas in the channel. Well above threshold, the gate leakage is small compared to the current in the channel and does not affect the response of the detector. However, for negative values of U_0, when electron concentration in the FET channel becomes exponentially small, the leakage current plays an important role and should be taken into account.

A new theoretical approach that takes into account the gate leakage current has been recently developed (see Ref. 21). Below we will discuss its main results.

The general equation for the electron concentration in the FET channel is given by

$$n = n^* \ln\left[1 + \exp\left(\frac{eU_0}{\eta k_B T}\right)\right]. \qquad (1)$$

Here $n^* = \dfrac{C\eta k_B T}{e^2}$, C is the gate capacitance per unit area and η is the ideality factor.

The response of the FET can be expressed as

$$\Delta u = \frac{eu_a^{\ 2}}{(4\eta k_B T)} \frac{1}{\left(1+\exp\left(-\frac{eU_0}{\eta k_B T}\right)\right)\left(1+\kappa\exp\left(-\frac{eU_0}{\eta k_B T}\right)\right)\ln\left(1+\exp\left(\frac{eU_0}{\eta k_B T}\right)\right)}$$

Where

$$\kappa = \frac{j_0 L^2 me}{2C\tau\eta^2 k_B^2 T^2}$$

is the dimensioneless parameter characterizing the leakeage current $j0$.

Results of the calculation of the detector response is shown in Fig.2

The temperature dependence of the photoresponse shown in Fig.2 represents a typical device behavior. One can see that the temperature evolution of the width and the amplitude of the photoresponse curves shown in Fig.2 qualitatively agree with the measured curves shown in Fig.1. (The photoresponse results in Fig.1 are shown as a function of the gate voltage U_g). At low temperatures a well defined maximum is observed and at high temperatures a broad maximum, or rather a step-like behavior near the device threshold voltage is observed. Generally, the shape of the photoresponse curve defined by Eq.2 is a function of two main parameters: $\eta k_B T$ that governs the carrier density in the subthreshold region (see Eq.1) and κ, which is a dimensionless parameter related to the leakage current.

The maximum value of photoresponse depends on $\eta k_B T$ only. The width and the position of the maximum depend both on $\eta k_B T$ and κ. For small κ, the photoresponse has a plateau (or step like) behavior near the threshold. For higher values of κ, the photoresponse has a well defined maximum. These rules, although approximate, allow to understand the photoresponse in most experimental situations. More detailed discussion of nonresonant detection of THz radiation by submicron FET's can be find in Ref. 21.

As can be seen in Fig.1 at lowest temperatures the resonant-like maximum is superimposed on the bacground. Region of the resonace response is shown in more details in Fig. 3. At the lowest temperatures (below 30 K) a well reproducible maximum - superimposed on the background was observed. The inset of Fig.2a shows the shape of the maximum after subtraction of the nonresonant background signal.

In Fig.2b, we plotted the plasma frequency versus gate voltage dependence for U_{th} =0.47. As can be seen, the resonant frequency of the transistor decreases with applied gate voltage from its maximum value ~1.8THz (at zero gate bias) to zero, which is reached when the gate voltage is equal to the threshold voltage. The horizontal line in figure 2b corresponds to 600GHz, which is the measurement frequency used in this work. One can see that the expected position of the resonance (marked by an arrow) is close to the one observed in the experiment. In order to get a more quantitative description of the results we use the following expressions for the drain response derived in Ref. 18:

$$U_{ds} = \frac{1}{4}\frac{U_a^2}{U_0} f(\omega) \qquad (1)$$

where

$$f(\omega) = 1 + \beta - \frac{1 + \beta\cos(2k_0^{'}L)}{\sinh^2(k_0^{"}L) + \cos^2(k_0^{'}L)} \quad , \quad \beta = \frac{2\omega\tau}{\sqrt{1 + (\omega\tau)^2}}$$

$$k_0^{'} = \frac{\omega}{s}(1/\beta + 1/2)^{1/2}$$

$$k_0^{"} = \frac{\omega}{s}(1/\beta - 1/2)^{1/2}.$$

Here U_a is the ac gate to drain voltage induced by external illumination. Using Eq. (1), we calculated the expected detector response using the device parameters determined from the I-V and magnetoresistance measurements for different temperatures. The results are shown in Fig.3. As can be seen from a comparison of Fig.2a and Fig.3 a good overall agreement of measured and calculated spectra was obtained. The temperature evolution of the background, as well as the position and the width of the resonant signal seen in the experiment (Fig.2a) are well reproduced in the calculated spectra (Fig.3).

In order to understand the temperature evolution of the spectra one has to consider the change of the $\omega\tau$ parameter. The measured mobility μ versus temperature dependence is plotted in the inset in Fig.3. As can be seen (right hand scale), the mobility increases with decreasing temperature. The momentum scattering time was calculated as $\tau=\mu m^*/e$. On the left hand side scale of the inset, we show the value of $\omega\tau$ for the frequency of 600 GHz. One can see that at temperatures above 50K $\omega\tau \ll 1$. At temperatures below 30K, one can approach the condition of resonant detection ($\omega\tau = 1$) . This explains why at higher temperatures only the $1/U_0$ –like non-resonant background is observed but at low temperatures, a resonant feature is superimposed on this background. This behavior is well seen in the experimental data (Fig.2a) and in the theoretical calculations (Fig.3). The value of the $\omega\tau$ parameter also determines the sensitivity of the detection. (The expected increase of the detector response when $\omega\tau$ slightly increases to 1.5 is shown by a dashed line in Fig.3). One expects a very high sensitivity of such detectors for high mobility GaAs FETs[18].

Resonant plasmons in grating gated double quantum well FET

Field effect devices are fabricated from modulation doped GaAs/AlGaAs double quantum well heterostructures grown on a semi-insulating GaAs substrate by molecular beam epitaxy. The quantum wells are 200Å wide and are separated by a 70Å barrier. The nominal electron densities in the quantum wells are $n_{upper}=1.7\times10^{11}cm^{-2}$ and $n_{lower}=2.57\times10^{11}$ cm^{-2}: the 4.2K mobility is ~1.7×10^6 cm^2/Vs. A 2x2mm mesa is defined by wet chemical etching and the ohmic contacts to both quantum wells are formed by evaporating and annealing NiAuGe over the edge and side of the mesa forming source and drain. A 700Å thick TiAu grating gate is evaporated with the lines of the grating parallel to the ohmic contacts, perpendicular to the current flow. Devices with grating periods of 4 and 8 μm were tested. (In both cases half of the period was metal the other half open.) The grating has three different functions 1) act as a gate by modulating the electron density in the quantum wells under the metallized part, 2) couple a component of the terahertz electric field normal to the layers and 3) select the wavevectors of the excited plasmon.

The terahertz sources used are the free-electron lasers at UCSB, which can cover a frequency range between 120 GHz and 4.8 THz, but the experiments were performed between 570 and 660 GHz. To measure the terahertz photoresponse, the samples are wire bonded and mounted on a fiberglass chip carrier inside a variable temperature cryostat. Radiation is focused onto the sample. A constant source-drain current of 100μA is applied and we study the photoconductive response of the double quantum wells as a function of gate voltage, terahertz frequency and temperature.[22]

Figure 1 shows the grating gate (4μm period) voltage dependent photoresponse at T=25K for different incident radiation frequency. For 570GHz there is a resonance in the photoresponse around $V_{gate}= -1.6V$. This resonant peak moves to lower gate voltage, higher electron density, as we increase the frequency of the incident radiation. This follows the expected dispersion relation for plasmon modes in 2DEG systems.[23,24,25] The response was a linear function of terahertz power and source drain current indicating a photoconductive response. The radiation is polarized perpendicular to the lines of the grating gate.

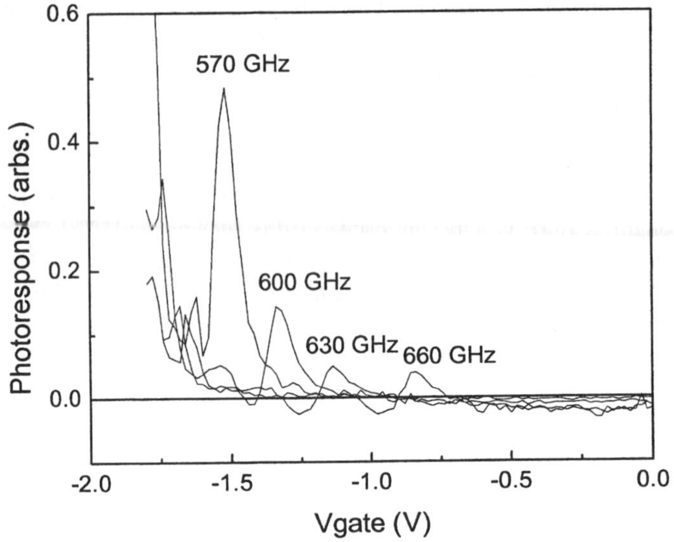

FIG. 1. Terahertz photoresponse as a function of gate voltage at four different frequencies. The temperature was T=25K and the source-drain current was I_{SD}=100μA. Grating period is 4μm.

To confirm that the observed resonant response is caused by the plasmon modes underneath the grating gate, we explored another device made from the same double quantum well material and processed in the same way but with an 8 μm grating period. The lower two plots in FIG. 2 show the photoresponse at 600 GHz as a function of gate voltage at T=25K for both the 4 μm (left) and the 8 μm (right) periods.

One can observe a larger number of resonances in the 8 μm period grating. The larger period, the greater the number of allowed standing waves in a frequency interval.

For a given excitation frequency, a larger number of resonances will appear in a gate voltage range.

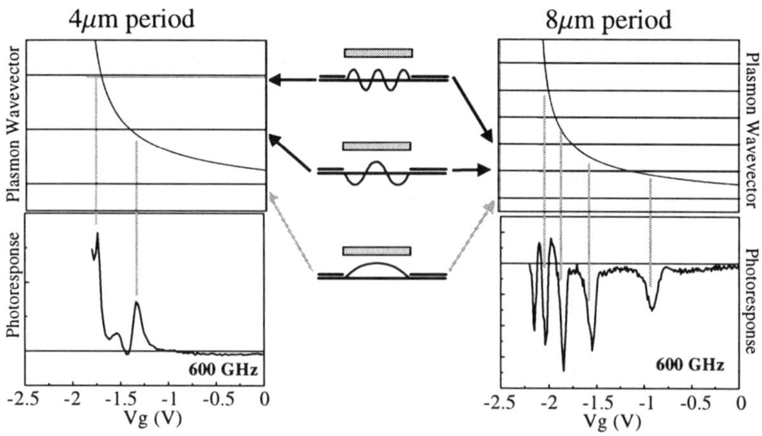

FIG. 2. Top: Model plasmon wavevector as a function of gate voltage at 600 GHz. Horizontal lines correspond to wavevectors where an odd integer number of plasmon ½ wavelengths fit under the metallic part of the gate. Bottom: Terahertz photoresponse as a function of gate voltage at 600 GHz. Middle column: Schematic representation of plasmon modes underneath the metal corresponding to the observed resonances. I_{SD}=100μA and T=25K.

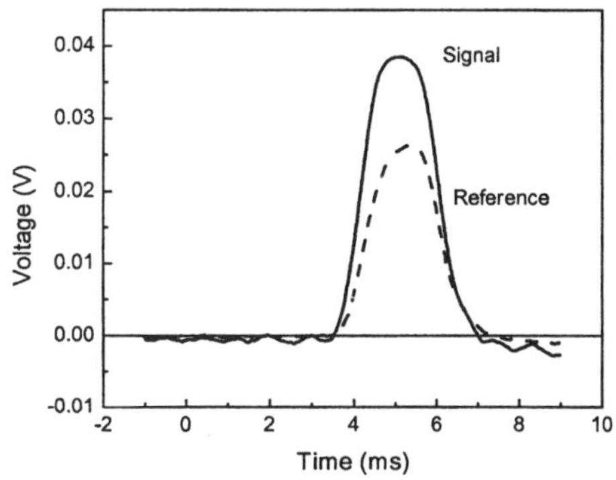

FIG. 3. Time response of the double quantum well field effect transistor at resonance (solid) and the reference pyroelectric detector (dotted) for frequency = 570GHz, I_{SD}=100μA and T=25K.

Figure 3 compares the time response of the double quantum well FET at resonance and the fast pyroelectric detector used as a reference for comparison. The frequency was 570GHz, I_{SD}=100μA and T=25K. The device is fast enough to detect the free electron lasers' rise and fall. The response time is faster than 700ns.

While it is clear that the terahertz photoconductance resonates at the standing plasmon resonance and its spatial harmonics under the grating metallization, the mechanism that produces a change in conductance is not clear. We note in Fig 2 that the sign of the conductance change depends on the grating period. We do note that a device similar in every respect to the above but with a single quantum well displayed no resonant terahertz photoconductance; the change in conductance at resonance with the standing plasmon is conditioned on the presence of the double quantum well in the ungated part of the device. We have also observed striking dependence on in plane magnetic fields which we do not discuss here. There are a number of issues that need to be addressed.

- What is the detection mechanism?
- What is the response time? Is it fast enough to be find use as a heterodyne detector?
- As an incoherent detector it would be useful to trade responsive time for sensitivity.
- What is the projected noise equivalent power (NEP)?

Discussion and prognosis

I. Plasma wave detectors in a gated FET.

Presented above results clearly demonstrate the resonant detection by the two-dimensional electron plasma confined in sub-micron field-effect transistors. The critical parameter that governs the physics of the detection problem is $\omega\tau$. We demonstrated that by increasing τ (by lowering the temperature) or increasing the detection frequency ω, one can reach a resonant detection of sub-THz radiation in a 0.15μm-gate length GaAs field effect transistor. Both the resonant and non-resonant parts of the observed photoresponse signal, as well as their evolution with temperature and frequency, were explained within the framework of a theoretical model. Theoretical estimates point to a possibility of achieving, in high mobility GaAs FETs, resonant detection up to several orders of magnitude more sensitive than that of conventional Schottky diodes[18]. There are however a number of issues that need to be addressed before such high sensitivity and tunable detectors can be realized :

- Is the submicron size compatible with requirement of the high mobility of the carriers?
- Can the limits of the device size and/or maximal frequency be changed by use of FET's based on other semiconductor heterojunctions like GaN/AlGaN or InAs/GaAlSb?
- How to incorporate efficient coupling antennas to the micron/submicron device?

II. Plasmon detection in grating gated double quantum well FET.

We have observed a resonant photoresponse in a 2DEG bilayer system corresponding to the excitation of standing plasma waves under the metallic part of a grating gate. At the moment, we understand that the tunable resonance is caused by the composite plasma oscillations of the double quantum well structure but we do not understand the actual mechanism that gives rise to the change in conductance at resonance. Directly related to our lack of understanding of this mechanism are other issues that need to be addressed - the sign reversal with grating period and lineshape variation with frequency.

From the data presented we can see that grating gated double quantum well FETs can be used as a tuneable detector. An estimate of the noise equivalent power, NEP and responsivity yield $6\mu W/Hz^{1/2}$ and $R=890\mu V/W$, respectively. Here we used a rough estimate of the power falling on the sample, not including terahertz losses through cryostat windows and detector coupling. These figures of merit are not competitive with good incoherent detectors. The ability to trade response time for sensitivity and NEP needs to be addressed.

On the other hand the response time has been measured to be no slower than 700 nsec. This opens the possibility of using the device as a heterodyne detector taking advantage of electronics integrated with the detector to amplify the IF and further process the signal.

Acknowledgments

The authors would like to thank D. Enyeart and G. Ramian at the Center for Terahertz Science and Technology. This work was supported by the ONR MFEL program, the DARPA/ONR THz Technology, Sensing and Satellite Communications Program and the ARO, Science and Technology of Nano/Molecular Electronics: Theory, Simulation, and Experimental Characterization. Sandia is a multiprogram laboratory operated by Sandia Corporation, a Lockheed Martin Company, for the United States Department of Energy under Contract DE-AC04-94AL85000.

References

1. M. Kroug, S. Cherednichenko, H. Merkel, E. Kollberg, B. Voronov, G. Gol'tsman, H. W. Huebers, H. Richter, IEEE Transactions on Applied Superconductivity **11**, 962 (2001).
2. P. J. Burke, R. J. Schoelkopf, D. E. Prober, A. Skalare, B. S. Karasik, M. C. Gaidis, W. R. McGrath, B. Bumble, H. G. LeDuc, J. of Appl. Phys., **85**, 1644 (1999).
3. B. S. Karasik, W. R. McGrath, M. E. Gershenson, A. V. Sergeev, J. of Appl. Phys., **87**, 7586 (2000).
4. B. Bumble, H. G. LeDuc, J. A. Stern, K. G. Megerian, IEEE Transactions on Applied Superconductivity **11**, 76 (2001).
5. N. N. Iosad, V. V. Roddatis, S. N. Polyakov, A. V. Varlashkin, B. D. Jackson, P. N. Dmitriev, J. R. Gao, T. M. Klapwijk, IEEE Transactions on Applied Superconductivity, **11**, 3832 (2001).
6. See Molectron's website at www.molectron.com
7. T. W. Crow, R. J. Mattauch, R. M. Weilke, U. V. Bhapkar in Compound Semiconductor Electronics, p.209 , World Scientific Publishing, ed.by M.Shur (1996).
8. S. M. Marazita, W. L. Bishop, J. L. Hesler, K. Hui, W. E. Bowen, T. W. Crowe, IEEE Transactions on Electron Devices **47**, 1152 (2000).
9. E. E. Haller and J. W. Beeman in Proceedings of Far-IR, Sub-mm & mm Detector Technology Workshop, April 2002, Monterey, CA, in press.
10. G. E. Stillman, C. M. Wolfe, J. O. Dimmock, Solid State Comm., **7**, 5 (1969), (Proceedings of the third international conference on photoconductivity, Stanford, CA, USA, 12-25 Aug. 1969.)
11. G. Strasser, K. Bochter, M. Witzany, E. Gornik, Infrared Physics **32**, (1991).
12. M. Dyakonov and M. S. Shur in "Terahertz Sources and Systems", Kluwer Academic Publishers-Netherlands ed. by R. E. Miles, p187-207 (2001).
13. M. Dyakonov and M. S. Shur, Phys. Rev. Lett. **71**, 2465 (1993). M. Dyakonov and M. S. Shur, IEEE Trans. on Elec. Dev. **43**, 380, (1996).
14. M. Dyakonov and M. S. Shur in "Terahertz Sources and Systems" Kluwer Academic Publishers-Netherlands ed. by R. E. Miles, p187-207 (2001)
15. S. J. Allen, Jr., D. C. Tsui, and R. A. Logan, Phys. Rev. Lett., **38**, 980 (1977)
16. D. C. Tsui, E. Gornik, and R. A. Logan, Solid State Comm., **35**, 875 (1980)
17. P.J.Burke et al Appl. Phys. Lett., **76**, 745 (2000)
18. M. Dyakonov and M.S. Shur, Phys. Rev. Lett., **71**, 2465 (1993) and M.Dyakonov and M.S.Shur , IEEE Transaction on Electron Devices. **43**, 380, (1996)
19. J.-Q. Lü, M. S. Shur, J. L. Hesler, L. Sun, and R. Weikle, IEEE Electron Dev. Lett., **19** (10), 373 (1998) and J.-Q. Lu and M.S. Shur, Appl.Phys. Lett., **78**, 2587(2001)
20. W. Knap , Y. Deng, S. Rumyantsev , J.-Q. Lü, M. S. Shur, C. A. Saylor, L. C. Brunel, Appl. Phys. Lett., May (2002) (accepted – in print)
21. W. Knap , V. Kachorovskii , Y. Deng, S. Rumyantsev , J.-Q. Lü, R. Gaska, M. S. Shur, G. Simin, X. Hu and M. Asif Khan, C. A. Saylor, L. C. Brunel, Journal of Applied Physics June (2002) (accepted – in print)
22. X. G. Peralta, S. J. Allen, M. C. Wanke, N. E. Harff, J. A. Simmons, M. P. Lilly, J. L. Reno, P. J. Burke, J. P. Eisenstein, "Terahertz photoconductivity and plasmon modes in double quantum well field effect transistors", submitted to Applied Physics Letters.
23. G. E. Santoro and G. F. Giuliani, Phys. Rev. B **37**, 937 (1981).
24. S. Das Sarma and A. Madhukar, Phys. Rev. B **23**, 805 (1981).
25. D. S. Kainth, D. Richards, A. S. Bhatti, H. P. Hughes, M. Y. Simmons, E. H. Linfield, and D. A. Ritchie, Phys. Rev. B **59**, 2095 (1999).

International Journal of High Speed Electronics and Systems, Vol. 12, No. 2 (2002) 501–510

THERMAL MANAGEMENT IN OPTOELECTRONICS

DANIEL K. JOHNSTONE
Air Force Office of Scientific Research
Physics and Electronics Directorate
Arlington, VA 22203-1977, USA

One of the primary considerations in lasers and detectors for infrared is the operating temperature and its effect on device performance. Several research areas are converging on the problem to cooperatively propel devices operating in the infrared to new levels. The fabrication of devices is advancing by better understanding of defects, and growth of the materials and structures is better able to control the desired crystallinity and composition. Also, the active region of the device may be shielded from the heat and phonon generation processes can be reduced by appropriate design of the structures and selection of materials. Furthermore, typically bulky coolers can be miniaturized and integrated with the device during the fabrication process. Each of these areas are being pursued with the expectation that some of the alternatives down each path will be mutually beneficial toward the goal of catapulting laser operating temperatures, and detection of weak infrared signals at longer wavelength in smaller packages. Better device fabrication and integrated cooling, both in device structure and added integrated coolers, should contribute to the proliferation and benefits of infrared lasers and detectors.

1. Introduction

Nobody would expect to see a cryostat on an automotive application. Nor would it be feasible to launch heavy cryocoolers that would be needed for many space applications, where the launch costs are currently $10,000/lb. Yet, infrared sensor applications are being demonstrated in automotive collision avoidance systems, and space sensor capabilities are continuously improving due in part to steady maturation of material growth that limits generation or recombination due to defects, and improvements in design that limit undesirable thermionic emission. Further advances are being made on other fronts. Cryocoolers have shown progress in recouping some of the potential efficiency, where the efficiency for typical coolers was only about 1% at the operating temperature. Miniaturization and integration of cooling near the heat generating active region of the device is another high-payoff area. Another consideration related to the device structure and material selection is the area of phonon generation and transport. However, phonon engineering has the difficult position of countering the current way of thinking that "phonons are there, and there's nothing we can do about them."

2. Device Design Examples

Several things can be done through design to increase the operating temperature or reduce the cooling requirements for various types of devices. The measures taken depend, among other things, on the type of device, the material system, and the growth method. For concrete examples of how the problem of heat can be addressed, detector and laser design and growth can be taken as somewhat representative. Each of the factors listed above can have a marked impact on the maximum allowable operating temperature.

2.1 Type II InSb detector

First, consider the limitations of detectors imposed by the operating temperature. It is most desirable for all of the current in the detector to be generated from the target scene. However, another source of current is from generation centers in the depletion region. Defects in the active area of the device can emit electrons and holes, depending on the characteristics of the deep level. A deep level defect that is near the conduction or valence band can emit an electron or hole previously trapped under bias conditions that bring the Fermi level above the defect level. If the detector bias brings the Fermi level back below the level of the defect, the trapped carrier can then be re-emitted. Discrete defects at the edge of the depletion region, where the Fermi level crosses the defect energy level, can contribute to the noise in the device by random emission and capture of carriers. A larger source of current would be a defect that is closer to the center of the band gap, where there would be a constant emission of an electron and hole. Deep level cross section and the density of states determine the rate of emission. In bulk material, or away from heterointerfaces, the defects are discrete states. At interfaces, the picture can be worse.

Interface defects are characterized by a distribution in defect energies, with an associated interface state density. The generation rate from interface states can be greater than for discrete states because the rate-limiting step from interface states can depend on lower energies, with trapped carriers passing through the near-continuum of interface states. Therefore, it is important to reduce the density of both interface states and discrete states in the active region of the detector.

One figure of merit for detectors is the specific detectivity, D*. A limiting factor for the detectivity is the noise current which has two components; the background from the scene, and the parasitic currents generated in the device from defects, as given by

$$I_n^2 = \frac{4kT}{R_0}\Delta f + 2q^2\eta A\Phi_b\Delta f \tag{1}$$

Where k is Boltzmann's constant, Δf is bandwidth, A is the detector electrical area, q is electron charge, λ is wavelength, Φ_b is background flux, and η is average quantum efficiency. R_0 is the differential resistance determined by the doping level and the carriers released by defects. The first term is the noise current from the device. The second term is noise associated with the random arrival of photons from the scene. Not much can be done about the random arrival of photons, but the resistance of the device can be increased by appropriate measures.

In this example, the InSb material for the device was chosen for its band gap, which allows detection out to 19um. Antimonides are good for this wavelength range, but they are notorious for Auger processes where carriers can be lost by any of several nonradiative processes. For example, InSb is particularly susceptible not only due to its narrow band gap, but also because the split-off band is separated from the valence band maximum by an energy approximately equal to the band gap. One way to mitigate the Auger process is to form a superlattice from layers of InAs and GaSb. The band offset is type II, meaning that the conduction band of one material is below the valence band of the adjacent material, separating the charges, thereby reducing the probability for Auger processes.

Other measures can be taken in selection of the starting material, and growth of the structure. Miscut of the substrate was found to be of importance in obtaining specular growth and limiting the scattering of electrons. Starting temperature for the growth of the superlattice is important to reduce the defects. The buffer is grown at 500°C but the growth of the superlattice is done below 400°C. The initial growth temperature of the superlattice is very vulnerable to the process of the superlattice deposition. The fluctuation of the temperature was suppressed to minimum to ensure high quality growth later on. Also, the design of the superlattice is important to reduce the relaxation of strain. Next, since absorption takes place mostly around the interfaces where there is maximum wavefunction overlap for type II heterojunctions, the composition of the junction on either side has a large effect on the noise current. It is most desirable to have InSb-like interfaces for best performance. However, the growth window is very narrow. Therefore, precise control of the growth temperature, to within $\pm$2°C over five hours, and appropriate III/V flux ratio, are critical. Outside of this growth window, dislocations can form and propagate through the rest of the grown layers. With proper selection of the material system, the starting substrate material, device architecture, and growth conditions, the background limited conditions for 12um can be as high as 80K.

2.2 Multi-quantum well InGaAsP/InP laser

Carriers injected into an InGaAsP/InP laser can be lost due to defects, but the primary loss in a mature material growth process is due to thermionic emission over heterojunction barriers in this example. This has the effect of raising the threshold current for lasing, and decreasing the slope efficiency. In a laser composed of multi-quantum wells in the InGaAsP/InP system the band offset is mostly in the valence band, which aggravates the confinement of electrons. Furthermore, the higher mobility of electrons compared to holes increases the loss rate of electrons over the heterobarriers. The problem of carrier loss can be addressed by adding to the barrier for thermionic emission by appropriate doping (Fig. 1). In this case, engineering the zinc doping profile on the p-type side of the device so that the edge of the Zn doping profile is moved toward the edge of the first quantum well increases the band bending both at the cladding layer and at the separate confinement heterojunction. In this manner, the leakage current density at an injection density of 4 kA/cm^2 at 80K decreased from 200 A/cm^2 down to 10 A/cm^2, more than an order of magnitude.

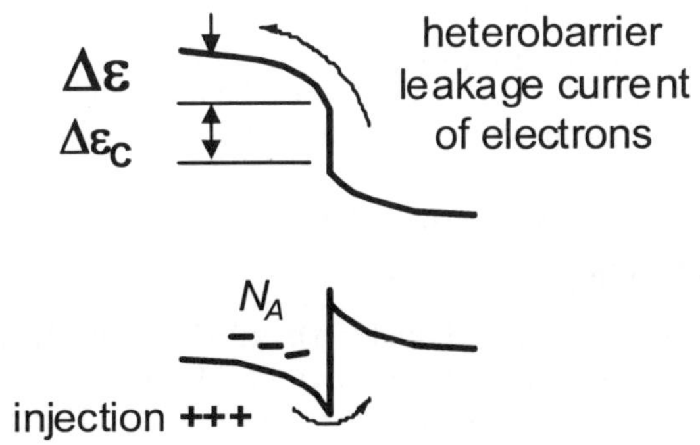

Fig. 1 Augmenting the thermionic emission barrier using the additional band bending from doping.

Attention to the details such as growth temperature stability, growth temperature, strain, and material for detectors, and doping profile control of the band bending for lasers are examples of measures that can be used to increase the operating temperature and stability in these types of devices. Minimal changes in growth are the most desirable methods of increasing the operating temperature. Design considerations are only partially effective for some applications, though. If greater detectivity or laser power is required, external cooling may be necessary.

3. Cryocoolers

For external cooling, several types of cryocooler are available based on Gifford-McMahon, Stirling, Joule-Thomson, Turbo-Brayton, pulse tubes, and in various combinations. Progress has been very measured in this field over the past few years. In 1998, the state of the art cryocooler had a specific mass (ratio of cryocooler mass to cooling power) of 1.5 kg/W, for a cooling power of 8 W, and specific power (ratio of power input to cooling power) of 14 W/W. Now, the specific mass has been reduced by 67% to 0.49 kg/W, the cooling power increased by 88% to 15W, and the specific power reduced by 20% to 11.2 W/W. Fig. 2 shows a pulse tube cryocooler capable of 10W of cooling at 95K. The improvements were based on a redesigned high capacity compressor. Next generation improvements are expected to be based on solid state and microelectromechanical technology (MEMs). One embodiment of solid state cooling would rely on the anti-Stokes process to absorb phonons, and eventually transfer the energy to fluorescence.[1] One program at the Defense Advanced Research Projects Agency (DARPA) sought to bring the cooling closer to the active area where the heat is being produced, reducing the cooling required for intervening materials.

Fig. 2 High efficiency space flight qualified cryocooler. Mass is less than 10kg.

4. Integrated Device Cooling

Several of the technologies pursued under DARPA's Heat Removal By Thermo-Integrated Circuits (HERETIC) project are based on MEMs technology, and others are based on semiconductor device physics. The objective of the HERETIC program is to develop solid-state and fluidic heat removal devices that are integrable with dense, high-performance electronics and photonics. These devices are envisioned to short-circuit the thermal resistance between heat sources and thermal sinks; they should also lead to a reduction in overall system volume and weight. It is expected that devices will be used for high efficiency operation in terms of work done to remove a unit of heat from a hot junction.

The universities and companies involved are investigating exciting and promising technologies. The following is a brief survey of some of the individual programs. Georgia Institute of Technology is developing micromachined synthetic jets for controllable heat convection and vibration induced droplet atomization (VIDA) where cooling uses phase-change heat transport. A prototype VIDA-based heat transfer cell is capable of removing 100W at 110°C and internal pressure of 4.5 psia. Rockwell Science Center is developing MEMs-based thermoacoustic refrigerators. Electrokinetic micro-coolers, including an electrokinetic pump for liquid/vapor phase change, shows promise at Stanford University. The inverse Nottingham effect is being developed by the University of North Carolina at Charlotte, where hot electrons are extracted from a high-field emitter, assisted by a resonant tunneling structure at the surface, and replaced by thermalized electrons in a low-field region.

4.1 Thermoelectric microcooler

We can look in more detail at one of the programs under the HERETIC project for an example of the efforts to integrate cooling with the device. NASA's Jet Propulsion Laboratory is developing thermo-integrated microcoolers (TEMC). The successful development of such monolithically integrated thermoelectric microcoolers with mid-IR lasers into a near room temperature thermal management package will free such lasers from the requirement of liquid cryogens and provide compact, low cost IR, and coherent IR sources for dual use applications. This project is focusing on the development of some key technologies: advanced thermal modeling tools, including a physical understanding of relevant thermionic and thermoelectric processes; development of miniaturized monolithically integrated thermoelectric microcoolers operating under transient conditions; ultra-high resolution thermal imaging microscopy for thermal characterization and modeling; direct integration of mid-IR lasers with near room temperature solid-state packaging and demonstration of significantly enhanced performance. The concept exploited is that cooling power scales proportional to the cross-sectional area divided by the length of the Bi_2Te_3 p-n thermoelectric leg. It should be possible to achieve 200 W/cm^2 cooling power at the reduced scale shown in Fig 3. Also, of interest for pulsed laser operation, transient deep

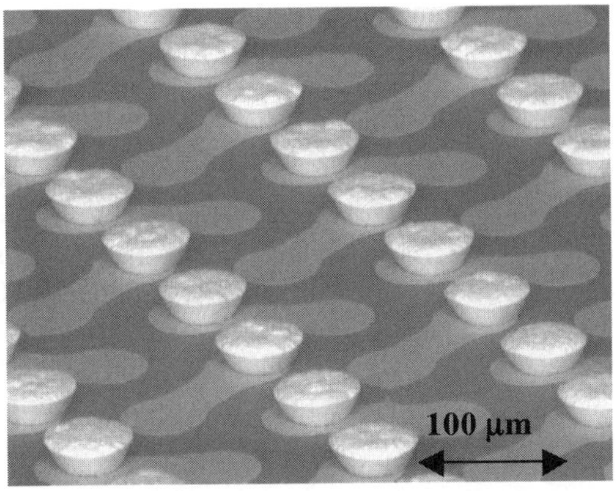

Fig. 3 Reduced size thermoelectric microcoolers based on Bi_2Te_3. Shortened legs, 10-50 μm, and pulsed operation will enhance the cooling capabilities.

spot cooling has been demonstrated where additional cooing from a pulsed current is achieved by delaying the Joule heat propagating to the cold junction. Under steady state conditions the cold side temperature can be as low as 230K. The maximum cooling is theoretically limited to a change of 80K. However, superimposing and shaping a transient pulse on the steady state current can result in temperatures as low as 100K.[2]

5. Phononics

Phonons are of course at the heart of the problem. Considering that phonons can be confined by quantum well structures, it is conceivable to change the device structure to control the generation and flow of phonons.[3] As an example system, in GaAs/AlAs superlattices or quantum wells, the optical phonon branches don't line up, resulting in confinement of the phonons.[4] With that as a basis, several groups have been working to characterize and model phonons, and develop methods of generating coherent phonons in order to influence the scattering processes. The modeling and tools are being used to enhance the operation of existing devices, and to come up with some new types of devices. An important part of the effort is focused on artificial semiconductor heterostructures such as quantum well, quantum dots and superlattices. Unlike bulk materials, the phonon properties of these low-dimensional systems and, in particular, the phonon frequency, group-velocity, spectral density as well as the strength of the interaction between phonons and carriers can be widely modified to improve the performance of devices. Existing devices that are being considered for modification by phonon participation by design are field-effect transistors and quantum-heterostructure lasers and detectors. New devices that will emerge will use phonons to assist in lasing and tunneling, and for light emission in indirect-gap materials. Specific examples are presented here in summary form on the effects of injected acoustic phonons on stimulated longitudinal optical (LO) phonon decay, a Si/Ge laser pumping by phonons, generation of coherent phonons associated with electron transport through a GaN/AlGaN field effect transistor (FET) structure, the phonon density of states in quantum wells and quantum wires, and generation of coherent phonons in a quantum dot diffraction grating for phonon annihilation.

5.1 Stimulated phonon decay

The spontaneous decay of LO phonons in many III-V materials is not very fast, typically on the order of 10-100 ps. However, injecting acoustic phonons can cause stimulated decay of the LO phonons. In InP the dominant decay channel of LO phonons are to convert LO phonons to transverse optical (TO) and combinations of longitudinal and transverse acoustic phonons. The lifetime of LO and TO phonons is 200 ps and 10 ps, respectively. If acoustic phonons could be injected to stimulate the decay, then more than an order of magnitude reduction in LO lifetime could, in principle, be achieved. From a theoretical analysis, there is a threshold acoustic phonon density,

$$N_c = \frac{2\pi f^2}{\tau_{TO} \upsilon^3}, \tag{2}$$

where f and v are the frequency and group velocity of the TO phonons, and τ_{TO} is the TO lifetime. For InP, the threshold is on the order of 10^{19} cm^{-3}, or an energy density of 8 mJ/cm^3. The dependence of the stimulated LO phonon lifetime, τ^*, on the density of injected LA phonons is

$$\frac{\tau_{LO}}{\tau^*} = \frac{1 + {N_C}/{N_{LA}}}{{\tau_{TO}}/{\tau_{LO}} + {N_C}/{N_{LA}}}. \tag{3}$$

From this equation, to obtain a ten-fold reduction in LO lifetime, the density of injected LA phonons must exceed N_c by a factor of 20. This phonon density corresponds to a flux density of 5x 10^{25}cm^{-2}s^{-1} for LA modes or, alternatively, to a phonon flux of 5x10^{19} phonons/second corresponding to delivering about 40 mW of power into a typical FET active region of 100 μm^2. The strain induced would be on the order of 0.13%, which can be obtained using ultrafast optical excitation. These are all realistic values and are currently being tested.

5.2 Phonon pumped laser

A prime example of a new device that would use phonons as a principle driver for device operation is the phonon pumped laser.[5] In the scheme shown in Fig. 4, a temperature gradient in the direction of growth along a superlattice waveguide structure is used to

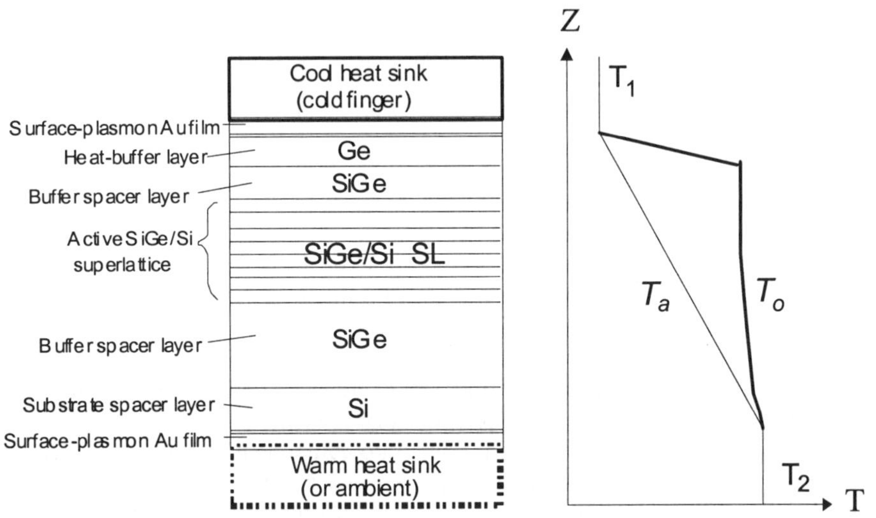

Fig. 4 Phonon pumped laser structure and temperature gradient of acoustic and optical phonons.

depopulate the lower laser level shown as HH1 in Fig. 5. In the Si/Ge system, phonon absorption can be a dominant factor if the laser photon energy is less than the optical phonon energy. The important aspect of this structure is that there is a mismatch in the phonon spectrum of the germanium heat buffer layers for optical phonons, but not for acoustic phonons. Optical phonons generated in the superlattice and supplied by the warm heat sink are reflected by the SiGe/Ge interface, while acoustic phonons are transmitted. This creates favorable conditions for optical phonons to excite a hole from HH1 to HH2, and relax to the superlattice zone boundary by emission of acoustic phonons. At the zone boundary, photon emission accompanies the transition from HH2 to HH1, followed by acoustic phonon emission as the carrier relaxes back to the minimum of HH1 at zone center. All of this depends on cooling of the acoustic phonons. Currently, methods of cooling the acoustic phonons are being sought. Sizable gain is expected for a reasonable temperature difference between the acoustic and optical phonons.

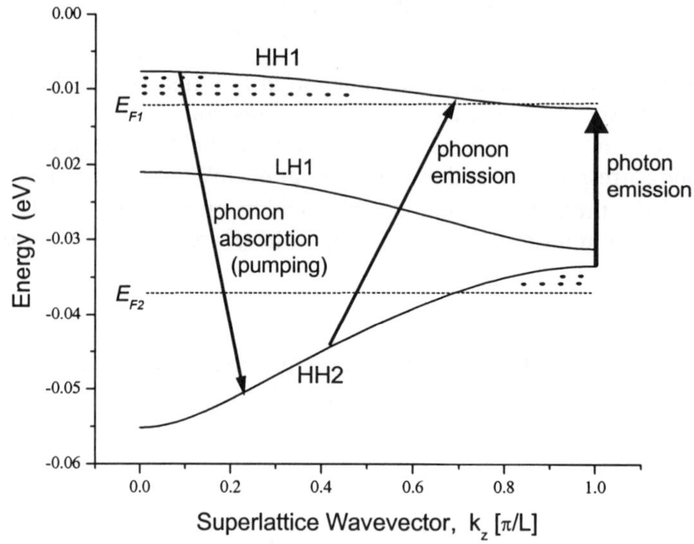

Fig. 5 Phonon pumping and THz lasing in SiGe/Si superlattice valence band.

5.3 Coherent phonons in GaN/AlGaN FETs

Under the strong piezoelectric effects and spontaneous polarization in AlN and GaN the dynamic effects when electrons are injected are expected to be a dominant effect. The band bending of these effects can be hundreds of millielectron volts. In a nitride quantum well structure, strong injection of electrons can result in large changes in the piezoelectric field and the spontaneous polarization by screening. The screening induced change then modifies the strain and hence, the confinement energies. The interesting result is that the strain redistribution launches acoustic phonons that are coherent.

The research into this effect breaks it down into component effects. One element is to study charge injected by ultrashort laser pulses into a GaN or InGaN strained QW at current levels realistic for an FET. The injection creates a large modulation in the piezofield by screening, leading to a large change in the local strain. The change in strain launches an acoustic phonon pulse, which is optically detected by the change in index of refraction. In this way, optical injection and detection experiments have generated 100GHz pulses. Next, FET devices will be used to look for and characterize coherent phonon emission originating from high frequency operation at high injection. Then, ultrafast optical techniques will be used to control and modulate GaN/AlGaN FETs at microwave frequencies. Several other approaches are also being investigated to generate coherent phonons and control the direction of propagation.

5.4 Modeling and phonon annihilation

Phonons will play an increasingly important role in the operation of devices with reduced feature sizes because of the generation of high heat densities. CFD Research, Huntsville, Alabama, is developing a simulation tool that can be used for design and optimization of heat removal procedures in sub-micron devices. Their approach is to combine the classical Boltzmann transport equation for phonons with a transfer matrix method to capture quantum effects. So far, the reduced size effects are found to be strong, but detrimental in nanoscale devices.[6] The simulation tools are being used to investigate new concepts of cooling including phonon waveguides, phonon bandgaps, wave annihilators, and spreaders.

One test-bed for numerical modeling is a CdS array of quantum dots formed in a porous anodized alumina film. This system was selected because formation of quantum dots in pores does not depend on a strain field being present, as it does for Stranski-Krastanow growth. Elimination of the strain simplifies the simulation, making it easier to discern confinement effects from strain effects. In a periodic structure such as this, the goal is to adjust the geometry of the device to act as a phonon filter or grating, rather than generating non-coherent acoustic phonons as is the usual case for bulk material. Under appropriate conditions, coherent or partially coherent phonon waves may form in some directions and be annihilated in other directions. With this structure, phonon carrier scattering can be tuned through periodicity, acoustic mismatch, and dot shape.

6. Conclusions

Studies that develop a deeper understanding of phonon generation and transport processes in the setting of reduced dimensional structures should enable new concepts for thermal management. The ability to generate or convert heat to coherent phonons is a valuable tool in controlling the heat in advanced devices. Phononics, or phonon engineering, is seen as secondary to electronics or photonics, but will play an increasingly important role especially as new types of devices, such as the phonon pumped laser, or existing but modified devices exhibit significantly enhanced performance.

As the pieces of the phonon investigations mature, using the physics of phonons will alleviate the burden on other thermal management methods and widen the trade space. The three methods can have a complementary nature. Reducing the defects in a detector allows a higher operating temperature, as does changing the doping profile to add band bending to the heterojunction band offset in a laser structure. New phonon management tools and methods can also alleviate the external cooling requirement, as well as leading to new types of devices.

Acknowledgements

This is a small piece of the broader body of hard work of several groups. Manijeh Razeghi, Northwestern University, is doing the work on detectors for cold body detection. Gregory Belenky from SUNY, Stony Brook, is working on improved lasers for communications, presented here. Thanks also to Barrett Flake, Air Force Research Lab's Space Vehicles Directorate, Kirtland Air Force Base, for the information on advanced cryocoolers. Jean-Pierre Fleurial, Jet Propulsion Laboratory, is responsible for development of the thermoelectric microcooler. Arto Nurmikko, Brown University, and Roberto Merlin, University of Michigan, are each leading multi-university teams studying the phonon-enhanced devices. Andrzej Przekwas is the principle investigator at CFDRC working on the modeling and demonstration of phonon annihilation methods, along with Alexander Balandin, University of California Riverside.

References

[1] S. Chu and T. R. Gosnell, "Laser cooling by 65K of Yb3+-doped ZBLANP glass," Quantum Electronics and Laser Scinece Conference, (2000) 179-180.

[2] J. P. Fleurial, G. J. Snyder, J. A. Herman, P. H. Giauque, W. M. Phillips, M. A. Ryan, P. Shakkottai, E. A. Kolawa and M. A. Nicolet, "Thick-Film Thermoelectric Microdevices," 18th international Conference on Thermoelectrics (1999) 294-300.

[3] X. Zianni, C. D. Simserides and G. P. Triberis, "Electron scattering by optical phonons in $Al_xGa_{1-x}As/GaAs/Al_xGa_{1-x}As$ quantum wells," Physical Review B, Vol 55, No 24, (1997) 16324-16330.

[4] D. J. Mowbray, M. Cardona and K. Ploog, "Confined LO phonons in GaAs/AlAs superlattices," Physical Review B, Vol 43, No 2, (1991) 1598-1603.

[5] G. Sun, R. Soref, J. Khurgin, "Phonon Pumped SiGe/Si Interminiband Terahertz Laser," IEEE Journal on Selected Topics in Quantuim Electronics, Vol. 7, No. 2, (2001) 376-380.

[6] J. Zou and A. Balandin, " Excess Heat Removal from Nanoscale Electronic Devices," Proceed. of Int'l Symposium of Quantum Confinement, Electrochemical Society Proceedings Vol 2001-19, (2001) 70-80.

International Journal of High Speed Electronics and Systems, Vol. 12, No. 2 (2002) 511–520

THE MULTIPLE-GATE MOS-JFET TRANSISTOR

BENJAMIN J. BLALOCK[1], SORIN CRISTOLOVEANU[2], BRIAN M. DUFRENE[3], F. ALLIBERT[2], & MOHAMMAD M. MOJARRADI[4]

[1]*Department of Electrical and Computer Engineering*
The University of Tennessee, Knoxville, TN 37996-2100; USA

[2]*Institute of Microelectronics, Electromagnetism, and Photonics (UMR CNRS)*
ENSERG, BP 257, 23 rue des Martyrs, F38016 Grenoble Cedex 1, France

[3]*Department of Electrical and Computer Engineering*
Mississippi State University, Mississippi State, MS; USA

[4]*Jet Propulsion Laboratory*
Pasadena, CA; USA

A new SOI device, the MOS-JFET, has been developed that combines two different transistors, JFET and MOSFET, superimposed in a single silicon island so that they share the same body. A unique attribute of the MOS-JFET is that it can be viewed as a four-gate transistor (two side junction-based gates, the top MOS gate, and the back gate activated by SOI substrate biasing). Each of these four gates can control the conduction characteristics of the transistor. This novel transistor's multiple gate inputs give rise to exciting circuit opportunities for analog, RF, mixed-signal, and digital applications. Measured results of MOS-JFET transistors, fabricated in a conventional partially-depleted SOI technology, demonstrate that the device is fully operational. From the experiments and systematic 2-D simulations, typical regions of operation are identified. These results indicate that optimum performance is reached when the MOS and junction field-effects are combined.

1. Introduction

The MOS-JFET is a new device that takes advantage of the unique flexibility and isolation capability of SOI technology to combine both MOS (Metal-Oxide-Semiconductor) field-effect and junction field-effect for conduction control within a single transistor body. Implementing, in bulk or SOI technology, an MOS transistor with more than 2 independent gates is difficult to imagine. By contrast, through the combination of MOS gates and junction-based gates, the MOS-JFET allows manipulation of multiple gates, from one to four, according to the application envisioned. Utilizing these gates separately, or in a combinatorial fashion, provides new circuit opportunities for analog, RF, mixed-signal, and digital applications.

In this paper, we describe the device structure and processing (section 2), and present experimental results (section 3). 2-D simulations (section 4) are used to clarify the mechanisms of operation, optimize the device architecture, and envisage pertinent applications (section 5) in micro- and nano-electronics.

2. The MOS-JFET Device Structure

The n-channel MOS-JFET (Figure 1) is constructed from a regular p-channel MOSFET that has two independent body contacts located on either side of the MOSFET. The body contacts are used as source and drain for the 'transversal' n-channel JFET (Junction Field-Effect Transistor). It is apparent from the MOS-JFET structure that no specialized processing is required to manufacture the device. The standard source and drain of the MOSFET are promoted as side-gates for the JFET and used to control the width of the conductive path. The maximum width roughly corresponds to the gate length of the original MOSFET, whereas the JFET channel length is defined by adjusting the 'width' of the MOSFET. Interestingly, when only the two side gates are operated, the MOS-JFET is somewhat similar to a double-gate MOS transistor that has recently attracted much interest [1]. An independent work used a similar device structure to implement either a MOSFET or a JFET, based on the demands of various digital applications [2]. The unique nature of our investigation stems from combining both MOS and junction field-effects to achieve current modulation through four-gate manipulation.

Also note that CMOS technology scaling should enhance the performance of the MOS-JFET transistor as MOS gate length shrinks (therefore reducing the lateral separation between the junction gates). This implies that it should be easier to pinch-off the conduction channel, particularly if body doping is optimized. Subsequently the MOS-JFET is highly amenable to modern trends in the microelectronics industry.

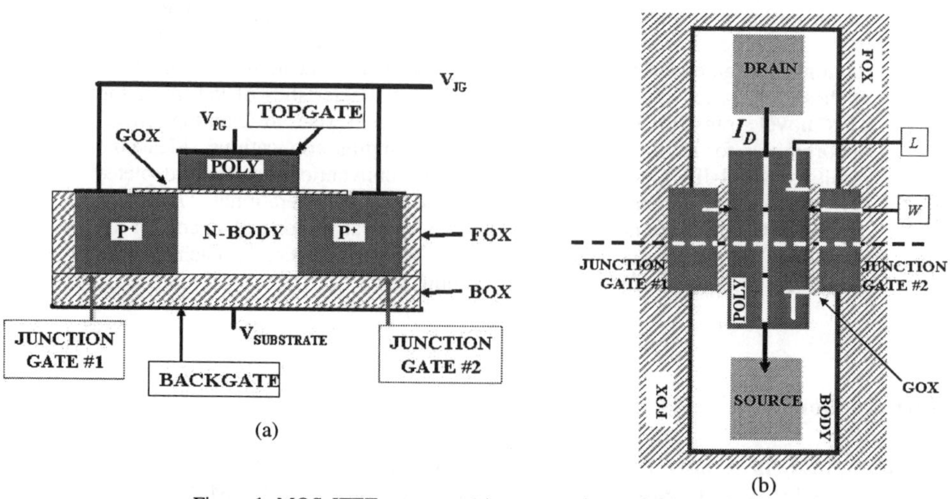

Figure 1: MOS-JFET structure: (a) cross-section and (b) top view.

3. Measured Characteristics

Extensive testing has demonstrated that the MOS-JFET is fully operational. Some of the measured characteristics will now be described to illustrate typical modes of operation. These measurements were taken from MOS-JFET devices fabricated in a conventional 0.35-micron partially-depleted SOI (PDSOI) technology with drawn features of L=1.5μm and W=0.35μm. For simplicity, both junction gates are tied together (Figures 2-10). In Figure 2, the measured I_D vs. V_{DS} characteristic is provided with substrate (or back-gate) voltage biased at V_{SUB} = 0V. V_{PG} is the top-gate (or poly-gate) voltage and V_{JG} is the junction-gate voltage. Curves corresponding to fixed V_{JG} and fixed V_{PG} are labeled accordingly. The fixed V_{PG} curves demonstrate the depletion-mode nature of the device. With V_{JG} set to 0V and V_{PG} swept in a positive fashion, the top-gate becomes accumulated and the drain current increases. For fixed V_{PG}, applying a negative junction bias (V_{JG} from 0 to –2 V) narrows the effective width, causing the current to decrease.

In Figure 2, V_{DS} is swept to 10V, implying that the MOS-JFET has a V_{DS} breakdown voltage in excess of 10V. Indeed this is true as the junction-to-body diode enters avalanche breakdown for V_{DS} near 15V. Given that these devices were fabricated in 3.3V technology, this result is quite encouraging from the standpoint of applying the MOS-JFET in higher voltage applications such as 5V interface circuits or MEMS-based sensors interface and control electronics.

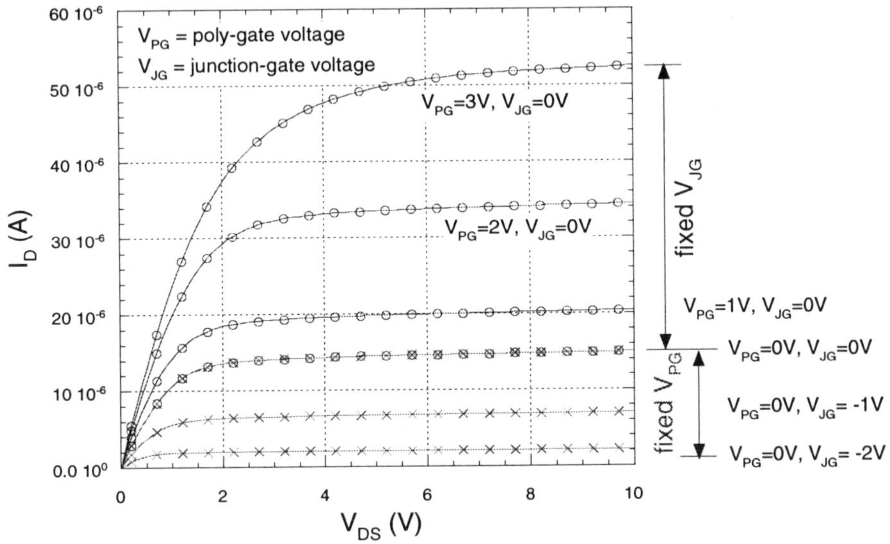

Figure 2: Measured I_D vs. V_{DS} for an N-channel MOS-JFET with L=1.5μm, W=0.35μm, and V_{SUB}=0V.

In Figure 3, the measured I_D vs. V_{JG} characteristic for ohmic operation (V_{DS} = 0.1 V) is shown with V_{PG} swept from –2V (depletion) to +1V (accumulation). The junction-gates provide cutoff control for the MOS-JFET over the entire range of top-gate biasing from depletion to weak accumulation. The subthreshold slope (Figure 3b) is improved for depletion at the front interface. When the top gate is accumulated (Figure 4: V_{PG} stepped from 0V to +3V), the MOS-JFET conductivity is enhanced at the expense of limiting the device cutoff and subthreshold swing.

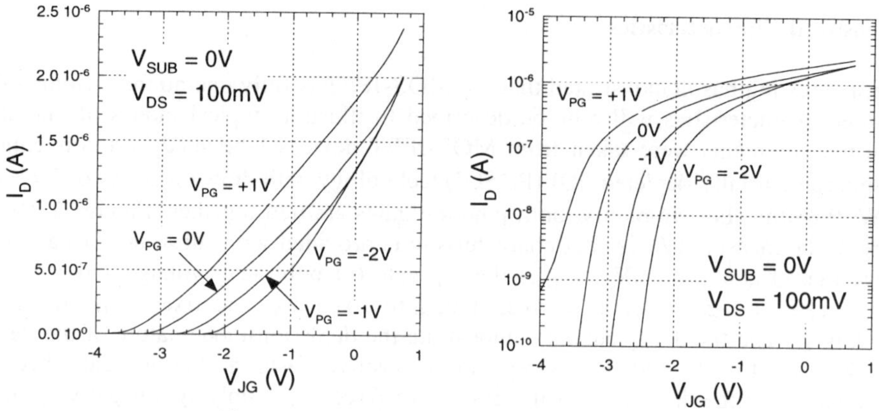

(a) (b)

Figure 3: Measured I_D vs. V_{JG} shown on (a) linear scale and (b) semilog scale (V_{SUB}=0V).

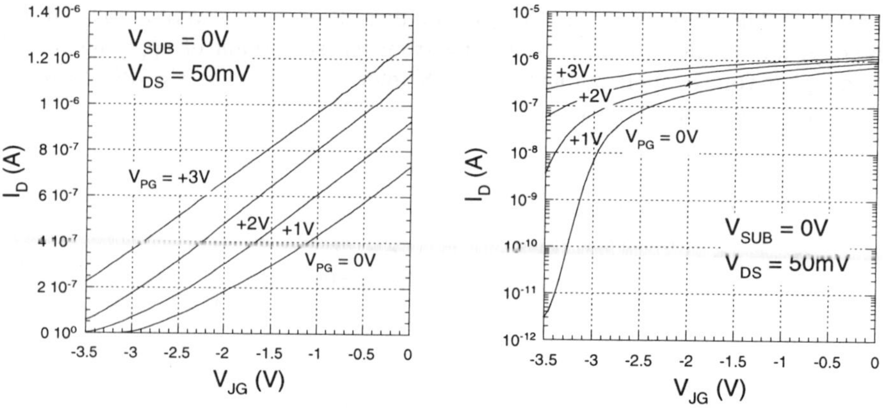

(a) (b)

Figure 4: Measured I_D vs. V_{JG} for top-gate accumulation shown on (a) linear scale and (b) semilog scale.

The reciprocal I_D vs. V_{PG} characteristics for linear operation are shown in Figure 5. Top-gate depletion can achieve cutoff, but only if the junction gates are also depleted. Note the junction-gate bias influence on the MOS-JFET's 'threshold' voltage V_T as seen from the top-gate: V_T increases with junction-gate depletion. More negative is V_{PG}, narrower is the channel, and less front-gate depletion is needed for controlling the subthreshold slope and V_T. From Figure 6, it is apparent that top-gate accumulation altogether prevents device cutoff, even for high-level depletion at the junction-gates.

Figure 7 provides the MOS-JFET's transconductance characteristic, g_m vs. V_{PG}, as a function of junction-gate bias during saturation-mode operation. Transconductance increases as the junction-gates become less depleted, again because the channel is wider. The dual-slope nature of g_m, for V_{PG} from 0V to approximately 0.75V, then from 0.75V to about 2V, is indicative of the conduction mode changing from neutral to accumulation as V_{PG} is increased. Also evident is the classical MOSFET-like roll-off of g_m for high-level top-gate accumulation (V_{PG} > 2.5V), induced by mobility degradation from surface scattering.

Figures 8 through 10 illustrate the role of the back-gate. With –20V applied to the substrate, the back-gate is heavily depleted and the channel thickness is reduced. Device cutoff is easily achieved for depletion at the junction-gates and back-gate, even while the top-gate is mildly accumulated. In addition, the current swing for the $V_{JG} = -1$V and -2V curves reveal attractive subthreshold characteristics.

Additional back-gate depletion data is shown in Figure 9 for I_D vs. V_{JG} and g_m vs. V_{JG} measurements. Complete device cutoff (full depletion) is readily achieved, with an excellent subthreshold swing (below 100 mV/decade, when $V_{PG} = -2$V, see Figure 9(a)). Figure 9(b) shows exotic double-hump transconductance curves.

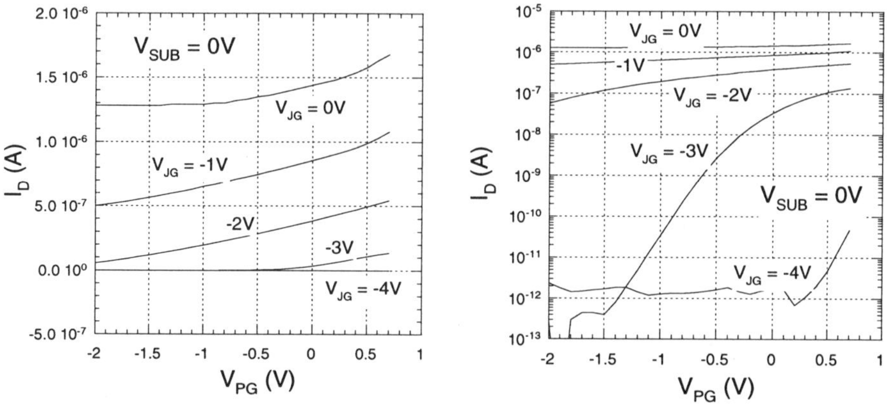

(a) (b)

Figure 5: Measured I_D vs. V_{PG} with V_{DS} = 100mV shown on (a) linear scale and (b) semilog scale.

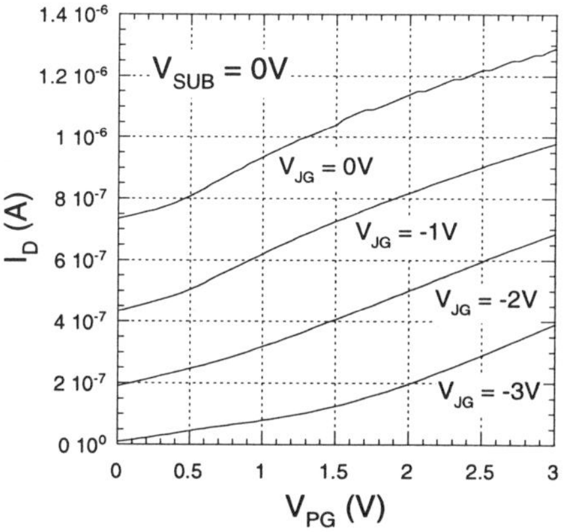

Figure 6: Measured I_D vs. V_{PG} shown on linear scale (V_{DS} = 50mV).

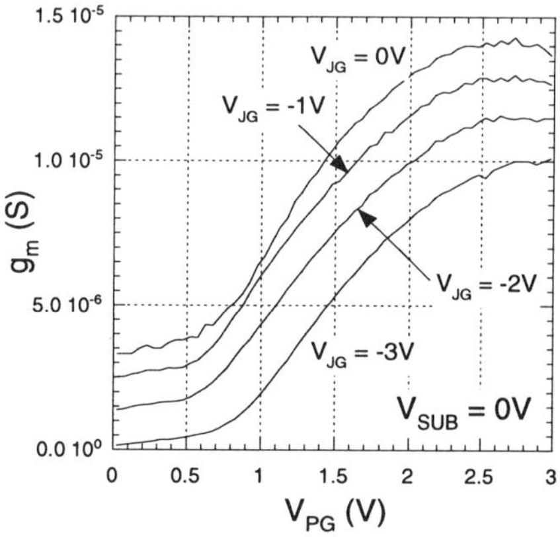

Figure 7: Measured g_m vs. V_{PG} for saturation operation (V_{DS} = 5V).

Figure 10 reproduces I_D vs. V_{PG} and I_D vs. V_{JG} characteristics while the back-gate is heavily accumulated (V_{SUB} = +20V). This condition prevents device cutoff and large currents are possible. Though they cannot force full depletion at the bottom of the film, the junction-gates still maintain good control of current magnitude.

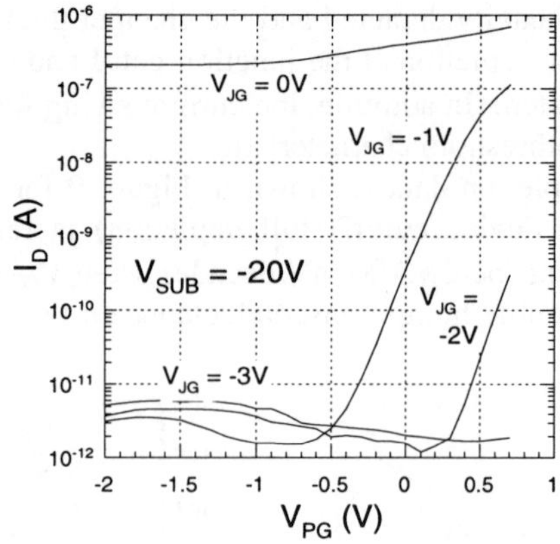

Figure 8: Measured I_D vs. V_{PG} demonstrating cutoff control using back-gate and junction-gates depletion.

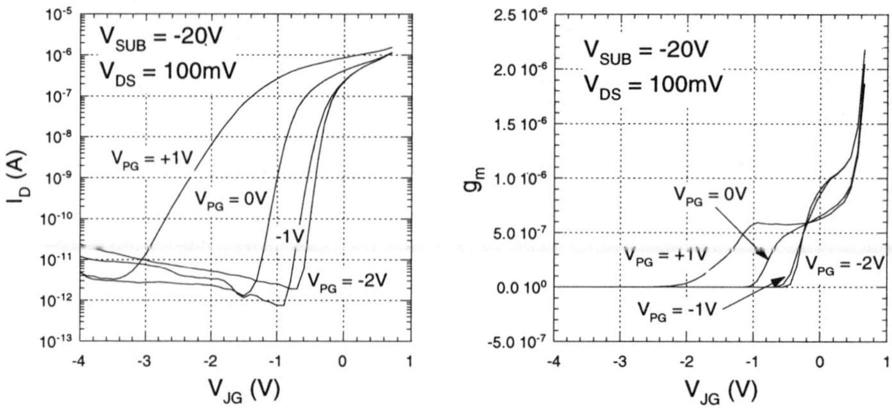

(a) (b)

Figure 9: Measured (a) I_D vs. V_{JG} (semilog scale) and (b) g_m vs. V_{JG} under back-gate depletion condition.

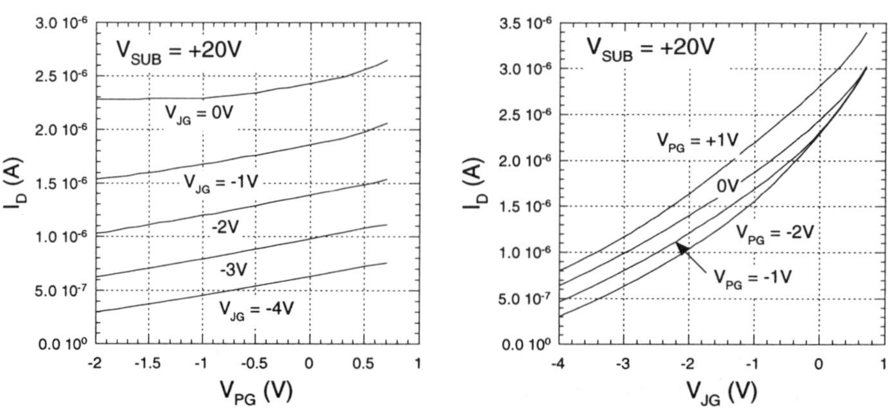

(a) (b)

Figure 10: Measured (a) I_D vs. V_{PG} and (b) I_D vs. V_{JG} under back-gate accumulation condition (V_{DS} = 100mV).

4. 2-D Simulations of MOS-JFET Cross-Section

Two-dimensional numerical simulations of the MOS-JFET cross-section are presented to help illustrate the conduction mechanisms under different gate bias conditions as well as the pattern for device optimization. Both wide-channel (W=10μm) and narrow-channel (W=0.3μm) are considered. Just as with the measured results, both junction-gates are tied together during the simulations. The results for various doping levels (selected through an iterative process) are shown but are only qualitatively indicative of the actual device described in the previous section.

4.1 Wide-channel simulation:

Figure 11 shows wide-channel simulation results under two different back-gate bias conditions: (a) V_{SUB} = 0V and (b) V_{SUB} = −10V. The status of various body regions is indicated with specific colors: accumulation (black), quasi-neutral (grey), and full depletion (white). The following parameters were selected and included in the Silvaco simulator: W = 10μm, $N_A = 2\times10^{20}$ cm^{-3}, $N_D = 5\times10^{17}$ cm^{-3}. With V_{PG} = −1V and V_{SUB} = 0V, only the top-channel is depleted in Figure 11(a), whereas both top and bottom channels are depleted for V_{SUB} = −10V in Figure 11(b). For this last case, a conductive plate with adjustable thickness is formed. A wide channel inhibits junction-gate control, and only the MOS field-effect (from the top and bottom) influence conduction.

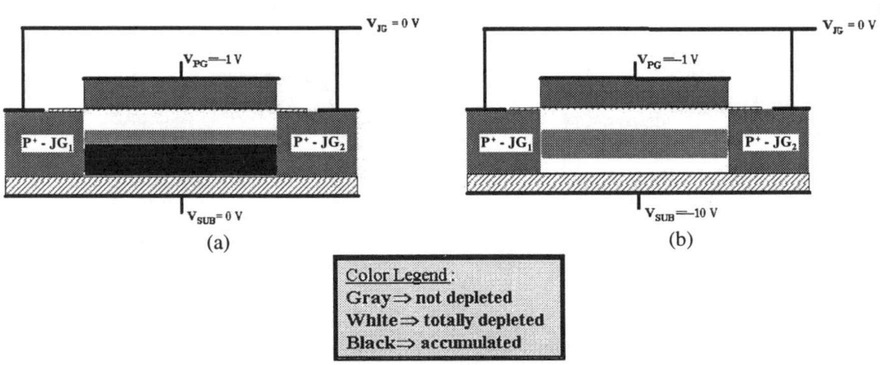

Figure 11: 2-D simulation of wide-channel MOS-JFET with (a) top-gate depletion and (b) top-gate and back-gate depletion (W = 10μm, $N_A = 2\times10^{20}$ cm^{-3}, $N_D = 5\times10^{17}$ cm^{-3}).

4.2 Narrow-channel simulation:

Figure 12 shows narrow-channel simulations with top-gate bias swept from (a) V_{PG} = +0.5V (accumulation), to (b) V_{PG} = 0V and (c) V_{PG} = −0.75V (depletion). These results correspond to W = 0.3μm, and grounded V_{PG} and V_{SUB}. With the top-gate in accumulation, conduction in the entire film (device body) is possible. By driving the top-gate into depletion, extremely narrow conductive 'wires' can be realized at the bottom of the film.

The additional impact of biasing the back-gate and junction-gates in depletion is illustrated in Figure 13. Just as with the measured results, depleting all four gates provides full film depletion (Figure 13(c)). And with bottom- and side-gate depletion, top-gate accumulation is used to maintain a conduction channel in the upper region of the film (Figures 13(a,b)).

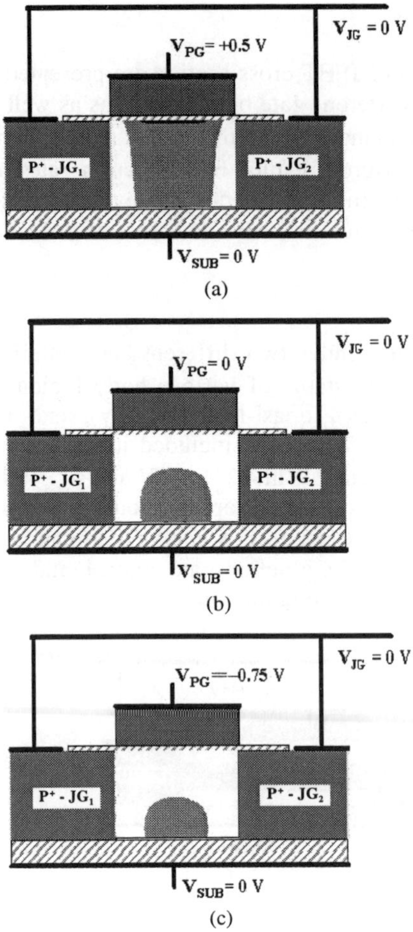

Figure 12: 2-D simulation of narrow-channel MOS-JFET with (a) top-gate accumulation and (b,c) top-gate depletion (W = 0.3μm, $N_A = 2\times10^{20}$ cm^{-3}, $N_D = 2\times10^{17}$ cm^{-3}).

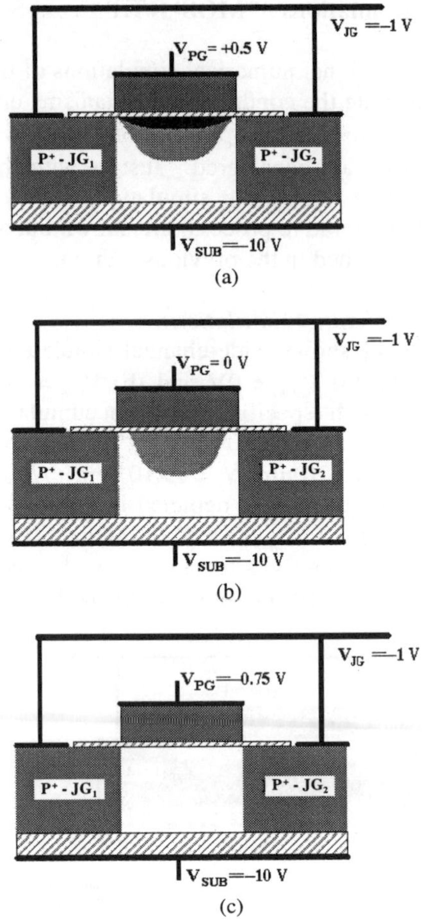

Figure 13: 2-D simulation of narrow-channel MOS-JFET with depleted back-gate and side-gates, and with top-gate biased in (a) accumulation, (b) neutral, and (c) depletion.

The simulation results provided in Figure 14 use the same V_{JG} and V_{PG} biasing levels as in Figure 13, but with V_{SUB} = +10V to accumulate the back-gate. With both top- and bottom-gates accumulated, high current is possible through channels formed at the top and bottom of the film, in addition to the film volume. Current modulation using V_{JG} should be possible, particularly when the top-gate is depleted.

With V_{SUB} and V_{PG} grounded, the simulation results shown in Figure 15 emphasize the influence of the junction-gates. As V_{JG} is taken more negative, the conduction path becomes narrower as the device approaches the off-state regime requiring full depletion. This is a quantum wire, located at the bottom of the film, whose width is controlled by the side-gates.

Figure 16 shows the unique four-gate modulation of conduction made possible by this device. In this simulation all four gates are depleted using V_{SUB} = –25V, V_{JG} = –1.75V, and V_{PG} = –2V. Under these conditions, the conduction "aperture" formed in the center portion of the transistor body is a 'floating' quantum wire. Furthermore, by varying each gate's level of depletion, the wire size and position is controlled within the film. Never before has the circuit designer been given so much control over conduction within a single transistor. Also by maintaining conduction away from both the top and bottom oxide interfaces, the MOS-JFET should achieve very low flicker noise. Interestingly, since the conduction aperture can be 'focused' to very fine dimensions, a quantum wire may be achieved and electrically modulated. This offers enhanced flexibility as compared with quantum wires defined by lithography.

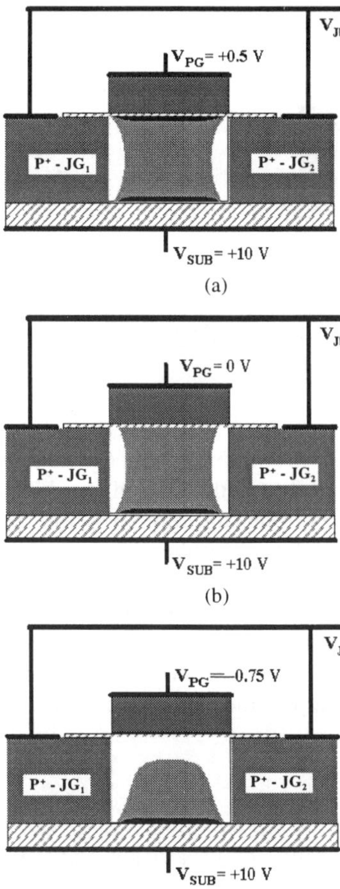

Figure 14: 2-D simulation of narrow-channel MOS-JFET with accumulated back-gate, depleted side-gates, and top-gate swept from (a) accumulation, (b) neutral, and (c) depletion.

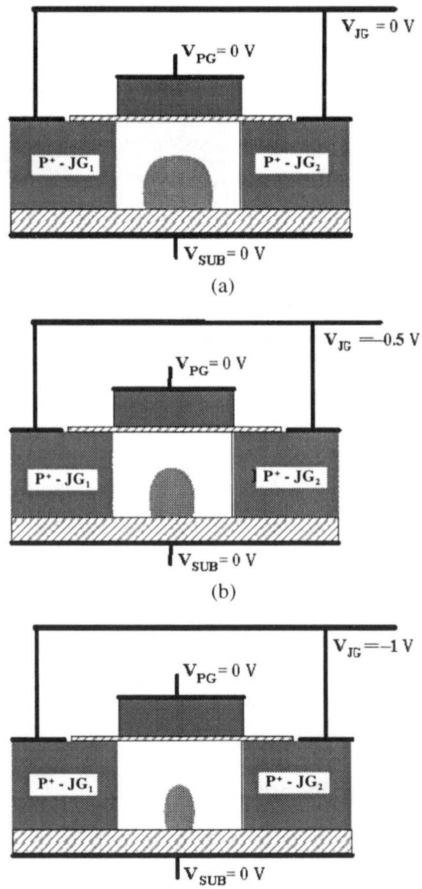

Figure 15: 2-D simulation of narrow-channel MOS-JFET showing junction-gate bias influence; junction-gate swept from (a) weak, (b) moderate, to (c) strong depletion.

5. Applications

The MOS-JFET provides many exciting new circuit opportunities. Given the four-gate modulation described in Figure 16, ultra low-noise high-speed analog circuits should be realizable. High-speed is expected based on the fact that bulk mobility is greater than surface mobility. The depletion-mode nature of the MOS-JFET makes it well suited for low-voltage analog circuits. Single-stage, low power, high frequency modulators and RF mixers are also conceivable given the multiple input terminals per transistor. Combinatorial manipulations of the MOS-JFET gates imply high-density (per transistor) digital logic. And given the V_{DS} breakdown of this device, the MOS-JFET is also suitable for 5V interface circuits and mixed-voltage systems-on-a-chip requiring MEMS control functions. Finally, through four-gate modulation, electrically controlled wires seem plausible using the MOS-JFET.

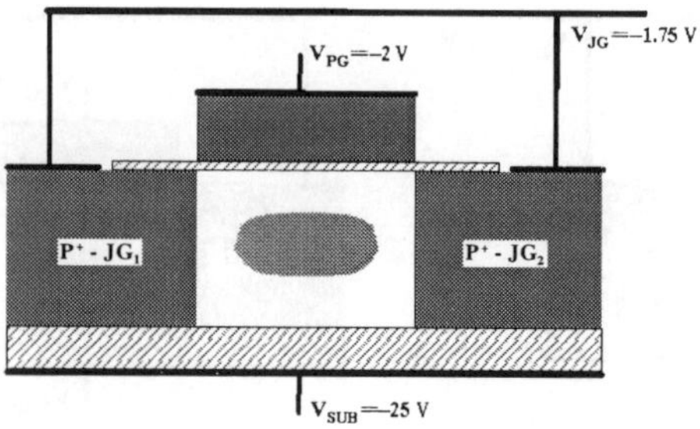

Figure 16: 2-D simulation of narrow-channel MOS-JFET demonstrating 4-gate modulation and formation of a floating wire (W = 0.3μm, $N_A = 2\times10^{20}$ cm^{-3}, $N_D = 5\times10^{17}$ cm^{-3}).

6. Conclusions

A unique four-gate device has been discovered and preliminary experimental data presented. The MOS-JFET takes advantage of the flexibility and isolation capability of SOI technology and is not feasible in bulk CMOS. This transistor combines both MOS field-effect and junction field-effect to modulate conduction. Furthermore, the MOS-JFET is highly amenable to technology scaling, as narrower channels will enhance pinch-off, particularly for lower channel doping. This exciting device can be used to modulate both high and low current levels. Four-gate modulation opportunities are provided, which open the door for many new circuit concepts in analog, RF, mixed-voltage, and digital applications. Much work is needed, however, to fully explore the capabilities of this new device.

Acknowledgements

The work described in this paper was carried out for the Jet Propulsion Laboratory, California Institute of Technology, under a contract with the National Aeronautics and Space Administration.

References

1. S. Cristoloveanu, T. Ernst, D. Munteanu and T. Ouisse, "Ultimate MOSFETs on SOI: ultra-thin, single gate, double gate, or ground plane," *Int. J. of High Speed Electronics and Systems, WOFE'99,* vol. 10, no. 1, pp. 217-230, 2000.
2. K. W. Golke and P. S. Fechner, "Integrated Circuit Impedance Device and Method of Manufacture Therefor," U.S. Patent 6 180 984, Jan. 30, 2001.

International Journal of High Speed Electronics and Systems, Vol. 12, No. 2 (2002) 521–529

ARCHITECTURE AND DESIGN OF AN 8-BIT FLUX-1 SUPERCONDUCTOR RSFQ MICROPROCESSOR

MIKHAIL DOROJEVETS

Department of Electrical and Computer Engineering
State University of New York, Stony Brook, NY 11794-2350, U.S.A.
midor@ece.sunysb.edu

The first single-chip superconductor FLUX-1 microprocessor has been designed in the Rapid Single Flux Quantum (RSFQ) logic and fabricated using 4 kA/cm^2, 1.75-μm Nb/AlOx/Nb Josephson junction technology as a result of the collaboration between SUNY Stony Brook and TRW, Inc. A FLUX-1 chip represents an 8-bit deeply pipelined microprocessor prototype with a target clock frequency of 17-20 GHz. A new parallel partitioned architecture has been developed in order to tolerate interconnect delays and fill long FLUX-1 processor pipelines with useful instructions. The processor includes the 16 x 32-bit pipelined instruction memory, 8 integer arithmetic-logic units interleaved with 8 registers, the branch unit, and I/O ports for 5-GHz chip-to-chip communication over Nb microstrip lines on a chip carrier. The FLUX-1 instruction set consists of ~25 arithmetic, logical, and control instructions. A FLUX-1 microprocessor chip contains 65,759 Josephson junctions on a 10.6 mm x 13.2 mm die with flip-chip packaging. First FLUX-1 chips fabricated in August 2001 are currently under testing at TRW, Inc.

1 Introduction

Several advances in superconductor technology and design made over the last years have established the foundation of the FLUX superconductor microprocessor project. Theoretical studies and experimental chip designs based on the Rapid Single Flux Quantum (RSFQ) logic have demonstrated a great potential of this logic for ultra-fast superconductor digital circuit implementation.[1,2] Circuit-level computer-aided design (CAD) and testing tools have been developed to cope with challenges of practical low & medium-scale RSFQ chip design and testing.[3,4,5,6] A new 4 kA/cm^2, 1.75-μm Nb/AlOx/Nb Josephson junction technology developed by TRW, Inc.[7] has allowed reproducible fabrication of relatively large chips containing tens of thousands junctions with tolerable variations of their technological parameters. The final contribution has come from the architectural and design studies of the SPELL RSFQ processors for a petaflops-scale system within the HTMT project.[8,9]

With these opportunities and support from our DoD sponsors available, we decided to proceed with the design and fabrication of the first single-chip RSFQ microprocessor prototype called FLUX.[10,11] From the beginning, we considered a FLUX-1 chip to be a technology and design driver that would allow us to discover challenges and problems in real-world superconductor processor design. The major goals of the FLUX project were formulated as follows: VLSI-scale complexity, high performance with a processor clock rate close to 20 GHz, chip-to-chip communication at a 5-GHz rate, design scalability, testability, and tolerance to process imperfections. We will show how and to what extent these objectives were achieved in the FLUX-1 microprocessor design when presenting its architecture and implementation in the following sections.

2 Superconductor Nb tri-layer technology

A good match between the FLUX architecture, superconductor technology, and RSFQ logic is crucial to achieving our goals. Among the key technological factors that had strong impact on the FLUX-1 design were: the critical current density, number of metal layers, chip area, circuit speed, and interconnection density.

Complexity of any superconductor chip design depends on the number of Josephson junctions (JJs) that can be laid out on a given die area. In order to use TRW's current fabrication techniques and get an acceptable die yield, the maximum area of a FLUX chip is limited to ~ 140 mm^2. With the technology providing the critical current density of 4 kA/cm^2, each 1.75-μm junction has to be overdamped with an external shunt resistor of ~ 3.8 Ω in order to have its Stuart-McCumber parameter of ~ 1. Due to fabrication requirements, Josephson junctions and shunt resistors, while occupying different layers, cannot overlap each other. The ratio of junction-to-shunt-resistor area is of ~ 1:5 for the current technological process. A dramatic increase in chip density is expected in future technologies featuring ~0.3-μm junctions and critical current density exceeding 150 kA/cm^2, when junctions become overdamped without any external shunt resistors.[12]

While shunt resistors occupy quire a large area compared to junctions, it is Nb passive microstrip and strip transmission lines who consume most of the FLUX-1 chip area because of the required line widths and the relatively small number of metal layers (four including one ground plane) available in TRW's technology.[7] All FLUX gates use at least two metal layers called TRIW and WIRA plus the ground plane, i.e., leaving only one metal layer (WIRB) unused. This WIRB level alone, however, cannot be used for transmitting any signals. In order to be reliable, all gate-to-gate communication has to be done via microstrip/strip transmission lines with one 'wiring' metal (WIRB/TRIW) and one/two additional grounded planes, respectively. These strip and microstrip lines can safely cross each other, being shielded from each other by the grounded planes. Special transceivers working as 'impedance bridges' have to be placed between the FLUX gates with their 1.9-Ω output resistance and transmission lines with a ~4.6-Ω impedance. While the minimum wire width and spacing required by the topological design rules are of 1.5 μm, the actual FLUX lines are much wider: 5 μm for the TRIW strip lines and 21 μm for the WIRB microstrip lines. While the drivers' output resistance and the line impedance are matched to each other, it is not possible to provide a perfectly matched load for transmission lines at the receiver's (termination) side due to the non-linear resistance of Josephson junctions.[13] This results in partial reflections off the receiver's connection of a transmission line, which limits the maximum data transfer rate over passive transmission lines and their length. However, even with these limitations, internal FLUX signals (represented by ~0.5-mV high and ~4-ps wide pulses) can be transmitted ballistically with very low signal attenuation and dispersion at a rate of 20 GHz over relatively long (up to 3-4 mm) Nb transmission lines.

Signals travel over the strip and microstrip lines at velocities of ~ 130 and ~ 90 μm/ps, respectively. It takes approximately two clock cycles for a signal to cross the FLUX-1 chip in a horizontal direction, and three cycles in a vertical one.

3 Rapid Single Flux Quantum Logic

The low-loss ballistic propagation of signals over transmission lines together with extremely high speed and low power consumption are the most attractive features of superconductor RSFQ circuits. In the meantime, the use of pulses rather than voltage levels to code signal values prevents RSFQ design elements from being connected to each other by buses with shared wires. Instead, binary signal distribution trees built of pulse splitters have to be used, which leads to quite a large area to be consumed by interconnect in RSFQ designs.

In the RSFQ logic, almost all Boolean functions can only be implemented with clocked gates. Theoretically, the maximum clock frequency of a pipelined RSFQ processor can be reached with only one logic gate (in other words, one level of logic) per pipeline stage. (Current CMOS microprocessors feature from 8 to 12 levels of logic per stage.) Thus, the maximum rate at which these clocked gates can operate imposes the upper limit on a FLUX-1 processor clock rate. For any clocked gate used in FLUX-1, the reciprocal of the sum of the gate's setup and hold times gives the maximum operating frequency (F_{max}) of the gate. Maximum speed is not the only requirement in the RSFQ gate design. RSFQ circuits must be operational with a given bit error rate and have relatively high tolerance to noise and technological process parameter variations. Table 1 shows key timing characteristics (accurate to ~10%) of major RSFQ gates, with the design of each of them optimized for a 20-GHz clock rate and noise margins of ~30%.[14]

Table 1 Timing characteristics of major FLUX-1 RSFQ gates (simulated with PSCAN[3]).

Library element	JJs	Delay (ps)	Setup (ps)	Hold (ps)	F_{max} (GHz)
JTL stage	1	3.0	n/a	n/a	n/a
Line driver	2	3.0	n/a	n/a	n/a
Line receiver	2	4.0	n/a	n/a	n/a
Splitter	3	6.5	n/a	n/a	n/a
Merger	5	4.5	n/a	n/a	n/a
D flip-flop (DFF)	6	7.5	6.5	7.5	71
Inverter	8	10.5	8.0	11.5	51
DFFC	9	14.0	10.0	17.5	36
D^2FF	10	9.5	3.0	9.5	80
NDRO					
a) Read/Write	10	10.0	17.5	25.5	23
b) Read Only	10	10.0	n/a	25.5	39
XOR	11	9.5	8.0	8.0	62
AND	13	15.0	0	18.0	55

A use of built-in flip-flops in RSFQ logic gates allows a superconductor processor to be built with very fine-grain processing pipelines and have a very high clock frequency. Unfortunately, due to the small amount work done per cycle in such fine-grain RSFQ pipelines (can be several times less than the amount of work done each cycle in current CMOS processors), they inevitably have to be long. Without special architectural support, it would be unrealistic to keep these pipelines filled with useful operations and achieve high sustained performance in superconductor processors.

Another challenge is that the read latencies even for small 16 x 32-bit memories or register files implemented with the RSFQ gates shown in Table 1 are of several cycles long because of the time spent in the address & word decoders (built with splitters and mergers) and the interconnect in these relatively large memory structures.

4 FLUX-1 architecture and organization

A new architectural design requiring localized data communication between processing elements and registers has been developed in order to address the realities of the current superconductor technology and RSFQ logic. Figure 1 shows the block diagram of a FLUX-1 microprocessor. The major design units are: the 16-word instruction memory with the embedded program counter and instruction fetch logic; the branch unit with five condition flags; the instruction register and dual decode/issue logic; eight bit-stream arithmetic logic units (ALUs) interleaved with eight 8-bit general-purpose integer registers (R0-R7); two 8-bit 5-GHz I/O ports; the clock controller; built-in scan path circuitry.

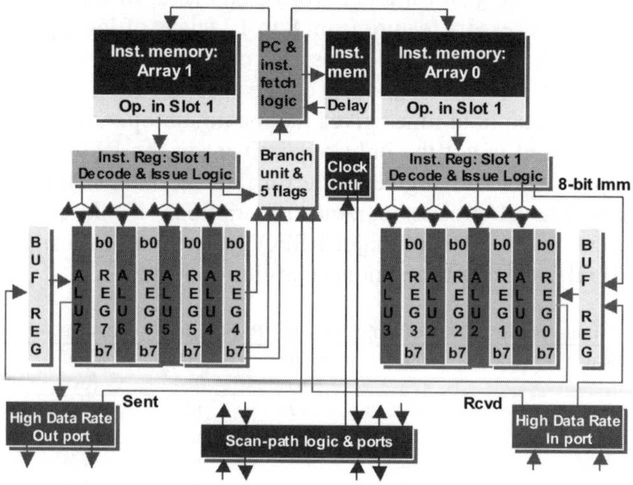

Figure 1. Block diagram of a FLUX-1 microprocessor. Each of the 8-bit ALUs shown here is a bit-stream ALU consisting of 8 single-bit pipelined ALUs that can work in parallel on data bits from integer registers.

The FLUX-1 instruction set consists of ~25 30-bit instructions, including conditional/unconditional branches, integer add/subtract, shift left/right, swap, or, invert, load an immediate value, etc. From a programmer's point of view, each bit-stream ALU operation has a two-register format, with the requirement that its source registers are to be adjacent and its destination register can be one of the source registers. After translation into binary code, ALU operations are represented by their codes consisting of a 2-bit ALU identifier and a 10-bit horizontal code field specifying the ALU operation to be performed on registers 'neighboring' the ALU.

An 8-bit FLUX-1 microprocessor implements a new parallel architecture that extends the synchronous dual-op Long-Instruction-Word (LIW) architecture with stream-like processing of data bits in integer registers. An 8-bit word in any of the registers is treated as a vector of eight single-bit elements starting from the least significant bit 0 (LSB) and ending with the most significant bit 7 (MSB). Each register bit has its own read and write ports. Each bit-stream ALU is built with eight single-bit ALUs. A single-bit ALU (bALU) has a 3-stage execution pipeline whose data inputs and outputs are connected to the corresponding read and write bit ports of the registers located left and right to the bALU by relatively short (~ 500 μm) wires. Due to short distances between ALUs and their source/destination registers, read & pre-process as well as write microoperations on each bit can be completed within one 50-ps cycle.

Two integer operations can be issued to two of the eight bit-stream ALUs and two other operations completed each cycle. In more detail, two operations of an instruction fetched from the instruction memory are placed into two slots of the instruction register. After two cycles spent in the first level of the pipelined instruction decoder, an ALU operation from each slot arrives at the ALU decoder specified by the operation's ALU-id field. Two cycles later, the pipelined ALU decoder initiates the specified register read and ALU operations by issuing properly delayed control signals to its least significant bALU and two registers surrounding it. Pipeline control logic embedded in each register bit and bALU transmits control, carry and data signals from one bit-stage to another until the completion of the operation. Four cycles after the decoded integer operation reaches the target ALU, the result's first bit (LSB) is calculated. (A FLUX processor with 16 or even 32 bits per register would have the same start-up delay of four cycles, which means that the FLUX-1 processing core design is perfectly scalable.) During the next eight cycles, 8 bits of the result (starting from LSB) will be written bit by bit into the operation's destination register by the eight single-bit bALUs comprising the target ALU. As a side effect of the operation on register R4, three control flags can be calculated and used in branch operations later.

What makes FLUX-1 different from any bit-sequential processor is that any operation dependent on the data to be calculated by some operation-in-progress can start working with the data as early as its first bit is ready, i.e., not waiting for the operation-in-progress to complete. Each bit-stream ALU can overlap execution of up to 12 independent or 3 serially-dependent instructions working with the same 8-bit integer register.

Synchronous LIW architectures often suffer from increase in code size (so-called code inflation) caused by the no-operation (NOP) instructions. This problem is solved in FLUX-1 by suppressing the NOP instructions using a special 4-bit fetch delay field in each instruction. With a corresponding encoding of this field, the processor can postpone fetching of the instruction following the current one for the specified number of cycles (up to four), during which the FLUX hardware creates and inserts the NOP instructions into the processor pipeline. Table 2 presents architectural parameters of a FLUX-1 microprocessor.

Table 2 Architectural characteristics.

Feature	Parameter
Architecture	2-way LIW with a processing-in-registers datapath and bit-level operation chaining (bit streaming)
Instruction set size	~ 25 (branches, add/sub, logical, I/O, etc.)
Register data size, bits	8
# of registers	8
# of functional units	8 bit-stream arithmetic-logic units 1 branch unit
I/O port width, bits	8 data + 2 control
Instruction length, bits	30 (+ 2 reserved for future extensions)
Inst. memory size, bits	16 x 32
Operations issue rate	2 operations/cycle
Peak performance	40 billion 8-bit operations/second (with a 20-GHz clock rate)

5 Ultra deep pipelining

FLUX-1 blocks are deeply pipelined with a typical pipeline stage consisting of one clocked gate (serving as a pipeline latch simultaneously with performing its Boolean function), a few non-clocked elements such as splitters/mergers (the latter performing non-clocked OR-like functions), and transmission lines with their drivers/receivers. Interconnect delays (including transceivers' ones) in critical paths are relatively small (~10-12 ps, i.e., 20-25% of the cycle time) thanks to the localized communication model discussed above.

A different technique, wave pipelining, is used to implement the instruction memory whose read latency is approximately three times larger than the clock cycle time of 50 ps. Wave pipelining is used in modern CMOS design to improve processor cycle time by pipelining functional units and caches without using intermediate pipeline latches with their additional synchronous clocking overhead resulting from flip-flop delay, setup time, and clock jitter/skew. The FLUX-1 instruction memory is pipelined across three stages with three waves of pulses propagating simultaneously one by one over non-clocked splitters/mergers and transmission lines in the memory address decoder. This allowed us to provide the memory bandwidth of one (sequential) dual-op instruction per cycle.

Besides the high-bandwidth memory, a FLUX-1 processor has very fast instruction select logic associated with its program counter (PC). In order to reach a fetch & issue rate of one instruction per cycle (in the absence of branches), PC and its update circuitry were 'fused' with the instruction memory word (row line) decoder. A 16-bit PC is implemented as a shift register (with additional inputs from the branch unit and scan path logic) holding such a 'linear' 16-bit address with only one non-zero bit corresponding to the position of the target instruction word in the memory arrays (so called 'one hot' encoding). As a result, the next PC address calculation can be done very quickly by shifting the contents of PC by one bit. For each conditional/unconditional branch operation, the branch unit converts a binary 4-bit target address encoded in the instruction into the linear 16-bit target address. Currently, the FLUX-1 cycle time is evaluated as ~54.5 ± 10% ps.

6 Physical chip design

Figure 2 shows a FLUX-1 chip photomicrograph. The chip is designed with a use of a full custom design flow. The development of a new RSFQ gate library[14] designed for TRW's technological process was accomplished with a use of the SUNY circuit-level design tools,[3,4,5] as well as TRW technological libraries, all integrated with the Cadence CAD tools. At the gate-level, FLUX-1 is built with a small set of basic building blocks, such as a one-bit ALU & register block, a 2 x 2 memory sub-array, etc. Each of these components is designed as a custom circuit macro optimized for high performance and small footprint. Placement and wiring of all elements (chip layout) is done by hand with neither synthesis nor automatic place & route tools used.

All FLUX-1 blocks operate synchronously with their clock supplied by the clock controller. The controller has several test ports (10 I/O pads total) connected to the external test equipment, allowing the processor clock frequency to be varied in a wide range from fractions of Hertz in an external clock mode, to more than 20 GHz in a normal mode when using an on-chip clock generator with the programmable number of

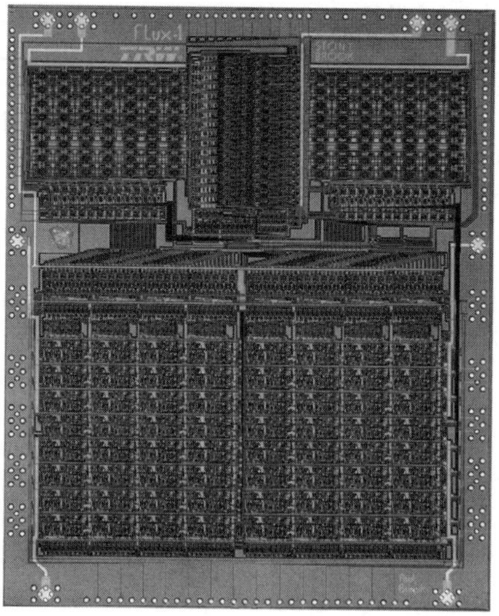

Figure 2. FLUX-1 chip photomicrograph. There are 65,759 Josephson junctions on a 140-mm^2 die. Photo courtesy of Lynn Abelson and George Kerber, TRW, Inc.

clock pulses to be generated. This clock is distributed throughout the chip using multiple-stage binary trees built of splitters.

Even if these trees were perfectly balanced (which is not the case for FLUX-1), clock skew caused by thermal fluctuations and to lesser extent technological process variations in these splitters would be quite large in any RSFQ processor with thousands of clocked gates. To solve the problem, FLUX-1 incorporates circuit-level clock delay adjustment capabilities. The clock controller drives four clock domain buffers associated with major blocks. These clock domain buffers are used for custom deskewing of clock signals. Each clock domain buffer represents the Josephson transmission line (JTL) consisting of 20 JJs in series. The bias currents of these JTLs are controlled from the outside, allowing the delays of individual junctions in these clock domain buffers to be changed in order to create the desired inter-domain clock skew of an order of ~5 ps. The elimination of signals generated by remote gates from the FLUX-1 critical path allowed us to keep the contribution from the data & clock skew relatively low (~ 8-10% of the cycle time).

The bias (DC) currents used for powering FLUX-1 RSFQ gates vary from ~ 0.2 mA for a transmission line driver/receiver up to ~1.7 mA for a TRS flip-flop. With the average bias current of ~0.1 mA per Josephson junction, the total bias current is ~7 Amps. The 8 DC pads available in a FLUX-1 chip distribute this current to eight 'power regions' each with an approximately equal number of junctions. The power dissipation of a FLUX-1 chip (without the power dissipated in the DC copper leads carrying the bias current into the cryostat) is a sum of the static power dissipated in bias registers (~ 14 mW at 2 mV) and the dynamic power dissipated in shunt resistors when junctions switch. With less than one fourth of all FLUX junctions switched per cycle and with ~ 0.06 mW dissipated per switching, the dynamic power consumption does not exceed 1 mW at 20 GHz.

FLUX-1 has full built-in testing (scan path) capabilities giving access to the contents of all registers and memory during testing. The chip has 10 test regions each with their separate scan path logic and I/O test ports. All bits of any register/memory word are connected to adjacent bits of the same or other registers/memory words, forming a sequential scan path of the region. Through these scan paths data can be shifted bit by bit into registers/memory and shifted out under control of clock signals generated by the test equipment. The 56 low-speed test pads available in FLUX-1 are used for scan path testing, clock deskewing, and I/O port tweaking. Table 3 summarizes physical characteristics of a FLUX-1 chip.

A cycle-accurate simulator was developed for the FLUX architecture verification, and partial VHDL simulation was used for functional verification of some FLUX-1 units. The full circuit-level verification of a FLUX-1 chip was not possible due to the current state of superconductor CAD tools and huge computational requirements of such a task. Instead, a multiple-chip approach including design, fabrication, and testing of several intermediate chips with increasing complexity was used. Twelve such chips containing individual FLUX-1 gates and components were designed before the first FLUX-1 chip went to fabrication after full layout-versus-schematics (LVS) and design rule checking (DRC) in June 2001. To be successful, however, any future design of an RSFQ microprocessor will require much more detailed functional and circuit-level verification.

Table 3 FLUX-1 chip physical data.

Feature	Parameter
Technology	4 KA/cm^2 Nb-trilayer
Min. junction size	1.75-μm (in diameter)
Metallization	4 Nb layers (incl. one ground)
Circuit family	Rapid Single Flux Quantum
Target clock rate	17-20 GHz
Max. I/O data rate	5 GHz via 8-bit I/O ports
Power dissipation	15 ± 20% mW at 4.2K
JJ count	65,759
Chip size	10.6 mm x 13.2 mm
Packaging	Flip-chip (solder bumps): 20 5-GHz I/O pads 56 low frequency test pads 8 DC pads
MCM	1.25 inch x 1.25 inch 2 Nb metal layers 56 signal + 16 DC pads

7 Current status

The first FLUX-1 'niobium' was delivered in August 2001. The final version of the FLUX-1 MCM carrier capable of passing the required bias current without overheating the MCM's DC pads was fabricated in December 2001. Currently, the processor is under testing at TRW, Inc. During testing, a cryogenic probe with a FLUX-1 chip flip-chip bonded to the MCM carrier is placed into a cryostat mounted on top of a liquid-helium storage Dewar.[15] Multi-channel Octopux system[16] connected the probe is used to apply test vectors, read-out and analyze test results, as well as control bias current sources. In order to significantly decrease the bit error rate currently observed in the first FLUX-1 chips, we consider the design and fabrication of next versions of the FLUX-1 chips with current recycling and other circuit-level improvements incorporated. In 2002, we start a FLUX-2 project whose goal is to demonstrate a multiple-chip testbed including floating-point and data storage RSFQ units with 20-GHz chip-to-chip communication over MCM.

Acknowledgements

This work is a result of the collaboration between the architecture, implementation and fabrication teams at SUNY Stony Brook and TRW, Inc. with the management support from the JPL (NASA). I thank the current members of our project, namely Paul Bunyk (circuit-level design including the final layout of the chip), Quentin Herr (high data rate chip-to-chip drivers/receivers), Mike Wire (MCM), Lynn Abelson and John Spargo (fabrication) from TRW, Inc., Arnold Silver and Alan Kleinsasser (project management) from the JPL (NASA) and Doc Bedard (project goals and funding) from DoD for their key contributions to the FLUX-1 project. This work was supported by DoD.

References

1. K. K. Likharev, and V. K. Semenov, "RSFQ logic/memory family: A new Josephson-junction digital technology for sub-terahertz–clock-frequency digital systems," *IEEE Trans. Appl. Supercond.* **1**, 3 (1991).
2 W. Chen, A.V. Rylyakov, V. Patel, and K. K. Likharev, "Rapid single-flux quantum T-flip-flop operating at 770 GHz," *IEEE Trans. Appl. Supercond.* **9**, 3212 (1999).
3. S. V. Polonsky, V. K. Semenov, and P. N. Shevchenko, "PSCAN – personal superconductor circuit analyzer," *Supercond. Sci. Techol.* **4**, 667 (1991).
4. S. Polonsky, P. Shevchenko, A. Kirichenko, et al., "PSCAN'96: new software for simulation and optimization of complex RSFQ circuits," *IEEE Trans. Appl. Supercond.* **7**, 2685 (1997).
5. P. I. Bunyk, and S. V. Rylov, "Automated calculation of mutual inductance matrices of multilayer superconductor integrated circuits," *Abstracts of Int. Supercond. Electronics Conf.*, NIST, Boulder, CO (1993).
6. K. Gai, Q.P. Herr, V. Adler, *et al.*, "Tools for the computer-aided design of multigigahertz superconducting digital circuits," *IEEE Trans. Appl. Supercond.* **9**, 18 (1999).
7. G. K. Kerber, L. A. Abelson, M. L. Leung, *et al.*, "A high density 4 kA/cm^2 Nb Integrated Circuit process," *IEEE Trans. Appl. Supercond.* **9**, 1061 (2001).
8. M. Dorojevets, P. Bunyk, D. Zinoviev, and K. Likharev, "Superconductor electronic devices for petaflops computing," FED Journal **10**, 3 (1999).
9. M. Dorojevets, "COOL Multithreading in HTMT SPELL-1 Processors," *Int. Journal on High Speed Electronics and Systems* **1**, 247 (2000).
10. M. Dorojevets, P. Bunyk, and D. Zinoviev, "FLUX Chip: Design of a 20-GHz 16-bit Ultrapipelined RSFQ Processor Prototype Based on 1.75-um LTS Technology," *IEEE Trans. Appl. Supercond.* **9**, 326 (2001).
11. M. Dorojevets, and P. Bunyk, "FLUX-1: The First Superconductor 20-GHz 8-bit RSFQ Microprocessor Built in 1.75-μm Nb-trilayer Technology," *Abstracts of Int. Workshop on Future Trends in Microelectronics: The Nano Millennium,* Ile de Bendor, France, 40 (2001).
12. Y. Naveh, D. Averin, and K. Likharev, "Physics of high-jc Josephson junctions and prospects of their RSFQ VLSI applications," *IEEE Trans. Appl. Supercond.* **9**, 1056 (2001).
13. P. Bunyk, P.-M. Koenig, and D. Zinoviev, "Passive interconnects: a revolutionary approach to RSFQ system design," *Abstracts of the 8th ISEC*, Osaka, Japan, (2001).
14. P. Bunyk, K. Likharev, and D. Zinoviev, "RSFQ technology: physics and devices," *Int. Journal High Speed Electronics and Systems* **1**, 257 (2001).
15. M.S. Wire, D.J. Durand, A.H. Silver, and M.K Wagner, "A multiple pin, flip-chip system for microwave and gigabit per second device testing at variable temperatures," *Rev. Sci. Instruments* **2**, 1542 (2001).
16. D. Zinoviev, and Yu. Polyakov, "Octopux: an advanced automated setup for testing superconductor circuits," *IEEE Trans. Appl. Supercond.* **7**, 3240 (1997).

International Journal of High Speed Electronics and Systems, Vol. 12, No. 2 (2002) 531–540

NOISE MEASUREMENT OF III–V COMPOUND DETECTORS FOR 2 μm LIDAR/DIAL REMOTE SENSING APPLICATIONS

M. NURUL ABEDIN
5 North Dryden Street, MS 468
NASA Langley Research Center, Hampton, VA 23681

TAMER F. REFAAT
Science and Technology Corporation
10 Basil Sawyer Drive, Hampton, VA 23666

UPENDRA N. SINGH
5 North Dryden Street, MS 468
NASA Langley Research Center, Hampton, VA 23681

Noise of a photodetector plays a vital role in determining the minimum detectable signal for lidar and DIAL receivers. A low noise trans-impedance amplifier circuit has been employed to examine the noise of III-V compound infrared detectors. These infrared detectors include InGaAs PIN diodes and newly developed InGaAsSb avalanche photodiodes (APDs) with separate absorption and multiplication (SAM) structure. The noise of these detectors are compared with well-established Si APDs. These measured noises are utilized to compute the figures-of-merit, such as noise-equivalent-power (NEP) and detectivity (D^*) of these devices and are presented in this paper.

1. Introduction

Light detection and ranging (lidar) is an active remote sensing technique that uses a pulsed laser and a collocated receiver to measure the density of atmospheric gases and aerosols as a function of range. In the differential absorption lidar (DIAL) technique, two laser pulses at slightly different wavelengths are used to measure the differential absorption feature of the gas. The transmitted laser pulses are subjected to absorption and scattering, thus the backscattered radiation collected by the instrument telescope contains information about the concentration of the atmospheric species, which can be retrieved using the lidar/DIAL equations.[1]

Recently, short wave infrared lidar/DIAL measurements draw attention due to their wide applications, especially in the 1.6-μm and 2-μm spectral regions. These measurement techniques are an attractive way to obtain carbon dioxide profiling and monitoring, aerosol, water vapor and cloud profiling and global wind and wind shear measurements. Ambrico *et al.* discussed the sensitivity of both the 1.6 and 2 μm DIAL measurement and emphasized the advantage of the 2 μm.[2] Ismail *et al.* simulated the measurement of CO_2 using the DIAL technique at 2 μm wavelength.[3]

The intensity of the lidar/DIAL return signal is inversely proportional to the square of the range. Therefore, photodetectors associated with the lidar/DIAL receiver play a major role in defining the instrument sensitivity and maximum range. High gain and low noise photodetectors are required for these applications to maximize the signal-to-noise ratio (SNR), which defines the system minimum detectable signal. This results in extending the measurement range, which is important in air and space-born applications.[1]

APDs are ideal detectors for lidar/DIAL measurements. Besides being rugged and compact, they have internal gain mechanism (avalanche multiplication) that allows increasing the output signal, which may increase their SNR.[1] On the other hand, the internal gain increases the noise, which increases the NEP of the device for low signals and introduces the excess noise factor for high signals as well slowing down the device. Research on different APD structures indicates some important design conditions that can minimize the device noise and further increases the SNR. Low noise APD requires an ionization ratio (defined as the ratio of the hole to the electron ionization coefficients) that is either much higher or much smaller than unity.[4] By satisfying this condition, avalanche multiplication is dominated by one charge carrier. When selecting an APD, key parameters, such as noise equivalent power (NEP), quantum efficiency, excess noise factor, spectral response, dynamic range, and linearity have to be considered.

Separate absorption and multiplication (SAM) regions structure is widely employed for 2 μm APDs, where photo-absorption occurs in the narrow bandgap InGaAsSb layer and multiplication occurs in the wide bandgap AlGaAsSb layer. Andreev *et al.* developed SAM APD based on III - V compound (InGaAsSb/AlGaAsSb) materials with low noise to operate in the wavelength range from 1.6 to 2.5 μm.[5-7] They also measured avalanche multiplication at room and liquid nitrogen temperatures for the InGaAsSb/AlGaAsSb heterostructures and reported the results.[5-6] Shellenbarger *et al.* developed InGaAsSb/AlGaAsSb SAM APD with cutoff wavelength of 2.3 μm.[8-10] They measured multiplication gain of 20 and detectivity of 1.85×10^8 $cmHz^{1/2}/W$ at room temperature.[8,10]

At NASA Langley Research Center, a few SAM APDs and pn diodes based on InGaAsSb/AlGaAsSb structures are acquired from Astropower, Inc. The main focus of this paper is to characterize the noise performances of these detectors and compare them to low noise InGaAs PIN diodes (Hamamatsu Corp.). The noise characteristics are also compared to Si APDs with reach-through (RTS) and super low ionization ratio (SLIK) structures.[11-12] The RTS and SLIK structures are well developed and are similar to the SAM structure. In this article, noise-equivalent-power (NEP) and detectivity (D^*) are discussed as well as the SNR variation with bias voltage.

2. Trans-Impedance Amplifier (TIA) Circuit

The conversion of the detectors output photocurrent signal into a voltage signal is an important step in lidar/DIAL detection system. This involves a current-to-voltage converter, or a trans-impedance amplifier (TIA) that virtually shorts the detector to its bias supply and defines its operating mode. The TIA choice is very critical, since it must have a minimal influence on the detected signal. One of the design aspects is the TIA noise contribution to the detected signal. The design must be in such a way that the TIA noise is small compared to the detector noise. During the detector noise measurement, the TIA noise must be well defined and separated.

Figure 1 shows a schematic diagram of the TIA circuit used in conjunction with the test detectors (except for the SLIK APD, since it has a built-in amplifier).[12] The OPA 627 operational amplifier has been utilized due to its low noise.[13] The value of the feedback resistor, R_f, is adjusted to amplify the detector signal, I_s, according to

$$V_o = -I_s \cdot R_f \,, \tag{1}$$

where V_o is the TIA output voltage. The resistor R and the capacitor C are designed to minimize the voltage ripples from the biasing source. The values and noise contributions of each element are given in table 1. Noise current is calculated by applying Johnson formula at room temperature for the resistors and by considering the operational amplifier input noise voltage and current. The total noise contribution from the circuit is estimated to be $1.4\text{x}10^{-12}$ A/Hz$^{1/2}$ based on noise power calculations referred to the TIA output. This value is verified experimentally by measuring the amplifier noise with replacing the detector by a short circuit. The detector noise current is separated from the amplifier noise current using the relation

$$I_n = \sqrt{I_t^2 - I_a^2} \,, \tag{2}$$

in which I_n is the detector noise current, I_a is the amplifier noise current, and I_t is the total noise current measured for the detector-TIA combination.

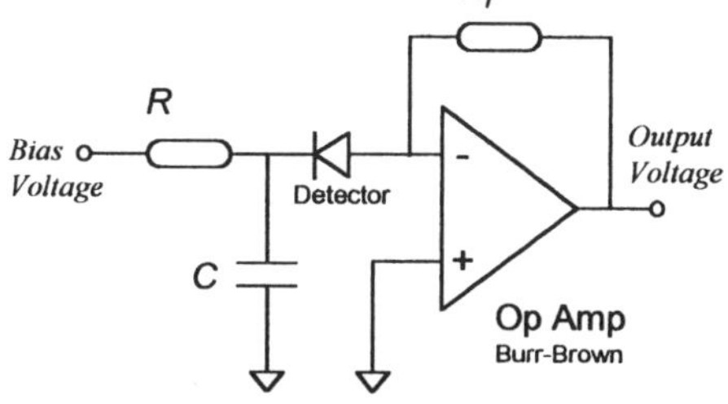

Figure 1. Schematic diagram of the TIA used with the detectors.

Table 1. Specifications of the TIA components.

Component	Value	Noise Current
R	10 kΩ	$1.3\text{x}10^{-12}$ A/Hz$^{1/2}$
C	1 μF	---
Rf	50 kΩ	$5.7\text{x}10^{-13}$ A/Hz$^{1/2}$
Op Amp	OPA627	$1.6\text{x}10^{-13}$ A/Hz$^{1/2}$

The TIA circuit as well as the detector is mounted on a printed circuit board (PCB) as shown in this figure 2. The PCB layout is designed to minimize noise pick-up by limiting the length of the conducting tracks with surface mount components. This PCB is mounted inside a cooling chamber, which controls the temperature of the detector and also provides proper shielding to the circuit.

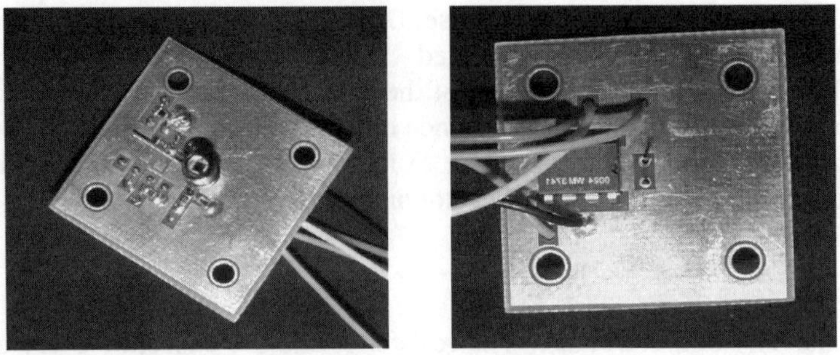

Figure 2. Printed circuit board of the TIA and the detector.

3. Detectors Noise Measurements

In lidar/DIAL detection systems, detector noise is a major concern, since it affects the instrument performance. Detectors noise is classified into shot noise, gain noise, thermal noise, and 1/f noise. Shot noise is the dominant noise source in optical detectors and it is due to the random arrival of photons to the device. This noise level increases due to the amplification of the noise current by the internal gain mechanism. Thermal noise or Johnson noise occurs due to the thermal excitation of charge carriers in the semiconductor and is proportional to the temperature. 1/f noise is a low frequency noise due to traps in the semiconductor material.[14] Other noise forms can be observed in the detected signal due to external sources, which are coupled via electromagnetic and electrostatic pick-up. These sources can be minimized by proper shielding and by signal processing techniques such as averaging and background subtraction.[15-16]

The influence of the total noise power on the detector output is of interest. Thus, the total detector noise is described as the root-mean-square (rms) deviation of the output from the average value. The total noise current spectral density, I_t, is calculated using the following equation with respect to the measured power spectral noise P_n (in dBm): [11]

$$I_t = \frac{\sqrt{10^{0.1P_n} \cdot R_L \cdot 10^{-3}}}{R_f}, \tag{3}$$

where R_L is a 50 Ω TIA load resistance. The power spectral noise is measured directly using SR785 spectrum analyzer and the mean noise is calculated in the 0.1 Hz to 10 kHz frequency ranges with 1 Hz normalization. Table 2 shows the results of the noise measurements of the tested detectors as well as the measurement conditions in this study.

Table 2 Detectors noise current spectral density results.

Detector	Part Number	Company	Diameter mm	I_n $A/Hz^{1/2}$	V_B V	T °C
Si RTS APD	C30955E	Perkin Elmer	1.5	2.0×10^{-12}	317.0	22.0
Si SLIK APD	C30649E	Perkin Elmer	0.5	2.7×10^{-13}	415.0	0.0
InGaAs pin	G5853-01	Hamamatsu	1.0	2.1×10^{-12}	0.0	22.0
InGaAs pin	G5852-01	Hamamatsu	1.0	4.5×10^{-13}	0.0	23.2
InGaAsSb pn	B17017A-25	Astro Power	0.2	4.0×10^{-12}	0.0	22.8
InGaAsSb APD	D19-K204-11#8	Astro Power	0.2	2.2×10^{-10} 4.8×10^{-12}	10.9 0.0	22.3

It is found that near infrared Si detectors have better noise performance as compared to others as shown in Table 2, while operating at high voltage. InGaAs short wave infrared detectors have slightly higher noise figures even at much lower operating bias voltage. InGaAsSb pn and unbiased SAM APD has similar noise performance close to the other devices. The device noise increases by a factor of 45 with a bias voltage of 10.9 V for SAM APD. This change in bias voltage only doubles the device responsivity.[17] This result indicates poor device performance characteristics for Lidar/DIAL measurement applications.

3.1. Noise-equivalent-power (NEP) calculation

The NEP of a detector is defined as the amount of optical power that will result in a unity signal to noise ratio in a 1 Hz bandwidth.[18] NEP is, therefore, an important parameter that is directly equivalent to the lidar/DIAL instrument minimum detectable signal after defining the system bandwidth. The noise current spectral density (table 2) is employed to evaluate the detector NEP by applying monochromatic radiation with measured responsivity, $\Re$, using

$$NEP = \frac{I_n}{\Re}. \qquad (4)$$

Figures 3 shows NEP measurements of the Si APDs as a function of wavelength in the spectral range 600 - 1100 nm. The measurements are obtained at 415V and 317 V bias voltage and 0°C and 22°C for the SLIK and the RTS structures respectively. SLIK APD has lower NEP than the RTS within this spectral range. This might be due to the smaller sensitive area of the device, as indicated in table 2. On the other hand, SLIK APD structure has a superior ionization ratio with respect to the RTS structure.[11, 19]

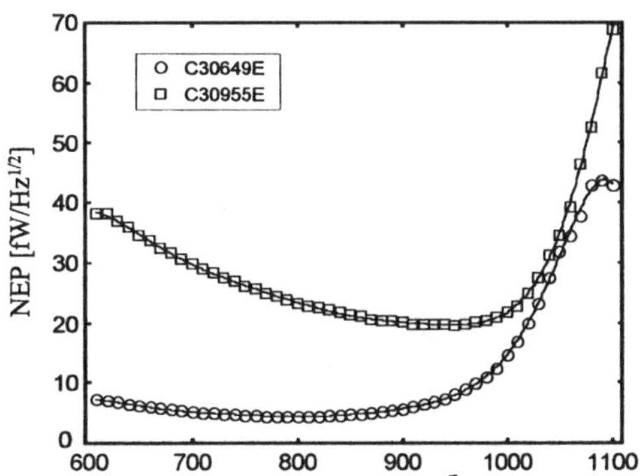

Figure 3. NEP of the SLIK APD structure (C30649E) at 415V and 0°C and RTS APD structure (C30955E) at 317V and 22°C.[11]

The NEP measurements of InGaAs PIN diodes as a function of wavelength in the spectral range 1560 - 1800 nm are shown in Figures 4. About one order of magnitude lower NEP for G5852-01 PIN detector is obtained with respect to G5853-01. This is due to the lower noise contents of the G5852-01 detector besides its relatively higher responsivity.[17] These results indicate that this detector is a good candidate for CO_2 lidar measurements.

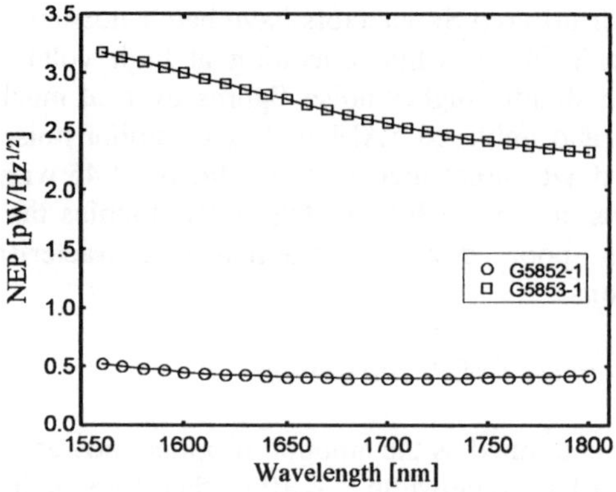

Figure 4. NEP of the InGaAs PIN diodes (G5852-01 and G5853-01) at 0V Bias and 23.2°C and 22.0°C respectively.

Figures 5 shows the NEP measurements in the spectral range 1560 - 1800 nm for the InGaAsSb pn detector at 0V bias and 22.8°C and SAM APD detector at 10.9V and 22.3°C. One order of magnitude higher NEP is obtained for InGaAsSb SAM APD as compared to a pn junction of the same material. Also, from the pn results, we can conclude that a higher NEP is measured in the InGaAsSb material system as compared to Si APD and InGaAs PIN diodes.

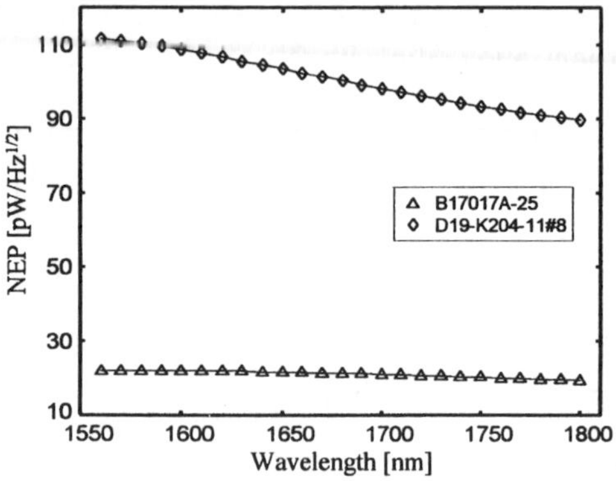

Figure 5. NEP of the InGaAsSb pn detector (B17017A-25) at 0V Bias and 22.8°C and the InGaAsSb SAM APD detector (D19-K204-11#8) at 10.9V Bias and 22.3°C.

3.2. Detectivity (D^*) calculation

Since the detector NEP is dependent on the sensitive area, false judgment on detectors can occur if we consider only this figure of merit. In other words, it is hard to compare different detectors with different areas only with this parameter, due to the fact that the noise content of a detector is directly related to its area. Thus, the detectivity is defined to normalize the detectors NEP with respect to the area, or in a mathematical form.[11]

$$D^* = \frac{\sqrt{A}}{NEP}, \tag{5}$$

where A is the detector active area in cm^2. Equation (5) is used to calculate the detectivity of the detectors as a function of wavelength at the same bias voltage and temperature as the responsivity calibration.[17] Figure 6 shows the D^* of the Si APDs. As indicated in figure 3, the SLIK APD structure has lower noise content up to 940 nm where the two curves intersect. RTS APD structure has better performance at longer wavelength regions.

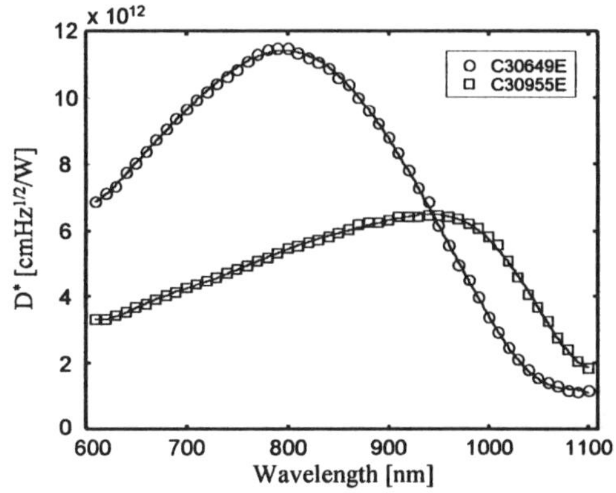

Figure 6. D^* of the SLIK APD structure (C30649E) at 415V and 0°C and RTS APD structure (C30955E) at 317V and 22°C.[11]

Figure 7 shows the D^* of the InGaAs PIN detectors. Results of this figure agree with the results obtained in figure 4. The G5852-01 detector has a better noise performance in this spectral range. Figure 8 shows similar curves for the InGaAsSb SAM APD and pn detector. Results indicate a higher noise content of this material structure in general and the SAM APD device in specific. These results might be due to the holes, which play dominant roles within charge carriers for the avalanche multiplication in that material system and is associated with higher noise levels than electrons.

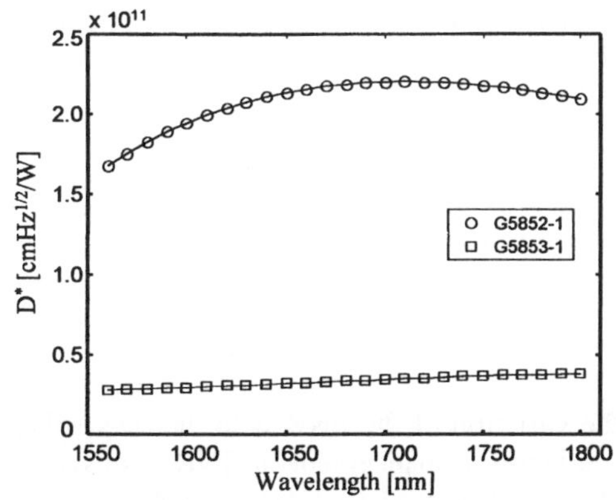

Figure 7. D^* of InGaAs PIN diodes (G5852-01 and G5853-01) at 0V bias and 23.2°C and 22.0°C respectively.

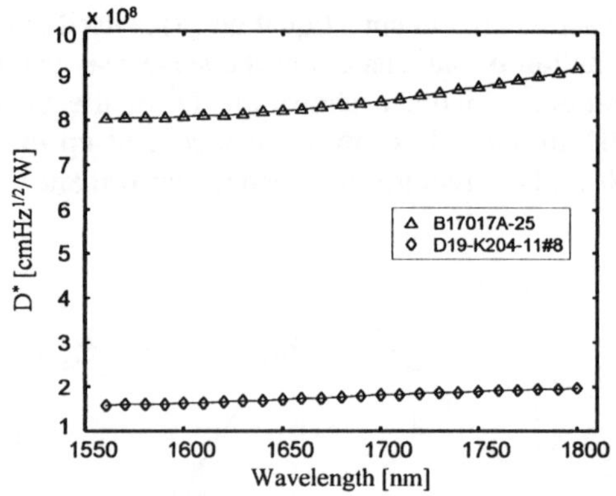

Figure 8. D* of the InGaAsSb pn detector (B17017A-25) at 0V Bias and 22.8ºC and the InGaAsSb SAM APD detector (D19-K204-11#8) at 10.9V Bias and 22.3ºC.

4. SNR Variation with Bias Voltage

SNR enhancement is the main focus of the noise characterization, since it is directly related to the sensitivity and the minimum detectable signal of the lidar/DIAL measurement. In general, raising the bias voltage of a detector increases the responsivity as well as the dark current and dark current noise. The SNR may increase or decrease depending on the individual variation of these two parameters. In this section, variation of the SNR with respect to the detector bias voltage is investigated. This is only applied to the InGaAs PIN detectors and the SAM APD.

The detector power SNR is defined as the square of the signal voltage divided by the square of the noise voltage. Experimentally, the SNR can be measured by applying a chopped optical signal. Assuming that an ideal optically chopped signal incident on a detector. The output of the device signal is displayed in figure 9, in which V_d is the dark current component corresponding to "light off" condition since it is blocked by the chopper and V_s is the signal current corresponding to the "light on" condition. The peak-to-peak optical signal, V_{pp}, is defined as

$$V_{pp} = \overline{V_s} - \overline{V_d}\,. \tag{6}$$

Therefore, the power signal-to-noise ratio will be given here with respect to the rms or the standard deviation of the signal, *std* (V_s):

$$SNR = \left(\frac{V_{pp}}{std(V_s)}\right)^2 . \tag{7}$$

Figure 10 shows the variation of the SNR with the bias voltage at 2 μm wavelength using constant source radiation. Enhancement in SNR is observed for InGaAs detectors as shown in figure 10(a) as compared to InGaAsSb APD shown in figure 10(b) with respect to bias voltage. Therefore, a significant discrepancy is observed for InGaAsSb APD with bias voltage and a peculiar behavior is seen in SNR ratio, i.e., SNR ratio decreases as a function of bias voltage. In remark, this APD needs more study to develop a high quality detector for the Lidar/DIAL measurement system.

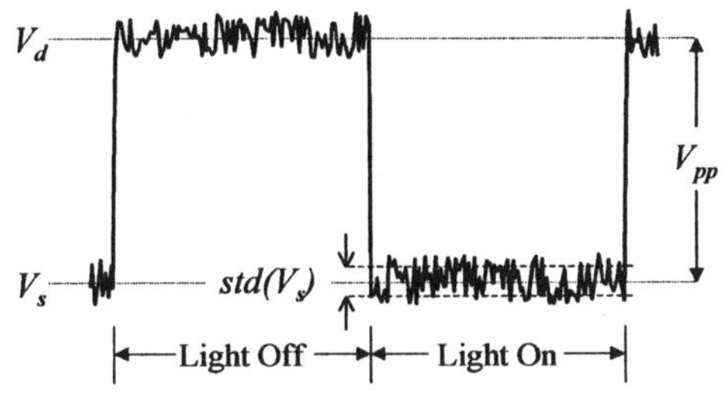

Figure 9. Signal-to-Noise Ratio Variation Measurements using optically chopped signal.

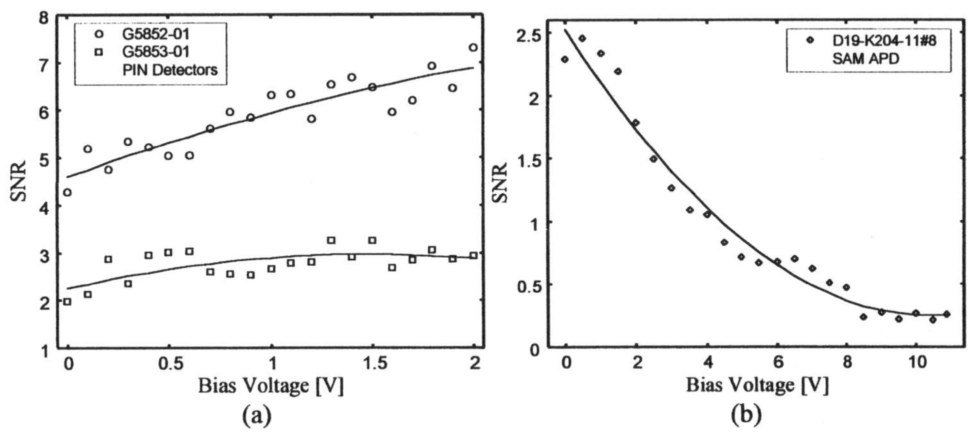

Figure 10. Signal-to-Noise Ratio Variation with Bias voltage at 2 μm Regime.

5. Conclusions

In summary, we have measured the noise of different types of detectors, such as Si-APD, InGaAs PIN, and InGaAsSb pn and SAM APD over different operating conditions. Comparison of different detector noise currents at certain bias voltage and temperature have been established in Table 2. The figures-of-merit, such as NEP and D^* of these detectors are determined as a function of wavelength. SNR of InGaAs PIN diodes and InGaAsSb APD are measured at 2 μm regime. No significant improvement in SNR is observed for the InGaAsSb SAM APD compared to InGaAs PIN photodiode. On the other hand, the SNR of the InGaAs PIN photodiodes have demonstrated improvements as a function of bias voltage. However, InGaAs PIN detector is a good candidate for 2 μm lidar application, but it has low responsivity. In conclusion, there is a need to in advance the technology to enhance the operation of SAM APD for infrared lidar detection.

Acknowledgment

The authors of this paper would like to thanks Earth Sciences Technology Office (ESTO) at NASA Goddard Space Flight Center for financial support of 2-micron detector development at NASA Langley Research Center. The authors would like to extend their acknowledgements to Dr. Syed Ismail and to Dr. Grady Koch for their suggestions and discussions on science of 2-μm Lidar/DIAL remote sensing for monitoring the atmospheric constituents. They also offer thanks to Mr. L. DiNetta of the AstroPower, Inc. for help with the InGaAsSb pn/SAM APD.

References

1. R. M. Measures, Laser Remote Sensing Fundamentals and Applications, John Wiley & Sons Inc, Toronto, 1984.
2. P. F. Ambrico, A. Amodeo, P. D. Girolamo and N. Spinelli, "Sensitivity analysis of differential absorption lidar measurements in the mid-infrared region", Appl. Opt. (2000) 6847-6865.
3. S. Ismail, E. V. Browell, J. C. Barnes, G. J. Koch, J. H. Hair and U. Singh, "Studies towards the development of a space-based 2-μm lidar system for profiling atmospheric CO_2", Int. Geosciences and Remote Sensing Society Symposium, 2001.
4. S. M. Sze, Physics of Semiconductor Devices, John Wiley and Sons, Taipei, 1983, p. 770.
5. I. A. Andreev, M. A. Afrailov, A. N. Baranov, M. A. Mirsagatov, M. P. Mikhailova and Y. P. Yakoviev, "Avalanche multiplication in photodiode structures using GaInAsSb solid Solutions", Sov. Tech. Phys. Lett. (1987) 199-201.
6. M. .P. Mikhailova, I. A. Andreev, A. N. Baranov, S. V. Mel'nikov, Y. P. Smortchkova and Y. P. Yakovlev, "Low-noise GaInAsSb/GaAlAsSb SAM avalanche photodiode in the 1.6-2.5 μm spectral range", SPIE Proc., vol. 1580 (1991) 308-312.
7. I. A. Andreev, M. A. Afrailov, A. N. Baranov, M. A. Mirsagatov, M. P. Mikhailova and Y. P. Yakoviev, "GaInAsSb/GaAlAsAb avalanche photodiode with separate absorption and multiplication regions", Sov. Tech. Phys. Lett. (1988) 435-437.
8. Z. A Shellenbarger, M. G. Mauk, M. I. Gottfried, J. D. Lesko and L. C. DiNetta, "GaInAsSb and InAsSbP photodetectors for mid-infrared wavelengths", SPIE Proc., vol. 2999 (1997) 25-33.
9. Z. A. Shellenbarger, M. Mauk, J. Cox, J. South, J. Lesko, P. Sims, M. Jhabvala and M. K. Fortin, "Recent progress in GaInAsSb and InAsSbP photodetetctors for mid-infrared wavelengths", SPIE Proc., vol. 3287 (1998) 138-144.
10. Z. A. Shellnbarger, M. Mauk, J. Cox, J. South, J. Lesko, P. Sims and L. C. DiNetta, "GaInAsSb and InAsSbP photodetectors for mid-infrared wavelengths", SPIE Proc., vol. 3379, (1998) 354-360.
11. T. F. Refaat, G. E. Halama and R. J. DeYoung, "Comparison between super low ionization ratio and reach through avalanche photodiode structures", Opt. Eng. (2000) 2642-2650.
12. H. Dautet, P. Deschamps, B. Dion, A. MacGregor, D. MacSween, R. McIntyre, C. Trottier and P. Webb, "Photon counting techniques with silicon avalanche photodiodes", Appl. Opt. (1993) 3894-3900.
13. Precision High-Speed Dift® Operational Amplifiers, Burr-Brown Corporation, 1998.
14. S. Nudelman, "The Detectivity of Infrared Photodetectors", Appl. Opt., (1962) 627-636.
15. H. W. Ott, Noise Reductions Techniques in Electronic Systems, John Wiley & Sons Inc., New York, 1988.
16. R. P. Areny, John G. Webster, Analog Signal Processing, John Wiley & Sons, Inc., New York (1999).
17. T. F. Refaat, M. N. Abedin and U. N. Singh, "Spectral Response measurements of short wave infrared detectors (SWIR)", to be published in Int. Journal of high speed electronics and systems (2002).
18. G. E. Stillman and C. M. Wolfe, "Avalanche Photodiodes", Semiconductors and Semimetals (1977) 291.
19. "Avalanche Photodiodes: A User's Guide", by EG&G Optoelectronics Canada.

International Journal of High Speed Electronics and Systems, Vol. 12, No. 2 (2002) 541–550

SPECTRAL RESPONSE MEASUREMENTS OF SHORT WAVE INFRARED DETECTORS (SWIR)

TAMER F. REFAAT
Science and Technology Corporation
10 Basil Sawyer Drive, Hampton, VA 23666

M. NURUL ABEDIN and UPENDRA N. SINGH
NASA Langley Research Center
5 North Dryden Street, MS 468, Hampton, VA 23681

Quantum detectors are critical components in infrared lidar receivers. They convert the optical return signal into electrical signal compatible with electronic data processing and storage devices. The detectors used in this study comprise InGaAs PIN diodes and InGaAsSb avalanche photodiodes (APDs) for short wave infrared applications and Si APDs, with different structures, for near-infrared applications. The spectral response of these infrared detectors utilized for lidar receivers was studied with respect to operating temperature and external bias voltage. Variation of these spectral responses as a function of bias voltage and temperature was determined. This variation is employed to estimate errors in the detected lidar return signal. Results of this research finding are reported in this article.

1. Introduction

Light detection and ranging (lidar) is an active remote sensing technique suitable for monitoring atmospheric species. A lidar system consists of a transmitter and a correlated receiver. The transmitter consists of a monochromatic laser source that transmits laser pulses through the atmosphere, spectrally tuned to an absorption feature of the measured species. The receiver consists of a telescope used to collect part of the backscattered radiation and focuses it onto a quantum detector. The detector converts the power of the return signal into photocurrent signal. The photocurrent signal is further converted into a voltage signal by a current-to-voltage converter or trans-impedance amplifier (TIA). Then the detected voltage signal is digitally processed and stored. Discrete time analysis of this signal retrieves the concentration profile of the molecular specie under study.[1-3]

Detectors used in lidar remote sensing applications must have high gain and low noise contents. This increases the instrument signal-to-noise ratio (SNR), which enhances the lidar sensitivity and range.[1] Depending on the operating wavelength, several types of detectors are used in lidar atmospheric applications.[4] Among these detectors, silicon avalanche photodiodes (APD) are used for water vapor detection near 820-nm wavelength.[5] Recently, measurements of atmospheric CO_2 profiles at 1.6 and 2.0 μm wavelength have gained interest.[6,7] Although several types of detectors might be suitable for this spectral range, high gain APDs are commercially unavailable. Some research efforts have focused on developing these devices. Avalanche multiplication process in InGaAsSb material was studied by Andreev *et al.*[8] Electron and hole impact ionization coefficients were evaluated,[9] and an ionization ratio up to 60 was reported.[10] A separate absorption and multiplication (SAM) regions GaInAsSb/ GaAlAsSb APD structure was studied by the same group. Room temperature multiplication factor of 10-20 in the 1.0 to 2.4 μm spectral range was achieved.[11] Benoit *et al.* emphasized on the constrain of the electric field value and profile for fast and sensitive device and discussed dark current reduction.[12] Development of pn and SAM APD structures were discussed by Shellenbarger *et al.* [13-15] Different characteristics were reported based on front and back sides illumination of the devices.

In this paper, the spectral response of short-wave infrared (SWIR) InGaAsSb pn and SAM APD detectors are discussed as well as InGaAs PIN detectors in the 1560 to 1800 nm spectral range. InGaAs PIN detectors are evaluated due to their sensible noise performance.[16] The results are compared to the well-developed near-infrared (NIR) Si APDs in the 600 to 1100 nm range. The Si APDs considered have super low ionization ratio (SLIK) and reach-through structures (RTS), [17,18] which are similar to the SAM APD structure. Effects of temperature and bias voltage are considered and the influence of changing these operating conditions on the lidar measurement is presented.

2. Detector Responsivity Deviation

The number density profile of an atmospheric specie, $N(r)$, as a function of the range, r, is given in terms of the laser transmitted power, P_o, and the backscattered power return signal, $P(r)$ in the lidar equation[2,3]

$$P(r) = P_o \frac{c\tau A\eta\beta(r)}{2r^2} exp\left\{-2\int_0^r [\sigma N(r) + \kappa(r)]dr\right\}, \tag{1}$$

where c is the speed of light, τ is the laser pulse duration, A is the receiver area, η is the receiver efficiency, $\beta(r)$ is the atmospheric volume backscatter coefficient, σ is the absorption cross section of the molecular specie and $\kappa(r)$ is the atmospheric extinction coefficient resulting from all other attenuation processes. At the receiver end, the optical detector and the TIA combination, convert this power return signal into a voltage signal, which is given by

$$V_d(t) = R_f \cdot \Re(\lambda, V, T) \cdot P(r), \tag{2}$$

in which R_f is the TIA feedback resistance and $\Re$ is the detector responsivity or gain. The time dependence of eq. (2) is converted to spatial (range) dependence using the relation

$$r = c \cdot t/2. \tag{3}$$

The detector responsivity is a function of operating wavelength λ, device operating bias voltage V, and temperature T. According to eq. (2), any deviation in the detector responsivity, $d\Re$, will results in a deviation in the detected lidar signal. The deviation of the detector responsivity can be expressed as

$$d\Re = \left.\frac{\partial\Re(\lambda)}{\partial\lambda}\right|_{V,T} \cdot d\lambda + \left.\frac{\partial\Re(V)}{\partial V}\right|_{\lambda,T} \cdot dV + \left.\frac{\partial\Re(T)}{\partial T}\right|_{V,\lambda} \cdot dT . \quad (4)$$

Wavelength deviation, $d\lambda$, is minimized by using narrow-bandpass optical filter at the lidar receiver, which is tuned to the narrow-linewidth transmitted laser. At the detector side, bias voltage and temperature deviations, dV and dT respectively, are limited by the performance of the detector electronic controllers.

3. Spectral Calibration

The detector spectral response is measured with respect to NIST (National Institute of Standards and Technology) calibrated reference detectors (Si PIN diode, Oriel, Model 70116, S/N 113, for NIR detectors and Ge pn diode, Newport, Model 818-IR, S/N 1127, for SWIR detectors). Table 1 lists the detectors considered in this study as well as some of the manufacturers' specifications. A photograph of the experimental setup is shown in figure 1 (detailed setup discussion is given in ref 18). A quartz tungsten halogen lamp is used as the radiation source. The lamp is supplied by a stabilized current source, to ensure stable spectrum and intensity. The output radiation is collimated and modulated at 200 Hz using an optical chopper. A monochromator is used to separate the light input into its spectral components. Monochromator grating and slit-widths are adjusted to obtain a 10-nm wavelength resolution. Optical long-pass filters with 600 nm and 1540 nm cut-on wavelengths are used at the instrument outlet, for NIR and SWIR detectors respectively, to prevent higher order dispersion of shorter wavelengths from being detected. The detectors are mounted on a circuit board with a TIA and biasing circuit, and placed inside a temperature controlled chamber.[16,18] For all the test detectors, the TIA gain is set at 50 kV/A with AC coupling and 50 Ω terminating resistance. The detector output is digitally measured and recorded for every wavelength step.

Table 1 Manufacturer and basic information of the investigated detectors.

Detector, Type	Part Number	Company	d mm	V_B V	λ_{peak} nm
Si, RTS APD	C30955E	Perkin Elmer	1.5	317	940
Si, SLIK APD	C30649E	Perkin Elmer	0.5	415	780
InGaAs, PIN	G5853-01	Hamamatsu	1.0	0	2300
InGaAs, PIN	G5852-01	Hamamatsu	1.0	0	1900
InGaAsSb, Pn	B17017A-25	Astro Power	0.2	10.9	2000
InGaAsSb, SAM APD	D19-K204-11#8	Astro Power	0.2	0	2000

Figure 2 shows the spectral response of the NIR Si APDs in the 600 to 1100 nm spectral range. Results are obtained at room temperature with the manufacturer recommended bias voltage, as given in Table 1. The RTS APD has higher responsivity than the SLIK within this range. Spectral shift of the peak responsivity between the two structures is mainly due to the difference in the devices' absorption depth. The SLIK APD package has a built-in thermoelectric cooler (TEC) with a thermistor for temperature control, as well as a TIA. The value of the TIA feedback resistance is 50 kΩ, which is mounted on the TEC to reduce the Johnson noise.

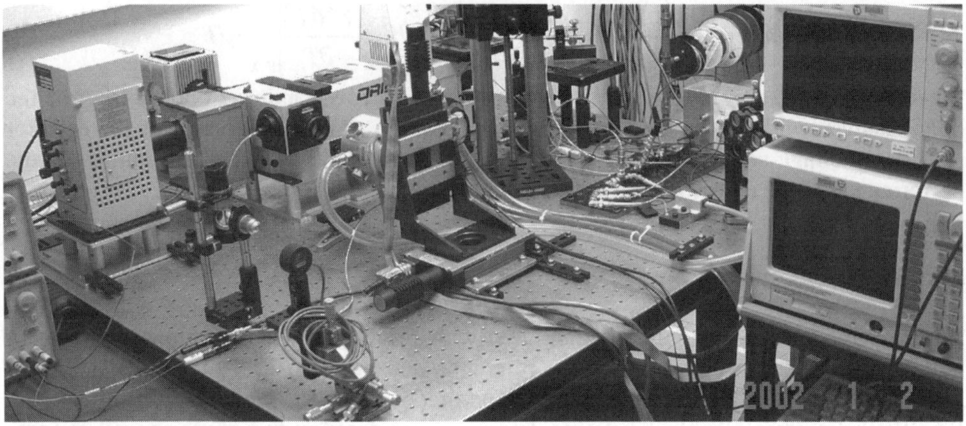

Fig. 1. Picture of the experimental setup used for spectral response measurements.

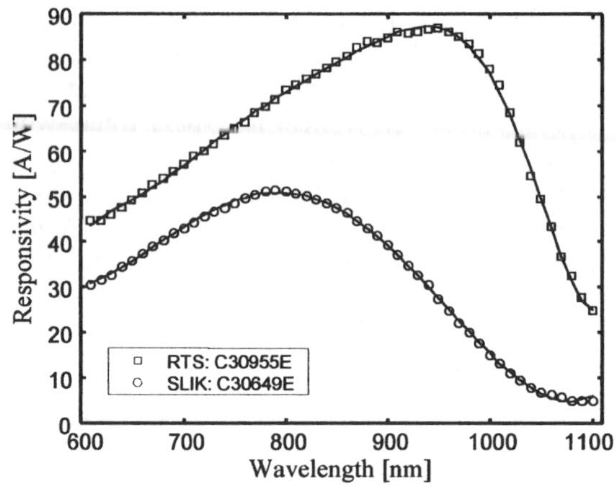

Fig. 2. Spectral response of NIR Si APDs, obtained at the manufacturer specified bias voltage and 22 and 23 °C temperature for the RTS and SLIK structures respectively.[18]

Figure 3 shows the spectral response of the SWIR detectors in the 1560 to 1800 nm spectral range. InGaAs pin diodes show small variation in this region, since it is far from their cut-off wavelengths (2.5 and 2.0 μm for G5852-1 and G5853-1 respectively).[19] To enhance their performance, a 1 V bias is applied to these detectors. Shown in the same figure are the spectral response curves of the InGaAsSb pn and SAM APD detectors. InGaAsSb SAM APD shows an order of magnitude higher responsivity compared to the pn device. A reverse bias of 10.9 V was selected close to the device breakdown voltage, which is confirmed by I-V characterization. The ripples in the SAM APD spectral response are due to an epoxy layer that covers the detector surface. Transmission characterization of 1 mm thick layer of this epoxy indicated that it has a strong non-uniform absorption in this spectral range.[14] Compared to the SWIR detectors the Si APDs have higher operating voltage, higher gain (multiplication gain is in the range of 150-200) and less noise contents specially the SLIK structure.[17, 18]

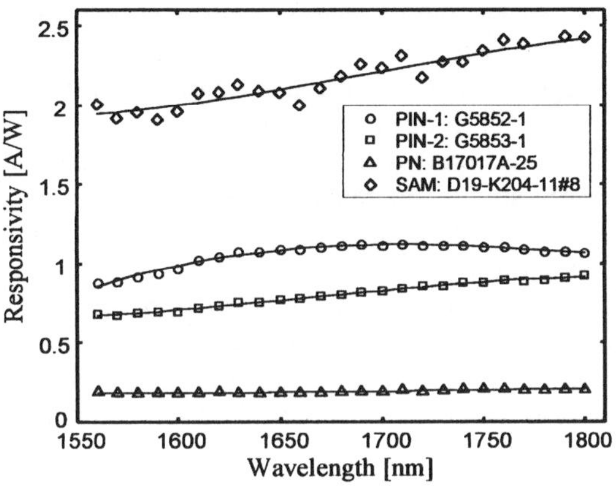

Fig. 3. Spectral response of SWIR detectors obtained at a temperature of 22.5 °C and bias voltages as stated in Table 1.

4. Temperature and Bias Voltage Effect

Figure 4 shows the responsivity variation with respect to temperature for the NIR detectors at constant wavelength and bias voltage. Responsivity of Si APDs increases by decreasing the temperature. The detector chamber allows temperature variation and stabilization within 1 °C in the range of 5 °C up to room temperature. For the SLIK APD, temperature as low as –3 °C is achievable because of the built-in TEC.

Due to setup limitation, no change in the SWIR detectors' responsivity is measured with respect to temperature. InGaAsSb detectors require lower temperature operation in order to increase their gain. Increasing the multiplication gain by a factor of 5 requires cooling the detector to liquid nitrogen temperature.[11] Temperature affects InGaAs PIN diodes by shifting the cut-off wavelength to shorter value when cooling the device.[19] Generally, low temperature operation of infrared detectors is preferable due to the reduction of the device noise, leading to increase in the SNR.

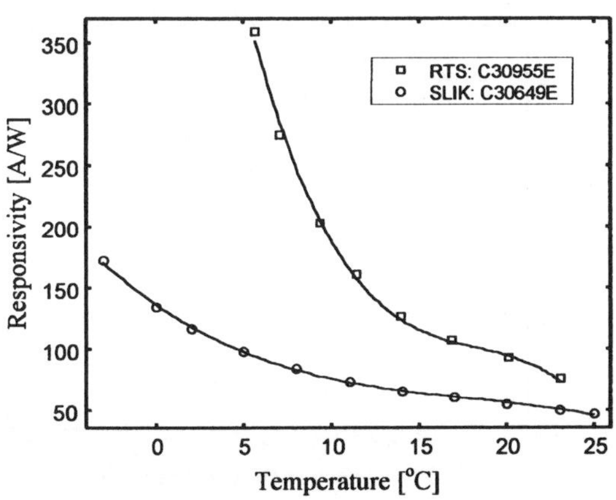

Fig. 4. Responsivity variation with temperature for NIR detectors, obtained at 820 nm wavelength with voltage bias of 317 and 415 V for RTS and SLIK structures respectively.[18]

Figure 5 shows the responsivity variation with bias voltage for the NIR detectors, obtained at constant wavelength and room temperature. The responsivity for Si APDs increases with bias voltage. This is due to the increase in the device electric field, which enlarge the impact ionization probability. Similar results are obtained, at 1800 nm wavelength, for the SWIR detectors as shown in figure 6. Increasing the bias voltage of the InGaAs PIN detectors slightly increases the responsivity and enhances the devices SNR.[16] For InGaAsSb SAM APD, increasing the bias voltage increases the responsivity more rapidly, but still at a lower rate than for the Si APDs. It appears from the figure that the responsivity tend to saturate near breakdown. On the other hand, high voltage operation of those devices increases the noise level tremendously as well as the dark current (0-10V increase in the device bias increases the dark current by two order of magnitudes).[16] This reduces the device SNR and limits its dynamic range.

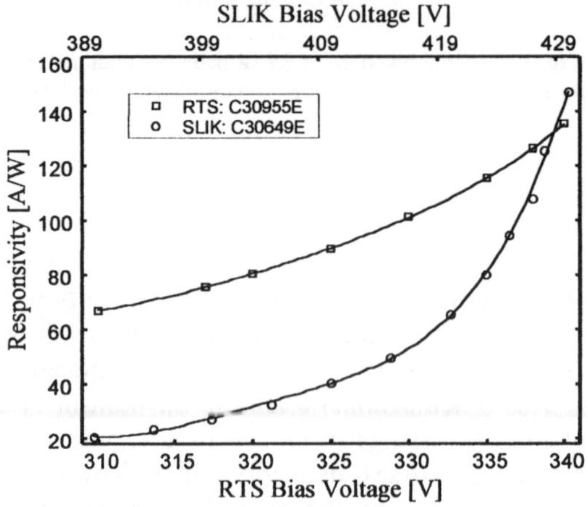

Fig. 5. Responsivity variation with bias voltage for NIR detectors, obtained at 820 nm wavelength with operating temperatures of 22 and 0 °C for RTS and SLIK structures respectively.[18]

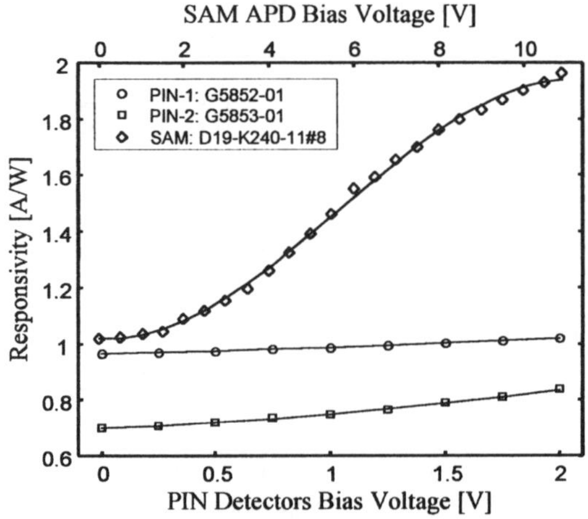

Fig. 6. Responsivity variation with bias voltage for SWIR detectors, obtained at 1800 nm wavelength with operating temperature of 22, 23.5 and 21 °C for PIN-1, PIN-2 and SAM detectors respectively.

5. Responsivity Deviation

The responsivity variation with temperature shown in figure 4 can be fitted using a polynomial fit in the form of

$$\Re(T)\Big|_{\lambda_i,V_j} = \sum_{m=0}^{M} a_m \cdot T^m . \tag{5}$$

where M is the curve fit order and a_m are the polynomial coefficients. Referring to eq. (4), the responsivity deviation due to a small change in temperature, can be obtained differentiating eq. (5) with respect to temperature, which is given by

$$\left.\frac{\partial \Re(T)}{\partial T}\right|_{\lambda_i,V_j} = \sum_{m=1}^{M} m \cdot a_m \cdot T^{(m-1)} . \tag{6}$$

Defining the responsivity relative error, $\Delta\Re$ by

$$\Delta\Re = \frac{\partial \Re}{\Re}, \tag{7}$$

and assuming a temperature deviation of 0.1 °C at the same operating conditions given in figure 4, the variation of the relative error with the detector operating temperature is shown in figure 7. This figure indicates that the SLIK detector has a better stability with respect to temperature than the RTS structure. Besides, operating temperature of about 17 °C will result in the minimum relative error in the responsivity.

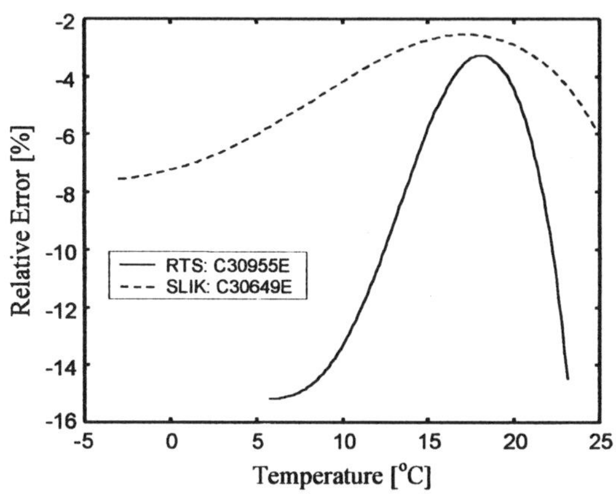

Fig. 7. NIR detectors responsivity relative error due to 0.1 °C temperature variation obtained at 820 nm wavelength.[18]

Similar analysis can be applied to the detectors bias voltage. The responsivity variation with voltage bias, given in figures 5 and 6, can be fitted using a polynomial fit in the form

$$\Re(V)\Big|_{\lambda_i,T_j} = \sum_{n=0}^{N} b_n \cdot V^n . \tag{8}$$

where N is the curve fit order and b_n are the polynomial coefficients. The responsivity deviation due to small change in bias voltage, will be given by

$$\left.\frac{\partial \Re(V)}{\partial V}\right|_{\lambda_i, T_j} = \sum_{n=1}^{N} n \cdot b_n \cdot V^{(n-1)}. \tag{9}$$

Applying eq. (7), the responsivity relative errors due to bias voltage deviations of 1 and 0.1 V are given in figures 8 and 9 for NIR and SWIR detectors respectively. RTS detector has higher stability with respect to bias voltage, as indicated by the lower relative error values in figure 8. Although it might have higher error (G5853-1), PIN detectors are more stable with respect to the bias voltage than the SAM APD. It was observed that the responsivity relative error curves, versus temperature and bias voltage, do not have strong dependence on the operating wavelength.[18]

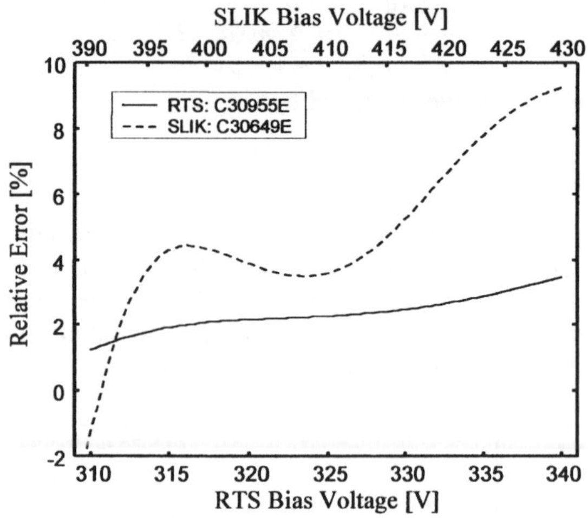

Fig. 8. NIR detectors responsivity relative error due to 1 V bias voltage variation obtained at 820 nm wavelength.[18]

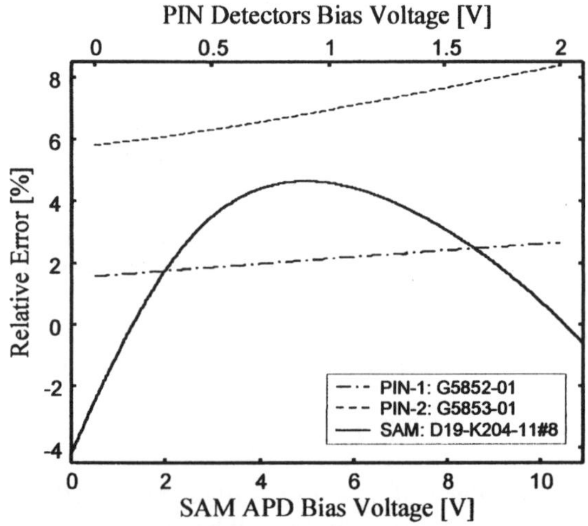

Fig. 9. SWIR detectors responsivity relative error due to 0.1 V bias voltage variation obtained at 1800 nm wavelength.

6. Conclusion

The spectral response of NIR and SWIR detectors in the 600-1100 nm and 1560-1800 nm wavelength range respectively, was presented in this paper. The variation of the responsivity with bias voltage and temperature and its influence on the lidar measurement was also discussed. NIR silicon detectors have high responsivity and operate at higher bias voltage compared to SWIR InGaAsSb SAM APDs. Within the SWIR region, InGaAs PIN diodes showed small gain with increasing the bias voltage, therefore they are more stable regarding bias voltage operation. Increasing the bias voltage for the SAM APDs has the disadvantage of increasing the device noise and dark current.

Due to their noise performance, InGaAs PIN detectors are good candidates for 1.6 μm CO_2 lidar detection, but they provide no internal gain. On the other hand, InGaAsSb APDs showed higher gain with larger noise figures. Research efforts are required to enhance the performance of these devices for the SWIR lidar detection.

Acknowledgements

The authors would like to thank Dr Syed Ismail for his consultations, which added in-depth to this article. Thanks also due Dr Grady Koch for his contribution to the experimental setup. The authors would like to extend their acknowledgement to the Earth Sciences Technology Office (ESTO) at NASA Goddard Space Flight Center for funding the development of 2 μm detectors. The authors also thank Louis DiNetta at AstroPower Inc. for the help with InGaAsSb detectors.

References

1. R. M. Measures, Laser Remote Sensing Fundamentals and Applications, John Wiley & Sons Inc, Toronto, 1984.
2. U. N. Singh, "Lidar for atmospheric remote sensing", Optical Measurement Techniques and Applications (P. K. Rastogi), Artech House Inc., Norwood, 1997, Chapter 13.
3. A. I. Carswell, "Lidar measurements of the atmosphere", Can. J. of Phys. (1983) 378-395.
4. G. E. Halama, J. McAdoo, M. D. Jhabvala, A. J. Ewin, M. K. Fortin, R. A, Bredthauer and M. N. Abedin, "SAGE III detector alternative", SPIE Proc., vol. 3756 (1999) 172-189.
5. T. F. Refaat, G. E. Halama and R. J. DeYoung, "Characterization of advanced avalanche photodiodes for water vapor lidar receivers", NASA/TP-210096, 2000.
6. P. F. Ambrico, A. Amodeo, P. D. Girolamo and N. Spinelli, "Sensitivity analysis of differential absorption liadr measurements in the mid-infrared region", Appl. Opt., (2000) 6847-6865.
7. S. Ismail, E. V. Browell, J. C. Barnes, G. J. Koch, J. W. Hair, U. Singh, "Studies toward the development of a space-based 2-μm lidar system for profiling atmospheric CO_2", International Geoscience and Remote Sensing Society Symposium, 2001.
8. I. A. Andreev, M. A. Afrailov, A. N. Baranov, M. A. Mirsagatov, M. P. Mikhailova and Yu. P. Yakovlev, "Avalanche multiplication in photodiode structures using GaInAsSb solid solutions", Sov. Tech. Phys. Lett. (1987) 199-201.
9. I. A. Andreev, M. P. Mikhailava, S. V. Mel'nikov, Yu. P Smorchkova and Yu. P. Yakovlev, "Avalanche multiplication and ionization coefficients of GaInAsSb", Sov. Phys. Semicond. (1991) 861-865.
10. M. P. Mikhailova, I. A. Andreev, A. N. Baranov, S. V. Mel'nikov, Y. P. Smortchkova and Y. P. Yakovlev, "Low-noise GaInAsSb/GaAlAsSb SAM avalanche photodiode in the 1.6-2.5 μm spectral range", SPIE Proc., vol. 1580 (1991) 308-312.

11. I. A. Andreev, M. A. Afrailov, A. N. Baranov M. A. Mirsagatov, M. P. Mikhailov and Yu. P. Yakovlev, "GaInAsSb/GaAlAsSb avalanche photodiode with separate absorption and multiplication regions", Sov. Tech. Phys. Lett. (1988) 435-437.
12. J. Benoit, M. Boulou, G. Soulage, A. Joullie and H. Mani, "Performance evealuation of GaAlAsSb/GaInAsSb SAM-APDs for high bit rate transmission in the 2.5μm wavelength region", J. Opt. Commun. (1988) 55-58.
13. Z. A. Shellenbarger, M. G. Mauk, M. I. Gottfried, J. D. Lesko and L. C. DiNetta, "GaInAsSb and InAsSbP photodetectors for mid-infrared wavelengths", SPIE Proc., vol. 2999 (1997) 25-33.
14. Z. A. Shellenbarger, M. G. Mauk, J. Cox, J. South, J. Lesko, P. Sims, M. Jhabvala and M. Fortin, "Recent progress in GaInAsSb and InAsSbP photodetectors for mid-infrared wavelengths", SPIE Proc., vol. 3287 (1998) 138-145.
15. Z. A. Shellenbarger, M. G. Mauk, J. Cox, J. South, J. Lesko, P. Sims, L. DiNetta, "GaInAsSb and InAsSbP photodetectors for mid-infrared wavelengths", SPIE Proc., vol. 3379 (1998) 354-360.
16. M. N. Abedin, T. F. Refaat and U. N. Singh, "Noise measurement of III-V compound detectors for 2μm lidar/DIAL remote sensing applications", to be published in Int. Journal of High Speed Electronics and Systems (2002).
17. H. Dautet, P. Deschamps, B. Dion, A. MacGregor, D. MacSween, R. McIntyre, C. Trottier and P. Webb, "Photon counting techniques with silicon avalanche photodiodes", Appl. Opt. (1993) 3894-3900.
18. T. F. Refaat, G. E. Halama and R. J. DeYoung, "Comparison between super low ionization ratio and reach through avalanche photodiode structures", Opt. Eng. (2000) 2642-2650.
19. Infrared Detectors, Hamamatsu Corporation, 2000.

International Journal of High Speed Electronics and Systems, Vol. 12, No. 2 (2002) 551–562

NUMERICAL ANALYSIS OF RANDOM DOPANT-INDUCED EFFECTS IN SEMICONDUCTOR DEVICES

I. D. MAYERGOYZ and P. ANDREI

Department of Electrical and Computer Engineering, University of Maryland, College Park, MD 20742, USA

A new approach to the analysis of random dopant-induced effects in semiconductor devices is presented. It is based on the "small signal analysis" (perturbation) technique. This approach is computationally much more efficient than the existing purely "statistical" techniques, and it yields the information that can be directly used for the design of dopant fluctuation-resistant structures of semiconductor devices. This approach is applied to the analysis of random dopant-induced fluctuations of threshold voltages and frequency response characteristics of MOSFET devices.

Keywords: MOSFETs; doping fluctuations; threshold voltage variances; small-signal parameters variance.

1. Introduction

The prevalent trends in mixed signal (analog–digital) circuit design are to incorporate higher functionality on a single silicon chip and appreciably increase the speed and operational frequency of digital and analog circuit components. This can be achieved by deeply scaling down semiconductor devices. It is well-known that such small devices will be very susceptible to random dopant fluctuations that are inevitably present due to the random nature of ion implantation and diffusion. For this reason, an accurate statistical analysis of random dopant-induced effects is very important for the further progress in semiconductor technology, in general, and in mixed signal electronics, in particular.

The existing approaches[1–3] to the analysis of random dopant-induced effects are based on generating random realizations (samples) of doping distributions and solving transport equations for each of such realizations. In this way, statistics of physical quantities of interest can be assembled. These statistics are then used for the evaluation of variances of the physical quantities of interest. These are purely statistical approaches that are computationally very expensive. Their accuracy is strongly dependent on the number of statistical realizations. In addition, the complete knowledge of doping statistics is needed for the implementation of these approaches. Moreover, randomly generated doping distributions are usually highly irregular in space, and this negatively affects the accuracy of numerical solutions of transport equations. This is especially the case when the same mesh is used for different doping realizations. Finally, in purely statistical approaches, the fluctuation of physical quantities are computed by substraction of approximately computed quantities. As a result, this approach is prone to numerical errors that are especially appreciable in the case of small fluctuations. It must be also noted that the analysis of random dopant-induced effects has been focused on computations of threshold voltage variances, while no effort has been made to analyze random dopant-induced fluctuations of frequency response characteristics.

In the paper, an entirely different approach[4] to the evaluation of variances of fluctuating physical quantities is presented. It is based on "small signal analysis" (perturbation) technique and completely circumvents the calculations for numerous doping realizations. This approach requires only the knowledge of variances of fluctuating doping concentrations and, in this sense, it is a "second-moment characterization" technique. As a result, this technique is computationally much more efficient than the existing purely statistical approaches. In addition, our method yields the information on sensitivity of the variances of the physical quantities of interest to doping locations. For this reason, our method can be instrumental in the design of dopant fluctuation-resistant structures of semiconductor devices. Our approach is universal in the sense that it is applicable to any transport model (drift-diffusion, semiclassical transport, quantum transport, etc.) and it can be used for the calculation of threshold voltage and terminal current variances, as well as for the analysis of fluctuations of frequency response characteristics.

The paper is organized as follows. In Sec. 2, the idea of the method is presented for the case of calculation of the threshold voltage variance. In Sec. 3, the numerical results obtained by using this method for the threshold voltage variance and fluctuations of frequency response characteristics are reported. These results are compared with those previously published in literature in the case of threshold voltage variance and with our numerical results obtained by using a purely statistical technique in the case of random dopant-induced fluctuations of frequency response characteristics. Conclusions are drawn in Sec. 4.

2. Technical Discussion

The basic idea of the proposed approach is described below for the calculation of the threshold voltage variance of MOSFET devices (see Fig. 1). The "current" definition of the threshold voltage is adopted in subsequent discussions. Namely, the threshold voltage is defined as such a gate voltage that guarantees the specified (small) source-to-drain current $I_{SD} = I_0$ for the specified source-to-drain bias voltage. It is clear from this definition that the calculations of the threshold voltage variance must be based on transport equations. These calculations will be described by assuming that the transport equations have been already discretized and represented in the algebraic form:

$$\boldsymbol{L}(\boldsymbol{X}, \boldsymbol{D}, V_G, V_{SD}) = 0\,. \tag{1}$$

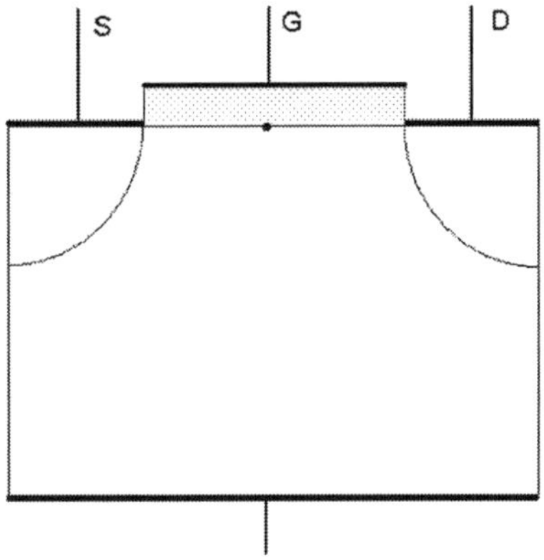

Fig. 1. MOSFET device.

Here, $\boldsymbol{L}$ is a nonlinear vector function of unknown "state" vector $\boldsymbol{X}$, vector $\boldsymbol{D}$ of mesh-point values of doping density as well as gate voltage V_G and source-to-drain voltage V_{SD}. If the transport model is the drift-diffusion model, then the state vector can be written as follows:

$$\boldsymbol{X} = \begin{pmatrix} \boldsymbol{\varphi} \\ \boldsymbol{n} \\ \boldsymbol{p} \end{pmatrix}, \tag{2}$$

where $\boldsymbol{\varphi}$ is the vector of mesh-point values of electric potential, while $\boldsymbol{n}$ and $\boldsymbol{p}$ are the vectors of mesh-point values of electron and hole densities, respectively. It can be remarked that, within the framework of the drift-diffusion model, quasi-Fermi potentials or Slotboom variables can also be chosen as "state" variables instead of electron and hole densities.[5–7]

If the semiclassical or quantum transport models are adopted then the state vector $\boldsymbol{X}$ has the form:

$$\boldsymbol{X} = \begin{pmatrix} \boldsymbol{\varphi} \\ \boldsymbol{F}_n \\ \boldsymbol{F}_p \end{pmatrix}, \tag{3}$$

where $\boldsymbol{F}_n$ and $\boldsymbol{F}_p$ are the vectors of mesh-point values of distribution or Wigner functions for electrons and holes, respectively.[8–10] By solving nonlinear discretized equations (1) with respect to $\boldsymbol{X}$, the source-to-drain current can be computed as some (nonlinear) functional $\mathcal{F}$ of $\boldsymbol{X}$:

$$I_{SD} = \mathcal{F}(\boldsymbol{X}). \tag{4}$$

As mentioned before, the threshold voltage is usually defined as such a gate voltage that the source-to-drain current achieves some small specified value I_0 that corresponds to the onset of conduction:

$$I_{SD} = I_0 \quad \text{for} \quad V_G = V_{th}. \tag{5}$$

Vector $\boldsymbol{D}$ is random and can be represented as follows:

$$\boldsymbol{D} = \boldsymbol{D}^0 + \boldsymbol{d}\,, \tag{6}$$

where $\boldsymbol{d}$ is the vector of mesh-point fluctuations of doping:

$$\bar{\boldsymbol{d}} = 0\,, \tag{7}$$

and symbol "-" stands for the expected (mean) value.

The doping fluctuations result in the shift $\boldsymbol{X}^{sh}$ and fluctuations $\boldsymbol{x}$ of state vector and this can be mathematically written as follows:

$$\boldsymbol{X} = \boldsymbol{X}^0 + \boldsymbol{X}^{sh} + \boldsymbol{x}\,, \tag{8}$$

where

$$\bar{\boldsymbol{x}} = 0\,. \tag{9}$$

This, in turn, leads to the shift V_{th}^{sh} and fluctuations v_{th} of the threshold voltage that are needed to keep the source-to-drain current the same and equal to I_0. Thus:

$$V_{th} = V_{th}^0 + V_{th}^{sh} + v_{th}\,, \tag{10}$$

where

$$\bar{v}_{th} = 0\,. \tag{11}$$

In Eqs. (8) and (10), $\boldsymbol{X}^0$ and V_{th}^0 stand for the value of state variables and threshold voltage in the case when the doping fluctuations can be neglected. This means that $\boldsymbol{X}^0$ and V_{th}^0 satisfy the following nonlinear equations:

$$\boldsymbol{L}(\boldsymbol{X}^0, \boldsymbol{D}^0, V_{th}^0, V_{SD}) = 0\,. \tag{12}$$

By assuming that shifts and fluctuations of $\boldsymbol{X}$ and V_{th} are relatively small, we can use the perturbation ("small signal analysis") technique and derive the following linearized equations for these shifts and fluctuations:

$$\hat{\boldsymbol{J}}(\boldsymbol{X}^{st} + \boldsymbol{x}) = \hat{\boldsymbol{A}}\boldsymbol{d} + (V_{th}^{sh} + v_{th})\boldsymbol{b}\,, \tag{13}$$

where $\hat{\boldsymbol{J}}$ is the Jacobian of $\boldsymbol{L}$ with respect to $\boldsymbol{X}$ evaluated at $\boldsymbol{X}^0$, $\boldsymbol{D}^0$, V_{th}^0 and V_{SD}:

$$\hat{\boldsymbol{J}} = \frac{\partial \boldsymbol{L}}{\partial \boldsymbol{X}}(\boldsymbol{X}^0, \boldsymbol{D}^0, V_{th}^0, V_{SD})\,, \tag{14}$$

while matrix $\hat{\boldsymbol{A}}$ and vector $\boldsymbol{b}$ are similarly defined by the equations:

$$\hat{\boldsymbol{A}} = -\frac{\partial \boldsymbol{L}}{\partial \boldsymbol{D}}(\boldsymbol{X}^0, \boldsymbol{D}^0, V_{th}^0, V_{SD})\,, \tag{15}$$

$$\boldsymbol{b} = -\frac{\partial \boldsymbol{L}}{\partial V_G}(\boldsymbol{X}^0, \boldsymbol{D}^0, V_{th}^0, V_{SD})\,. \tag{16}$$

By taking average on both sides of Eq. (13) and taking into account Eqs. (7), (9) and (11), we find:

$$\hat{J}\mathbf{X}^{sh} = \hat{\mathbf{A}}\boldsymbol{d} + V_{th}^{sh}\boldsymbol{b}. \tag{17}$$

By subtracting Eq. (17) from Eq. (13), we arrive at the following equation for fluctuations:

$$\hat{J}\boldsymbol{x} = \hat{\mathbf{A}}\boldsymbol{d} + v_{th}\boldsymbol{b}. \tag{18}$$

Fluctuations $\boldsymbol{x}$ result in fluctuations i_{SD} of the source-to-drain current. By using the perturbation technique, fluctuations i_{SD} can be represented as a linear functional of $\boldsymbol{x}$:

$$i_{SD} = f(\boldsymbol{x}). \tag{19}$$

The specific mathematical form of f can be derived from the specific form of $\mathcal{F}$ by using the equation $i_{SD} = \nabla\mathcal{F}|_{X_0} \cdot \boldsymbol{x}$. Since Eq. (18) and functional f are linear, fluctuations i_{SD} must be a linear combination of fluctuations v_{th} and $\boldsymbol{d}$:

$$i_{SD} = \alpha v_{th} + \sum_{i,j} \lambda_{i,j} d_{i,j}. \tag{20}$$

Fluctuations of the threshold voltage are such that they compensate for doping fluctuations and keep the source-to-drain current the same. This means that:

$$i_{SD} = 0. \tag{21}$$

The last equation leads to the following expression for the fluctuations of the threshold voltage in terms of doping fluctuations:

$$v_{th} = -\sum_{i,j} \frac{\lambda_{i,j}}{\alpha} d_{i,j}. \tag{22}$$

By assuming that doping densities at different locations are independent random variables, from Eq. (22) we derive:

$$\sigma_{v_{th}}^2 = \sum_{i,j} \left(\frac{\lambda_{i,j}}{\alpha}\right)^2 \sigma_{D_{i,j}}^2. \tag{23}$$

To find $\sigma_{D_{i,j}}^2$, we remark that it is usually assumed (with some justification) that the total number $N_{i,j}$ of dopant atoms in volumes $\Delta V_{i,j}$ are random variables with Poisson distributions. In the case of Poisson distributions, the expected value and variance coincide. Thus, we have:

$$\sigma_{N_{i,j}}^2 = N_{i,j}^0 = D_{i,j}^0 \Delta V_{i,j}. \tag{24}$$

By using this fact, we derive:

$$\sigma_{D_{i,j}}^2 = \left(\frac{1}{\Delta V_{i,j}}\right)^2 \sigma_{N_{i,j}}^2 = \frac{D_{i,j}^0}{\Delta V_{i,j}}. \tag{25}$$

Next, by using Eq. (25), we arrive at:

$$\sigma_{v_{th}}^2 = \sum_{i,j} \left(\frac{\lambda_{i,j}}{\alpha}\right)^2 \frac{D_{i,j}^0}{\Delta V_{i,j}}. \tag{26}$$

Thus, the problem of calculation of the threshold voltage variance is reduced to the calculation of "superposition" coefficients α and $\lambda_{i,j}$ in Eq. (20). It can be observed that, according to Eq. (18), the α-coefficient is equal to the value of i_{SD} when $v_{th} = 1$, $d_{i,j} = 0$. Similarly, λ_{i_0,j_0} is equal to the value of i_{SD} when $d_{i_0,j_0} = 1$, $v_{th} = 0$ and $d_{i,j} = 0$ for all mesh points (i,j) except (i_0, j_0). The above observations suggests that the superposition coefficients α and $\lambda_{i,j}$ can be found by solving the linearized equations (18) with respect to x for many different right hand sides and then computing the values of i_{SD}. It turns out that there exists a technique that can substantially simplify the above computations and reduce them to the solution of certain linear equations for only *one* right hand side. To explain this technique, we first remark that the linear functional $f(x)$ in Eq. (19) can be represented as an inner product:

$$i_{SD} = \langle \boldsymbol{g}, \boldsymbol{x} \rangle, \tag{27}$$

where the vector $\boldsymbol{g}$ can be always deduced from the mathematical form of the linear functional $f(\boldsymbol{X})$, namely, it is equal to $\nabla\mathcal{F}|_{X_0}$.

Now, we consider linear equations:

$$\hat{\boldsymbol{J}}^t \boldsymbol{c} - \boldsymbol{g}, \tag{28}$$

where $\hat{\boldsymbol{J}}^t$ is the transpose matrix of Jacobian $\hat{\boldsymbol{J}}$. By substituting Eq. (28) into Eq. (27), we obtain:

$$i_{SD} = \langle \hat{\boldsymbol{J}}^t \boldsymbol{c}, \boldsymbol{x} \rangle = \langle \boldsymbol{c}, \hat{\boldsymbol{J}} \boldsymbol{x} \rangle. \tag{29}$$

By using Eq. (18), we arrive at:

$$i_{SD} = \langle \boldsymbol{c}, \hat{\boldsymbol{A}} \boldsymbol{d} + v_{th} \boldsymbol{b} \rangle. \tag{30}$$

From the last equation and the described relations between i_{SD} and the superposition coefficients, we find:

$$\alpha = \langle \boldsymbol{c}, \boldsymbol{b} \rangle, \tag{31}$$

$$\lambda_{i,j} = \langle \boldsymbol{c}, \boldsymbol{a}^{(i,j)} \rangle, \tag{32}$$

where $\boldsymbol{a}^{(i,j)}$ is the vector obtain by multiplying matrix $\hat{\boldsymbol{A}}$ by the vector whose component corresponding to the mesh point (i,j) is equal to one, while all other components are equal to zero.

Now, the algorithm of threshold voltage variance calculation can be summarized as follows. First, we solve nonlinear discretized equations (12) subject to the condition $I_{SD} = \mathcal{F}(X_0) = I_0$. Second, we solve linear equations (28) and then compute the superposition coefficients α and $\lambda_{i,j}$ by using Eqs. (31) and (32). Finally, the threshold voltage variance is computed by using Eq. (26).

Coefficients $(1/\Delta V_{i,j})(\lambda_{i,j}/\alpha)^2$ can be viewed as coefficients that provide information on how the threshold voltage variance is sensitive to doping locations. For this reason, these sensitivity coefficients can be very instrumental in the design of random dopant-fluctuation resistant structures.

The presented technique has been extended to the analysis of random dopant-induced fluctuations of frequency response characteristics of semiconductor devices. This far-reaching extension is mathematically more involved than the discussion presented above and its details will be published elsewhere. Here it is worthwhile emphasizing that this extension of our approach has been used for the calculation of variances of admittance matrix elements in a wide frequency range. In the case of MOSFET device, the admittance matrix relates a.c. components of gate and drain currents to a.c. components of gate and drain voltages:

$$\boldsymbol{i} = \hat{\boldsymbol{Y}}\boldsymbol{v}\,, \tag{33}$$

where

$$\boldsymbol{i} = \begin{bmatrix} i_G \\ i_D \end{bmatrix}, \quad \boldsymbol{v} = \begin{bmatrix} v_G \\ v_D \end{bmatrix}, \tag{34}$$

$$\hat{\boldsymbol{Y}} = \begin{bmatrix} y_{GG} & y_{GS} \\ y_{SG} & y_{SS} \end{bmatrix}. \tag{35}$$

Random dopant fluctuations may result in shifts and fluctuations of the values of admittance matrix elements. Since the admittance matrix representation (33) leads to the a.c. equivalent circuit of MOSFET devices, the above random dopant-induced effects may have important implications for analog circuit designers. However, this issue has hardly been studied (or even mentioned) in the existing literature because the most of the effort has been directed towards the understanding of random dopant-induced effects that are relevant to digital electronics.

3. Computational Results

The technique described in the previous section has been numerically implemented and extensively used for the computation of the threshold voltage and terminal current variances, as well as for the analysis of fluctuations of frequency response characteristics. Some of these results are presented below.

It has been proposed[10,11] to use a thin epitaxial layer adjacent to the oxide interface (see Fig. 2) as a means to suppress random dopant-induced threshold voltage fluctuations. Figure 3 presents the results of calculations of $\sigma_{V_{th}}$ as a function of the thickness d_{epi} of the epitaxial layer for the MOSFET device with $L = 0.05$ μm, $w = 0.05$ μm, $\Delta = 3$ nm, $D_0^{epi} = 10^{15}$ cm^{-3} and $D_0 = 4.3 \cdot 10^{17}$cm^{-3}. This figure also presents the results obtained in Ref. 11. It is clear that our results are in a reasonably good agreement with those published in Ref. 11.

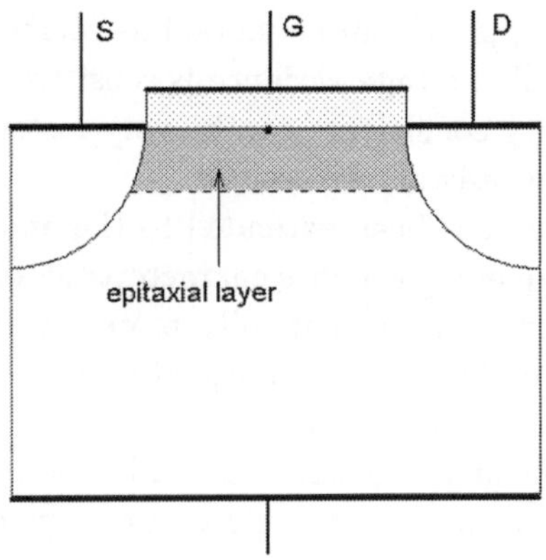

Fig. 2. MOSFET device with epitaxial layer.

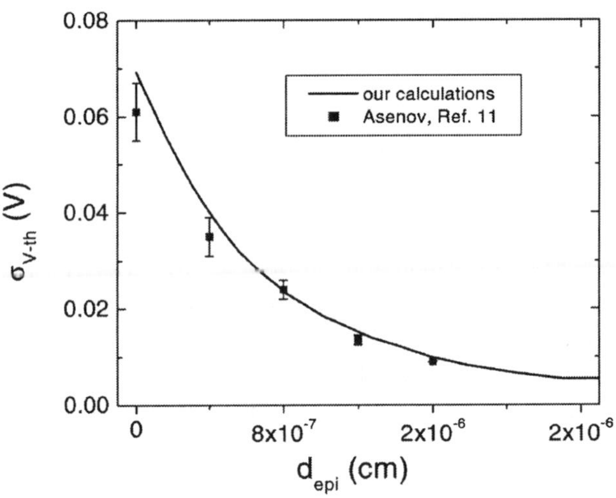

Fig. 3. Dependence of $\sigma_{V_{th}}$ on the epitaxial layer thickness.

Equation (26) can be represented in the form:

$$\sigma_{V_{th}}^2 = \sum_{i,j} \xi_{i,j} n_{i,j}^0 \, , \tag{36}$$

where $n_{i,j}^0$ are the expected numbers of doping atoms in the discretization volumes $\Delta V_{i,j}$:

$$n_{i,j}^0 = D_{i,j}^0 \Delta V_{i,j} \, , \tag{37}$$

and $\xi_{i,j}$ can be interpreted as "sensitivity" coefficients. It is important to stress that our approach yields the values of these coefficients that provide the information on the sensitivity of the threshold voltage variance to discretization volume locations. As demonstrated by Fig. 4, the value of these "sensitivity" coefficients goes down

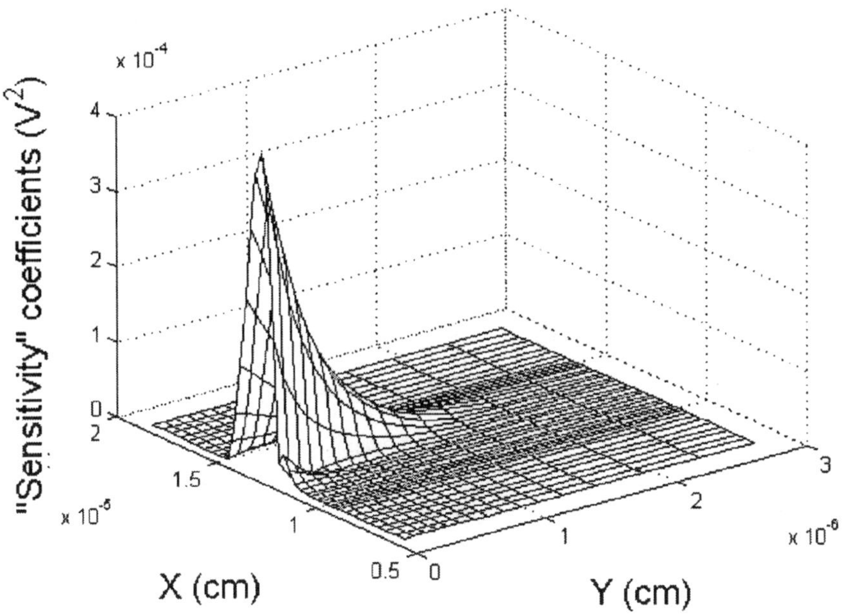

Fig. 4. The spatial distribution of sensitivity coefficients for $\sigma_{V_{th}}$.

as the distances of mesh points to the oxide interface are increased. We have found that the sensitivity coefficients are "weakly" dependent on the actual value of the bulk doping (Fig. 5). These facts suggest that the computed sensitivity coefficients can be immediately used to estimate the appropriate thickness of the epitaxial layer. In other words, our technique yields the information that can be very instrumental in the design of random dopant fluctuation-resistant structures.

Table 1 presents the expected values of admittance matrix elements computed by neglecting random dopant fluctuations ("classical method") and by averaging calculations performed for 100 different realizations of random doping ("statistical method"). The computations were carried out for a MOSFET device with $L = 0.05\ \mu$m, $W = 0.05\ \mu$m, $\Delta = 3$ nm, and $D_0 = 10^{18}$ cm^{-3}. Symbols y^r and y^i stand for real and imaginary parts of the corresponding admittances. It is evident from Table 1 that there is no appreciable shift in the expected values of the admittance matrix elements.

Table 2 presents the results of calculation of standard deviation of y^i_{GG} and y^r_{SG} performed by using our perturbation technique and the "brute force" purely statistically approach. The computational results are presented for these two admittance matrix elements because they are very important for the calculation of current gain cutoff frequency. In addition, y^i_{GG} is directly related to the gate capacitance, while y^r_{SG} is directly related to the transconductance. It is evident from Table 2 that there is a fairly good agreement between the above two techniques, which deteriorates to some extent for very high frequencies close to the cutoff frequency.

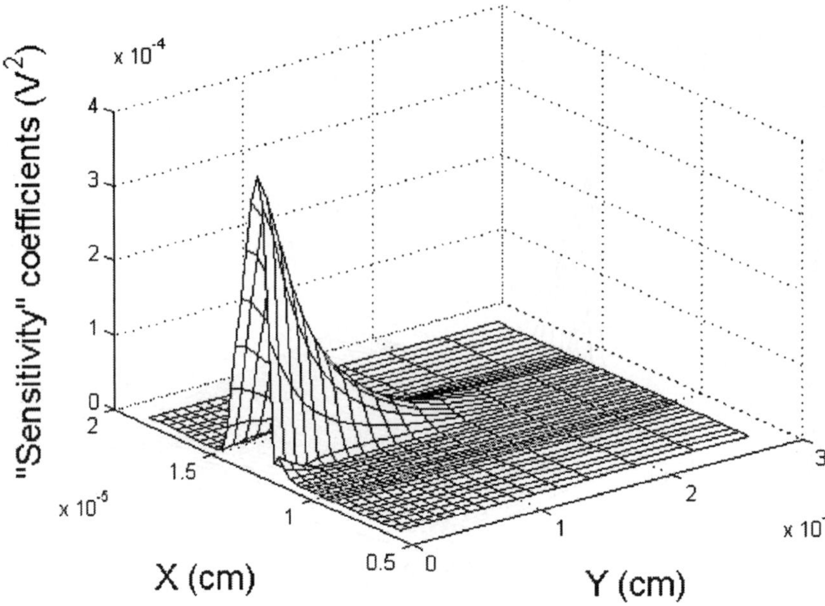

(a)

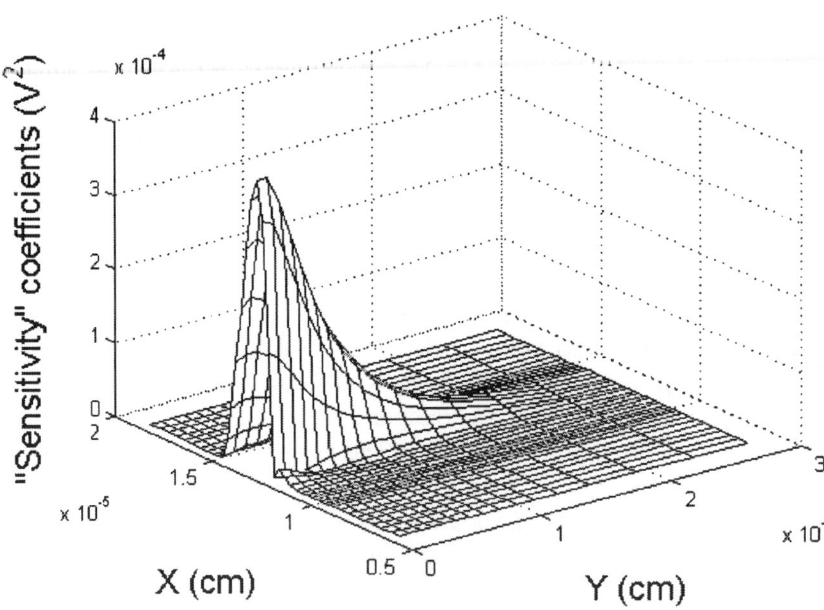

(b)

Fig. 5. The spatial distribution of sensitivity coefficients for $\sigma_{V_{th}}$ for (a) $d_{epi} = 10^{-6}$ cm and (b) $d_{epi} = 2 \times 10^{-6}$ cm.

Table 1. Expected value of admittance elements ($V_G = 0.95$ V, $V_D = 1.0$ V, and $V_S = V_B = 0.0$ V.

	$f = 10^8$ Hz		$f = 10^{10}$ Hz		$f = 3 \times 10^{11}$ Hz	
(A/V)	Classical analysis	Statistical method	Classical analysis	Statistical method	Classical analysis	Statistical method
G	5.21×10^{-13}	5.25×10^{-13}	5.21×10^{-9}	5.25×10^{-9}	4.09×10^{-6}	4.09×10^{-6}
G	-1.33×10^{-8}	-1.32×10^{-8}	-1.33×10^{-6}	-1.32×10^{-6}	-3.78×10^{-5}	-3.76×10^{-5}
G	2.14×10^{-14}	2.07×10^{-14}	2.14×10^{-10}	2.07×10^{-10}	1.83×10^{-7}	1.82×10^{-7}
G	2.94×10^{-9}	2.95×10^{-9}	2.94×10^{-7}	2.95×10^{-7}	8.74×10^{-6}	8.74×10^{-6}
S	3.32×10^{-5}	3.30×10^{-5}	3.32×10^{-5}	3.30×10^{-5}	3.01×10^{-5}	2.96×10^{-5}
S	7.73×10^{-9}	7.75×10^{-9}	7.73×10^{-7}	7.75×10^{-7}	2.18×10^{-5}	2.17×10^{-5}
S	8.13×10^{-6}	8.17×10^{-6}	8.13×10^{-6}	8.17×10^{-6}	1.06×10^{-5}	1.07×10^{-5}
S	-1.08×10^{-8}	-1.08×10^{-8}	-1.08×10^{-6}	-1.08×10^{-6}	-3.19×10^{-5}	-3.18×10^{-5}

Table 2. Standard deviation of admittance matrix elements ($V_G = 0.95$ V, $V_D = 1.0$ V, and $V_S = V_B = 0.0$ V.

	$f = 10^8$ Hz		$f = 10^{10}$ Hz		$f = 3 \times 10^{12}$ Hz	
(A/V)	Perturbation technique	Statistical method	Perturbation technique	Statistical method	Perturbation technique	Statistical method
G	3.32×10^{-10}	3.85×10^{-10}	3.32×10^{-8}	3.85×10^{-8}	8.95×10^{-7}	1.09×10^{-6}
S	2.38×10^{-6}	2.40×10^{-6}	2.38×10^{-6}	2.40×10^{-6}	2.15×10^{-6}	2.36×10^{-65}

4. Conclusions

In the paper, a novel technique for the calculation of random dopant-induced effects is presented. This technique completely circumvents the computations of numerous doping realizations and, for this reason, it is very computationally efficient. For instance, the overhead computational cost of evaluation of threshold voltage variance is only about 10% of the computational cost of finding the threshold voltage. In addition, this technique yields the information on sensitivity of variances of the physical quantities of interest to doping locations. For this reason, this technique can be very instrumental in the design of dopant fluctuation — resistant structures of semiconductor devices. The developed technique is conceptually universal in the sense that it is applicable to various transport models, and it is naturally extended to take into account random fluctuations of oxide thickness and oxide trapped charges. The above technique has been numerically implemented, and computational results are reported and compared with the results previously published in the literature or obtained by using a purely statistical technique.

References

1. A. Asenov, *IEEE Trans. Electron. Devices* **45** (1998) 2505.
2. H. S. Wong and Y. Taur, *IEEE IEDM Tech. Dig.* (1993) 705.
3. A. Asenov, A. R. Brown, J. H. Davies, and S. Saini, *IEEE Trans. Electron. Devices* **18** (1999) 1558.
4. I. D. Mayergoyz and P. Andrei, *J. Appl. Phys.* **90** (2001) 6.
5. S. Selberherr, *Analysis and Simulation of Semiconductor Devices*, Springer-Verlag, 1984.
6. I. D. Mayergoyz, *J. Appl. Phys.* **59** (1986) 195.
7. C. E. Korman and I. D. Mayergoyz, *J. Appl. Phys.* **68** (1990) 1324.
8. M. Lundrstrom, *Fundamentals of Carrier Transport*, Addison-Wesley, 1990.
9. S. P. Singh, N. Goldsman, and I. D. Mayergoyz, *Solid-State Electron.* **39** (1996) 1695.
10. P. A. Markowich, C. A. Ringhofer, and C. Schmeiser, *Semiconductor Equations*, Springer-Verlag, 1990.
11. A. Asenov and S. Saini, *IEEE Trans. Electron. Devices* **46** (1999) 1718.

International Journal of High Speed Electronics and Systems, Vol. 12, No. 2 (2002) 563–572

A SIMULATION-BASED PREVIEW OF EXTREMELY SCALED DOUBLE-GATE CMOS DEVICES AND CIRCUITS

JERRY G. FOSSUM

Department of Electrical and Computer Engineering
University of Florida
Gainesville, FL 32611-6130, U.S.A.

This paper gives a simulation-based preview of device-design issues and performance of extremely scaled DG CMOS. A suite of simulation tools, including a 2D numerical device simulator, a 1D numerical Poisson-Schrödinger solver, and a generic, physics/process-based DG MOSFET compact model in Spice, is applied to both asymmetrical- and symmetrical-gate DG CMOS devices and circuits to provide physical insight at both the device and circuit levels. The results give added motivation as well as preliminary guidance for the development of DG CMOS.

1. Introduction

Because of their inherent immunity to short-channel effects (SCEs), double-gate (DG) MOSFETs [1] in very thin (<10nm) Si films, as illustrated in Fig. 1, could possibly constitute mainstream CMOS integrated circuits when the end of the SIA roadmap (ITRS) [2] (where channel lengths of <10nm are intimated) is approached in 10-15 years. In addition to their scalability, DG MOSFETs could yield superior (relative to the single-gate counterparts) device and circuit figures of merit, including near-ideal subthreshold slope, extremely high I_{on}/I_{off} ratios, and very short propagation delays. Whereas the development of a viable DG CMOS fabrication process is clearly a formidable task, so is the optimization of DG device design, which will require good understanding of the complex underlying physics, for example, the quantum mechanics of carriers confined in thin Si-film channels and the quasi-ballistic, or ballistic, carrier transport from source to drain. This paper gives a simulation-based preview of device-design issues and performance of extremely scaled DG CMOS.

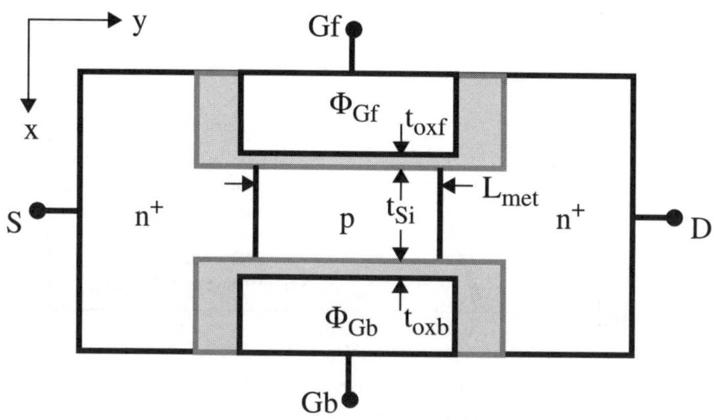

Fig. 1. Generic DG nMOSFET structure.

2. Inherent Features of DG MOSFETs

The inherent gate-gate charge coupling enabled by the thin Si film results in advantageous features of DG MOSFETs, having asymmetrical or symmetrical gates [3] as reflected by Fig. 2. The I_{DS}-V_{GS} (with $V_{GfS} = V_{GbS} = V_{GS}$) characteristics shown were predicted by MEDICI [4], a 2D tool based on hydrodynamic carrier-transport equations (which if not properly calibrated can overestimate I_{on} because of excessive velocity overshoot and no accounting for the ballistic limit [5], [6]). The DG nMOSFETs simulated have L_{met} = 50nm with symmetrical (n^+ polysilicon) and asymmetrical (n^+ and p^+ polysilicon) gates; the Si-film body is lightly (and insignificantly) doped and thin (t_{Si} = 10nm), and the gate oxides are relatively thick ($t_{oxf} = t_{oxb} = t_{ox}$ = 3nm). Note first the near-ideal subthreshold slopes for both DG devices, contrasted to those of the single-gate (SG: back gate grounded) counterparts. This inherent DG feature is obvious from the 1D Poisson equation applied (in x) to the Si film; it yields [3] $dV_{GS}/d\psi_{sf} = dV_{GS}/d\psi_{sb} = 1$, where ψ_{sf} and ψ_{sb} are the front and back surface potentials. Hence, the intrinsic gate swing factor, for any DG MOSFET, is (at T = 300°K)

$$S = \left(\frac{kT}{q}\ln(10)\right)\frac{dV_{GS}}{d\psi_{sf}} = 60mV \quad . \tag{1}$$

Note further in Fig. 2 that, whereas the threshold voltage of the asymmetrical DG nMOSFET is reasonable, that for the symmetrical device is negative. When we assume an "exotic" gate material having a work function (near midgap) that yields the same I_{off} as in the asymmetrical device, interestingly MEDICI predicts virtually the same I_{on} as well [3]. However, because of the equivocation due to the classical approximations of MEDICI, as well as the noted uncertainty in I_{on} predictions, we further checked this result using SCHRED-2 [7], a 1D (in x) self-consistent solver of the Poisson and Schrödinger equations. SCHRED-predicted areal electron charge densities (Q_i, which correlates with the channel current) versus V_{GS} in both devices are plotted for V_{DS} = 0 in Fig. 3. Although the quantum-mechanical (QM) electron distributions across the Si film, shown

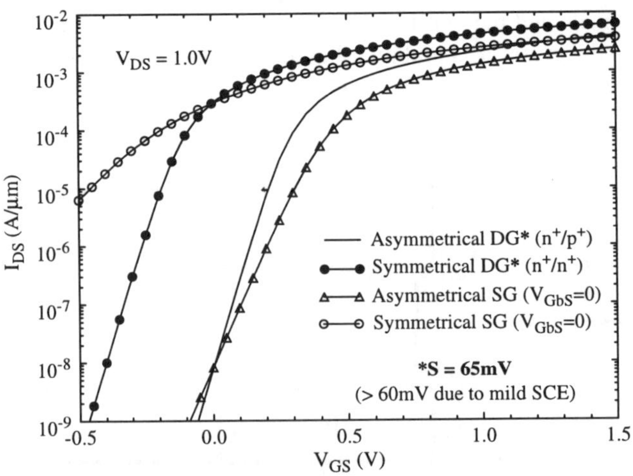

Fig. 2. MEDICI-predicted current-voltage characteristics of DG nMOSFETs and their SG counterparts.

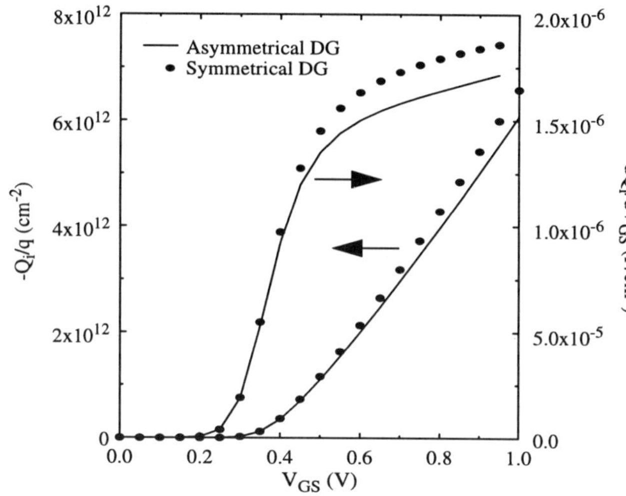

Fig. 3. SCHRED-predicted integrated electron densities versus gate voltage; $V_{DS} = 0$. The corresponding derivatives (C_G) are also plotted.

in Fig. 4, differ noticeably from the classical results, Q_i is nearly the same in both devices, implying nearly equal currents as predicted by MEDICI.

The near-equivalence of Q_i and I_{DS} in the symmetrical and asymmetrical DG MOSFETs can be explained by integrating the 1D Poisson equation, in which the charge density comprises both electron (n(x)) and dopant-ion (low N_A) terms. For the symmetrical device, integration over half the Si film yields [8]

$$Q_{i(sym)} = -2C_{Gf(sym)}(V_{GS} - V_{tf(sym)}) \tag{2}$$

where $C_{Gf(sym)}$ is the total front-gate (or back-gate) capacitance, which is less than the oxide capacitance ($C_{of} = \varepsilon_{ox}/t_{oxf}$) due to finite inversion-layer capacitance (or depth), and $V_{tf(sym)}$ is a nearly constant threshold voltage for strong-inversion conditions. The factor of 2 in (2) reflects two identical channels and gates. For the asymmetrical device, we integrate Poisson's equation over the entire Si film and derive [3]

$$Q_{i(asym)} \cong -C_{Gf(asym)}(1 + r)(V_{GS} - V_{tf(asym)}) \tag{3}$$

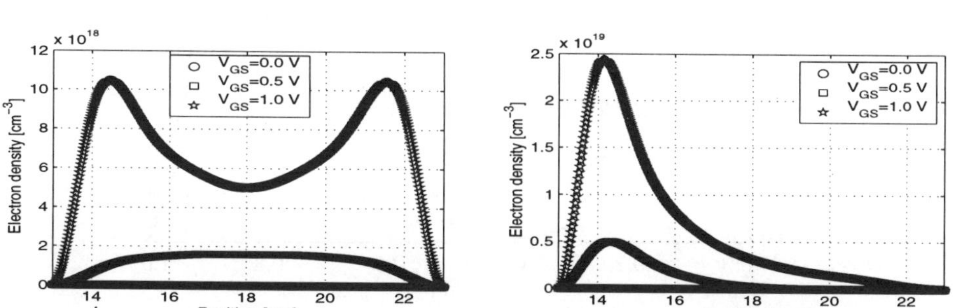

Fig. 4. SCHRED-predicted n(x) vs. V_{GS} across the Si film of the symmetrical (left) and asymmetrical (right) DG nMOSFETs; $V_{DS} = 0$.

where $r \cong 3t_{oxf}/(3t_{oxb}+t_{Si})$ is a gate-gate charge-coupling factor,

$$C_{Gf(asym)} \cong \frac{C_{of}}{1 - r\dfrac{t_{i(asym)}}{3t_{oxf}}} \tag{4}$$

in which $t_{i(asym)}$ is the average depth of the inversion layer, and $V_{tf(asym)}$ is a nearly constant threshold voltage for strong-inversion conditions. Note in (3) that both $r > 0$ and $C_{Gf(asym)} > C_{Gf(sym)}$ define the total two-gate capacitance ($C_G = -dQ_i/dV_{GS}$), thereby revealing that *extended* (relative to that for symmetrical gates) gate-gate charge coupling and a *reverse* inversion-layer capacitance effect in the asymmetrical DG MOSFET underlie its near-equivalence with the symmetrical device as shown in Fig. 3. Note, however, that this equivalence depends on r, and hence could be undermined by variations in t_{oxf}, t_{oxb}, and t_{Si}.

As illustrated in Fig. 4, the asymmetrical device has one predominant channel, but with Q_i nearly equal to the total charge in both identical channels of the symmetrical device. The degree of on-state electron degeneracy is thus higher in the former device, as is the surface electric field (although the field is substantially lower than that in a bulk-Si counterpart because of the absence of significant depletion charge [6]). The degeneracy can be beneficial since it tends to increase the thermal injection velocity at the source [5], [6], which defines the ballistic-limit current. The higher field implies more confinement, which leads to more occupation of states in the lowest-energy subband where the effective electron mass is lower, implying even higher injection velocity [5], [6], [9]. Further, the higher Q_i per channel in the asymmetrical device suggests higher $V_{DS(sat)}$ and hence more I_{on}. Note for the symmetrical device in Fig. 4 the condition of "volume inversion" [10] at low V_{GS}. This condition, which tends to obtain as t_{Si} is thinned as depicted by the predicted wavefunctions (which define n(x)) in Fig. 5, can result in higher

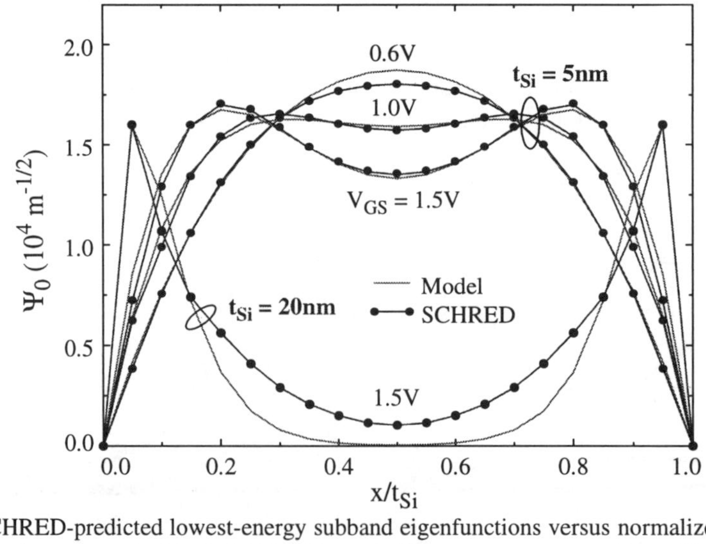

Fig. 5. SCHRED-predicted lowest-energy subband eigenfunctions versus normalized position in the Si film for symmetrical DG nMOSFETs with different t_{Si} at varying V_{GS}. (Also shown are predictions from a physics-based compact model (UFDG) [11].)

carrier mobility [11], which in turn tends to support ballistic transport. We stress, however, that t_{Si} cannot be thinned to less than ~2nm because of severe QM-confinement effects [11].

The carrier transport in the 50nm devices is actually quasi-ballistic, meaning that one population of carriers flows ballistically and another is scattered. Such transport is characterized in MEDICI [4] via the hydrodynamic transport equations, and it can be analytically described by an average velocity-overshoot model, $v_{sat(eff)} > v_{sat}$, derived from the first and second moments of the Boltzmann transport equation [12]. The overshoot depends on high carrier mobility, facilitated by relatively low transverse electric field. This is another beneficial feature of DG MOSFETs that results from the noted gate-gate coupling [6].

The threshold voltages in (2) and (3) [3] are virtually independent of the low N_A, and are defined mainly by the gate workfunctions. However, the QM-confinement effects, which are dependent on t_{Si} (mildly for t_{Si} > ~2nm) as well as the transverse electric field [11], result in, relative to the classical MEDICI results, higher V_t, due to the electron-energy quantization, and lower C_G, due to the deeper electron distribution in the Si film (see Fig. 4). Thus, both $V_{tf(sym)}$ and $V_{tf(asym)}$ depend on t_{Si}, but the latter dependence is stronger because of its dependence on r [3].

The main benefit of DG MOSFETs, in addition to the doubling (or more so) of the on-state current (per device area) implied by (2) and (3), is the inherent control of SCEs resulting from the gate-gate coupling and the associated control of the transverse electric field in the thin Si film. For t_{Si} scaled to 5nm and t_{ox} = 1.5nm in the symmetrical DG nMOSFET, for example, the gate-length dependences of V_t and S shown in Fig. 6 are predicted by self-consistent solutions of the 2D Poisson and 1D Schrödinger equations [13]. These results clearly reveal the scalability of DG devices to less than 20nm.

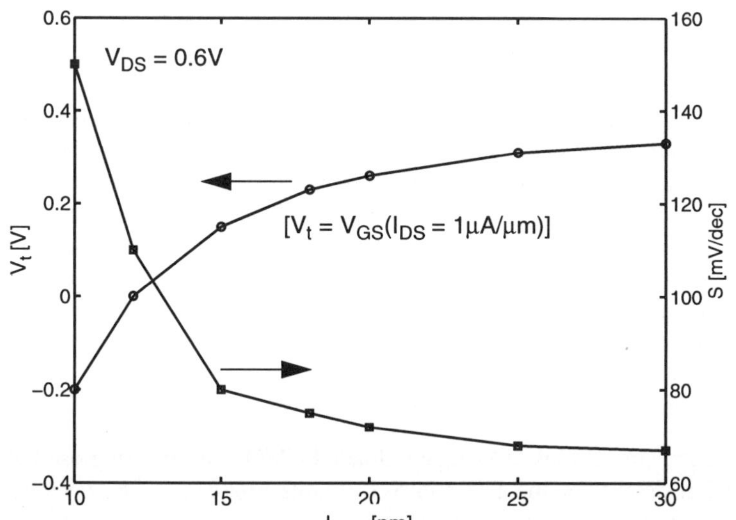

Fig. 6. 2D Poisson/1D Schrödinger-predicted threshold voltage and gate swing factor versus channel length for a symmetrical DG nMOSFET; t_{Si} = 5nm, t_{ox} = 1.5nm.

Additional 2D simulations with MEDICI reveal that the asymmetrical device has virtually the same superior (relative to bulk-Si MOSFETs) immunity to SCEs as the symmetrical device [14].

3. Projected Performance of DG CMOS

Irrespective of the potential near-ideal features of DG MOSFETs discussed in Sec. 2, commensurate DG CMOS performance cannot be inferred directly. The technology, when developed, will be complex, and substantial device parasitics such as excessive gate-overlap capacitance may be unavoidable. Hence, as has been frequently wondered with reference to the CV/I speed metric, if the DG device gives twice the current but with twice the gate capacitance relative to the SG (e.g., bulk-Si) counterpart, then, whereas for large loads (e.g., interconnect) DG should be faster, for small loads (e.g., low fanout) no speed improvement is implied, and the DG parasitics may render inferior speed. Meaningful performance projections are needed, and they will require predictive circuit simulations using a physics/process-based compact model for the DG MOSFET. We are developing such a model (UFDG [11], [15]), focusing on the crucial physics of the extremely scaled device: gate-gate charge coupling, inversion-carrier distribution throughout the Si-film channel defined by the QM confinement with 2D density of states and Fermi-Dirac statistics (see Fig. 5), 2D SCEs, quasi-ballistic carrier transport based on velocity overshoot [12] and a mobility model dependent on t_{Si} as well as transverse electric field, and associated terminal charge modeling for transient simulations. The UFDG model is generic, applicable to asymmetrical or symmetrical DG devices, and, via Newton-Raphson iterative solutions, is an effective compact self-consistent Poisson-Schrödinger solver. The UFDG model, and in particular its predictive accounting for the gate-gate charge coupling, have been supported by measured characteristics of DG MOSFETs as well as by numerical device simulations [15].

We used UFDG in Spice3 to simulate unloaded DG CMOS-inverter ring oscillators [16]. The nominal DG device structure we assumed, for both nMOSFETs and pMOSFETs, has L_{met} = 50nm, with $t_{oxf} = t_{oxb} = t_{ox}$ = 3nm and t_{Si} = 10nm; the Si film doping density is low ($10^{15}cm^{-3}$). The electron mobility was calibrated based on Monte Carlo transport simulations [17], and the hole mobility was assumed to be a factor-of-two lower. Reasonable source and drain (specific) series resistances were presumed ($R_S = R_D$ = 80Ω-μm for nMOSFETs, and twice that value for pMOSFETs), and gate resistance was neglected ($V_{GfS} = V_{GbS} = V_{GS}$). We simulated DG MOSFETs with both asymmetrical (n^+/p^+ polysilicon) and symmetrical (workfunction defined to yield I_{off} equal to that of the asymmetrical device) gates. To benchmark the DG MOSFETs against an SG counterpart, we simply used the asymmetrical DG device structure with its back (p^+ in the nMOSFET, n^+ in the pMOSFET) gate grounded. Indeed, this SG (i.e., FD/SOI with thin BOX) device emulates the bulk-Si counterpart well [16].

UFDG-aided simulations of 9-stage unloaded CMOS-inverter ring oscillators yielded the propagation delays for the DG and SG circuits plotted in Fig. 7 versus V_{DD}. The simulations included gate-source/drain overlap capacitances defined by the oxide thickness and assumed front and back overlaps of 40% of L_{met}. Note that the asymmetrical and symmetrical DG devices yield about the same delays (e.g., about 5ps at

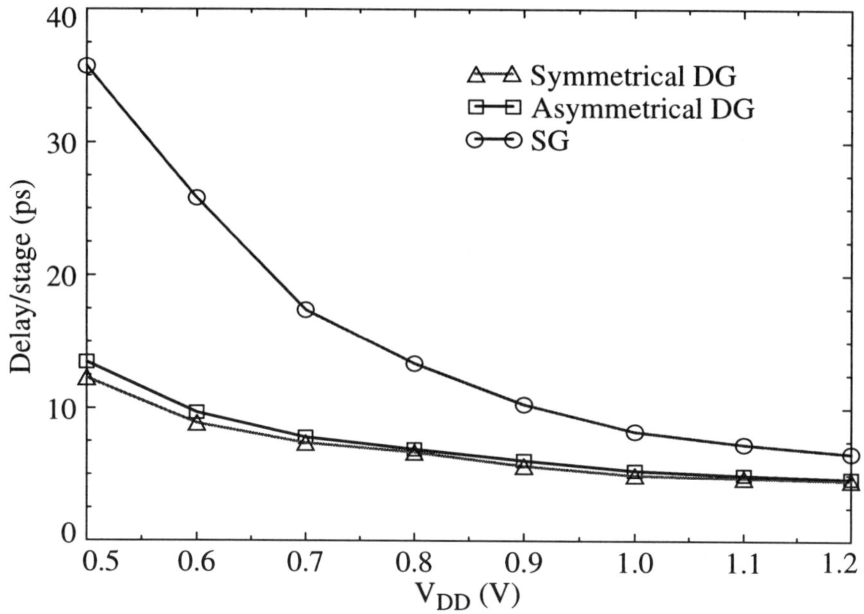

Fig. 7. Predicted propagation delays versus supply voltage from UFDG-aided simulations of unloaded 9-stage CMOS-inverter ring oscillators comprising asymmetrical DG, symmetrical DG (workfunction defined for I_{off} equal to that of the asymmetrical counterpart), and SG (asymmetrical DG structure with back gate grounded) MOSFETs; L_{met} = 50nm. Gate-source/drain overlaps of 40% of L_{met} were assumed for all gates, front and back.

V_{DD}=1.0V, lengthened from ~1.5ps by the noted gate overlaps), in accord with their comparable characteristics as discussed previously, and that the delays are substantively shorter than that of the SG CMOS, e.g., by about 40% at V_{DD}=1.0V. The predicted DG CMOS speed superiority is significantly enhanced when V_{DD} is reduced below 1V; e.g., at V_{DD}=0.5V, it is almost three-times faster.

The DG speed superiority can be partially explained by the near-ideal S illustrated in Fig. 2, which, for equal I_{off} and low V_{DD}, results in the DG on-state current being greater than that of the SG device by more than a factor of two. UFDG-predicted I_{on} versus V_{DD} is plotted in Fig. 8 for the DG and SG nMOSFETs. For V_{DD}=1.0V, I_{on} of the symmetrical DG device is 2.6 times that of the SG device; and this ratio increases to 3.7 when V_{DD} is reduced to 0.5V. Similar ratios obtain for the asymmetrical DG device as shown in the Fig. 8 inset.

The DG speed superiority is fully explained by considering the intrinsic gate capacitance, in addition to I_{on}. The UFDG-predicted $C_G(V_{GS})$ plots shown in Fig. 9, in accord with the SCHRED predictions in Fig. 3, reveal differences between DG and SG gate capacitances that underlie the faster DG speeds. Most significantly, for V_{GS} decreasing toward zero, C_G of the DG devices decreases dramatically, becoming less than that of the SG device and approaching zero in weak inversion. In contrast, the low-V_{GS} C_G of the SG device is finite and substantive, like that of a bulk-Si MOSFET. (Note in Figs. 3 and 9 that for moderately high V_{GS}, C_G of the asymmetrical DG nMOSFET is

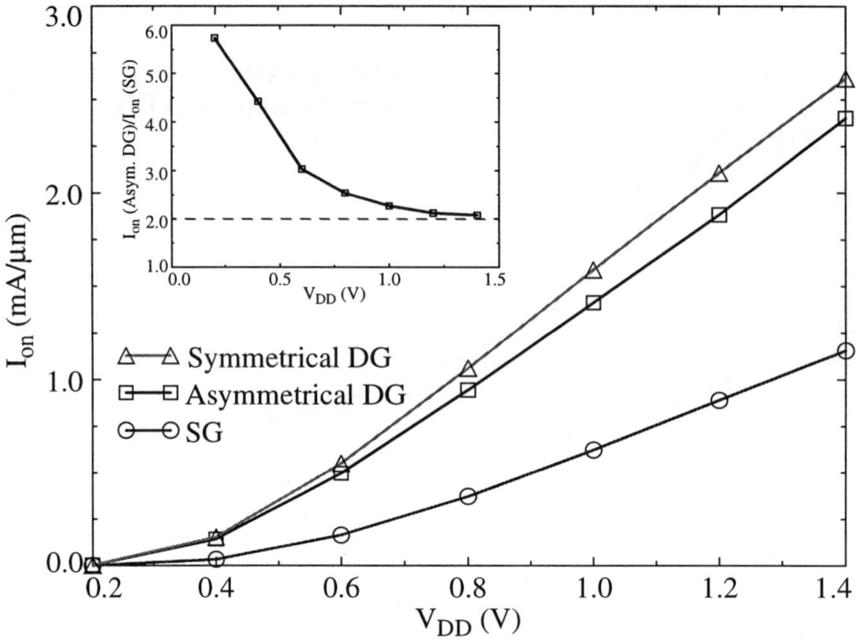

Fig. 8. UFDG-predicted on-state currents versus supply voltage for the asymmetrical DG, symmetrical DG, and SG nMOSFETs in the ring oscillators of Fig. 7. The asymmetrical DG-to-SG current ratio plotted in the inset clearly shows it, and hence that for the symmetrical DG device, to be substantially greater than two for V_{DD} less than 1V.

significantly less than that of the symmetrical device, even though its I_{on} is comparable. This difference is due to the different inversion-charge distributions in the respective devices as shown in Fig. 4. For higher V_{GS}, the two devices are virtually identical.) The predicted $C_G(V_{GS})$ plots also show that for increasing V_{DS}, C_G decreases more for the DG devices than for the SG device, implying smaller dynamic loading. Thus, whereas the

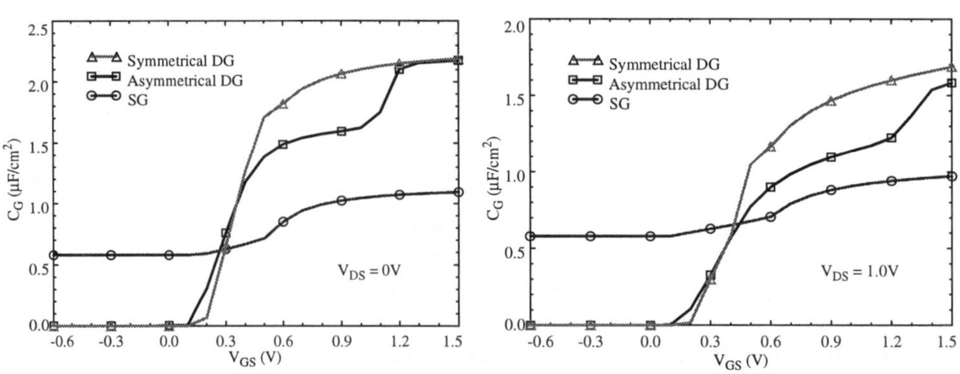

Fig. 9. UFDG-predicted intrinsic gate capacitances (without the source/drain overlaps) versus gate voltage, at low (left) and high (right) V_{DS}, for the asymmetrical DG, symmetrical DG, and SG nMOSFETs in the ring oscillators of Fig. 7.

ratio of the symmetrical DG-to-SG capacitances at V_{DS}=0 is about two for V_{GS}=1.0V, it is less than 1.7 at V_{DS}=1.0V.

4. Summary

A suite of simulation tools, including a 2D numerical device simulator, a 1D numerical Poisson-Schrödinger solver, and a generic DG MOSFET physics/process-based compact model in Spice, were applied to both asymmetrical- and symmetrical-gate DG CMOS devices and circuits, providing physical insight and previewing design and performance issues and at both the device and circuit levels. Simulation results, and explanations thereof, revealed how the SCEs are suppressed and why the subthreshold slope is near ideal in thin-film DG MOSFETs, how the performances of DG MOSFETs having asymmetrical and symmetrical gate structures can be comparable, how the thin Si film affects the carrier energy quantization and mobility, why the ballistic-limit current could possibly be reached in DG MOSFETs, and how and why DG CMOS, even with substantive parasitics, can yield significantly superior speed performance relative to the single-gate counterpart, for unloaded or loaded circuits operating at low supply voltages (<1V). While the development of a viable DG CMOS technology is a formidable task, the results presented, suggesting that moderate parasitics will not be showstoppers because of the inherent near-ideal features of the intrinsic DG MOSFET, give added motivation as well as preliminary guidance for the development of the technology, which is now only in very early stages in a few of the major semiconductor companies.

Acknowledgments

This work was supported, in part, by Motorola and the Semiconductor Research Corporation. I acknowledge contributions made by Meng-Hsueh Chiang, Keunwoo Kim, Lixin Ge, Zhibin Ren, and Mark Lundstrom.

References

1. D. J. Frank, S. E. Laux, and M.V. Fischetti, "Monte Carlo simulation of a 30nm dual-gate MOSFET: How short can Si go?," *IEDM Tech. Dig.*, pp. 553-556, Dec. 1992.
2. Semiconductor Industry Association, "International Technology Roadmap for Semiconductors," 2001.
3. K. Kim and J. G. Fossum, Double-gate CMOS: Symmetrical- versus asymmetrical-gate devices," *IEEE Trans. Electron Devices*, vol. 48, pp. 294-299, Feb. 2001.
4. *Avant! - MEDICI 4.0: Two-Dimensional Device Simulation Program.* Fremont, CA: Avant! Corp., 1999.
5. F. Assad, Z. Ren, D. Vasileska, S. Datta, and M. Lundstrom, "On the performance limits for Si MOSFET's: A theoretical study," *IEEE Trans. Electron Devices*, vol. 47, pp. 232-240, Jan. 2000.
6. K. Kim and J. G. Fossum, "Achieving the ballistic-limit current in Si MOSFETs," submitted to *Solid-State Electron.*, July 2001.
7. D. Vasileska and Z. Ren, "SCHRED-2.0 Manual," Arizona State University, Tempe, AZ, and Purdue University, West Lafayette, IN, Feb. 2000.
8. J. A. López-Villanueva, P. Cartujo-Cassinello, F. Gámiz, J. Banqueri, and A. J. Palma, "Effects of inversion-layer centroid on the performance of double-gate MOSFET's," *IEEE Trans. Electron Devices*, vol. 47, pp. 141-146, Jan. 2000.

9. J. G. Fossum, Z. Ren, K. Kim, and M. Lundstrom, "Extraordinarily high drive currents in asymmetrical double-gate MOSFETs," *Superlattices and Microstructures*, vol. 28, pp. 525-530, Nov./Dec. 2000.
10. F. Balestra, S. Cristoloveanu, M. Benachir, J. Brini, and T. Elewa, "Double-gate silicon-on-insulator transistor with volume inversion: A new device with greatly enhanced performance," *IEEE Trans. Electron Devices Lett.*, vol. EDL-8, pp. 410-412, Sep. 1987.
11. L. Ge and J. G. Fossum, "Analytical modeling of quantization and volume inversion in thin Si-film double-gate MOSFETs," *IEEE Trans. Electron Devices*, vol. 49, pp. 287-294, Feb. 2002.
12. L. Ge, J. G. Fossum, and B. Liu, "Physical compact modeling and analysis of velocity overshoot in extremely scaled CMOS devices and circuits," *IEEE Trans. Electron Devices*, vol. 48, pp. 2074-2080, Sept. 2001.
13. Z. Ren, R. Venugopal, S. Datta, M. Lundstrom, D. Jovanovic, and J. G. Fossum, "The ballistic nanotransistor: A simulation study," *Tech. Dig. 2000 IEEE Internat. Electron Devices Meeting*, pp. 715-718, Dec. 2000.
14. J. G. Fossum, "Physical insights on double-gate MOSFETs" *Dig. Government Microcircuit Appl. Conf.*, pp. 322-325, Mar. 2001.
15. M.-H. Chiang and J. G. Fossum, "A process-based compact model for double-gate MOSFETs," *Proc. Tenth Internat. Symp. on SOI Technology and Devices*, ECS Vol. 2001-3, pp. 421-426, Mar. 2001.
16. J. G. Fossum, L. Ge, and M.-H. Chiang, "Speed superiority of scaled double-gate CMOS," *IEEE Trans. Electron Devices*, vol. 49, 2002.
17. F. Gámiz, J. B. Roldan, J. A. Lopez-Villanueva, P. Cartujo-Cassinello, J. E. Carceller, and P. Cartujo, "Monte Carlo simulation of electron transport in silicon-on-insulator devices," *Proc. Tenth Internat. Symp. on SOI Technology and Devices*, ECS Vol. 2001-3, Mar. 2001.

International Journal of High Speed Electronics and Systems, Vol. 12, No. 2 (2002) 573–582

FULL-CHIP POWER-SUPPLY NOISE: THE EFFECT OF ON-CHIP POWER-RAIL INDUCTANCE

C.W. FOK
Department of Physics and Astronomy, University of British Columbia
Vancouver, BC V6T1Z4, Canada

D.L. PULFREY
Department of Electrical and Computer Engineering, University of British Columbia
Vancouver, BC V6T1Z4, Canada

The importance of on-chip power-rail inductance in generating delta-I power-supply noise is examined in this paper using systematic circuit simulation of the complete integrated-circuit power net. This source of noise is compared to the resistive IR drop in the net, and to the delta-I noise due to both high-inductance- and low-inductance-bonding packages. Results are presented for a typical on-chip power net in 0.18 μm CMOS technology, and it is demonstrated that the inductance of this on-chip power net is the dominant contributor to the full-chip power-supply noise. The simultaneous switching events which produce the triggering current transients for the delta-I noise are taken to arise from core-logic switching; the mitigating, de-coupling role of the capacitance of non-switching gates within the core-logic block is considered.

1. Introduction

As silicon integrated-circuit sizes increase, the long, on-chip power lines become important sources of resistance and self- and mutual-inductance. Also, as silicon MOSFET sizes decrease, the circuit density increases, with the result that larger currents are demanded and larger current transients appear. These two size-related phenomena, coupled with the trend of decreasing power-supply voltage, lead to the possibility of significant power-supply noise due to both resistive voltage drops and inductive voltage changes. In this paper, the full-chip power-supply noise is considered, and particular attention is paid to determining the importance of the contribution to this noise of the inductance of the on-chip power net. The intention is to resolve the issue of whether the on-chip inductance is really important, in view of the presence of significant inductance in the bonding package: there is presently some difference of opinion on this [1,2,3].

Our approach is to perform systematic circuit simulations of a full chip comprising a representative, 0.18 μm CMOS power system, with either wire-bond- or flip-chip-packaging. The impedance of the full-chip power network is simulated in its entirety, and also by considering each of the impedance components separately, namely: package inductance, on-chip resistance, and on-chip inductance. This approach reveals that it is necessary to consider the inductance of the on-chip power net if the power-supply noise (PSN) is not to be seriously underestimated.

The ΔI component of PSN originates in the massive, simultaneous switching of many on-chip devices. Here, we consider this switching to take place in core-logic blocks, and we use a Monte-Carlo approach to generate a switching pattern that gives rise to off-chip, power-supply currents that are typical of modern processors. Within the logic blocks considered in this work, only 20% of the gates switch in a given cycle. Consideration is given to the de-coupling role that the capacitance of the non-switching gates within the block might play in providing some of the current demanded by the switching gates, thereby reducing the current that needs to be delivered via the inductive power-supply net.

2. Methodology

Figure 1 illustrates the system considered: it comprises an off-chip power supply, bonding-package inductances, an inductively coupled on-chip power net, and blocks of logic gates. The capacitance of the power net is also considered, but it has been found to have negligible effect on the simulated power-supply noise.

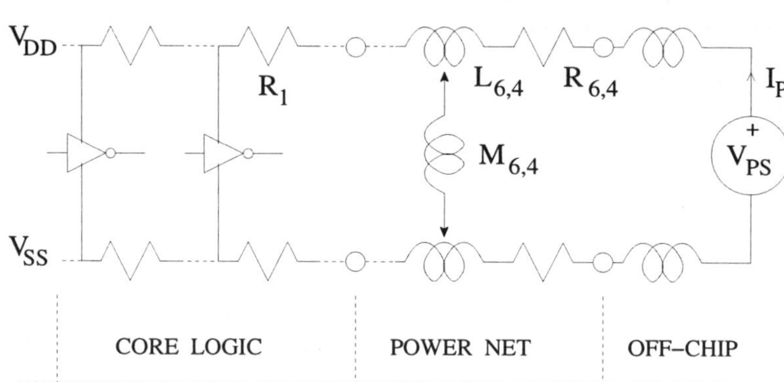

Fig. 1. Full-chip schematic. $R_{6,4}$ and $L_{6,4}$ refer, respectively, to the resistance and self-inductance of power lines in Metals 6 and 4. $M_{6,4}$ is the mutual inductance of the network of Metal 6 and Metal 4 lines. R_1 is the resistance of the Metal 1 lines.

In this work, the network size is large enough to ensure a statistically meaningful result, but it is also small enough to allow the simulation of PSN to be accomplished in a reasonable time (several hours on a SUN Ultra 10 workstation). Specifically, wire- and flip-chip-bonding schemes are applied to 15 pairs of V_{DD}/V_{SS} lines arranged in a parallel fashion in Metal 6. These lines are connected at the appropriate crossover points to 15 pairs of orthogonal lines in Metal 4 which, in turn, are connected to 40 pairs of orthogonal, resistive power lines in Metal 1. This is a typical power net arrangement [4], and, with the wire specifications listed in Table 1, leads to a chip area of 36.9 mm^2.

Table 1. Metal layer specifications.

Metal	Resistivity (Ω/sq.)	Width (μm)	Height (μm)	Spacing (μm)
6	0.036	16	0.99	180
4	0.078	16	0.53	180
1	0.078	1.5	0.53	4.1

The inductance, resistance and capacitance of the power-supply net are computed using RAPHAEL, in which the Partial Equivalent Electric Circuits (PEEC) model is employed. The vias connecting Metals 6 and 4 occur at intervals along the power rails of 360 μm; we accommodate two partial self-inductance elements in this length. Thus, each of Metals 6 and 4 is represented by an inductance matrix of 930$\times$930 elements. Clearly, some simplification is needed for a solution to be achieved using a circuit simulator. In this work, the partial-inductance matrix is successively reduced by iteratively subtracting the smallest matrix term from all the other, non-zero matrix terms. For symmetrical matrices of the type considered here, this procedure is known to not affect the circuit solution [5]. We cease the reduction process when no more than four mutual inductance components are present in any column. As Fig. 2 shows, the eigenvalues of the full- and reduced-matrices match very well, except at very large eigenvalues. However, the differences have an insignificant effect on the circuit solution because the large eigenvalues correspond to the inductances of very large current loops, which carry insignificant current in the presence of smaller current loops [5].

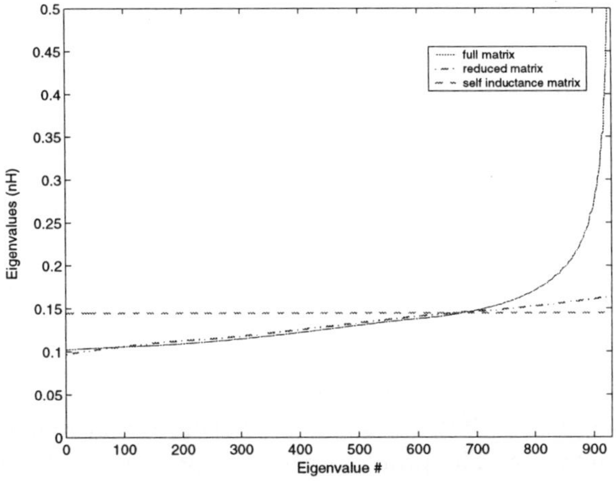

Fig. 2. Comparison of eigenvalues for the inductance matrices.

The resulting net list, which includes 12516 resistors, 1860 partial self-inductors, and 7196 partial mutual-inductors, is incorporated into HSPICE, along with a description of the logic gates. Blocks of 10,000 gates (minimum-size inverters) are

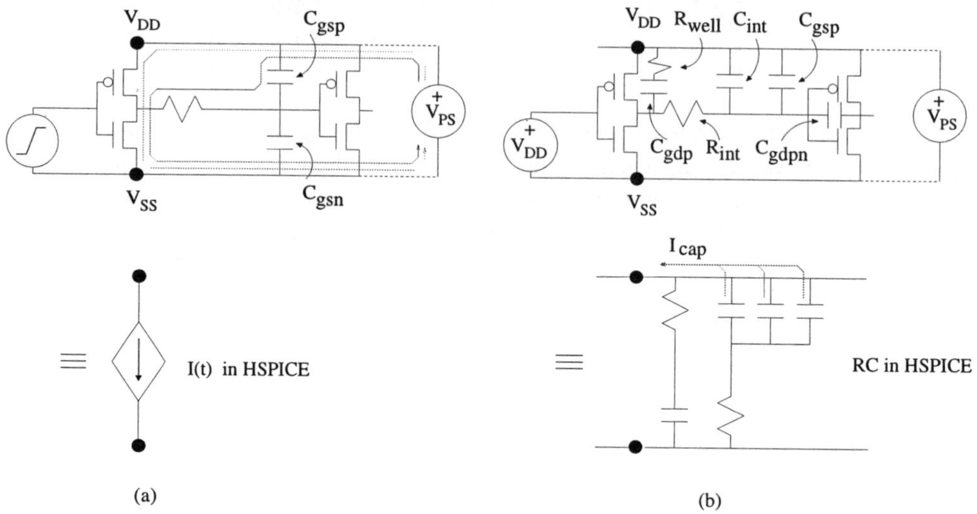

Fig. 3. Gate equivalent circuits: (a) switching gate (LO-HI transition), (b) idle gate (input HI).

placed between the Metal 1 power rails. Each gate is represented by either a current source or a series RC branch [6], as shown in Fig. 3. The current source is activated if a gate is switching. Fig. 3(a) shows three switching currents, two of which must be supplied from outside of the gate. It is these two currents that are represented in HSPICE by a time-dependent current source. The RC branch is activated if a particular gate is not switching. This description allows the current de-coupling effect of the capacitance of the idle gates to be evaluated, as illustrated in Fig. 3(b). In the equivalent-circuit representation of the idle gate, the gate-drain capacitance of the p-channel transistor in the first inverter of the gate is not considered, in view of its association with an n-well resistance R_{well}, which is large compared to the interconnect resistance R_{int} that couples the other capacitors to the power-rail node [6].

The current-source waveform for a single gate is determined from HSPICE simulations, using 0.18 μm CMOS rules, by simulating the switching of one inverter driving a similar inverter. The profile of the supply-current (I_{PS} in Fig. 1) is then generated by arranging for 20% (2000) of the gates within a block to switch at appropriate time intervals. The instant at which a gate switches is determined by Monte-Carlo procedures [2], and the locations of the switching gates are randomly distributed within the 10,000-gate block. Switching events are allowed every 50 ps, and each gate switches in 85 ps, *i.e.*, during the assumed rise-time of the local clock.

The overall switching period is 1 ns. To achieve profiles for I_{PS} which are typical of recent microprocessors [7], the number of gates forced to switch within a prescribed period (485-685 ps) is varied. In the examples shown in Fig. 4, the higher profile is produced by 1869 gates switching within the stated period, while the number for the lower profile is 1239. The respective peak currents are 43 and 25 mA, and the average currents are 4.66 and 4.59 mA. Thus, the peak-to-average ratio is about 9 for the higher-profile, which is consistent with values suggested for modern, low-power, chip design [8].

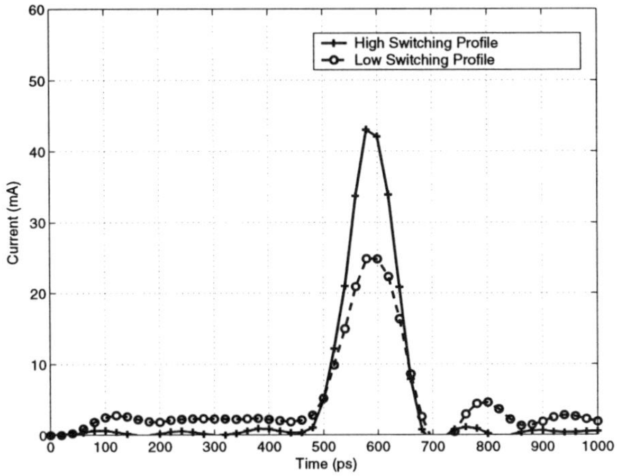

Fig. 4. Power-supply current waveforms for high- and lower-numbers of gates (1869 and 1239, respectively) switching in a 200 ps period.

Results are presented below for PSN due to various instances of switching within one core-logic block. The major objective of determining the role played by the on-chip power-net inductance is achieved by systematically increasing the complexity of the on-chip power-net impedance. First, the on-chip power net is represented merely as a resistive network. Next, the on-chip inductance is then included by separately introducing self- and mutual-inductance into the network description. For each of these cases, simulations are performed for two types of bonding, thereby allowing the relative importance of the bonding-package inductance to be assessed. Additionally, the importance of the de-coupling capacitance of the idle gates is determined by performing simulations with and without this capacitance present. Finally, by determining the PSN at a central spot in the chip due to switching in a block which can be located at various distances from this reference point, the effect of multiple blocks switching simultaneously is determined by superposition.

3. Results and Discussion

A block of 10,000, minimum-sized inverters in 0.18 μm technology occupies an area of 0.286 mm^2 [9]. When considering switching within a block, we evaluate the PSN at one corner of the block. In all results reported here, the high-switching current profile depicted in Fig. 4 is used. For the lower-switching case portrayed in that figure, we found that the PSN scaled approximately with the magnitude of the power-supply current. This is not surprising, as a lower magnitude also means a lower rate-of-change of current in our prescription for switching. Two bonding packages were used: wire-bond and flip-chip. These represented cases of high- and low-package inductance, respectively (see below).

To illustrate the nature of the PSN, typical disturbances to V_{DD} and V_{SS}, due to switching within one block, are shown in Fig. 5. Flip-chip bonding was used in this case, and the figure shows the effect of modeling the on-chip power net as either resistive (R), resistive plus self-inductive (RL), or resistive plus self- and mutual-inductive (RLM). Clearly, the presence of the power-line inductance causes significant overshoot in both V_{DD} and V_{SS}, following the usual "droop" of V_{DD} and "bounce" of V_{SS}.

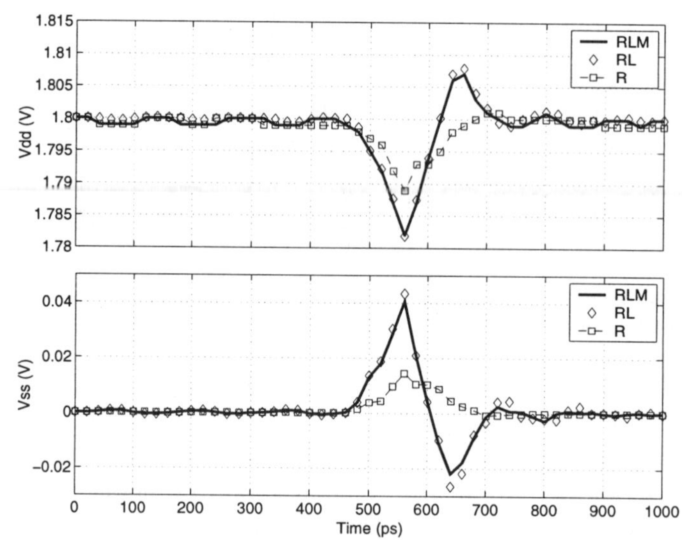

Fig. 5. Voltage waveforms: the effect of modeling the on-chip power net as either resistive (R), resistive plus self-inductive (RL) or resistive plus self- and mutual-inductive (RLM). The bonding is flip-chip and the high-switching-activity case is used.

For comparison purposes, we quantify the power-supply noise by:

$$\mathrm{PSN} = \mathrm{MAX}\{\frac{\mathrm{V_{PS}} - \mathrm{V_{DD}(t)} + \mathrm{V_{SS}(t)}}{\mathrm{V_{PS}}}\}, \tag{1}$$

where V_{PS} is the off-chip supply voltage. Thus, the data in Fig. 5 is summarized in Fig. 6, along with the results from employing a wire-bonded package. In the wire-bonded case, each bond pad is assumed to have an inductance of 2.3 nH and a capacitance of 27 fF [10]. In the flip-chip case, the corresponding values per bond are 35 pH and 20 fF [10]. The flip-chip bonding scheme used here employs diagonally routed, metal lines in an additional layer (Metal 7) to connect the solder bumps. This arrangement produces five bumps per line and gives a minimum spacing between adjacent V_{DD}- and V_{SS}-bonds of 438 μm.

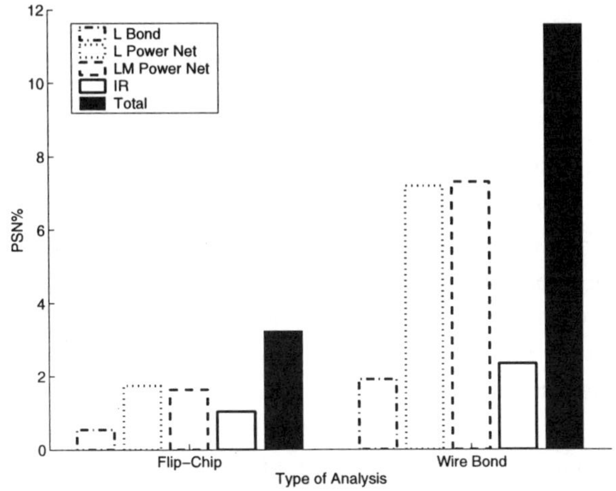

Fig. 6. PSN: the effect of the various components of the full-chip impedance network.

Figure 6 is the key figure in this paper as it shows the contribution to the PSN of each component of the full-chip impedance network. The IR case refers to the ideal situation when there is neither bonding-package inductance nor on-chip power-net inductance. The other cases shown refer to situations in which various combinations of inductance are present in a network with zero resistance. For each type of bonding package, the case labeled "total" refers to the full-chip impedance case when all resistances and inductances are modeled. It is immediately clear that the inductance of the on-chip power net is the major contributor to the PSN. This is true even in the wire-bonded case where the package inductance is high. For the flip-chip case, there is a significant ($\approx 2\times$) increase in PSN. Interestingly, the absolute increase in PSN caused by considering the power-net inductance in the wire-bond case is larger than in the flip-chip case. This indicates an interactive effect between the package- and rail-inductances that would be missed if a complete-system simulation were not performed. Thus, the PSN is higher in the wire-bond case not only because the package inductance is higher, but also because the current is forced by the end-placement of the bonds to pass through a longer length of the on-chip power net.

Regarding the effect of mutual coupling, Fig. 6 shows that when the mutual inductance components of the network are considered, there is very little change in PSN for both instances of bonding package. While some cancellation of the voltages induced by different mutual inductances is to be expected, because the coupling between inductances in lines carrying similarly directed currents gives voltages that are of the opposite sense to those induced by coupling between lines carrying oppositely directed currents, we emphasize that the almost complete cancellation observed here is probably due to the particular choice of length for the partial self-inductance elements. The chosen length of 180 μm, which also applies to a partial mutual-inductance element that couples between adjacent segments on a given power line, is the same as that of the partial mutual-inductance elements coupling most directly between adjacent V_{DD}- and V_{SS}-rails. Thus, significant cancellation of voltages induced by these two sets of partial mutual-inductances can be expected. This is a very desirable situation because it means that a reasonable estimation of PSN can be obtained by considering only the partial self-inductance elements. This greatly reduces the computation time.

Having established the importance of considering power-rail inductance, we disabled the parasitic capacitance of the idle gates (see Fig. 3(b)) and repeated the above simulations with the on-chip power net represented by its resistive and inductive components. The PSN increased by $\approx$ 25%, illustrating the significant de-coupling role that is played by the capacitance of the non-switching gates within a block.

Next, we investigated the effect on PSN at our centrally located reference point within the chip of the placement of the switching block. The previous results are for the switching block located with one corner at the reference point. Fig. 7 shows results for the PSN induced at the reference point by a single switching block placed at different locations. Clearly, the influence of a switching block falls off quite rapidly

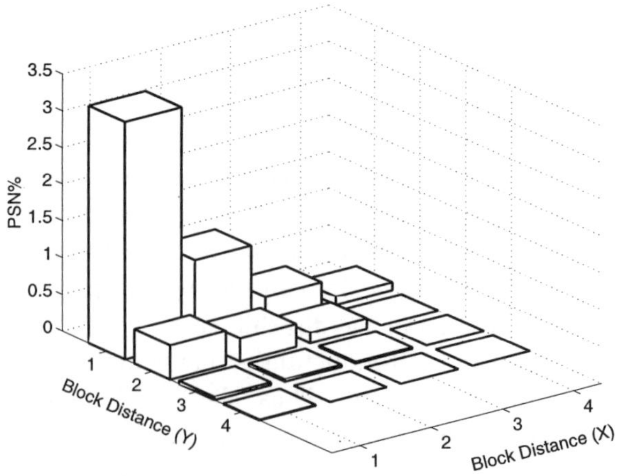

Fig. 7. PSN: the effect at the reference point (0,0.5) of the location of a switching block.

with distance from the reference point. Nevertheless, the cumulative effect of many blocks switching simultaneously cannot be ignored, as Fig. 8 reveals. For example, four blocks, each with a corner at the reference point, conspire to produce a PSN of about 13%. This would already be unacceptable for situations with stringent PSN budgets [7,11]. Moving further out from the reference point and including the next twelve surrounding blocks causes the PSN to increase to 20%, a figure that would probably be universally unacceptable. Clearly, if such close-proximity switching is unavoidable, then the idle-gate de-coupling capacitance considered here must be augmented by additional, explicit capacitative elements in order to reduce the noise to acceptable levels [7]. Fig. 8 also shows the effect of not considering the power-rail inductance. The indications are that this would be a serious omission. For example, a PSN budget of 10% would be predicted to allow about 20 blocks to switch simultaneously, whereas, when the power-rail inductance is considered, this number drops to the much more restrictive value of three. Note that the predictions

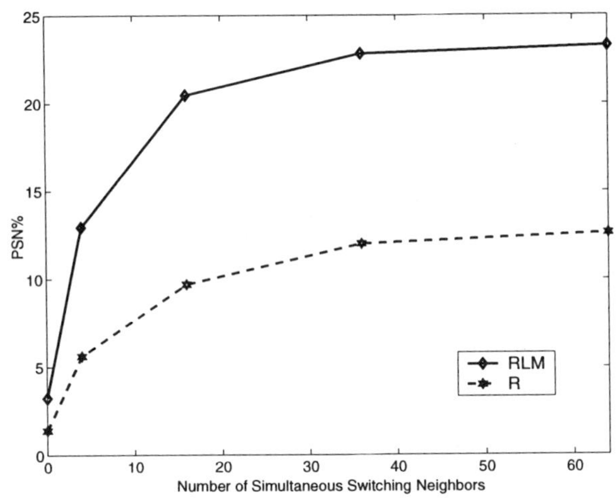

Fig. 8. PSN: the cumulative effect at the central-chip reference point of the location of additional switching blocks. The on-chip power net is modeled as either resistive (R) or resistive and inductive (RLM).

of PSN in excess of 10% should be considered with caution. This is because the current drawn by a particular block is dependent on the supply voltage, and as the voltage drops due to switching at other blocks, the current drawn by the block also decreases, thereby reducing its impact on the noise. However, for voltage variations less than about 10%, superposition should still give an accurate estimate of PSN [12].

4. Conclusions

From this systematic study by simulation of power-supply noise (PSN) it can be concluded that:

(i) PSN depends on both the bonding-package inductance and the power-rail inductance, but even in the wire-bonded case, when the package inductance is large, the power-rail inductance is the dominant contributor to the noise;

(ii) for the flip-chip packaging case considered here, the presence of power-rail inductance increases PSN by a factor of about 2;

(iii) the de-coupling capacitance of non-switching gates in a logic block has a significant effect on PSN, but it is not likely to be large enough to realize reasonable noise budgets if more than 2 or 3 contiguous logic blocks of the type considered here are active simultaneously.

Acknowledgements

The authors are grateful to Sameer Khushal of PMC-Sierra, Inc. for valuable advice and information. Financial support from the Natural Sciences and Engineering Research Council of Canada helped make this work possible.

References

1. P. Zarkesh-Ha and J.D. Meindl, "Optimum On-Chip Power Distribution Networks for Gigascale Integration (GSI)", *Proc. IEEE Intl. Interconnect Technology Conf. (IITC)*, 2001, pp. 125–127.
2. R. Panda, D. Blaauw, R. Chaudhry, V. Zolotov, B. Young, and R. Ramaraju, "Model and Analysis for Combined Package and On-Chip Power Grid Simulation", *Proc. Intl. Symp. on Low Power Electronics and Design (ISLPED)*, 2000, pp. 179–184.
3. P. Heydari and M. Pedram, "Analysis and Optimization of Ground Bounce in Digital CMOS Circuits", *Proc. IEEE Intl. Conf. on Computer Design (ICCD)*, 2000, pp. 121–126.
4. W.S. Song and L.A. Glasser, "Power Distribution Techniques for VLSI Circuits", *IEEE J. Solid-State Circuits*, **SC-9** (1986) 150–156.
5. B. Krauter and L.T. Pileggi, "Generating Sparse Partial Inductance Matrices with Guaranteed Stability", *IEEE/ACM Intl. Conf. Computer Aided Design (ICCAD)*,1995, pp. 45-52.
6. R. Chaudhry, R. Panda, T. Edwards and D. Blaauw, "Design and Analysis of Power Distribution Networks with Accurate RLC Models", *Proc. 13th International Conference on VLSI Design*, 2000, pp. 151-155.
7. H.H. Chen and J.S. Neely, "Interconnect and Circuit Modeling Techniques for Full-Chip Power Supply Noise Analysis", *IEEE Trans. Comp., Packag., Manufact. Technol.*, **21** (1998) pp. 446-452.
8. M. Benoit, S. Taylor, D. Overhauser and S. Rochel, "Power Distribution in High-Performance Design", *Proc. Intl. Symp. on Low Power Electronics and Design (ISLPED)*, 1998, pp. 274-278.
9. D. Sylvester and K. Keutzer, "Getting to the Bottom of Deep Submicron", *IEEE/ACM Intl. Conf. Computer Aided Design (ICCAD)*, 1998, pp. 203-211.
10. S. Gong, H. Hentzell, S.-T. Persson, H. Hesselbom, B. Lofstedt and M. Hansen, "Packaging impact on switching noise in high-speed digital systems", *IEE Proc.-Circuits Devices Syst.*, **145** (1998) pp. 446-452.
11. S. Lin and N. Chang, "Challenges in Power-Ground Integrity", *IEEE/ACM Intl. Conf. Computer Aided Design (ICCAD)*, 2001, pp. 651-654.
12. S. Bobba and I.N. Hajj, "Simultaneous Switching Noise in CMOS VLSI Circuits", *Southwest Symp. on Mixed-signal Design*, 1999, pp. 15-20.

International Journal of High Speed Electronics and Systems, Vol. 12, No. 2 (2002) 583–592

QUANTUM DOT SUPERLATTICES IN A CONSTANT ELECTRIC FIELD: LOCALIZATION AND BLOCH OSCILLATIONS

ROBERT A. SURIS and IVAN A. DMITRIEV

A.F.Ioffe Physical-Technical Institute, 194021, Polytechnicheskaya 26

St.Petersburg, Russia

An analysis has been presented of electron localization in ideal 2D and 3D quantum dot superlattices (QDSL) in a homogeneous dc electric field and of Bloch oscillations in such structures. A very strong dependence of the ideal QDSL spectrum and wave functions on the field orientation is demonstrated. Bloch oscillations (BO) in QDSL are performed at two (or three for 3D QDSL) main Stark frequencies, which can be independently tuned by variation of field value and orientation. Due to the strong dependence of spectrum on field orientation, in QDSL intraminiband scattering processes can be almost totally suppressed. That allows to enlarge the BO lifetime by two orders of magnitude as compared with conventional quantum well superlattices.

1. Introduction

Technology of manufacturing ordered quantum dot arrays have been intensively developed during last decade. As purely discrete spectrum of such structures makes them very attractive for applications in electronics and optoelectronics.[1] Here a further development of these ideas is performed, which concerns carrier localization and Bloch oscillations in quantum dot superlattices (QDSL) in a constant electric field. We deal with an ideal QDSL, supposing the electric field strength and tunneling matrix elements between quantum dots (QD) small enough to describe QDSL in a single miniband approach.

First of all, we give a picture of the electron spectrum and wave functions in the QDSL. We demonstrate that at some orientations of the electric field with respect to the basic QDSL vectors, the spectrum is purely discrete and forms 2D (or 3D for 3D QDSL) Stark ladder. Otherwise, the spectrum is continuous and electrons are not localized in the direction(s) normal to the electric field. In the last case, the energy width of a transverse miniband depends on the field orientation exponentially.

We demonstrate that in contrast to conventional quantum well superlattices (QWSL), Bloch oscillations (BO) in QDSL take place at two (or three for 3D QDSL) main Stark frequencies and, with exponentially smaller amplitudes, at their various combinations. The main frequencies can be independently tuned, varing the field value and orientation.

It is well known that the main obstacle for observation and practical applications of BO in QWSL is their short lifetime due to the scattering of carriers by phonons and defects. We show that in QDSL due to the strong dependence of the spectrum on the field orientation it is possible almost totally to suppress these scattering processes and thus significantly enlarge the BO lifetime.

2. The Spectrum and Localization Area of Carriers in DC biased QDSL

The spectrum of an ideal QDSL looks unusualy owing to its very strong dependence on the field orientation. We demonstrate this for a square 2D QDSL with a basic vectors $\boldsymbol{a}_1$, $\boldsymbol{a}_2$. The generalization to the 3D case or to the cases of other geometry is obvious.

Let us call the orientation of field the irrational one, if the ratio $e\boldsymbol{F}\boldsymbol{a}_1 / e\boldsymbol{F}\boldsymbol{a}_2$ is an irrational number. For such field directions, the electrostatic potential of all QDs is different (Fig. 1) and the spectrum can be determined from the translation symmetry of the system:

$$E_{N1,N2} = - N_1\, e\mathbf{F}\cdot\mathbf{a}_1 - N_2\, e\mathbf{F}\cdot\mathbf{a}_2\,. \tag{1}$$

The spectrum is purely discrete, forming the 2D (3D for 3D QDSL) Stark ladder. Electron is localized in all three dimensions.

For the rational directions of field (when the ratio of field projections on the basic vectors of QDSL is rational), chains of QDs (or planes in 3D QDSL) with the same electric potential appear in the direction(s) normal to the field (Fig.1). Neglecting the carrier tunneling between QDs in these chains (planes) leads to degeneracy (see Eq. (1)). The carrier tunneling lifts off the degeneracy and leads to creation of transverse minibands of widths exponentially depending on the field orientation with respect to the QDSL axes: the larger the separation of QDs in the chain (plane) the less is the miniband width (Fig. 2):

$$E_N(k_\perp) = \boldsymbol{\Delta}_\perp/2\cos(k_\perp d_\perp) - N\, eF\, d_\parallel\,, \tag{2}$$

where $d_\parallel$, $d_\perp$ are the distances between chains and QDs in chains respectively, $\boldsymbol{\Delta}_\perp/4$ is the tunneling matrix element between neighbor QDs in a chain.

We underline the nontrivial fact that only the transverse tunneling matrix elements $\boldsymbol{\Delta}_\perp/4$ lift off the degeneracy though they are exponentially smaller than ones

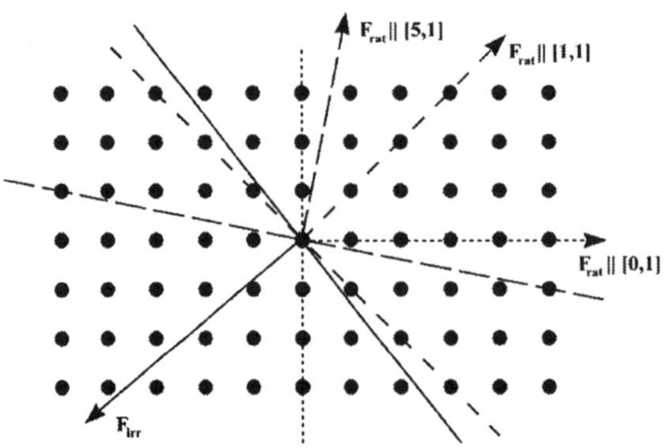

Fig. 1. Rational and irrational orientations of field in 2D QDSL.

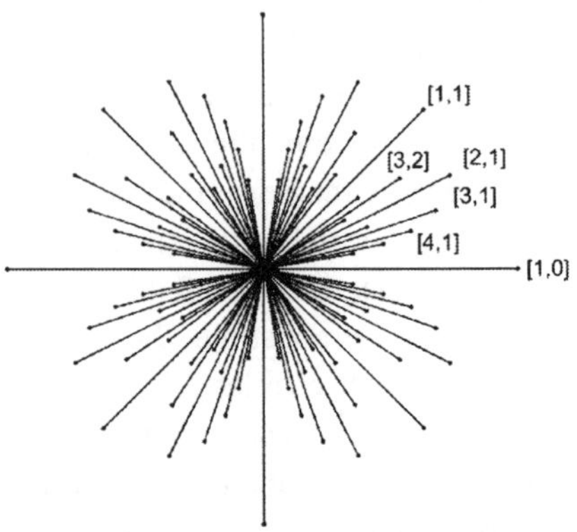

Fig. 2. The transverse miniband width dependence on rational field orientation (in logarithmic scale).

binding the nearest QDs. So, the first neighbor approximation can be used for the QDSL description only in special cases, for instance, when the field orientation is normal to one of basic QDSL vectors $\boldsymbol{a}_1$, $\boldsymbol{a}_2$. Taking into account all tunneling matrix elements, we came to the following rigorous expressions for the spectrum and the wave functions:[2]

$$E_R = -eF\cdot R,$$
$$\Psi_R(r) = \sum_\rho C^R_\rho \varphi(r-\rho),$$
$$C^R_\rho = V_B^{-1}\int dk\, \exp\Big(ik(\rho - R) + \tfrac{1}{4}\sum_{F\cdot\rho'\neq 0}\lambda_{\rho'}\exp(ik\cdot\rho')\Big) \tag{3}$$

for irrational field orientations, and

$$E_R(k_\perp) = \sum_{F\cdot\rho=0}\Delta_\rho/2\cos(k_\perp\cdot\rho) - eF\cdot R,$$
$$\Psi_R^{k_\perp}(r) = \sum_\rho C^R_\rho(k_\perp)\varphi(r-\rho),$$
$$C^R_\rho(k_\perp) = l_{norm}\int dk_{||}\exp\Big(ik(\rho - R) + \tfrac{1}{4}\sum_{F\cdot\rho'\neq 0}\lambda_{\rho'}\exp(ik\cdot\rho')\Big) \tag{4}$$

for rational field orientations.

Here $\{\mathbf{R}, \rho\} = n\boldsymbol{a}_1 + m\boldsymbol{a}_2$ are QDSL vectors; $\mathbf{k} = \mathbf{k}_\perp + \mathbf{k}_{||}$ is the electron wave vector, consisting of its components normal and parallel to the field direction and lying within the QDSL Brillouin zone of volume V_B; $\varphi(\boldsymbol{r}-\rho)$ are the Wannier functions of QDSL miniband, which in the limit of noninteracting QDs are the wave functions of the basic level of corresponding QDs; the ratios $\lambda_\rho = \Delta_\rho/[e\mathbf{F}\cdot\rho]$ determine the zone of electric field localization.

Let us analyze the above expressions for certain field orientations. For the irrational orientation, (3) gives the following expressions in the nearest - neighbor approximation:

$$E_{N_1,N_2} = -N_1 eF\cdot a_1 - N_2 eF\cdot a_2,$$
$$C^{N_1,N_2}_{n_1,n_2} = \boldsymbol{J}_{n_1-N_1}(\lambda_1/2)\boldsymbol{J}_{n_2-N_2}(\lambda_2/2), \tag{5}$$

where $\boldsymbol{J}_\mathbf{n}$ are the Bessel functions of the first kind. The localization lengths along two basic directions (measured in QDSL periods) are $\lambda_{1,2} = \boldsymbol{\Delta}_{[1,0]}/[\mathbf{eF}\cdot\mathbf{a}_{1,2}]$ until, as the field increases, $\lambda_{1,2}/2$ becomes less than unity and the electron localizes mainly at one QD (Fig. 3).

Now let us keep the field value constant and change its orientation near the [1,1] direction, so we should take into account the tunneling matrix element between QDs in that direction (Fig. 1). When the angle between the field direction and [1,1] $\delta\phi$ becomes less then the ratio $\Delta_{[1,1]}/\Delta_{[1,0]}$, the localization length in the transverse direction rapidly increases until electron delocalizes while $\delta\phi$ becomes zero (Fig. 4). For the delocalized electron, (4) gives

$$E_N(k_\perp) = \Delta_{[1,1]}/2\cos(k_\perp d) - N eFd/\mathbf{2},$$
$$C^N_{n_\perp,n_{||}}(k_\perp) = \exp(ik_\perp n_\perp d)\boldsymbol{J}_{n_{||}-N}\Big(\frac{2\Delta_{[1,0]}}{eFd}\cos(k_\perp d/2)\Big), \tag{6}$$

where $\boldsymbol{d}$ is the distance between QDs in transverse chains. A remarkable feature of this wave function is the fact that the localization length along field direction depends on the transverse momentum.

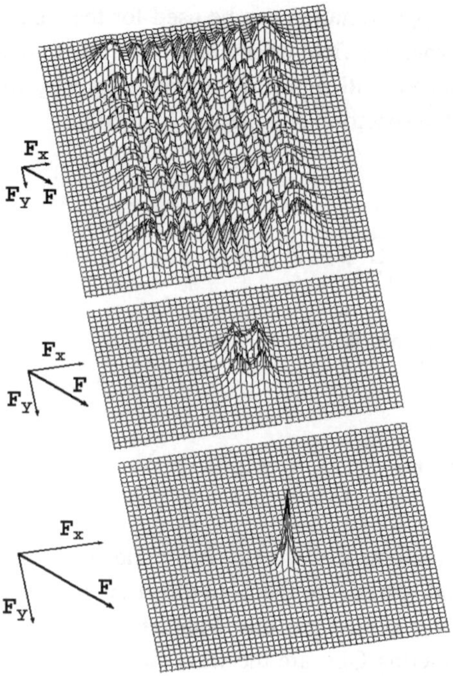

Fig. 3. Localization area vs. field value under the fixed irrational field orientation.

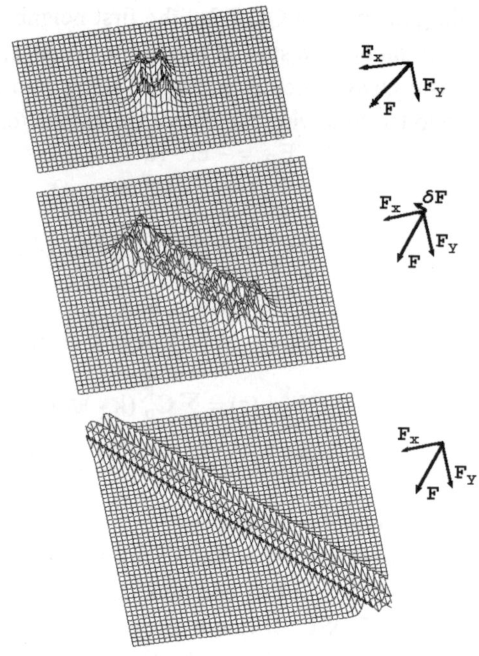

Fig. 4. Localization area near the rational field orientation under fixed field value.

3. Bloch Oscillations

In this section, we describe the Bloch oscillations arising in QDSL after the abrupt switching-on of the electric field. For such initial conditions, the semiclassical description appears to be correct for the whole field range. Applying semiclassical dynamics equations to the QDSL miniband spectrum

$$\frac{d}{dt}\hbar\mathbf{k}(t) = e\mathbf{F}, \quad \mathbf{v}(t) = \frac{1}{\hbar}\,\mathrm{grad}_{\mathbf{k}}E(\mathbf{k}), \quad E(\mathbf{k}) = \sum_{\mathbf{R}}\frac{\Delta_{\mathbf{R}}}{2}\cos(k\cdot R), \tag{8}$$

we arrive at the following expression for the BO current:

$$j = \sum_{R}\frac{1}{2\hbar}e\,n_e\Delta_R R < \cos(k_0\cdot R) > \sin(\Omega_R t), \quad \hbar\Omega_R = eF\cdot R = n\,eF\cdot a_1 + m\,eF\cdot a_2. \tag{9}$$

As far as the tunneling matrix elements $\Delta_{\mathbf{R}}/4$ decrease with the distance between QD exponentially, BO take place at two (or three in 3D QDSR) main frequencies, which can be independently tuned by varying the applied field in value and orientation. The amplitudes of the subharmonics are exponentially smaller.

The oscillations are periodic for the rational field orientations, when the ratio of frequencies is a rational number, and quasi-periodic for the irrational field orientations (because any irrational number is close to a rational one) (Fig. 5).

Fig. 5. Field projection of BO current
for rational (A) and irrational (B) field orientations.

4. Bloch Oscillation Damping

Observation of the BO in QWSL is a difficult problem due to fast dephasing processes. In Fig. 6, one can see that the wide transverse spectrum of QWSL provides strong phonon and elastic scattering of electrons both inside and between the transverse minibands at an arbitrary field value.

In QDSL, we can, in addition to possibility to change distance between adjacent steps of Stark ladder, vary the transverse miniband width by an appropriate choice of field orientation and thus significantly reduce phonon scattering (Fig. 7). To eliminate the intraminiband optical phonon scattering, it is enough to make the transverse miniband width smaller than the optical phonon energy. The interminiband optical phonon scattering between different transverse minibands is also impossible if the phonon frequency is not close to any integer number of Stark frequencies and the transverse miniband width is small enough.

Furthermore, the acoustic phonon scattering can also be strongly suppressed. The matrix element of the Stark electron interaction with a bulk acoustic phonon has the form

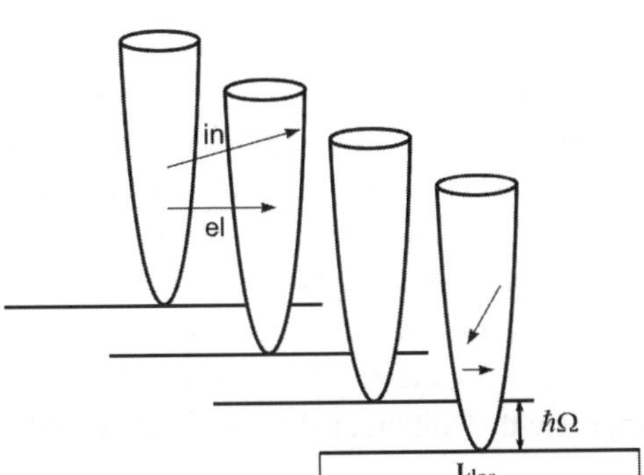

Fig. 6. Elastic and inelastic scattering processes in QWSL.

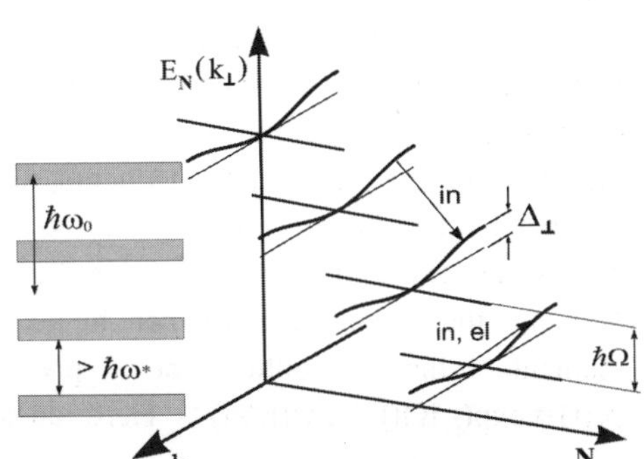

Fig. 7. Scattering processes in QDSL.

$$<R',k';0_q\,|\,V\,|\,R,k;1_q> = iGq\sqrt{\frac{\hbar}{2\omega_q V_0 \rho_a}}\; Q(q)\, S^{kk'}_{RR'}(q)$$

$$Q(q) = \int d^3r\,|\,\varphi(r)\,|^2\, e^{iq\cdot r}, \quad S^{kk'}_{RR'}(q) = \sum_{\rho} C^R_\rho(k)\, C^{R'}_\rho(k')\, e^{iq\cdot\rho}. \tag{10}$$

Here G is the constant of deformation potential; V_0, ρ_a are the volume of elementary cell and the mass density of QDSL material; the amplitudes C are defined in (4). The formfactor $Q(q)$ describes fine structure of the wave functions. Because the Wannier function $\varphi(r)$ exponentially decreases outside QD, it is obvious that the formfactor is small for phonon wavelengths less than the QD size $2R_d$. If the electron effective mass is different inside and outside QD, it is easy to show that the formfactor decreases as $(q/q^*)^{-3}$ as $q > q^* = \pi/R_d$. For an infinitely deep spherical QD or in the case of equal effective mass and exponential decrease outside the QD, the formfactor decreases as $(q/q^*)^{-4}$. So, the interminiband acoustic phonon scattering is strongly reduced if $\hbar\,\Omega - 4\Delta_\perp > \hbar\,\omega^* = \hbar s\,\pi/R_d$, where s is the sound velocity (Fig. 7).

Under the above conditions, only intraminiband electron transitions with absorption and emission of acoustic phonons still remain. The theory of the BO damping due to this scattering mechanism is described in the next section.

5. The Bloch Oscillation Damping due to Acoustic Phonon Intraminiband Scattering

Following the procedure developed by Kohn and Luttinger [3] and then, in particular, applied for the study of carrier kinetics in QWSL,[4] one can derive the quantum kinetic equations describing BO damping in the density matrix formalism under the condition of weak scattering $\tau_{eff} >> \Omega^{-1}$. Generally, these equations are too complicated to be analysed here. It is why we limit ourselves to the simplest case of abrupt switching-on of the electric field directed along one of the base vectors of rectangular 2D QDSL or tetragonal 3D QDSL. This leads to a closed form of equations for the density matrix elements. We explore the dependence of BO damping rate on the localization length $\lambda = \Delta_\parallel/eFa$ (measured in QDSL periods)[a] and on the transverse miniband width $\Delta_\perp$.

As an initial conditions for these equations, one should take the density matrix that is diagonal in the single miniband k-representation. We suppose the Bolzman distribution for electrons in the initial state. Due to spatial homogeneity, the density matrix elements $\rho_{NN+n}(k, k',t)$ do not depend on N (here ***k*** and ***k'*** are the transverse miniband wavenumbers). In the nearest-neighbor approximation, there survives the only matrix element with **n** = 1 describing the transitions between neighbor Stark ladder states. Namely, the current density can be expressed in terms of the single density matrix element that could be presented as $\rho_{NN+1}(k,k,t) \equiv \rho_0(t) g_k(t)$ with $\rho_0(t) = \exp(-i\Omega t) < \cos(k^0_\parallel a) >$. Here an average (with Boltzman distribution) quantity

[a] For the sake of siplicity, in our calculation we suppose the QDSL period a to be the same both along the electric field direction and the transverse direction, and, at the same time, allow $\Delta_\parallel$ and $\Delta_\perp$ to be different. It is a reasonable simplification because of very strong (exponential) dependence of tunneling matrix elements on the distance between QDs.

$<\cos(k^0_{\parallel}a)> = [I_1(\Delta_{\parallel}/2T)/I_0(\Delta_{\parallel}/2T)]$, I_0, I_1 are modified Bessel functions. In order to simplify the equations we restrict ourselves to a high temperature limit, $T >> \hbar\,\omega^*$. This allows the spontaneous emission of phonons to be ignored. Under the above conditions, one arrives at the following equations:

$$-\frac{\partial}{\partial t} g_k(t) = \sum_{k'} J_{kk'}\, g_{k'}(t), \qquad J_{kk'} \equiv \delta_{k',k} \sum_{k''} W^{out}_{kk''} - W^{in}_{kk'},$$

$$W^{out}_{kk'} = \frac{\pi}{\hbar} \sum_K \int d^3q\, \nu_q (|V^q_{N+1N+1}|^2 + |V^q_{NN}|^2) \delta(|\varepsilon_k - \varepsilon_{k'}| - \hbar\omega_q) \delta(k' - k - q_\perp + K),$$

$$W^{in}_{kk'} = \frac{2\pi}{\hbar} \sum_K \int d^3q\, \nu_q \operatorname{Re}[V^q_{N+1N+1}(V^q_{NN})^*] \delta(|\varepsilon_k - \varepsilon_{k'}| - \hbar\omega_q) \delta(k' - k - q_\perp + K),$$

$$j(t) = j_0(t) \sum g_k(t), \quad j_0(t) = \frac{e n_e \Delta_{\parallel} a}{\hbar} < \cos(k^0_{\parallel} a) > \sin(\Omega_R t), \quad g_k(0) = f_k. \qquad (11)$$

Here $\nu_q \approx T/\hbar\,\omega_q$ is the phonon distribution function in the high-temperature limit; $\varepsilon_k = \Delta_\perp/2\cos(k\cdot a)$ denotes the transverse miniband spectrum; summing over $K=2\pi n/a$ describes Umklapp processes in the transverse miniband; $j_0(t)$ is the current density in the absence of scattering (9), f_k is the Boltzman distibution function in the transverse miniband.

Let us now construct a formal solution of (11), making use of the initial conditions and eigenvalues η_i and normalized eigenvectors g^i_k of the integal operator $\boldsymbol{J}$:

$$g_k(t) = \sum_i (\sum_{k'} g^i_{k'} f_{k'}) g^i_k f_k e^{-\eta_i t}, \quad j(t) = j_0(t) \sum_i (\sum_{k'} g^i_{k'} f_{k'})^2 e^{-\eta_i t}. \qquad (12)$$

We see that generally damping the BO current is not exponential. Let the damping rate of current be introduced as $\Gamma(t) = -\partial/\partial t \ln[j(t)/j_0(t)]$. From (12) we notice that $\Gamma(t)$ monotonically decreases with time, starting from the initial value $\Gamma(0)$ and approaching $\Gamma(\infty) = \eta_{min} \equiv \min\{\lambda_i\}$ at $t\to\infty$.[a]

Essentially, the BO are the quantum beatings arising when the initial state of electrons just after applying of the electric field is a coherent mixture of Stark eigenstates. The coherence becomes degraded due to scattering processes that results in BO damping. Therefore, the equation for $g_k(t)$ describes degradation of the coherence between the neighboring Stark states. The elements of the matrix W^{out}, describing outcome processes, are positive, what leads to destruction of the coherence between neighbor Stark ladder states. In contrary, the elements of income, $\boldsymbol{W}^{in}$, may be positive or negative depending on the phonon phase incursion a superlattice period. Respectively, the income can enhance or reduce the coherence between the states.

Besides, it is easy to see that $\boldsymbol{W}_{kk'}{}^{out} \geq |\boldsymbol{W}_{kk'}{}^{in}|$. Then, as was shown by Gershgorin[5], all eigenvalues η_i of $\boldsymbol{J}$ are not negative (η_i are real as $\boldsymbol{J}$ is a real and symmetric matrix), and the following inequality is true:

$$\eta_{min} \geq \Lambda \equiv \min_k \{\sum_{k'} W^{out}_{kk'} - \sum_{k'\neq k} |W^{in}_{kk'}|\} \geq 0. \qquad (13)$$

[a] Actually, the spectrum of $\boldsymbol{J}$ is continuous and there should be the logarithmic correction to the asymptotic behavior of damping rate, always arising for a decay processes in a continuous spectrum. For our analysis, it is not of a principal value.

Further numerical calculation of eigenvalues shows that Λ serves as a good estimate for $\Gamma(t \to \infty)$. For a whole parameter range it gives $\Lambda/\Gamma(\infty)= 0.7 \div 1$.

For a better understanding of dephasing process, it is useful to study equation (10) also in a time domain. Let us rewrite it in the form

$$(\partial / \partial t + \gamma_k) g_k(t) = \sum_{k'} W^{in}_{kk'}(g_{k'}(t) - g_k(t)),$$
$$\gamma_k = \frac{\pi}{\hbar} \int d^3 q v_q \,|\, V^q_{N+1N+1} - V^q_{NN} \,|^2\, \delta(|\,\varepsilon_k - \varepsilon_{k+q_\perp}\,| - \hbar\omega_q). \quad (14)$$

As $t = 0$, the right-hand side of (14) approaches zero under chosen initial conditions. It means that for γ_k=const the right-hand side is identically zero and damping is exponential with the decrement γ. (Specifically, oscillations do not fade at all for γ_k=0). For $\gamma_k \neq$const, the right-hand side describes mixing in the transverse miniband, which leads to a nonexponential decay of g_k. One can see that γ_k includes a specific combination of matrix elements: γ_k vanishes if the phonon phase variation on the QDSL period is zero. From (12) and (14), we easily obtain the initial value of current damping rate $\Gamma(t = 0) = \langle\gamma\rangle = \Sigma\, f_k \gamma_k$.

In Figs. 8, 9, we present the dependencies of $\Gamma(0) = \langle\gamma\rangle$ and $\Gamma(t \to \infty) = \eta_{min}$ on the electron localization length $\lambda=\Delta_\parallel/eFa$ and the transverse miniband width $\Delta_\perp$. The parameters of 2D QDSL, used in the calculation, were: T = 300 K, a = 100 Å, R_d = a/4; s = 5,22·10^5 cm/sec, G = 6 eV (GaAs); $\hbar\,\omega^*=\hbar s\, \pi/R_D$=4,3 meV, $2\hbar\, s/a$ = 0,6 meV (see eq. (10) and comments below). For the reference, we notice that scattering rate in a bulk GaAs is about 10^{11} sec^{-1}.

One can see that the dependencies of damping rate both on λ and $\Delta_\perp$ are very strong, so we have rich possibilities to control dephasing rate by varying the field value and orientation.

The damping rate dependence on the localization length is mainly controlled by the localization length dependence of scattering matrix elements. Direct calculation gives

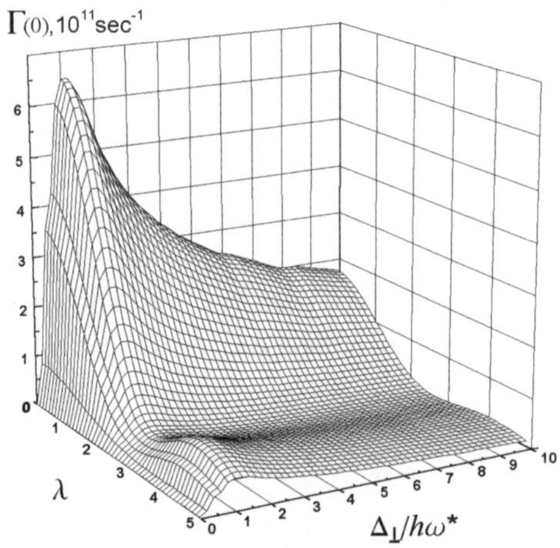

Fig. 8. Localization length and transverse miniband width dependence of the initial value of BO current damping rate.

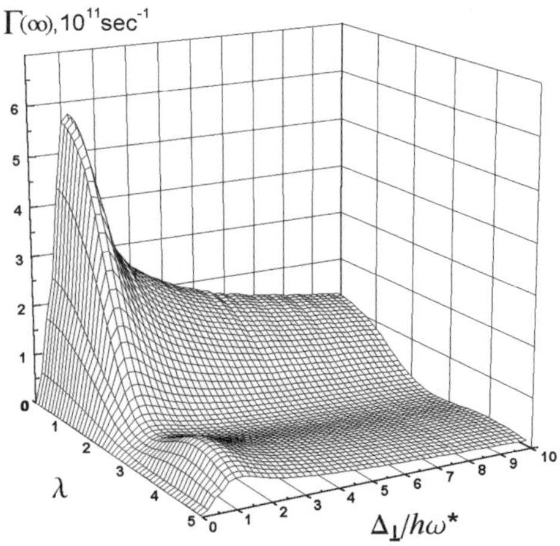

Fig. 9. Localization length and transverse miniband width dependence of the final value of BO current damping rate.

(see eq. (4), (9))

$$| V^{q}_{N+1N+1} - V^{q}_{NN} |^2 = 2(1-\cos(q_{||}a)) | V^{q}_{NN} |^2 = f(\omega_q)(1-\cos(q_{||}a)) J_0^2(\lambda \sin(q_{||}a/2)) \quad (15)$$

One can conclude from (14) and (15) that γ_k differs from a usual scattering rate in the transverse miniband by the factor (1- $\cos(q_{||}a)$), describing on the effect of phonon phase variation $q_{||}a$ on the QDSL period. One can see in Fig. 8 and 9 that the maxima and minima of damping rate dependence on λ coincide with those of the squared Bessel function. As for the impact of the factor on integral in (14), here we only notice that in the case of full localization $\lambda = 0$, the Bessel function is equal to unity, so the scattering rate has a maximum. The damping rate at this point is equal to the scattering rate, while a mean cosine gives zero. With a rise of localization length, the damping rate decreases faster than the scattering rate. At $\lambda \approx 2{,}7$ the damping rate has the first minimum where the order of magnitude is already smaller than that at $\lambda = 0$ and about three times smaller than the scattering rate.

The overall run of damping decrement dependence on the transverse miniband width can also be clearly appreciated. In the wide transverse minibands $\Delta_\perp > \hbar\omega^*$, the energy of actual phonons is limited by formfactor (10) at $\hbar\omega^*$, so, the scattering is quazielastic. Respectively, the damping rate together with the density of states in the tranverse miniband depend on the miniband width inversely. On contrast, for $\Delta_\perp < \hbar\omega^*$, the energy of participating phonons is limited by miniband edges, and the rise of miniband width results in a rise of damping rate. At last, in a very narrow minibands $\Delta_\perp < 2\hbar s/a$, the group velocity of electron in the transverse miniband becomes less than the sound velocity, and the energy and momentum conservation laws cannot be obeyed together. Damping in this case is possible only due to the interminiband scattering processes not taken into account in this calculation.

The strong suppression of interminiband scattering is possible only for $\hbar\Omega - \Delta_\perp > \hbar\omega^*$ (Fig. 7) what for the chosen QDSL parameters means that the Bloch frequency should exceed 10^{12} Hz. So, we see that the damping time in QDSL can consist of hundreds of oscillation periods. For comparison,[6] in QWSL with similar parameters, the observed BO lifetime at room temperature is 10^{13} sec^{-1}.

In 3D QDSL, the dependencies analogous to those presented in Fig. 8, 9 are quite similar. However, the damping decrement in 3D case is about three times greater than in 2D QDSL as a result of a different density of states in the 2D transverse miniband.

6. Conclusions

In conclusion, we would like to stress that QDSL application enables us to control the scattering processes. It has been shown that the damping rate of Bloch oscillations can be reduced at least by two order of magnitude as compared with conventional QWSL. The possibility of such drastic reduction originates in a variety of new attractive capabilities of the spectrum, wave function and scattering managing in QDSL, where we deal not only with the electric field value but also with its direction to QDSL basic vectors.

Namely, varying the electric field value and orientation in QDSL allows to:

(i) qualitatively change the spectrum of electrons, from the continuous one to the purely discrete one;

(ii) control the transverse miniband width in a wide range since it exponentialy depends on the field orientation;

(iii) totaly suppress the optical phonon scattering which is the main mechanism of scattering in QWSL;

(iv) almost totally suppress the acoustic phonon scattering between transverse minibands;

(v) significally reduce the intraminiband acoustic phonon scattering by choosing appropriate electric field strength and miniband width.

In addition, in contrast to QWSL, the Bloch oscillations in QDSL operate at two (or three in 3D QDSL) main frequencies, and one has an attractive possibility to tune these frequencies independently, varying the electric field value and orientation.

All these peculiarities make QDSL very promising for applications in electronics.

Acknowledgements

The work was supported by Russian Foundation for Basic Research (Grant 99-02-16796), by the Federal Program on Support of Promient Scientists and Leading Scientific Schools (Grant 00-15-96812) and by the Program "Physics of Solid State Nanostructures".

References

1. R. A. Suris, “Prospects for Quantum Dot Applications in Electronics and Optoelectronics”, in *Future Trends in Microelectronics, Reflections on the Road to Nanotechnology*, ed. S. Luryi, J. Xu, and A. Zaslavski, NATO ASI Series, Ser. E: Applied Sciences, Kluwer Academic Publishers, Dordrecht/ Boston/ London, 1996, p. 197.
2. I. A. Dmitriev, R. A. Suris, “Electron localization and Bloch oscillations in quantum-dot superlattices under a constant electric field”, *Semiconductors* **35** (2001) 212-219.
3. W. Kohn, J. M. Luttinger, “Quantum theory of Electrical Transport Phenomena”, *Phys. Rev.* **108** (1957) 590-610.
4. R. F. Kazarinov, R. A. Suris, “Electric and electromagnetic properties of semiconductors with a superlattice””, *Sov. Phys. Semicond.* **6** (1972) 120-131; R. A. Suris, B. S. Shchamkhalova, “Heating of electrons in superlattice semiconductors”, *Sov. Phys. Semicond.* **18** (1984) 738-742
5. S. A. Gershgorin, “Ueber die Abgrenzung der Eigenwerte einer Matrix”, *Izv. Akad. Nauk SSSR ser. Phys.-Mat.* (1931) 749-754.
6. T. Dekorsy, R. Ott, H. Kurz, K. Köhler, “Bloch oscillations at room temperature”, *Phys. Rev. B* **51** (1995) 17275.